Food-Derived Bioactive Peptides

Food-Derived Bioactive Peptides

Green Preparation, Activity Mechanisms, and Health Applications

Shao-Yun Wang, Xu Chen, and Xixi Cai
The College of Biological Science and Engineering
Fuzhou University
Fuzhou, Fujian
China

WILEY

Registered Offices
John Wiley & Sons, Inc., 111 River Street, Hoboken, NJ 07030, USA
John Wiley & Sons Ltd, New Era House, 8 Oldlands Way, Bognor Regis, West Sussex, PO22 9NQ, UK

For details of our global editorial offices, customer services, and more information about Wiley products visit us at www.wiley.com.

The manufacturer's authorized representative according to the EU General Product Safety Regulation is Wiley-VCH GmbH, Boschstr. 12, 69469 Weinheim, Germany, e-mail: Product_Safety@wiley.com.

Wiley also publishes its books in a variety of electronic formats and by print-on-demand. Some content that appears in standard print versions of this book may not be available in other formats.

Library of Congress Cataloging-in-Publication Data
Names: Wang, Shao-Yun author | Chen, Xu author | Cai, Xixi author
Title: Food-Derived Bioactive Peptides : Green Preparation, Activity Mechanisms, and Health Applications / Shao-Yun Wang, Xu Chen, Xixi Cai.
Description: Hoboken, NJ : Wiley, [2026] | Includes bibliographical references and index.
Identifiers: LCCN 2025042817 (print) | LCCN 2025042818 (ebook) | ISBN 9781394349487 cloth | ISBN 9781394349500 adobe pdf | ISBN 9781394349494 epub | ISBN 9781394349517 obook
Subjects: LCSH: Peptides | Functional foods | Bioactive compounds | Food industry and trade
Classification: LCC QP552.P4 W35 2026 (print) | LCC QP552.P4 (ebook)
LC record available at https://lccn.loc.gov/2025042817
LC ebook record available at https://lccn.loc.gov/2025042818

Cover Design: Wiley
Cover Image: © Shao-Yun Wang

Set in 9.5/12.5pt STIXTwoText by Straive, Pondicherry, India

Contents

Preface

"Bioactive peptides"—a term that felt obscure just two decades ago—now stands firmly in the spotlight of human nutrition, precision medicine, and sustainable food innovation. Once dismissed as mere short-chain amino acids—"by-products of protein degradation"—they are now being redefined as signaling molecules capable of taming oxidative stress, modulating immunity, delaying fatigue, enhancing flavor, and even intervening in the aging process. Functional peptides are moving from laboratory test tubes to the public's dinner plate, hailed as a key lever for extending health span and reshaping the food industry.

Yet, the path from "bioactivity discovery" to "ingredient commercialization" is far from smooth. What's missing is not bioactivity data, but an integrated road map that bridges algorithms, processing, clinical validation, and market access.

This book was born to chart that road map. Drawing on over a decade of R&D and commercialization experience—both successes and setbacks—I have joined forces with colleagues at home and abroad to compile this volume. Centered on the theme of "green preparation—mechanism of action—health applications," it offers a systematic review of advances in food-derived bioactive peptides. Chapters 1 and 2 lay the foundation—concepts, preparation, and evaluation. Chapters 3–11 spotlight nine-star peptide categories (antifreeze, antioxidant, antimicrobial, metal-chelating, antiaging, anti-fatigue, immunomodulatory, flavor-enhancing, and self-assembling), dissecting their molecular mechanisms alongside state-of-the-art process parameters and cost–benefit assessments. Each chapter covers activity-assay methods, molecular mechanisms, safety evaluation, applications, and current limitations and challenges.

Designed as a desktop reference rather than a conventional academic monograph, this book offers graduate students a step-by-step protocol backed by sufficient theoretical context to grasp why each step matters. Industry professionals are provided with ready-to-use economic-model templates, supplier lists, and regulatory checklists that can be dropped straight into a business plan. Investors and policymakers are provided with forward-looking essays on the IP landscape, safety early-warning systems, and the socioeconomic implications of replacing animal proteins with precision-fermented peptides.

May every reader, while turning these pages, capture the elegance of microscopic peptide chains and perceive the complexity of the macroscopic systems they inhabit—ultimately reenter the grand experiment of reengineering humanity's relationship with food, armed with evidence and curiosity.

This book is dedicated to all fellow travelers who believe that "small molecules can change the big world."

Golden Autumn, 2025

Shao-Yun Wang, Xu Chen, and Xixi Cai

1

Introduction

Food-derived functional peptides are bioactive molecules obtained from dietary proteins, which exist in food proteins through specific amino acid sequences. Upon protein degradation, these peptides are released and acquire physiological activity during the process. With abundant natural protein resources and industrial scalability, food-derived peptides have gained significant attention due to their high safety profile.

Recent scientific advancements have revealed that these bioactive peptides, as functional fragments of proteins, not only supply essential nutrients for growth and development but also exhibit unique physiological characteristics compared to intact proteins:

1) **Enhanced bioavailability**: Food-derived peptides demonstrate superior absorption efficiency with low or no antigenicity. Current research indicates that small peptides (2–3 amino acids) can be directly absorbed into the circulatory system without further digestion or energy expenditure. Their absorption, transformation, and utilization processes are both efficient and complete. Compared to proteins, peptides have shorter residence times in the body, minimizing contact with antibody-producing cells and reducing immune stimulation.
2) **High biological activity**: These peptides exhibit potent effects even at low concentrations. Their actions span multiple biological levels, from cellular processes to tissue and organ functions.
3) **Structural modifiability**: The molecular structure of food-derived peptides is easily amenable to chemical modifications and resynthesis for targeted functional development.
4) **Nutritional balance regulation**: Unlike excessive protein intake, these peptides effectively regulate metabolic balance without causing nutrient overload.

In biological systems, endogenous peptides are essential for normal physiological functions. Bioactive peptides (or functional peptides) are defined as compounds formed by two or more α-amino acids linked through peptide bonds. The diverse biological structures and functions of peptides arise from variations in amino acid composition, quantity, and sequence arrangement. Based on their origin, peptides can be classified as endogenous (naturally produced within organisms) or exogenous (dietary-derived).

Food-Derived Bioactive Peptides: Green Preparation, Activity Mechanisms, and Health Applications, First Edition. Shao-Yun Wang, Xu Chen, and Xixi Cai.

Extensive research has characterized food proteins as a rich source of functional peptides. These peptides demonstrate a wide range of biological activities, including antihypertensive effects (e.g., ACE inhibition by soy peptides), antioxidant properties (e.g., wheat germ peptides), anticancer potential (e.g., marine peptides), anti-inflammatory actions (e.g., casein-derived peptides), immune modulation (e.g., immunostimulatory peptides from egg proteins), and cholesterol-lowering capabilities (e.g., oat peptides).

As highlighted in recent studies (e.g., Food & Fermentation Industries 2024), these bioactive peptides represent promising therapeutic agents for human health. Their application in the food industry as functional ingredients offers significant potential for improving consumer health outcomes, particularly in addressing chronic metabolic disorders and enhancing nutritional status across diverse populations. Current research emphasizes their role in precision nutrition and personalized dietary interventions, aligning with the growing demand for science-based functional foods.

The *Handbook of Foodborne Bioactive Peptides* offers a comprehensive overview of bioactive peptides. This volume focuses on the source, preparation, activity evaluation, mechanism consideration, and practical application, as well as the classifications of major bioactive peptides such as antifreeze peptides, antioxidant peptides, antimicrobial peptides, metal chelate peptides, antiaging peptides, antifatigue peptides, immune regulatory peptides, flavor peptides, and self-assembling peptides. Moreover, the new technology, future research direction, and emphasis of functional peptides were summarized.

2

Preparation of Bioactive Peptides

2.1 Introduction

In recent years, food-derived bioactive peptides have attracted substantial attention from researchers in the food and pharmaceutical industries globally due to their abundant raw materials, high product safety, and notable biological functions. With the rapidly growing demand for functional peptides, efficient and precise preparation, separation, and purification of target bioactive peptides have become critical research topics in food science and biotechnology. Traditional peptide preparation methods, typically involving sequential separation and purification procedures, are time-consuming, costly, and yield limited peptide quantities, thus hindering the industrialization of bioactive peptides. Therefore, developing efficient, controllable, and cost-effective peptide preparation and purification techniques holds considerable theoretical significance and practical value for advancing the functional food industry.

Currently, preparation methods for food-derived bioactive peptides primarily include protein hydrolysis, chemical synthesis, recombinant DNA technologies, and chemical modification [1, 2]. Among these, enzymatic hydrolysis techniques have been widely adopted in the food industry due to the mild reaction conditions and scalability. Recent studies have demonstrated that optimization of hydrolysis conditions and selective use of specific proteases can yield peptides with targeted molecular weight distributions and functional characteristics [3, 4]. Microbial fermentation, which utilizes microbial metabolic pathways to convert food proteins into specific bioactive peptides, has also shown significant promise, especially in dairy products and fermented soybean foods [5, 6]. Nonetheless, these methods still face challenges such as complex peptide mixtures and difficulties in purifying target peptides. To overcome the limitations of traditional approaches, computer-assisted technologies, including molecular docking, quantitative structure–activity relationships (QSARs), and bioinformatics analyses, have been increasingly introduced into the screening and preparation processes for food-derived bioactive peptides [2, 7, 8]. These technologies rapidly and efficiently predict target peptide sequences through virtual simulation of enzymatic hydrolysis and targeted bioactive peptide screening.

Overall, this chapter systematically reviews recent advances and technological methods for the preparation, separation, and purification of food-derived bioactive peptides, with a

Food-Derived Bioactive Peptides: Green Preparation, Activity Mechanisms, and Health Applications, First Edition. Shao-Yun Wang, Xu Chen, and Xixi Cai.

particular focus on the theoretical basis, practical application, and emerging trends of computer-assisted screening technologies. It aims to provide theoretical and practical guidance for the efficient preparation and industrial development of food-derived functional peptides.

2.2 Preparation Methods for Bioactive Peptides

Currently, bioactive peptides are primarily prepared through four methods. The first method involves protein hydrolysis, including enzymatic hydrolysis, microbial fermentation, and chemical hydrolysis. The second method comprises chemical synthesis based on known amino acid sequences. The third method employs recombinant DNA technology, and the fourth involves chemical modification of existing peptides or proteins.

2.2.1 Protein Hydrolysis

2.2.1.1 Enzymatic Hydrolysis

Enzymatic hydrolysis refers to a process in which proteins are hydrolyzed by specific or nonspecific proteases to produce small peptides. Since proteases exhibit substrate specificity and distinct cleavage sites, hydrolyzing the same protein substrate using different enzymes can yield peptides varying considerably in size and sequence [9, 10]. Currently, proteases widely employed in enzymatic hydrolysis include trypsin, chymotrypsin, pepsin, alkaline protease, neutral protease, acid protease, papain, and complex proteases [3, 6]. Furthermore, reaction conditions significantly influence the hydrolysis process; thus, industrial production typically involves optimizing reaction parameters such as temperature and pH. Selection of proteases is guided by parameters including the degree of hydrolysis (DH) and the desired bioactivity of peptides.

The general process for producing bioactive peptides via enzymatic hydrolysis is illustrated in Figure 2.1. In some cases, initial separation steps are followed by further purification to obtain peptides with targeted properties and enhanced functional performance.

Research on protein enzymatic hydrolysis processes has primarily focused on four areas. Firstly, studies on developing novel protein resources through optimized hydrolysis processes are conducted, aiming to produce functional additives or ingredients with superior processing properties for the food industry. Secondly, novel proteases or innovative hydrolysis techniques are utilized to generate peptides with narrower molecular weight distributions. For instance, Pintado et al. [11] reported an optimized hydrolysis process to produce whey protein hydrolysates concentrated predominantly within two

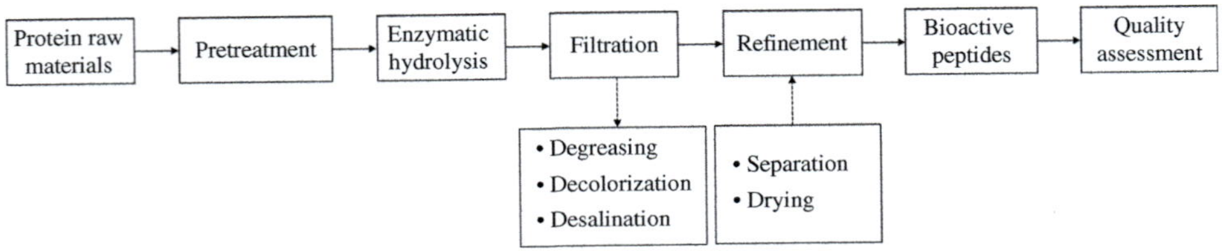

Figure 2.1 Steps involved in the production of bioactive peptides via enzymatic hydrolysis.

molecular weight ranges (7.5–8.0 and 4.0–4.5 kDa). Thirdly, it further improves the yields of peptides with specific bioactivities through optimized hydrolysis processes, such as antioxidant properties, from diverse protein sources. Antioxidant peptides, for example, can be obtained from both plant proteins (soybean, cereals) and animal-derived proteins (milk, blood, fish). Lastly, research on novel enzymatic hydrolysis processes, such as combining enzymatic hydrolysis with membrane separation technology to establish continuous hydrolysis processes or employing immobilized enzymes to achieve enhanced efficiency and process stability.

Selection of Proteases Protease selection requires consideration of enzyme specificity and the avoidance of bitterness that often accompanies higher degrees of hydrolysis. Depending on the protein substrate, different protease combinations are utilized to achieve optimal proportions of amino acids, dipeptides, and peptides. Typically, a mixture of trypsin and chymotrypsin is selected because trypsin specifically targets lysine and arginine residues, while chymotrypsin only hydrolyzes peptide bonds involving tyrosine, phenylalanine, and tryptophan residues [2, 12]. Additionally, to maximize the yield of bioactive peptides, exopeptidases should generally be avoided, except when necessary for specific modifications for a short duration, as their inclusion typically increases operational complexity. Previous studies indicated that the generation of certain inhibitory peptides required hydrolysis by two or more proteases [13]. Thus, the selection of proteases compatible with production requirements should align with the structural and functional properties of the targeted bioactive peptides.

Proteases can be classified into plant, animal, or microbial proteases based on their source. These enzymes have distinct active sites and chemical properties, enabling them to cleave internal peptide bonds under specific pH and temperature conditions. Combining multiple proteases may yield better enzymatic hydrolysis effects compared to a single protease [14, 15]. For instance, the combination of alkaline protease (Alcalase), chymotrypsin, trypsin, pepsin, and proteases derived from certain bacteria and fungi has demonstrated enhanced protein utilization and improved hydrolysis degrees.

Enzymatic Hydrolysis Methods for Specific Bioactive Peptides Extensive studies have focused on collagen hydrolysates due to their notable antifreeze activities. Wang et al. [16] prepared purified peptides via enzymatic hydrolysis of shark skin (*Carcharodon carcharias*) collagen. The peptide significantly enhanced the viability of *Lactobacillus bulgaricus* after freezing when added at a concentration of 250 mg/mL. Cao et al. [17] extracted collagen peptides exhibiting notable thermal hysteresis and recrystallization inhibition activities from porcine skin hydrolysates, effectively increasing the glass transition temperature of ice cream. Du et al. [18] investigated the protective mechanism of chicken protein hydrolysates on actin during freeze–thaw cycles, finding significant inhibition of ice crystal growth. Additionally, Wu et al. [19] utilized alkaline protease to hydrolyze porcine skin collagen to produce antifreeze peptides (AFPs), elucidating their protective mechanism on *Streptococcus thermophilus*. These peptides preserved the intracellular enzymes β-galactosidase and lactate dehydrogenase, forming protective coatings on the cell surface to maintain cellular integrity. Collagen peptides are abundant in glycine, alanine, proline, and hydroxyproline residues. Previous studies have established that antifreeze activities correlate with glycine, alanine, threonine, aspartic acid, and serine residues [20]. Consequently, these amino acids, along with the alkyl side chains of proline, hydroxyproline, and alanine residues, create a partially nonpolar environment conducive to stabilizing hydrogen bonds between AFPs and water molecules, thereby inhibiting ice crystal formation.

Two primary enzymatic hydrolysis methods for preparing oligopeptides with high *F*-values are single-enzyme hydrolysis and multienzyme hydrolysis [21, 22]. Single-enzyme hydrolysis, mainly employing trypsin, is currently limited to oyster protein and pearl oyster protein, but requires subsequent activated carbon adsorption and gel filtration purification to obtain the final high *F*-value oligopeptides. Multienzyme hydrolysis, the predominant method for high *F*-value oligopeptides, involves a two-step process. The initial step employs enzymes such as pepsin, trypsin, alkaline protease, and chymotrypsin to cleave peptide bonds formed by aromatic amino acids, exposing these residues and degrading proteins into peptide mixtures. The subsequent step typically utilizes papain or flavorzyme to further hydrolyze the exposed aromatic residues into free amino acids [23]. This method requires only a single de-aromatization step to yield high *F*-value oligopeptides. Table 2.1 summarizes selected

Table 2.1 Selected proteases and their hydrolysis conditions and specificities for the preparation of high *F*-value oligopeptides.

Protease type	Substrate	Hydrolysis conditions	Hydrolysis specificity
Basophilic protease	Zein	$S = 1\%$, $E/S = 0.2\%$, $T = 37\,°C$, $t = 4\,h$, pH 11.0	Hydrolyze peptide bonds formed by AAA
Kerase	Sunflower protein	$S = 1\%$, $E/S = 1\%$, $T = 37\,°C$, $t = 6\,h$, pH 6.5	Hydrolyze peptide bonds formed by hydrophobic macromolecular amino acid residues
Chymotrypsin	Zein	$S = 3\%$, $E/S = 2\%$, $T = 37\,°C$, $t = 5\,h$, pH 8.0	Hydrolyze peptide bonds formed by AAA or amino acid residues with large nonpolar side chains
Pepsin	Whey protein	$S = 3\%$, $E/S = 2\%$, $T = 37\,°C$, $t = 5\,h$, pH 2.0	Hydrolyze peptide bonds formed by the carboxyl or amino groups of AAA or other hydrophobic amino acids
Alkaline protease	Zein	$S = 3\%$, $E/S = 2\%$, $T = 45\,°C$, $t = 5\,h$, pH 10.0	Hydrolyze peptide bonds formed by AAA
Actin protease	Zein	$S = 1\%$, $E/S = 1\%$, $T = 37\,°C$, $t = 6\,h$, pH 6.5	Release AAA
Pronase	Whey protein	$S = 3\%$, $E/S = 1\%$, $T = 37\,°C$, $t = 7\,h$, pH 9.0	Release AAA
Papain	Whey protein	$S = 3\%$, $E/S = 1\%$, $T = 37\,°C$, $t = 2\,h$, pH 6.5	Release AAA

Note: S represents substrate concentration, E/S represents enzyme-to-substrate ratio, T represents reaction temperature, and t represents reaction time.

proteases and their specificity and reaction conditions [24]. The hydrolysis process is typically controlled by monitoring the DH. A low DH indicates incomplete protein hydrolysis, whereas excessively high DH may lead to extensive hydrolysis, resulting in abundant free amino acids present in the mixture.

2.2.1.2 Microbial Fermentation

Microbial fermentation leverages microbial metabolism to convert large-molecule proteins from plant sources into small-molecule bioactive peptides. Through the modulation of fermentation conditions and microbial metabolic pathways, peptides with varied amino acid sequences and molecular weights are generated [25, 26]. It represents an effective strategy for producing bioactive peptides and food-grade protein hydrolysates. The obtained peptides vary according to controlled fermentation conditions and can be directly absorbed by the human digestive system. Compared to enzymatic hydrolysis, peptides produced via microbial fermentation exhibit higher edible safety but have limited industrial applications.

Numerous microbial enzymes have been employed to identify angiotensin converting enzyme (ACE)-inhibitory peptides derived from milk proteins. Solieri et al. [27] demonstrated that milk fermented with *Lactobacillus casei* PRA205 yielded stronger ACE inhibitory peptides (Val-Pro-Pro and Ile-Pro-Pro) compared to milk fermented by *Lactobacillus rhamnosus* PRA331, likely due to intracellular peptidase activity, cell-envelope-associated protease specificity, or enhanced peptide transport systems in lactobacilli. Kefir peptides (Lys-Ala-Val-Pro-Tyr-Pro-Gln, Asn-Leu-His-Leu-Pro-Leu-Pro, Ser-Lys-Val-Leu-Pro-Val-Pro-Gln, and Tyr-Gln-Lys-Phe-Gln-Tyr) have also demonstrated ACE inhibitory activity [28]. Co-culturing the specific microorganisms (lactic acid bacteria and yeasts) enhanced ACE-inhibitory peptide production due to synergistic growth [29]. Dairy products also serve as carriers for yeast isolates, with the potential to increase the production of ACE-inhibiting peptides during fermentation. Consequently, microbial co-cultures potentially offer improved ACE-inhibitory peptide yields through mutual growth enhancement. Hydrolysates derived from whey proteins represent a major source of ACE-inhibitory peptides [30], whereas meat proteins typically exhibit lower hydrolysis rates during microbial fermentation due to the limited proteolytic activity of lactic acid bacteria in meat-based substrates [31].

The application of microbial fermentation to produce immunologically active peptides has been increasingly reported. Chi et al. [32] employed defatted soybean meal as a substrate and, by optimizing fermentation conditions, achieved a peptide conversion rate of up to 79.33%. Zhang et al. [33] conducted mixed fermentation of fresh camel milk using lactic acid bacteria and yeast, followed by ultrafiltration and gel chromatography to isolate three peptide fractions. In vitro lymphocyte assays revealed that two of these fractions significantly stimulated murine splenic lymphocytes. High-performance liquid chromatography-tandem mass spectrometry (HPLC-MS/MS) analysis further identified three peptide sequences within the third fraction that matched regions of glycosylation-dependent cell adhesion molecule-1, whey acidic protein, and α-lactalbumin. Sun et al. [34] identified immunologically active peptides glycine-leucine-proline-asparagine (GLPN) from *Lactobacillus helveticus*-fermented milk using 3-(4,5-dimethylthiazol-2-yl)-2,5-diphenyltetrazolium bromide (MTT) assay, neutral red uptake, and nitric oxide assays.

GLPN was found to significantly enhance macrophage activity at 0.1 g/L without inducing excessive cytokine expression at 0.5 g/L under lipopolysaccharide (LPS) stimulation. Liu et al. [35] utilized *Bacillus subtilis*, *Aspergillus oryzae*, and *Aspergillus niger* to ferment ovine placental residues. After membrane-based purification, the peptide fractions were evaluated for the immunostimulatory potential. At a concentration of 100 μg/mL, peptides derived from *A niger* and *B. subtilis* fermentation yielded stimulation indices of 24.89% and 22.81%, respectively.

2.2.1.3 Chemical Hydrolysis

Chemical hydrolysis involves treating proteins with acid or alkaline solutions under controlled temperature to cleave peptide bonds and disrupt tertiary protein structure, thereby yielding low-molecular-weight peptides [36]. Common reagents include hydrochloric acid, phosphoric acid, and sodium hydroxide. This approach is particularly suitable for structural proteins rich in collagen or keratin [37]. Advantages of this method include procedural simplicity and low cost. However, it often leads to degradation of sensitive amino acids, thereby reducing the nutritional value of peptides [38, 39]. For example, acid hydrolysis destroys tryptophan, partially degrades methionine, and converts glutamine and asparagine to the corresponding acids. Alkaline hydrolysis can result in the extensive degradation of most amino acids [2]. Furthermore, the hydrolysis sites are difficult to control, and the resulting peptide quality can be inconsistent. Additional steps are required to remove residual acids or bases, limiting the applicability of chemical hydrolysis in producing immunologically active peptides.

2.2.2 Chemical Synthesis

Chemical synthesis is commonly used in a laboratory for preparing peptides with specific amino acid sequences. This category includes solution-phase synthesis, solid-phase synthesis, and recombinant genetic techniques [40]. Solution-phase synthesis, a traditional method, involves peptide assembly in homogeneous liquid environments. However, after each coupling step, extensive purification is required to remove unreacted reagents and byproducts, making this approach time-consuming and technically demanding [41].

Solid-phase peptide synthesis (SPPS) typically proceeds from the C-terminus to the N-terminus. A major advantage of SPPS is the ability to perform all reactions in a single vessel, where unreacted materials and byproducts can be easily removed by washing. This greatly simplifies purification and eliminates the need to isolate intermediates [42]. To minimize side reactions, the amino acid side chains are protected and the carboxyl group is activated prior to reaction. SPPS is currently the most effective method for synthesizing peptides comprising 10–100 residues at a small scale [43].

2.2.3 Recombinant Expression Technologies

With the rapid development and widespread application of biotechnology, genetic engineering has emerged as a promising strategy for the large-scale production of bioactive peptides. By introducing exogenous bioactive peptide genes into suitable host organisms, these peptides can be efficiently expressed and secreted [44, 45]. This approach is particularly relevant for AFPs, which have seen broad application potential.

2.2.3.1 Genetic Engineering of Fish-Derived Antifreeze Proteins or Peptides

Among various expression systems, *Escherichia coli* remains the most widely used prokaryotic host for expressing cloned AFPs due to its advantages such as rapid growth, short cultivation cycles, simple culture requirements, low production costs, and high yield [46]. The expressed AFPs may exist either as soluble active proteins or as inactive inclusion bodies. Chen et al. [47] successfully constructed a recombinant plasmid (pET22b-His6-*AFP IV*) by splicing the type IV *AFP* gene from Atlantic salmon with the pET-22b vector and transformed it into *E. coli* BL21 (DE3). Protein expression was induced by isopropyl β-D-1-thiogalactopyranoside, and the recombinant protein was purified using Chelating Sepharose Fast Flow affinity chromatography. Yeh et al. [48] expressed *AFP I* in a lactic acid bacteria system and demonstrated that its inclusion in frozen meat and dough products effectively reduced water loss and improved product quality. Zhang et al. [49] heterologously expressed the antifreeze protein *SF-P* in *Lactococcus lactis*.

While prokaryotic systems are convenient, they often yield inactive inclusion bodies and lack post-translational modification capabilities. To overcome these limitations, yeast-based eukaryotic expression systems have been widely adopted. Compared with prokaryotic expression systems, it not only features fast growth, genetic stability, and simplified purification but also enables proper folding and post-translational modifications [50, 51]. Li et al. [52] used the secretion vector pGAPZα A to express a C-terminal His-tagged herring *AFP II* in *Pichia pastoris* strains X-33 and SMD1168H. The recombinant protein was secreted directly into the culture medium and exhibited functional equivalence to native herring AFPs. Furthermore, low-temperature induction significantly enhanced expression yields, potentially due to improved protein folding efficiency and increased cellular activity.

2.2.3.2 Genetic Engineering of Insect-Derived Antifreeze Proteins

Insect-derived AFPs have also gained traction as targets for genetic engineering. Various vectors and host organisms have been utilized, including yeast systems capable of expressing disulfide-rich AFPs [53, 54]. Zhao et al. [55] isolated the *MpAFP5* gene from *Microdera punctipenis dzunarica*, a desert beetle, and cloned it into pGEX4T-1 and pGAPZα A vectors. These constructs were transformed into *E. coli* BL21 and *Pichia pastoris* SMD1168, respectively, yielding high-activity recombinant AFPs. Liu et al. [56] synthesized an *AFP* gene based on the sequence of *Dendroides canadensis* from GenBank and expressed it in *E. coli* using the pGEX-4T-1 vector. The resulting fusion protein exhibited significant cryoprotective effects. Gauthier et al. [57] successfully expressed *AFP 337* from the spruce budworm in a prokaryotic system. Leinala et al. [58] further expressed *AFP 501* in a prokaryotic host, refolding it from inclusion bodies into a functional protein. Liou et al. [59] reported the successful expression of an 8.5 kDa *AFP* from the yellow mealworm using an *E. coli* expression system.

2.2.3.3 Genetic Engineering of Plant-Derived Antifreeze Proteins

Early genetic engineering focused primarily on fish- and insect-derived AFPs. However, plant AFPs, which are taxonomically distant from these two groups and exhibit differences in gene structure, present challenges in heterologous expression. It has been shown that endogenous plant *AFP* genes are more suitable for functional expression within plant systems [60]. In a seminal study, Worrall et al. [61] identified and cloned an *AFP* gene from carrot, marking the first reported plant-derived AFP. The purified protein displayed significant thermal hysteresis and recrystallization inhibition activity. When expressed in transgenic tobacco, the thermal hysteresis value of the homogenate was 0.35 °C, demonstrating successful functional expression. Griffith et al. [62] identified two *AFP* components from winter rye, corresponding to CHT9 (31.7 kDa) and CHT46 (24.8 kDa). Both genes were successfully expressed in *E. coli*, and both fusion proteins exhibited antifreeze activity. Since then, *AFP* genes have been isolated and characterized from various plant species, including *Hydrangea macrophylla*, winter wheat, *Ammopiptanthus mongolicus*, cabbage, *European spruce*, peach, and *Rhodiola tangutica*, enriching the genetic resources available for plant antifreeze engineering [63].

2.2.3.4 Chemical Modification

The chemical modification method involves the process of introducing certain functional groups into proteins or inducing polymerization or cleavage reactions in the side chains of amino acid residues or polypeptide chains through chemical means, thereby altering the physical and chemical properties and functional characteristics of proteins [64]. The common chemical modification strategies include alkali and acid treatments, acylation, deamidation, phosphorylation, glycosylation, and thiolation [65]. Wang et al. [66] reported that a protein-dextran mixture at a 1:4 ratio, partially glycosylated in phosphate buffer at various pH values under 70–80 °C, exhibited an 8-fold increase in thermal stability after glycosylation.

Antifreeze glycoproteins (AFGPs) are highly glycosylated and the carbohydrate moieties are crucial for antifreeze activity. However, due to the challenges associated with their extraction and purification, chemically synthesized or chemically modified AFGP analogs have been explored to elucidate the structure–function relationship of these biomolecules [67]. Recent research has focused on developing glycosylation-based analogs of antifreeze proteins. For instance, a synthetic AFP composed of four Ser-Gly-Gly tripeptide repeats with a galactosyl group attached to the C-terminal serine residue demonstrated strong recrystallization inhibition activity [68]. Furthermore, it has been shown that the type of sugar residue, the glycosylation site on the peptide backbone, and the nature of adjacent amino acid residues significantly influence antifreeze activity. For example, AFP analogs modified with galactose exhibited superior recrystallization inhibition compared to those modified with mannose [69]. Additionally, the presence of proline residues, which restrict backbone flexibility and promote secondary structure formation in solution, enhances low-temperature protective activity in AFP analogs [70].

2.3 Separation and Purification Techniques of Peptides

Bioactive peptides are amphiphilic compounds derived from the hydrolysis of proteins. Consequently, separation and purification strategies commonly used for proteins are also applicable to peptides. Table 2.2 summarizes the standard purification methods used for proteins [3, 71, 72]. Currently, the most widely employed methods for separating and purifying bioactive peptides include ultrafiltration, ion exchange chromatography, gel filtration chromatography, reversed-phase high-performance liquid chromatography (RP-HPLC), capillary electrophoresis (CE), and immobilized metal affinity chromatography (IMAC).

Table 2.2 Classification of common methods for protein separation and purification.

Category	Separation methods	Separation basis	Applications
Centrifugation	Differential centrifugation	Density, sedimentation coefficient	Solid–liquid separation
Extraction	Supercritical extraction	Solubility	Separation of small-molecule proteins
	Aqueous two-phase extraction	Solubility	Protein separation
Membrane processes	Fractional salting	Solubility	Protein separation
	Dialysis	Molecular weight	Solvent exchange, desalting
	Ultrafiltration	Molecular weight	Protein concentration
	Microfiltration	Particle size	Solid–liquid separation
	Electroosmosis	Electric charge	Desalting
Electrophoresis	Molecular sieve electrophoresis	Molecular weight	Protein separation
	Moving boundary electrophoresis	Electromobility	Protein separation
	Continuous flow electrophoresis	Electromobility	Protein separation
	Isoelectric focusing	Isoelectric point differences	Protein separation
	Gel filtration	Molecular weight	Protein separation
	Ion exchange	Charge	Protein separation
Chromatography	Affinity chromatography	Specific binding	Separation of antigens, antibodies, and certain enzymes
	Reversed-phase chromatography	Hydrophobicity	Peptide or protein separation
	Normal-phase chromatography	Surface non-specific binding	Protein separation
	Hydrophobic chromatography	Hydrophobicity	Protein separation
	Chromatographic aggregation	Isoelectric point differences	Protein separation

2.3.1 Ultrafiltration

Ultrafiltration is a physical separation process that utilizes membranes with varying pore sizes to selectively filter solutes based on molecular size under pressure [73]. The molecular weight cut-off typically ranges from 1 to 500 kDa, with pore sizes between 10 and 100 nm. Macromolecular solutes and particulates, such as starch or undigested proteins, are retained, while small peptides and solvents pass through the membrane. Ultrafiltration membranes can be used continuously and maintain stable separation efficiency and throughput over time. Compared with traditional methods, ultrafiltration enhances product purity, reduces reagent consumption, enables continuous operation, and shortens production cycles.

2.3.2 Ion Exchange Chromatography

Ion exchange chromatography is a column-based method that exploits the differential affinity of ions for charged groups on ion exchange resins. The separation is achieved by reaching equilibrium between exchangeable ions on the resin and target ions in the sample matrix [74]. This method features high sensitivity, good reproducibility and selectivity, and rapid analysis, making it one of the most frequently used chromatographic techniques. Although ion-exchange chromatography effectively removes most impurities, fractions exhibiting similar charge properties often co-elute with the target active fractions. Moreover, buffer salts and residual NaCl are typically difficult to eliminate completely, which may interfere with subsequent assays of antioxidant activity [75].

2.3.3 Gel Filtration Chromatography

Gel filtration chromatography, also known as size-exclusion chromatography, molecular sieve chromatography, or gel permeation chromatography, is a separation technique that exploits differences in molecular size by passing a mixture through a porous gel matrix [71]. Sephadex G-50 gel filtration chromatography, for instance, offers a defined molecular weight fractionation range, enabling the resolution of solutes into three distinct categories: First, molecules with molecular weights exceeding the upper limit of the fractionation range are entirely excluded from the pores of the gel particles. These molecules migrate vertically through the interstitial spaces between the gel beads, encountering minimal resistance and following the shortest elution path, thereby eluting first from the column. Second, molecules whose molecular weights fall within the fractionation range partially penetrate the gel matrix to varying extents according to their size, allowing for efficient separation within this molecular weight window. Third, molecules with molecular weights below the lower limit of the fractionation range can fully access the internal pore structure, resulting in the longest retention times and greatest resistance, and thus are eluted last. Based on this principle, samples containing molecules of different molecular weights can be effectively separated [76, 77], as illustrated in Figure 2.2.

One of the key advantages of gel filtration chromatography is its high recovery rate, with minimal loss of biological activity. The principal types of gels employed include cross-linked

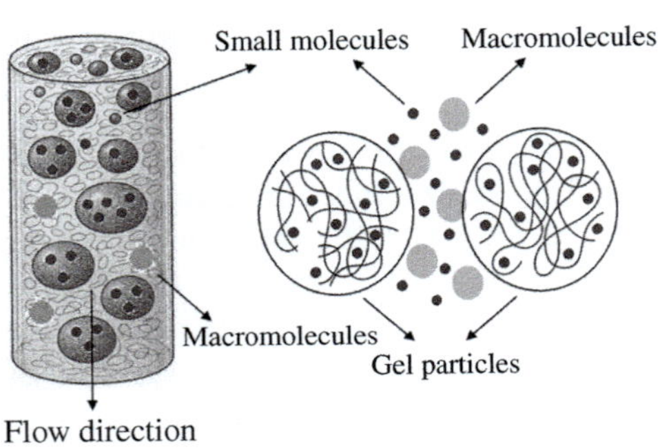

Figure 2.2 Schematic diagram of gel filtration chromatography principle.

dextran gels, agarose gels, and polyacrylamide gels. Over several decades of practical application and methodological development, gel filtration chromatography has evolved into a robust and reliable technique for both the separation and purification of biomolecules and the determination of their molecular weights. Its procedural simplicity, ease of instrumentation, and straightforward data analysis have contributed to its widespread adoption in biochemical research [3, 77].

2.3.4 Reversed-Phase High-Performance Liquid Chromatography

Since the mid-1970s, RP-HPLC has been widely applied in peptide research for purification, qualitative and quantitative analysis, and molecular weight determination. By the 1980s, it had become the method of choice for peptide separation [78]. RP-HPLC offers several advantages: it primarily uses water-based mobile phases, which are compatible with the physicochemical properties of peptides, and it provides higher resolution and broader applicability than other separation methods. Therefore, RP-HPLC is now a standard technique in the separation and purification of peptides.

2.3.5 Capillary Electrophoresis

CE is a modern analytical technique characterized by high resolution, speed, sensitivity, and automation. CE is especially suited for analyzing proteins, peptides, and other biomolecules [79]. It has been employed in the analysis of peptide and protein composition, identification of genetic engineering products, pharmaceutical development, food analysis, agricultural biotechnology, water treatment, and polymer chemistry.

High-performance capillary electrophoresis combines electrophoretic and mass spectrometry, offering high efficiency, fast analysis, minimal sample requirements, and ease of automation. Its use in the separation and analysis of peptides, proteins, and nucleic acids has grown rapidly in recent years [80].

2.3.6 Immobilized Metal Affinity Chromatography

IMAC utilizes immobilized transition metal ions chelated to a solid-phase matrix to selectively bind peptides containing specific amino acid residues. The target peptides are then eluted using competitive agents [81]. IMAC is noted for its high affinity and selectivity, biocompatibility, and reusability, making it a common first-step purification technique for metal-chelating peptides. Additionally, hydroxyapatite chromatography has gained popularity due to its high resolution. Hydroxyapatite interacts with calcium ions and binds strongly to specific peptides, making it suitable for purifying calcium-chelating peptides [3].

2.4 Computer-Aided Screening Technologies for Bioactive Peptides

Traditional methods for screening bioactive peptides typically rely on sequential purification and separation processes, which are complex, time-consuming, costly, and yield low peptide recovery rates. Consequently, these limitations significantly impede the practical production and industrial development of bioactive peptides. Computer-aided peptide screening technology has recently emerged as a cutting-edge interdisciplinary technique, integrating molecular docking, molecular dynamics simulations, high-resolution mass spectrometry, QSAR modeling, chemometric methods, bioinformatics, artificial intelligence techniques, and peptidomics database mining [2, 82, 83]. This method enables efficient and accurate identification of bioactive peptides from food-derived proteins by combining *in silico* predictions with experimental validations, thereby overcoming the limitations of traditional methods and substantially enhancing research efficiency and precision [84].

Computational peptide screening begins with retrieving amino acid sequence data from proteomics and peptidomics databases such as BIOPEP-UWM, UniProtKB, and the NCBI Protein database. Next, computational tools, including BIOPEP and PeptideCutter, predict potential proteolytic cleavage sites, enabling the virtual (*in silico*) enzymatic digestion of proteins and subsequent prediction of possible peptide sequences. The resultant peptides are then screened for biological activity, solubility, toxicity, and allergenicity using specialized prediction tools such as PeptideRanker, Pepcalc, ToxinPred, and AllerTop. For instance, Martini et al. [85] integrated peptidomics with computational simulations to identify a potent DPP-IV inhibitory peptide, SAPI ($IC_{50} = 57.7\,\mu M$), from common bean protein hydrolysates. Similarly, Gomez et al. [86] employed BIOPEP simulations to predict potential ACE and DPP-IV inhibitory peptides from oyster proteins, experimentally validating significant DPP-IV inhibitory activity ($44.37\% \pm 0.09\%$) at 2 mg/mL concentration. This research highlights that integrating computational predictions with experimental verification substantially improves bioactive peptide screening outcomes.

Molecular docking and QSAR modeling play crucial roles in elucidating peptide-target interaction mechanisms. Molecular docking rapidly identifies peptides with high binding affinity to target proteins (such as enzymes or receptors) by simulating interaction modes and binding energies, thereby providing preliminary insights into activity mechanisms [87]. Zhao et al. [88] utilized docking to screen three DPP-IV inhibitory peptides (ADF, MIR, and FGR) from egg protein hydrolysates, with peptide MIR exhibiting the strongest activity ($IC_{50} = 4.86\,mM$), largely attributed to hydrogen bonding interactions with the DPP-IV active site. QSAR models mathematically correlate peptide structural properties with biological activities, thus improving screening accuracy. Nongonierma et al. [89] developed a QSAR model based on 56 competitive DPP-IV inhibitory peptides characterized by amino acid descriptors, achieving high agreement between predicted and experimental values. Similarly, Wang et al. [90] successfully applied QSAR modeling to predict six potent tripeptides with DPP-IV inhibitory activity from black bass protein hydrolysates, confirming the method's practical accuracy.

Despite significant advantages, computational screening techniques exhibit certain limitations. First, peptides predicted through computational methods do not always correlate fully with actual enzymatic digestion products. Nongonierma et al. [91] reported that only approximately 60% of peptides predicted from α-lactalbumin hydrolysates were experimentally detectable. Second, current computational tools may yield false-positive predictions due to limited accuracy in predicting cleavage sites for proteases with low specificity. Additionally, molecular docking simulations may inadequately replicate true biological interactions, being constrained by structural resolution limitations and dynamic conformational changes of target proteins. Therefore, current research emphasizes integrating computational predictions with experimental validations, especially high-throughput peptidomics sequencing, to enhance screening accuracy and reliability.

References

1 Liao, A.M., Xu, T., Thakur, K. et al. (2025). Recent updates on plant-derived anti-fatigue peptides as novel energy boosters: expansion from mechanisms of action and beyond. *Trends in Food Science & Technology* 163: 105160.

2 Liang, S., Wang, X., Zheng, T. et al. (2025). Advancing bioactive peptide research: a comprehensive evaluation framework bridging computational prediction, organoid models, and clinical translation for functional innovation. *Journal of Agricultural and Food Chemistry* 73: 16673–16696.

3 Sun, J., Chen, Y., Yang, Y. et al. (2025). Research progress on bioactive peptides from animal sources: a comprehensive review. *Food Chemistry* 490: 145006.

4 Zheng, Z., Yang, X., Liu, J. et al. (2017). Effects of wheat peptide supplementation on anti-fatigue and immunoregulation during incremental swimming exercise in rats. *RSC Advances* 69 (7): 43345–43355.

5 Kul Lysaar, T., Songisepp, E., Mikelsaar, M. et al. (2003). Antioxidative probiotic fermented goats' milk decreases oxidative stress-mediated atherogenicity in human subjects. *British Journal of Nutrition* 90: 449–456.

6 López-García, G., Dublan-García, O., Arizmendi-Cotero, D., and Gómez Oliván, L.M. (2022). Antioxidant and antimicrobial peptides derived from food proteins. *Molecules* 27 (4): 1343.

7 Vidal-Limon, A., Aguilar-Toala, J.E., and Liceaga, A.M. (2022). Integration of molecular docking analysis and molecular dynamics simulations for studying food proteins and bioactive peptides. *Journal of Agricultural and Food Chemistry* 70 (4): 934–943.

8 Lin, W.W., Wang, Y.J., Ko, C.W. et al. (2020). Cyclic peptide inhibitors of the Tsg101 UEV protein interactions refined through global docking and Gaussian accelerated molecular dynamics simulations. *Polymers* 12 (10): 2235.

9 An, G., Wei, M., Wang, Z. et al. (2022). Effects of enzymolysis method on the preparation of peptides from wheat flour. *Food Bioscience* 49: 101956.

10 Huang, T., Bu, G., and Chen, F. (2018). The influence of composite enzymatic hydrolysis on the antigenicity of β-conglycinin in soy protein hydrolysates. *Journal of Food Biochemistry* 42 (5): e12544.

11 Pintado, M.E., Pintado, A.E., and Malcata, F.X. (1999). Controlled whey protein hydrolysis using two alternative proteases. *Journal of Food Engineering* 42 (1): 1–13.

12 Liu, L., Li, S., Zheng, J. et al. (2020). Safety considerations on food protein-derived bioactive peptides. *Trends in Food Science & Technology* 96: 199–207.

13 Maqsoudlou, A., Mahoonak, A.S., Mora, L. et al. (2019). Controlled enzymatic hydrolysis of pollen protein as a promising tool for the production of potential bioactive peptides. *Journal of Food Biochemistry* 43: e12819.

14 Mardani, M., Badakne, K., Farmani, J., and Aluko, R.E. (2023). Antioxidant peptides: overview of production, properties, and applications in food systems. *Comprehensive Reviews in Food Science and Food Safety* 22 (1): 46–106.

15 Ishak, N.H. and Sarbon, N.M. (2018). A review of protein hydrolysates and bioactive peptides deriving from wastes generated by fish processing. *Food and Bioprocess Technology* 11 (1): 2–16.

16 Wang, S.Y., Zhao, J., Xu, Z.B., and Wu, J.H. (2011). Preparation and partial isolation of antifreeze peptides from fish gelatin with hypothermia protection activity. *Applied Mechanics and Materials* 140: 411–415.

17 Cao, H., Zhao, Y., Zhu, Y.B. et al. (2016). Antifreeze and cryoprotective activities of ice-binding collagen peptides from pig skin. *Food Chemistry* 194: 1245–1253.

18 Du, L. and Betti, M. (2016). Chicken collagen hydrolysate cryoprotection of natural actomyosin: mechanism studies during freeze–thaw cycles and simulated digestion. *Food Chemistry* 211: 791–802.

19 Wu, J.H., Rong, Y.Z., Wang, Z.W. et al. (2015). Isolation and characterisation of sericin antifreeze peptides and molecular dynamics modelling of their ice-binding interaction. *Food Chemistry* 174 (1): 621–629.

20 Duman, J.G. and DeVries, A.L. (1972). Freezing behavior of aqueous solutions of glycoproteins from the blood of an Antarctic fish. *Cryobiology* 9 (5): 469–472.

21 He, H.L., Chen, X.L., Sun, C.Y. et al. (2006). Analysis of novel angiotensin-I-converting enzyme inhibitory peptides from protease-hydrolyzed marine shrimp *Acetes chinensis*. *Journal of Peptide Science* 12 (11): 726–733.

22 Lin, L. (2003). Preparation of high Fischer ratio oligopeptides from protease hydrolysate of corn gluten meal and application. Master's thesis, Northeast Agricultural University, China. (in Chinese)

23 Yang, H., Hou, W., Zhao, L. et al. (2019). Optimization on preparation of high Fischer ratio oligopeptides by two-step enzymolysis from corn yellow powder. *Journal of Food Science and Technology* 37 (6): 100–107.

24 Zheng, M. (2013). Preparation of corn high Fischer ratio oligopeptide and physiological function study. Master's thesis. University of Jinan, China. (in Chinese)

25 Ma, J., Wu, W., Liao, J. et al. (2024). Preparation of *Dendrobium officinale* polysaccharide by lactic acid bacterium fermentation and its protective mechanism against alcoholic liver damage in mice. *Journal of Agricultural and Food Chemistry* 72 (31): 17633–17648.

26 Peng, Q., Zheng, H., Li, J. et al. (2024). Impact of *Bacillus subtilis* on Chinese yellow rice wine (*Huangjiu*) fermentation: method variations and flavor analysis. *Food Chemistry* 460: 140658.

27 Solieri, L., Rutella, G.S., and Tagliazucchi, D. (2015). Impact of non-starter lactobacilli on release of peptides with angiotensin-converting enzyme inhibitory and antioxidant activities during bovine milk fermentation. *Food Microbiology* 51: 108–116.

28 del Mar Contreras, M., Carrón, R., Montero, M.J. et al. (2009). Novel casein-derived peptides with antihypertensive activity. *International Dairy Journal* 19 (10): 566–573.

29 Chen, L., Wang, L., Li, J., and Shu, G. (2021). Antihypertensive potential of fermented milk: the contribution of lactic acid bacteria proteolysis system and the resultant angiotensin-converting enzyme inhibitory peptide. *Food & Function* 12: 11121–11131.

30 Tagliazucchi, D., Shamsia, S., and Conte, A. (2016). Release of angiotensin converting enzyme-inhibitory peptides during in vitro gastrointestinal digestion of camel milk. *International Dairy Journal* 56: 119–128.

31 Lee, S.Y. and Hur, S.J. (2017). Antihypertensive peptides from animal products, marine organisms, and plants. *Food Chemistry* 228: 506–517.

32 Chi, X., Zhang, T., and Wang, N. (2012). The fermentation condition of soybean polypeptide by microbial fermentation method. *Food Science and Technology* 37 (2): 69–72. (in Chinese).

33 Zhang, Y., Yu, L., Xiao, X. et al. (2015). Isolation and identification of immune active peptide from Xinjiang fermented camel milk. *Dairy Industry* 43 (11): 12–14. (in Chinese).

34 Sun, G., Zhan, D., Ma, L. et al. (2013). Study on immune activity of bioactive oligopeptides derived from *Lactobacillus helveticus* fermented milk. *Dairy Industry* 41 (9): 4–7. (in Chinese).

35 Liu, W., Hou, Y., Wang, H. et al. (2015). Comparative study on activities of placenta peptide produced by different microbiological strains ferment goat placenta residue. *Food Industry* 36 (8): 149–153. (in Chinese).

36 Ananey-Obiri, D., Matthews, L.G., and Tahergorabi, R. (2019). Proteins from fish processing by-products. In: *Proteins: Sustainable Source, Processing and Applications*, 163–191. Elsevier.

37 Hou, Y., Wu, Z., Dai, Z. et al. (2017). Protein hydrolysates in animal nutrition: industrial production, bioactive peptides, and functional significance. *Journal of Animal Science and Biotechnology* 8 (3): 513–525.

38 Corrêa, J.A.F., de Melo Nazareth, T., da Rocha, G.F., and Luciano, F.B. (2023). Bioactive antimicrobial peptides from food proteins: perspectives and challenges for controlling foodborne pathogens. *Pathogens* 12 (3): 477.

39 Nirmal, N.P., Santivarangkna, C., Rajput, M.S. et al. (2022). Valorization of fish byproducts: sources to end-product applications of bioactive protein hydrolysate. *Comprehensive Reviews in Food Science and Food Safety* 21 (2): 1803–1842.

40 Jakubczyk, A., Karas, M., Rybczynska-Tkaczyk, K. et al. (2020). Current trends of bioactive peptides: new sources and therapeutic effect. *Food* 9 (7): 846.

41 Shimodaira, S., Asano, Y., Arai, K., and Iwaoka, M. (2017). Selenoglutathione diselenide: unique redox reactions in the GPx-like catalytic cycle and repairing of disulfide bonds in scrambled protein. *Biochemistry* 56 (42): 5644–5653.

42 Rossino, G., Marchese, E., Galli, G. et al. (2023). Peptides as therapeutic agents: challenges and opportunities in the green transition era. *Molecules* 28 (20): 7165.

43 Mijalis, A.J., Thomas, D.A. III, Simon, M.D. et al. (2017). A fully automated flow-based approach for accelerated peptide synthesis. *Nature Chemical Biology* 13: 464–466.

44 Götze, S., Arp, J., Lackner, G. et al. (2019). Structure elucidation of the syringafactin lipopeptides provides insight into the evolution of nonribosomal peptide synthetases. *Chemical Science* 10: 10979–10990.

45 Loureiro, C., Medema, M.H., van der Oost, J., and Sipkema, D. (2018). Exploration and exploitation of the environment for novel specialized metabolites. *Current Opinion in Biotechnology* 50: 206–213.

46 Chen, X., Wu, J., Yang, F. et al. (2023). New insight into the mechanism by which antifreeze peptides regulate the physiological function of *Streptococcus thermophilus* subjected to freezing stress. *Journal of Advanced Research* 45: 127–140.

47 Chen, Y., An, H., Zhang, Q. et al. (2017). Expression and purification of *salmo salar* Type IV antifreeze protein. *Biotechnology* 27 (3): 218–222. (in Chinese).

48 Yeh, C.M., Kao, B.Y., and Peng, H.J. (2009). Production of a recombinant type I antifreeze protein analogue by *Lactococcus lactis* and its applications on frozen meat and frozen dough. *Journal of Agricultural and Food Chemistry* 57 (14): 6216–6223.

49 Zhang, L., Huang, X., Ma, T. et al. (2017). Expression of recombinant antifreeze peptide in *Lactococcus lactis* and its antifreeze activity. *Food & Machinery* 33 (10): 123–127.

50 Kim, H.J., Shim, H.E., Lee, J.H. et al. (2015). Ice-binding protein derived from *Glaciozyma* can improve the viability of cryopreserved mammalian cells. *Journal of Microbiology and Biotechnology* 25: 1989–1996.

51 Hashim, N.H.F., Bharudin, I., Nguong, D.L.S. et al. (2013). Characterization of Afp1, an antifreeze protein from the psychrophilic yeast *Glaciozyma antarctica* PI12. *Extremophiles* 17: 63–73.

52 Li, Z., Xiong, F., Lin, Q. et al. (2001). Low-temperature increases the yield of biologically active herring antifreeze protein in *Pichia pastoris*. *Protein Expression and Purification* 21: 438–445.

53 Graham, L.A., Walker, V.K., and Davies, P.L. (2000). Developmental and environmental regulation of antifreeze proteins in the mealworm beetle *Tenebrio molitor*. *European Journal of Biochemistry* 267 (21): 6452–6458.

54 Zhang, X., Qi, H., Yang, J. et al. (2023). Development of low immunogenic antifreeze peptides for cryopreservation. *Industrial & Engineering Chemistry Research* 62 (31): 12063–12072.

55 Zhao, Q., Ma, J., Xue, N. et al. (2005). Cloning of a cDNA encoding antifreeze protein in *Microdera punctipennis dzunarica* (Coleoptera: Tenebrionidae) and its activity assay. *Acta Entomologica Sinica* 48 (5): 667–673. (in Chinese).

56 Liu, Z., Zhang, F., Mao, X., and Wang, Y. (2004). Study on expression of heterologous proteins using *Pichia pastoris*. *Biotechnology* 14 (1): 56–58.

57 Gauthier, S.Y., Kay, C.M., Sykes, B.D. et al. (1998). Disulfide bond mapping and structural characterization of spruce budworm antifreeze protein. *European Journal of Biochemistry* 258 (2): 445–453.

58 Leinala, E.K., Davies, P.L., Doucet, D. et al. (2002). A β-helical antifreeze protein isoform with increased activity. *Journal of Biological Chemistry* 277 (36): 33349–33352.

59 Liou, Y.C., Daley, M.E., Graham, L.A. et al. (2000). Folding and structural characterization of highly disulfide-bonded beetle antifreeze protein produced in bacteria. *Protein Expression and Purification* 19 (1): 148–157.

60 Dong, X., Wei, J., Cui, J. et al. (2023). Research progress in plant antifreeze protein. *Chinese Bulletin of Botany* 58 (6): 966–981.

61 Worrall, D., Elias, L., Ashford, D. et al. (1998). A carrot leucine-rich-repeat protein that inhibits ice recrystallization. *Science* 282 (2): 115–117.

62 Griffith, M. and Yaish, M.W. (2004). Antifreeze proteins in overwintering plants: a tale of two activities. *Trends in Plant Science* 9 (8): 399–405.

63 Deng, S.Y. (2010). Cloning, prokaryotic expression and transgenic study of antifreeze protein gene from ryegrass. Master's thesis, Central South University of Forestry and Technology, China. (in Chinese)

64 Horvat, Š. and Jakas, A. (2004). Peptide and amino acid glycation: new insights into the Maillard reaction. *Journal of Peptide Science* 10 (3): 119–137.

65 Shadrack, S.M., Wang, Y., Mi, S. et al. (2025). Enhancing bioavailability and functionality of plant peptides and proteins: a review of novel strategies for food and pharmaceutical applications. *Food Chemistry* 485: 144440.

66 Wang, Q. and Ismail, B. (2012). Effect of Maillard-induced glycosylation on the nutritional quality, solubility, thermal stability and molecular configuration of whey protein. *International Dairy Journal* 25 (2): 112–122.

67 Bang, J.K., Lee, J.H., Murugan, R.N. et al. (2013). Antifreeze peptides and glycopeptides, and their derivatives: potential uses in biotechnology. *Marine Drugs* 11 (6): 2013–2041.

68 And, S.L. and Ben, R.N. (2005). C-linked galactosyl serine AFGP analogues as potent recrystallization inhibitors. *Organic Letters* 7 (12): 2385–2388.

69 Tam, R.Y., Ferreira, S.S., Czechura, P. et al. (2008). Hydration index—a better parameter for explaining small-molecule hydration in inhibition of ice recrystallization. *Journal of the American Chemical Society* 130 (51): 17494–17501.

70 Heggemann, C., Budke, C., Majer, Z. et al. (2010). Antifreeze glycopeptide analogues: microwave-enhanced synthesis and functional studies. *Amino Acids* 38 (1): 213–222.

71 Wang, X., Yu, H., Xing, R., and Li, P. (2017). Characterization, preparation, and purification of marine bioactive peptides. *BioMed Research International* 2017 (1): 9746720.

72 Zhang, J., Li, M., Han, Z., and Shao, J.H. (2025). A review of insect peptides: from extraction and preparation to bioactivities. *Current Opinion in Food Science* 63: 101301.

73 Kong, Y., Huang, S.L., Shao, Y. et al. (2013). Purification and characterization of a novel antithrombotic peptide from Scolopendra subspinipes mutilans. *Journal of Ethnopharmacology* 145 (1): 182–186.

74 Wallace, R.G. and Rochfort, K.D. (2023). Ion-exchange chromatography: basic principles and application. *Methods in Molecular Biology* 2699: 161–177.

75 Wu, R., Wu, C., Liu, D. et al. (2015). Overview of antioxidant peptides derived from marine resources: the sources, characteristic, purification, and evaluation methods. *Applied Biochemistry and Biotechnology* 176: 1815–1833.

76 Huang, D., Li, Y., Cui, F. et al. (2016). Purification and characterization of a novel polysaccharide–peptide complex from Clinacanthus nutans Lindau leaves. *Carbohydrate Polymers* 137: 701–708.

77 Wang, P., Zhang, Y., Hu, J., and Tan, B.K. (2024). Bioactive peptides from marine organisms. *Protein and Peptide Letters* 31 (8): 569–585.

78 Huang, H., Cen, J., Yang, D. et al. (2022). Isolation and characterization of antioxidant peptides from oyster (*Crassostrea rivularis*) protein enzymatic hydrolysates. *Food Science & Nutrition* 11 (1): 261–273.

79 Scriba, G.K. (2016). Separation of peptides by capillary electrophoresis. *Methods in Molecular Biology* 1483: 365–391.

80 Benavente, F., Pero-Gascon, R., Pont, L. et al. (2018). Identification of antihypertensive peptides in nutraceuticals by capillary electrophoresis–mass spectrometry. *Journal of Chromatography A* 1579: 129–137.

81 Liu, P., Lan, X., Yaseen, M. et al. (2020). Immobilized metal affinity chromatography matrix modified by poly(ethylene glycol) methyl ether for purification of angiotensin I-converting enzyme inhibitory peptide from casein hydrolysate. *Journal of Chromatography B: Analytical Technologies in the Biomedical and Life Sciences* 1143: 122042.

82 Jia, W., Peng, J., Zhang, Y. et al. (2023). Exploring novel ANGICon-EIPs through ameliorated peptidomics techniques: can deep learning strategies serve as a core breakthrough in peptide structure and function prediction? *Food Research International* 174: 113640.

83 Fernando, I.P.S. and Wu, J. (2025). Food-derived bioactive peptides: the gateway to reach the full potential of food proteins for human health. *Trends in Food Science & Technology* 157: 104896.

84 Chang, J., Wang, H., Su, W. et al. (2025). Artificial intelligence in food bioactive peptides screening: recent advances and future prospects. *Trends in Food Science & Technology* 156: 104845.

85 Martini, S., Cattivelli, A., Conte, A., and Tagliazucchi, D. (2021). Application of a combined peptidomics and in silico approach for the identification of novel dipeptidyl peptidase-IV-inhibitory peptides in in vitro digested pinto bean protein extract. *Current Issues in Molecular Biology* 44 (1): 139–151.

86 Gomez, H.L.R., Peralta, J.P., Tejano, L.A., and Chang, Y.W. (2019). In silico *and* in vitro *assessment of Portuguese oyster (Crassostrea angulata) proteins as precursor of bioactive peptides. International Journal of Molecular Sciences* 20 (20): 5191.

87 Yu, Y., Xu, S., He, R., and Liang, G. (2023). Application of molecular simulation methods in food science: status and prospects. *Journal of Agricultural and Food Chemistry* 71 (6): 2684–2703.

88 Zhao, W.Z., Zhang, D., Yu, Z.P. et al. (2020). Novel membrane peptidase inhibitory peptides with activity against angiotensin converting enzyme and dipeptidyl peptidase IV identified from hen eggs. *Journal of Functional Foods* 64: 103649.

89 Nongonierma, A.B. and Fitzgerald, R.J. (2016). Learnings from quantitative structure–activity relationship (QSAR) studies with respect to food protein-derived bioactive peptides: a review. *RSC Advances* 6 (79): 75400–75413.

90 Wang, K., Yang, X.X., Lou, W.Y., and Li, N. (2020). Discovery of dipeptidyl peptidase 4 inhibitory peptides from largemouth bass (Micropterus salmoides) by a comprehensive approach. *Bioorganic Chemistry* 105: 104432.

91 Nongonierma, A.B., Paolella, S., Mudgil, P., and Sharma, M. (2018). Identification of novel dipeptidyl peptidase IV (DPP-IV) inhibitory peptides in camel milk protein hydrolysates. *Food Chemistry* 244: 340–348.

3

Antifreeze Peptides

3.1 Introduction

The high-water content inherent in fresh foods creates an environment conducive to microbial proliferation and accelerated chemical reactions, ultimately leading to spoilage [1]. While cryopreservation is widely recognized as the most effective strategy for long-term food preservation [2–5], it introduces structural challenges. Ice crystal formation and subsequent recrystallization during freezing and storage induce mechanical damage to cellular and tissue structures, significantly compromising product quality [6, 7]. This damage is exacerbated by temperature fluctuations, which perpetuate ice crystal growth and recrystallization cycles, further degrading cellular integrity and diminishing sensory and nutritional attributes [8–10]. The cumulative impact of these phenomena—not only causing quality deterioration but also incurring substantial economic losses—has prompted significant research interest in mitigating ice crystal-related damage [11, 12]. Consequently, the development of strategies to control ice crystal morphology and dynamics during freezing and storage remains a critical priority for maintaining the quality of cryopreserved food products [13].

The incorporation of cryoprotectants into frozen food systems represents a highly effective strategy for mitigating quality degradation during freezing and frozen storage [5, 13]. Among these agents, antifreeze peptides (AFPs)—low-molecular-weight proteins or hydrolyzed protein derivatives—demonstrate exceptional capacity to mitigate cold-induced damage under both freezing and supercooled conditions [14]. Based on their production methodology, AFPs are categorized into three distinct classes: enzymatic AFPs, natural AFPs, and synthetic AFPs. These compounds exhibit the dual advantages of safety and high cryoprotective efficacy. Compared to conventional cryoprotectants such as sucrose-sorbitol mixtures, AFPs offer distinct nutritional benefits by reducing sweetness intensity, salt content, and caloric density in frozen food products. This unique profile positions AFPs as a promising non-sweet alternative for cryoprotective applications in frozen food manufacturing. Furthermore, growing research interest in AFPs has expanded their potential applications to functional food ingredient development and food additive innovation, reflecting their multifaceted value in the food science domain.

Food-Derived Bioactive Peptides: Green Preparation, Activity Mechanisms, and Health Applications, First Edition. Shao-Yun Wang, Xu Chen, and Xixi Cai.

Despite the growing interest in antifreeze peptides (AFPs), a comprehensive systematic review integrating their production methodologies, activity assessment protocols, and structure–function correlations remains lacking. To address this gap, this review systematically examines the current technologies and strategies employed in AFP production, purification, functional characterization, and structural analysis. A key focus is placed on elucidating the molecular mechanisms underlying AFP activity and their structure–function relationships, alongside recent advancements in industrial applications within the frozen food sector. Additionally, practical approaches for incorporating AFPs into food matrices are critically evaluated. The review culminates in a discussion of existing challenges and future research directions, including potential strategies for optimizing AFP performance. By synthesizing these findings, this work aims to deepen the mechanistic understanding of AFPs, thereby stimulating innovative research avenues and serving as a foundational reference for the discovery of novel ice-binding modulators in frozen food preservation technologies.

3.2 Production and Purification of AFPs

3.2.1 Evaluation Methods of Antifreeze Activity

The detection of AFP activity is inherently challenging due to the supercooling phenomenon, which occurs when a solution remains in liquid form despite its temperature reaching or falling below the freezing point under constant pressure [15]. This behavior arises primarily from the solution's purity: in highly pure systems, ice nuclei are numerous yet small, delaying spontaneous crystallization. However, alterations in solution purity or external conditions (e.g., temperature gradients, impurity introduction) can promote the formation of critical-sized ice nuclei, triggering solidification [7, 15, 16]. This section systematically examines methodologies for quantifying and characterizing AFP activity, addressing both technical limitations imposed by supercooling and strategies to mitigate these challenges.

3.2.2 Nanoliter Osmometer

The morphology of individual ice crystals can be visualized using a nanoliter osmometer coupled with a microscope. This system comprises four core components: a refrigeration unit, capillary loading mechanism, computerized image processing module, and high-resolution camera system [17], with a schematic diagram presented in Figure 3.1. As a precision analytical instrument, the freezing point osmometer measures solution freezing points by dynamically modulating temperature at variable rates [18]. The nanoliter osmometer, a specialized variant of this technology, offers distinct advantages over conventional systems, including reduced sample volume requirements (nanoliter scale) and enhanced measurement accuracy [18–20]. Temperature regulation in this system leverages the Peltier effect [21], enabling precise control (±0.01 °C) within the operational range of 0 to −9 °C. Critical thermal parameters are defined as follows: the temperature inducing slow ice crystal nucleation corresponds to the freezing point, while the temperature promoting crystal dissolution represents the melting point. The thermos hysteresis (THA) is quantified as the temperature gap

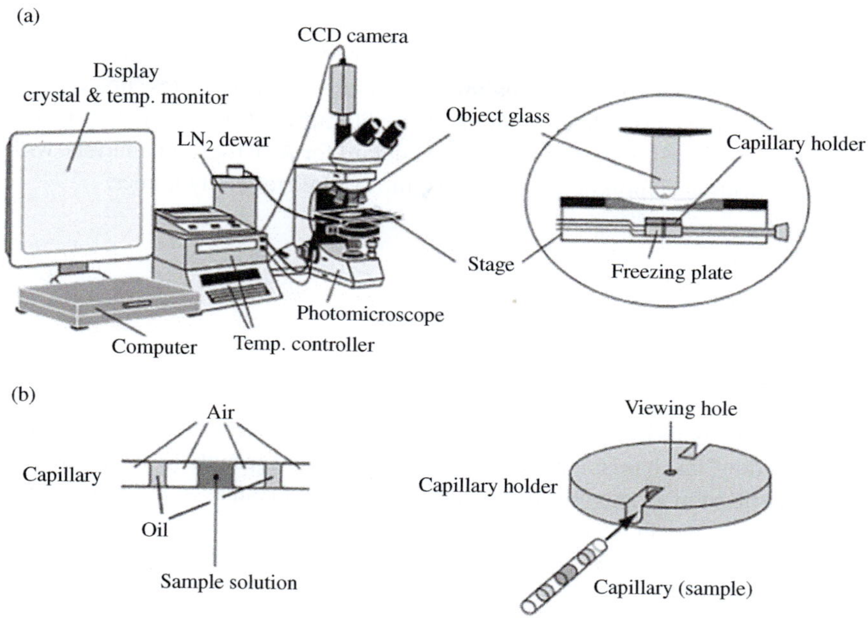

Figure 3.1 Nanoliter osmometer. (a) The photosystem is composed of a temperature controller, a color video charge-coupled device (CCD) camera, a photomicroscope, and an image-processing computer. (b) The capillary cell containing a drop of sample solution and setting of the sample into the holder. *Source:* [17]/John Wiley & Sons.

between these two points [22]. Beyond basic characterization, this methodology facilitates detailed morphological analysis of single ice crystals and enables systematic evaluation of antifreeze protein (AFP) effects on crystal growth patterns [19, 23, 24].

3.2.3 Differential Scanning Calorimetry (DSC)

Differential scanning calorimetry (DSC) is a widely adopted technique for determining the thermal hysteresis activity (THA) of AFPs. This method quantifies the actual crystallization initiation temperature through precise measurement of exothermic and endothermic enthalpy changes during phase transitions. The DSC approach offers objective and accurate analytical results, enabling researchers to determine ice crystal content with high precision [10, 25–27].

A key advantage of the optimized DSC protocol lies in its ability to overcome supercooling challenges during liquid–solid phase transitions. Through controlled thermal cycling—including initial cooling to supercooling points, equilibrium attainment, and subsequent heating—the method systematically captures critical parameters: melting point™, freezing point (T_0), and ice crystal content (Φ). The operational workflow involves: (i) controlled cooling to achieve supercooling, (ii) equilibrium heating for Tm and enthalpy of melting (ΔHm) determination, and (iii) repeat cycles with temperature hold (T_h) to measure enthalpy of freezing (ΔH_f) and T_0 when re-freezing occurs.

This technique establishes a reliable strategy for AFP analysis by calculating ice crystal fractions using Eq. ((3.1) and enabling quantitative assessment of mixed samples through precise $\Delta H_m/\Delta H_f$ measurements. The annealing process effectively mitigates supercooling artifacts, allowing accurate THA determination even without nucleating agents. Current applications demonstrate DSC's status as the gold standard for AFP thermal analysis, with thermal profiles providing comprehensive insights into crystallization dynamics.

$$\Phi\,(\%) = \left(1 - \frac{\Delta H_f}{\Delta Hm}\right) \times 100 \tag{3.1}$$

3.2.4 Splat-Cooling

AFPs exhibit ice recrystallization inhibition (IRI) by binding to ice surfaces through the "adsorption-inhibition" mechanism, which suppresses ice crystal growth. Recrystallization involves the redistribution of growth among pre-existing ice crystals: some grow larger while others shrink, and smaller ice crystals often aggregate into larger ones. This process, typically triggered by temperature fluctuations, results in increased ice crystal numbers and structural integration.

The most widely employed method for assessing IRI activity is the "splat cooling" technique. This approach generates ice crystals via rapid freezing and monitors their growth dynamics by cycling the sample's temperature near its freezing point using a cold stage microscope (Figure 3.2). The procedure involves sandwiching 2 μL of a 30% sucrose solution between two circular coverslips to form a thin film. High sucrose concentrations are critical to minimize interfacial inhibition effects. The assembly is first immersed in heptane cooled to −80 °C with dry ice, then transferred to a −6 °C circulating-cooled glass box. After 30–60 minutes of equilibration, micrographs are captured [29–31].

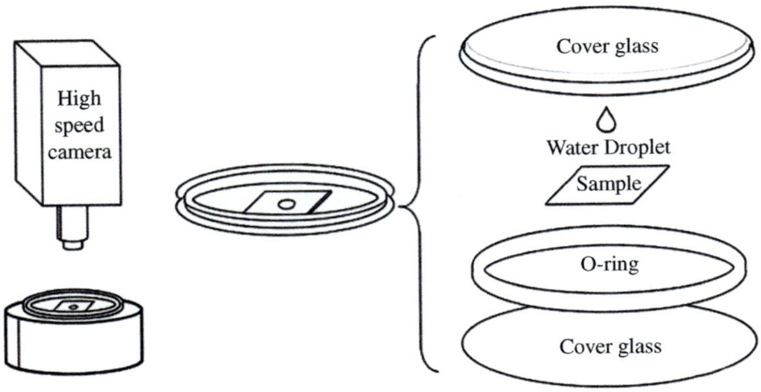

Figure 3.2 Schematic representation of the experimental apparatus used to study the effect of the surface on ice nucleation. The sample cell on the cryostage is composed of a rubber O-ring sandwiched between two glass coverslips. The right panel illustrates the process of the formation of water microdroplets through the evaporation and condensation process. *Source:* Reproduced with permission from He et al. [28]/American Chemical Society.

This low-temperature protocol presents significant operational challenges. Technological advancements have since introduced improved methodologies, such as polarizing microscopes integrated with cold stages. Modern systems employ liquid nitrogen-programmed temperature control and nitrogen-protected image microprocessing systems to collect high-resolution ice crystal micrographs [28, 32–34].

3.2.5 Low-Field Nuclear Magnetic Resonance (LF-NMR)

LF-NMR is a rapid, non-destructive analytical technique with advantages including high sensitivity, environmental compatibility, and fast testing speed. It has been extensively applied in food quality assessment, seed breeding, petroleum exploration, life sciences, and the characterization of rubber cross-linking density [35]. The methodology primarily relies on hydrogen protons as molecular probes to investigate the physical properties of samples, given the ubiquity of hydrogen atoms in natural systems and the strong, easily detectable NMR signals they produce [36]. Key parameters measured in LF-NMR include longitudinal relaxation time (T_1), transverse relaxation time (T_2), self-diffusion coefficients, and imaging modalities such as T_1-weighted, T_2-weighted, and proton density-weighted imaging [9].

The differential relaxation behaviors of water molecules in samples enable rapid and precise characterization of their physicochemical properties. This principle allows water to function as a dynamic probe for studying sample microstructures [9, 37]. In the context of antifreeze proteins (AFPs), LF-NMR has been particularly valuable for analyzing THA, where AFPs lower the freezing point relative to the melting point of a solution. By monitoring proton signals during phase transitions, LF-NMR can quantify the spatial distribution of mobile water and distinguish between liquid and solid states: moving water generates detectable signals, while solid ice remains signal-free [10].

Studies utilizing LF-NMR have demonstrated that AFPs significantly accelerate ice melting rates in frozen solutions [10, 29, 38–40]. This technology has further advanced understanding of AFP-induced water migration and redistribution in frozen foods, offering insights into cryopreservation mechanisms [5, 9, 41, 42].

3.2.6 Bio-Valuation Model

The conventional methods for quantifying antifreeze activity (AF-A) typically require sophisticated instrumentation and intricate procedural protocols, prompting researchers to develop simplified, low-cost alternatives [8, 19, 27, 43–46]. A critical challenge arises when cellular environments drop below the physiological freezing point: extracellular ice crystal formation induces mechanical stress, accelerates apoptosis, and compromises cell viability [43–46]. To address this, a bio-valuation model was established to evaluate the cryoprotective efficacy of antifreeze proteins (AFPs). This model quantifies post-cold-stress cell survival rates, directly correlating with AFP functionality [47].

For microbial systems, the turbidimetric method has emerged as a practical tool. This approach measures optical density (OD) values to assess bacterial growth dynamics, where the OD value reflects relative biomass under standardized culture conditions [26, 29, 48]. Wu et al. [25] demonstrated its utility in assessing sericin-derived AFPs' cryoprotective activity on *Lactobacillus delbrueckii subsp. bulgaricus* LB340 LYO. Validation studies further confirmed the reliability of the OD method by correlating it with metabolic activity, membrane integrity, and microstructural analyses of *Streptococcus thermophilus* post-cold stress [10, 26].

3.2.7 Novel Technology in Utilization of the Evaluation of Antifreeze Activity

Advances in computational and imaging technologies have revolutionized the study of AFPs. Molecular dynamics simulations, quantum mechanics, and computational chemistry are now indispensable tools in elucidating atomic-scale mechanisms and modeling AFP-ice/water interactions [25, 49, 50]. Quantum mechanical calculations of binding energies between AFPs and ice/water interfaces enable predictive modeling of antifreeze activity. Concurrently, fluorescence-based imaging systems—such as confocal laser scanning microscopy and fluorescence microscopy—leverage intrinsic AFP fluorescence or exogenous labeling to visualize AFP activities [51].

For structural analysis of frozen systems, a multi-modal approach combining environmental scanning electron microscopy (ESEM), atomic force microscopy (AFM), and micro-slicer image processing systems (MSIPS) provides critical insights into ice crystal morphology and AFP-mediated cryoprotection in food matrices. Non-invasive techniques like X-ray micro-computed tomography (CT), hyperspectral imaging, and Raman spectroscopy, alongside nuclear magnetic resonance (NMR) and magnetic resonance imaging (MRI), have gained prominence in recent years [2].

While AFP preparation technologies are mature, optimizing production parameters remains resource-intensive. A breakthrough strategy involves virtual peptidomics databases to predict antifreeze activity of peptide fragments, significantly accelerating experimental screening. For instance, Raus et al. [52] employed *in silico* proteolytic digestion simulations to analyze molecular weight, amino acid composition, isoelectric point, and hydrophobicity of hydrolyzed peptides. This computational hydrolysis approach enables high-throughput screening of food-derived protein precursors, reducing both time and cost burdens.

Modern peptidomics integrates high-resolution mass spectrometry, structure–function relationship models, chemometrics, and database management to decode peptide interactions and activities [53]. Coupled with advanced instrumentation like liquid chromatography-time-of-flight (LC-TOF) tandem mass spectrometry and quadrupole linear ion trap mass spectrometry, and data analysis tools such as ProteinPilot, these innovations have dramatically shortened structural characterization cycles for bioactive peptides [54]. This technological synergy addresses critical challenges in AFP structural identification, advancing both fundamental research and industrial applications.

3.3 Molecular Characteristics and Structure–Function Relationships of AFPs

The structural-functional relationship of AFPs has been extensively investigated, with particular focus on how amino acid composition, molecular mass, and peptide sequence influence antifreeze activity [33, 55, 56]. Key structural features include tripeptide repeat motifs (-–Gly-Pro-X-, –Gly-X-Pro-, –Gly-X-X-, or -T-X-T-), conserved sequences such as GTPG- and GPP(OH)G-, and high abundances of glycine (Gly), alanine (Ala), and threonine (Thr). These residues, particularly Gly and Thr, are critical for IRI, as their hydroxyl groups form hydrogen bonds with ice surfaces, while their hydrophobic interactions (via Pro, Hyp, and Ala side chains) stabilize AFP-water interfaces [33, 47, 57].

Structural studies (Table 3.1) reveal that AFPs with molecular masses of <3,000 Da and 10–20 amino acid residues exhibit optimal antifreeze activity. Smaller peptides demonstrate enhanced hydrophilicity, flexibility, and thermal stability, along with superior cell penetration and ice crystal adsorption capabilities compared to larger proteins [10, 25, 55]. For example, snow flea-derived AFPs (sfAFPs) display a -Gly-Z-X- repeat structure with 45% Gly content, establishing a template for synthetic AFP design [43]. Similarly, Nguyen et al. [59] identified a collagen-derived peptide (GLLGPLGPRGL, 1163.8 Da) with sequence homology and functional similarity to sfAFP, achieving 70% ice crystal growth inhibition in 23% sucrose systems.

Comparative analyses of AFPs from diverse sources further highlight structure–activity correlations. Du and Betti [32] isolated chicken skin-derived peptides (2,147.1 Da: GSPGADGPIGAHypGTHypGPQGIAGQR; 2,532.2 Da: GNDGAHypGAAGPHypGPTGPA GPHypGFHypGAAGAK) that mimic sfAFP functionality. Shark skin collagen-derived SsC-AFP (GAIGPAGPLGP, 906 Da, pI <5.0) exhibits 64% sequence homology to sfAFP [47], while fish gelatin-derived cationic peptides (SP2, 1,850.82 Da: KDGTPGQFGP(OH) PGAPGKGN(OH)H; 2,036.88 Da: NEGTPGTGPAGPP(OH)GFHTPK(OH)W) show superior antifreeze activity compared to anionic counterparts (SP1) [33]. Despite dissimilar sequences, SP2 peptides share conserved motifs (-GTPG-, -GPP(OH)G-) and hydroxyl group content (3–5 residues), underscoring the importance of structural fingerprints over strict sequence identity in determining antifreeze efficacy.

3.4 Action Mechanisms of AFPs

AFPs exhibit two primary antifreeze activities: thermal hysteresis (TH) and IRI [14, 25]. TH activity arises from a temperature differential between the freezing and melting points of a solution containing AFPs. This non-colligative effect allows AFPs to depress the freezing point without significantly altering the melting point [43, 44]. For instance, when an AFP-containing ice-water solution reaches 0 °C, ice crystal growth is suppressed until the temperature drops to the depressed freezing point. Conversely, ice crystals do not melt upon warming to the growth initiation temperature but instead begin melting at ~0 °C (the equilibrium melting point of ice). This phenomenon is thermodynamically explained by the Kelvin effect, where the curvature of ice surfaces increases the energy barrier for water molecule incorporation, thereby lowering the freezing point [63].

IRI activity, on the other hand, involves AFP adsorption at the ice-water interface to inhibit or modify ice crystal growth. Recrystallization—a process where larger ice crystals grow at the expense of smaller ones due to interfacial energy minimization—can cause severe cellular damage. AFPs mitigate this by adsorbing onto ice prismatic faces via hydrogen bonding between hydrophilic AFP side chains and ice crystal surfaces (Figure 3.3b). This interaction exposes hydrophobic regions with high surface free energy, effectively halting further crystal growth. As a result, ice crystals grow only on their base surfaces,

Table 3.1 The sources, structure–function relationship and characteristics of antifreeze peptides.

Source of peptides	Peptide sequence	Preparation	Activity	Molar mass (Da)	References
Tilapia scales	NA	Trypsin	Cell protection	< 2,000	[10]
silver carp	GVDNPGHPFIM	Protamex	TH, IBA	1,183.35	[58]
Tilapia scales	NA	Trypsin	TH, Cell protection	< 2,000	[10]
porcine skin	GLLGPLGPRGLL	Alkaline	TH, IRI	1,163.80	[59]
fish gelatin	KDGTPGQFGP(OH)PGAPGKGN(OH)H	Alcalase	IRI	1,850.82	[33]
fish gelatin	NEGTPGTGPAGPP(OH)GFHTPK(OH)W	Alcalase	IRI	2,036.88	[33]
Pig skin	GLLGPLGPRGLL	Alkaline	TH, IRI	1,162.80	[19]
Chicken skin	GSPGADGPIGAHypGTHypGPQGIAGQR	Trypsin	IRI	2,147.1	[32]
Chicken skin	GNDGAHypGAAGPHypGPTGPAGPHypGFHypGAAGAK	Trypsin	IRI	2,532.2	[32]
Pigskin collagen	NA	Alcalase	Cell protection	150–2,000	[48]
Shark skin	GAIGPAGPLGP	Acid Protease	Cell protection	906	[47]
Shark skin	NA	Alcalase	IRI	< 3,000	[55]
Bovine gelatin	NA	Alcalase	IRI	1,600–2,400	[30]
Gelatin	NA	Papain	IRI	2,000–5,000	[55]
Snow flea	-Gly-Gly-x- or –Gly-x-x-	Natural	IRI, TH	6,500	[43]
Type I AFPs	-DTASDAAAAAAL-	Natural	IRI, TH	1,047	[60]
Winter flounder	TAANAKAAAELTAANAAAAAAATAR	Natural	IRI, TH	2,213.44	[61]
Synthesis	STSTATATGR	Synthesis	IRI, TH	951.99	[13]
Synthesis	RGTATTTATGT	Synthesis	IRI, TH	1,037.09	[62]
Synthesis	DTASDAKAAAEL	Synthesis	IRI, TH	1,162	[41]
Synthesis	DTASDAFAAAAL	Synthesis	IRI, TH	1,123	[41]
Synthesis	GAGP[(GVGVP)(GEGVP)9]2GWPH6	Synthesis	IRI	10,193	[56]
Synthesis	GAGP[GVGVP (GKGVP)9]2GWPH6	Synthesis	IRI	10,176	[56]

NA: relevant information was not available in the original references.

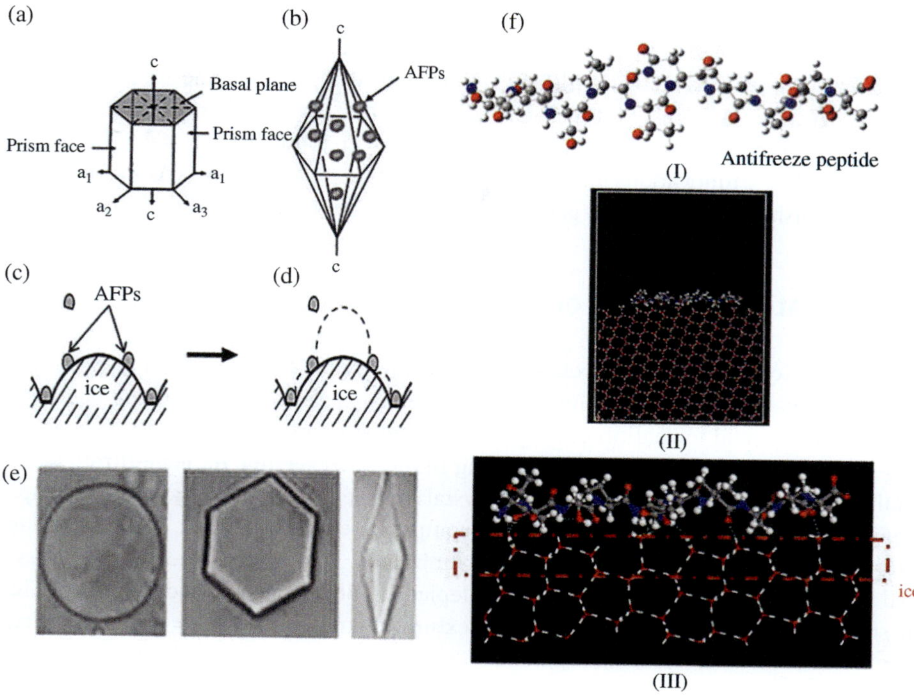

Figure 3.3 Mechanism of antifreeze peptides. (a) The general crystal growth direction; (b) Adsorption model of antifreeze peptides combined with the prismatic surface of ice crystals. *Source:* Adapted from [10]; (c) and (d) Kelvin effect. *Source:* [17]/John Wiley & Sons; (e) Different morphologies observed for single ice crystals; (f) Adsorption model of antifreeze peptide and ice crystals. I, the optimized molecular structure of antifreeze peptide shown in a ball and stick model; II, snapshot of the lowest energy ice/peptide system; III, interception of the surface area of the ice/peptide system. Red, blue, gray, and white balls represent oxygen, nitrogen, carbon, and hydrogen atoms, respectively, and blue dashed lines represent hydrogen bonds. *Source:* [25]/with permission of Elsevier.

forming biconical or plate-like structures (Figure 3.3e) [64]. Notably, AFPs with molecular masses of <3,000 Da (e.g., collagen-derived peptides) exhibit enhanced IRI activity due to their flexibility and ability to form rigid, collagen-like helices that align with ice oxygen triad planes [32, 55, 59]. Larger AFPs (>3,000 Da) may lose efficacy due to steric hindrance, which disrupts optimal ice surface interactions.

Despite both activities being common to most AFPs, a strong correlation between high TH and IRI activity is rarely observed [25]. This dissociation stems from their distinct mechanistic underpinnings: TH is assessed during freezing (when AFPs depress the freezing point below the melting point), whereas IRI is evaluated during post-freezing temperature fluctuations (when AFPs prevent ice crystal restructuring). The adsorption-inhibition theory, proposed by Raymond and DeVries [64], remains the most robust framework for explaining these mechanisms. According to this model, AFPs selectively bind to ice prismatic faces (Figure 3.3b), altering growth kinetics and producing irregular crystal morphologies (Figure 3.3d–e). Computational studies using ChemSite Pro and MOPAC further reveal that hydrophilic-hydrophobic balance in AFPs—mediated by residues like Pro and Ala—optimizes ice-water interface interactions, enhancing IRI efficacy [25].

Recent advances highlight the potential of AFPs in cryopreservation and biomaterials. For example, plant-derived AFPs (e.g., from *Astragalus adsurgens* in China) demonstrate unique structural features, such as a 50 kDa β-sheet-rich protein with ~0.35 °C thermal hysteresis. Meanwhile, synthetic AFP-inspired hydrogels developed by Fu et al. [65] exhibit intrinsic anti-freezing properties, toughness, and transparency, enabling applications in smart optical sensing. These innovations underscore the importance of elucidating AFP structure–function relationships to design next-generation cryoprotectants with tailored activity.

3.5 Application Advances of AFPs in Frozen Food Industry

High water activity (aw) in food accelerates spoilage by promoting microbial growth and enzymatic degradation. While freezing technology extends shelf life by slowing bacterial proliferation, ice crystal formation during freezing and storage remains a critical quality-limiting factor. For most food systems, freezing induces irreversible damage through ice recrystallization—a process where small ice crystals merge into larger, structurally disruptive aggregates. This phenomenon particularly compromises cellular foods, as intracellular ice formation during thawing ruptures cell membranes, leading to increased drip loss, reduced water-holding capacity, and nutrient depletion [66, 67]. Even foods consumed in frozen states, such as ice cream, experience textural degradation due to progressive ice crystal growth [68]. Notably, certain perishables like tomatoes and strawberries undergo severe structural collapse when frozen, necessitating storage at non-freezing temperatures to minimize quality loss [69].

AFPs have emerged as natural ice crystal growth inhibitors in frozen food preservation. Their dual mechanisms—lowering freezing point depression and inhibiting recrystallization—make them ideal for maintaining the structural integrity of frozen fish, meats, desserts, and cellular produce (Figure 3.4). At low concentrations (<10 μg/mL), AFPs significantly preserve nutritional profiles and sensory attributes, as evidenced by reduced ice crystal size in ice cream matrices treated with gelatin-based AFPs [33]. For instance, Damodaran and Wang [33]

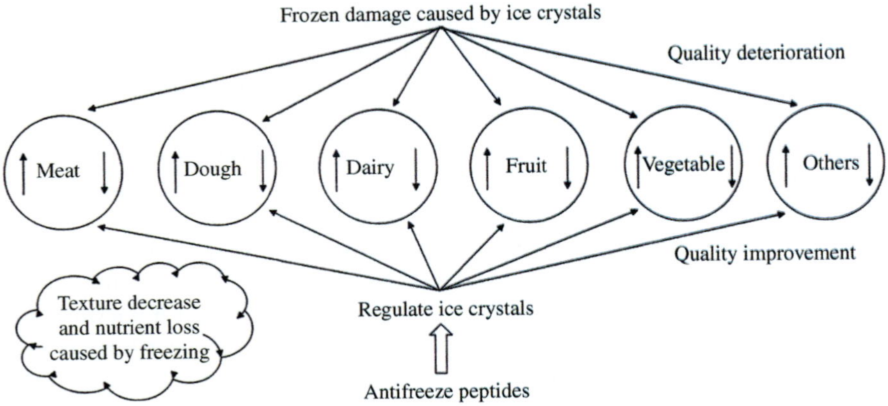

Figure 3.4 Application of antifreeze peptides in frozen food industry.

demonstrated that AFP-modified ice cream exhibited smaller ice particles (mean diameter 25 μm vs. 40 μm in controls), correlating with enhanced textural stability. In cellular foods like meat, AFPs mitigate ice-induced cell damage by preferentially binding to ice crystal surfaces, thereby reducing nutrient leaching [5, 70, 71].

3.5.1 Edible Safety of AFPs

AFPs employed in frozen food preservation are primarily sourced from collagen, gelatin, and food-derived protein hydrolysates. Notably, studies on collagen/gelatin hydrolysates have consistently demonstrated significant antifreeze activity in collagen-derived peptides [14, 19, 59]. This is primarily attributed to the broad availability of collagen/gelatin across animal and plant sources, as well as their established safety profile as food ingredient. Furthermore, enzymatic hydrolysis of proteins into peptides is a widely adopted strategy to mitigate allergenic risks. The molecular weights of AFPs typically range below 3,000 Da, with hydrolysis effectively disrupting epitope structures and molecular conformations that trigger pathological immune responses [67, 72].

Recent advancements in AFP research highlight the diversity of protein sources beyond traditional animal-derived materials. For instance, plant-based AFPs, such as the 50 kDa ice-structuring protein isolated from *Amorpha fruticos*, exhibit THA of ~0.35 °C at 5 mg/mL, comparable to fish-derived AFPs like those from winter flounder (*Pseudopleuronectes americanus*). Structural analyses using circular dichroism (CD) reveal that plant AFPs contain 11% α-helices and 34% antiparallel β-sheets, suggesting distinct folding mechanisms compared to fish AFP. This structural diversity underscores the potential of alternative protein sources, such as wheat-derived AFPs developed by Nanjing Anfei Biotechnology, which effectively inhibit ice recrystallization in frozen foods like ice cream and preserved fruits.

The safety of AFPs as ice crystal regulators is further supported by their non-antigenic properties post-hydrolysis. Enzymatic digestion not only reduces molecular weight but also eliminates immunogenic epitopes through conformational changes, minimizing the risk of hypersensitivity reactions [67]. This aligns with regulatory frameworks that prioritize non-allergenic ingredients for food applications.

3.5.2 Methods of Introducing AFPs into Food Products

The incorporation of AFPs into frozen food systems requires strategic design considerations. Given their extracellular mode of action, AFPs can be directly applied to food matrices using mechanical methods such as vacuum infiltration, injection, mixing, or soaking. These physical approaches are well-established for surface or bulk incorporation of functional additives in food processing.

Beyond mechanical strategies, genetic engineering offers an alternative pathway for AFP integration into food systems. Theoretically, stable transgenic organisms expressing AFPs could confer inherent cold resistance to both parent and progeny populations, as demonstrated in studies involving transgenic plants and microorganisms [70]. However, this approach faces significant technical hurdles. Successful gene transfer necessitates precise DNA targeting to ensure expression of functional peptides, which requires resolving multiple interdependent challenges: (i) accurate gene integration into host genomes to avoid off-target effects; (ii) proper post-translational folding of AFPs to maintain β-sheet or α-helix structures critical for ice-crystal inhibition (as observed in fish AFPs like *Pseudopleuronectes americanus*); (iii) environmental stability of expressed proteins under food storage conditions; (iv) resistance to proteolytic degradation during food processing; (v) optimal intracellular distribution; and (vi) targeted extracellular localization.

These challenges are inherently dependent on the selected target organism and AFP compatibility with the host system. For instance, while plant-based AFPs (e.g., from *Amorpha fruticosa* with 34% β-sheets) exhibit distinct folding patterns compared to fish-derived AFPs, their integration into food-grade organisms like yeast or *E. coli* requires tailored expression systems to preserve functional conformation. Additionally, regulatory approval for genetically modified AFP-containing foods remains a critical barrier, necessitating extensive safety assessments for allergenicity and unintended metabolic interactions.

3.5.3 Application in Frozen Dough

Frozen dough technology extends dough shelf life while enabling on-demand fresh bread production. However, freezing-induced alterations in gluten-starch networks and yeast activity significantly compromise product quality, manifesting as reduced specific volume (by ~15–20% compared to fresh dough) and increased crumb firmness (measured via Kramer shear test at 1.2–1.5 N vs. 0.8–1.0 N for fresh bread) [73, 74]. Ice crystal formation during freezing further exacerbates structural degradation through (i) protein denaturation (e.g., glutenin aggregation as shown by SDS-PAGE analysis) and (ii) starch particle separation, leading to macroscopic defects like dough shrinkage (12–18% volume reduction post-thaw) and surface cracks (observed via SEM imaging) [75].

While conventional additives (e.g., emulsifiers, hydrocolloids) mitigate these issues, they often alter flavor profiles (notably reducing volatile esters like ethyl hexanoate by 30–40%) and raise safety concerns over synthetic compound residues [76]. Antifreeze peptides (AFPs) offer a biocompatible alternative with dual functional benefits. At optimal concentrations (0.5–1.5% w/w), AFPs:

1) **Enhance cryoprotection** by reducing freezable water content (by 18–25% as measured by DSC) through hydrogen bonding with ice surfaces and inhibition of recrystallization [14, 77];
2) **Preserve dough microstructure** by modulating water distribution (NMR T2 relaxation time shifts from 25 ms to 42 ms in AFP-treated dough) and maintaining yeast viability (survival rates increased from 65% to 88% post-four-week storage) [9, 77];
3) **Improve sensory attributes** without flavor modification, as demonstrated by Zhang et al. [78] who found no alteration in volatile compounds (GC–MS analysis showed stable levels of 2-pentylfuran and 3-methylbutanal) and reported a Michelia alba-like aroma enhancement [78].

Mechanistic studies reveal that collagen-derived AFPs (CoAFPs) at 1% w/w specifically:

- Stabilize water mobility during freeze–thaw cycles (reducing ice crystal size by 40% as shown by XRD analysis)

- Maintain dough softness (texture profile analysis showed 22% lower hardness vs. controls) by limiting freezable water content to <35%
- Preserve bread volume (retaining 85–90% of fresh bread volume vs. 70–75% in controls) [9].

Sericin-based AFPs (SP) demonstrate complementary mechanisms:

- Structural matching with ice surfaces (confirmed via AFM imaging)
- Non-covalent interactions (hydrogen bonds and π-π stacking) that suppress crystal growth
- Yeast cryoprotection through reduced cell membrane rigidity (lipid order parameter decreased from 0.82 to 0.65) [77].

3.5.4 Application in Frozen Meat Products

The high nutrient content of meat products creates an ideal environment for spoilage microorganisms and pathogenic bacteria, rendering these foods highly susceptible to microbial degradation and posing significant health risks to consumers [79, 80]. While freezing is a critical preservation method for muscle-based products like meat, fish, and surimi, ice crystal formation during freezing and storage remains a major challenge. Slow freezing conditions promote intracellular ice crystal growth, which ruptures cellular membranes during thawing, leading to irreversible structural damage and nutrient loss (e.g., 20–30% reduction in protein solubility and 15–25% moisture loss) [10].

Antifreeze proteins (AFPs) have emerged as a promising solution to mitigate these issues through their IRI activity. In studies involving AFP-treated frozen meat, two key improvements were observed:

1) **Reduced ice crystal size** (from 50–100 μm in controls to 20–40 μm with AFPs)
2) **Decreased tissue damage** (histological analysis showed 40–50% reduction in cell membrane disruption) [66, 81]

Kittiphattanabawon et al. [82] demonstrated the application of shark skin-derived gelatin AFPs in surimi preservation. Their freeze–thaw cycle experiments revealed:

- **Progressive quality decline in control samples**:
  - Ca^{2+}-ATPase activity decreased by 35–45% with each cycle
  - Total sulfhydryl content dropped by 25–30%
  - Surface hydrophobicity increased by 20%
  - Disulfide bond content rose by 15–20% (measured via Ellman's assay and UV–Vis spectroscopy)
- **Significant protective effects with 10% AFP treatment**:
  - Ca^{2+}-ATPase activity stabilized at 90–95% of initial levels
  - Surface hydrophobicity reduced by 30% (determined via fluorescence spectroscopy)
  - Disulfide bond accumulation slowed to 8–10% per cycle (confirmed by SDS-PAGE analysis)

These results indicate that AFPs not only inhibit ice crystal growth but also preserve protein functionality by maintaining thiol-disulfide balance and membrane integrity. The mechanism appears to involve AFP binding to ice surfaces (validated by AFM imaging) and hydrogen bonding with water molecules, which suppresses recrystallization and reduces mechanical stress on muscle fibers [82].

3.5.5 Application in Frozen Fruit and Vegetable Products

Low-temperature storage is a fundamental strategy for maintaining the quality of fruits and vegetables. Research indicates that a 1 °C reduction in storage temperature can significantly extend the shelf life of perishable produce [83]. However, freezing processes present critical challenges: physical damage to cell membranes and structures—such as moisture loss (up to 20–30%) and cell wall rupture—represents the most severe quality degradation mechanism [84, 85]. Post-freeze–thaw cycles, cellular integrity is frequently compromised, leading to dehydration (loss of 15–25% water content) and irreversible softening of tissues [86].

The primary driver of frozen food quality decline is temperature-induced ice recrystallization, which enlarges ice crystals and exacerbates cellular damage. Antifreeze proteins (AFPs) have emerged as a transformative solution, demonstrating dual functionality during freezing, storage, and thawing:

1) **Ice recrystallization inhibition (IRI)**: Reduces crystal size from 50–100 μm (untreated) to 20–40 μm (AFP-treated) [8]
2) **Cellular protection**: Maintains tissue firmness by preserving cell membrane integrity and reducing structural deformation

Experimental evidence from Song et al. [69] validates these benefits in vegetables:

- **Tenebrio molitor AFP treatment** on cucumber, carrot, zucchini, and onion showed:
- 40–50% reduction in post-thaw softening (measured via texture profile analysis)
- 13-day storage at 0 °C maintained 95–98% structural integrity compared to 70–80% in controls
- **Kong et al.** [13, 62] demonstrated antifreeze peptide effects on:
  - **Carrots**: 25% reduction in drip loss and 30% improvement in color retention (L*a*b* colorimetry)
  - **Cherries**: 18% smaller ice crystal size (SEM imaging) and 20% lower volatile compound loss (GC–MS analysis)
  - Innovative application methods further enhance efficacy:
- **Drimys angustifolia AFPs** applied via vacuum infiltration to starfruit (Averrhoa carambola) achieved:
  - 60-day storage at −20 ± 2 °C with 92% firmness retention
  - 55% reduction in drip loss compared to non-treated controls [34]

3.5.6 Application in Dairy Products

Inhibiting ice recrystallization in frozen dairy products, such as ice cream, is another application of AFPs [68, 87–89]. AFPs reduce the size of ice crystals in ice cream, which is one of the most important factors in keeping ice cream smooth and creamy [33, 90]. However, if the temperature fluctuates during storage or transportation, recrystallization will inevitably occur, resulting in a rough texture and lowered ice cream quality. A small amount of AFPs was added to an ice cream sample frozen at −80 °C and stored at −6 °C to −8 °C. The results showed that there was almost no growth of ice crystals in the sample containing AFPs after one hour of storage, whereas the size of ice particles in control group increased significantly [88].

A study employing cryogenic microscopy investigated the impact of AFPs and fish collagen-derived ice-structure peptides on ice crystal formation in ice cream. The experimental protocol involved placing 5 μL ice cream samples (with/without additives) on glass slides, which were then subjected to a Linkman cryostage at −40 °C for five minutes. The temperature was subsequently raised to −14 °C at 1 °C/min, followed by cyclic temperature fluctuations between −14 °C and − 12 °C every three minutes. This thermal regime induced irruptive recrystallization at −14 °C, generating numerous ice nuclei that further developed during the thermal cycles.

AFP-treated samples demonstrated significantly reduced ice crystal count and size after multiple cycles, confirming AFPs' ability to inhibit ice crystal growth. Parallel research by Damodaran and Wang [33] revealed that ice-structure peptides (1,600–2,400 Da molecular weight) from fish collagen similarly reduced ice crystal size. When simulating cold-chain temperature variations via a cold-heat circulation system, these peptides effectively suppressed recrystallization. Notably, collagen-based AFPs also elevated the glass transition temperature and enhanced melting resistance in ice cream [90, 91]. These findings highlight the potential of biomacromolecules in optimizing ice cream microstructure stability during storage.

3.5.7 Application in Cryobiology

A systematic analysis of cryopreservation challenges in microbial cultures reveals that long-term continuous cultivation of bacterial strains or cells presents significant practical limitations in terms of temporal investment and operational efficiency [92]. To address this, industrial and academic sectors predominantly adopt ultralow-temperature preservation and freeze-drying technologies, despite the inherent cellular risks associated with cold stress. During cryopreservation, complex physicochemical transformations occur that compromise cellular integrity—particularly membrane destabilization, reduced survival rates, and metabolic activity suppression. Notably, direct freeze-drying can induce mortality exceeding 90% in bacterial populations [92], with cryopreservation recovery rates further constrained by oxidative stress pathways activated during cold exposure [8].

Traditional cryopreservation protocols require 10–40% organic solvents as cryoprotectants, which paradoxically introduce cytotoxic risks and potential irreversible cellular damage [93, 94]. This critical limitation has been effectively mitigated through the strategic incorporation of AFPs, which demonstrate multiple protective mechanisms. Extensive experimental evidence (10,28,29) confirms AFPs' capacity to enhance freeze survival rates while maintaining cellular metabolic vitality. The molecular mechanism involves hydrogen bonding interactions between AFP hydroxyl groups and phospholipid bilayer head groups [10, 48, 95], which preserves membrane fluidity and minimizes intracellular substance leakage [48]. Additionally, AFPs can translocate into cells to mitigate intracellular ice crystal formation [10].

Chen et al.'s [8] investigation on tilapia scale antifreeze peptides (TSAPP) in Streptococcus thermophilus cryopreservation provides mechanistic insights. TSAPP application resulted in: (i) ATP activity enhancement, (ii) cell membrane potential hyperpolarization mitigation, (iii) intracellular Ca^{2+} homeostasis regulation, and (iv) significant suppression of reactive oxygen species (ROS) and malondialdehyde (MDA) accumulation. These interventions yielded a 10.85-fold increase in post-cryopreservation viability, with preserved total antioxidant capacity and elevated catalase/peroxidase activities. This study underscores the multifaceted cytoprotective effects of AFPs in cryopreservation systems.

3.6 Future Direction

The present review aims to provide a comprehensive analysis of AFPs across four critical dimensions: biosynthesis pathways, purification methodologies, structure–function correlations, and molecular mechanisms of action, with particular emphasis on their transformative applications in the frozen food industry. Current research has established that ice nucleation and recrystallization processes are primary drivers of structural degradation in frozen foods, leading to texture deterioration and nutritional loss. As demonstrated through extensive experimental validation, AFPs exhibit exceptional ice crystal growth inhibition properties, establishing their status as next-generation cryoprotectants. Their multifunctional attributes—including non-toxicity, bioavailability, and nutritional compatibility—position AFPs as ideal food additives for cryogenic food preservation and tissue engineering applications.

While preliminary studies have confirmed their efficacy in enhancing freeze–thaw stability of dairy products (e.g., ice cream recrystallization control) and seafood preservation, significant technological barriers remain. First-generation AFPs derived from fish and insects predominantly exist in research laboratories, with industrial-scale production constrained by purification challenges (e.g., chromatographic costs) and limited structural elucidation of key binding domains. This knowledge gap impedes the development of structure-optimized variants through protein engineering.

Safety evaluation frameworks for novel AFPs require rigorous implementation, particularly for vulnerable populations. Current introduction methods rely on mechanical incorporation or recombinant expression in microbial systems. Although genetic engineering enables direct gene integration into target organisms (e.g., transgenic yeast expression of AFP genes), practical limitations persist: (i) limited success in functional protein expression, (ii) regulatory hurdles for genetically modified organisms (GMOs), and (iii) cost inefficiencies in large-scale production.

To advance translational research, three priority areas are identified:

1) **Structural characterization**: Advanced cryo-EM and molecular docking simulations are needed to decipher AFP-ice crystal interactions at atomic resolution.
2) **Safety assessment**: Long-term toxicological studies and allergenic potential evaluations must precede commercialization.
3) **Process optimization**: Development of cost-effective expression systems (e.g., plant-based transient expression) and scalable purification protocols.

The field is also evolving toward real-time monitoring technologies for ice crystal dynamics, necessitating interdisciplinary collaboration between food scientists, material engineers, and analytical chemists. These advancements will require not only technological innovation but also robust regulatory frameworks to ensure safe and sustainable integration of AFPs into food production systems.

References

1. Schafer, W. (2017). The science of freezing foods. USA: University of Minnesota Extension. Available from: https://the-science-of-freezing-foods/.
2. Dalvi-Isfahan, M., Hamdami, N., and Le-Bail, A. (2016). Effect of freezing under electrostatic field on the quality of lamb meat. *Innovative Food Science & Emerging Technologies* 37: 68–73.
3. Maity, T., Chauhan, O.P., Shah, A. et al. (2011). Quality characteristics and glass transition temperature of hydrocolloid pre-treated frozen pre-cut carrot. *International Journal of Food Properties* 14 (1): 17–28.
4. Parniakov, O., Bals, O., Lebovka, N., and Vorobiev, E. (2016). Pulsed electric field assisted vacuum freeze-drying of apple tissue. *Innovative Food Science & Emerging Technologies* 35: 52–57.
5. Zhu, Z., Zhou, Q., and Sun, D.-W. (2019). Measuring and controlling ice crystallization in frozen foods: a review of recent developments. *Trends in Food Science & Technology* 90: 13–25.
6. Elliott, G.D., Wang, S., and Fuller, B.J. (2017). Cryoprotectants: a review of the actions and applications of cryoprotective solutes that modulate cell recovery from ultra-low temperatures. *Cryobiology* 76: 74–91.
7. Petzold, G. and Jose, M.A. (2009). Ice morphology: fundamentals and technological applications in foods. *Food Biophysics* 4 (4): 378–396.
8. Chen, X., Li, L., Yang, F.J. et al. (2020). Effects of gelatin-based antifreeze peptides on cell viability and oxidant stress of *Streptococcus thermophilus* during cold stage. *Food and Chemical Toxicology* 136: 111056.
9. Chen, X., Wu, J.H., Li, L., and Wang, S.Y. (2017). The cryoprotective effects of antifreeze peptides from pigskin collagen on texture properties and water mobility of frozen dough subjected to freeze–thaw cycles. *European Food Research and Technology* 243 (7): 1149–1156.
10. Chen, X., Wu, J.H., Li, L., and Wang, S.Y. (2019). Cryoprotective activity and action mechanism of antifreeze peptides obtained from Tilapia scales on *Streptococcus thermophilus* during cold stress. *Journal of Agricultural and Food Chemistry* 67 (7): 1918–1926.
11. Sun, D.W. (2005). *Emerging Technologies for Food Processing*, 468–536. Elsevier Academic Press.
12. Hartel, R. (2001). *Crystallization in Foods*, 192–231. Maryland: Aspen Publishers Inc.
13. Kong, C.H.Z., Hamid, N., Ma, Q. et al. (2017). Antifreeze peptide pretreatment minimizes freeze-thaw damage to cherries: an in-depth investigation. *LWT—Food Science and Technology* 84: 441–448.
14. Chen, X., Cai, X.X., Wang, S. et al. (2019). Recent progress and application prospects of antifreeze peptides in food industry. *Food Science* 40 (17): 337.

15 Drori, R., Davies, P.L., and Braslavsky, I. (2015). When are antifreeze proteins in solution essential for ice growth inhibition? *Langmuir* 31 (21): 5805–5811.

16 Sun, D.W. (2011). *Handbook of Frozen Food Processing and Packaging*, 2nde. Boca Raton, Florida, USA: CRC Press/Taylor & Francis.

17 Takamichi, M., Nishimiya, Y., Miura, A., and Tsuda, S. (2007). Effect of annealing time of an ice crystal on the activity of type III antifreeze protein. *The FEBS Journal* 274 (24): 6469–6476.

18 Bayer-Giraldi, M., Sazaki, G., Nagashima, K. et al. (2018). Growth suppression of ice crystal basal face in the presence of a moderate ice-binding protein does not confer hyperactivity. *Proceedings of the National Academy of Sciences* 115 (29): 7479–7484.

19 Cao, L., Huang, Q., Wu, Z. et al. (2016). Neofunctionalization of zona pellucida proteins enhances freeze-prevention in the eggs of Antarctic notothenioids. *Nature Communications* 7 (1): 12987.

20 Vorontsov, D.A., Sazaki, G., Titaeva, E.K. et al. (2018). Growth of ice crystals in the presence of type III antifreeze protein. *Crystal Growth & Design* 18 (4): 2563–2571.

21 Peltier, J.-C.-A. *Peltier effect*. Encyclopædia Britannica, inc.

22 Au-Braslavsky, I. and Au-Drori, R. (2013). LabVIEW-operated novel Nanoliter Osmometer for ice binding protein investigations. *Journal of Visualized Experiments* 72: e4189.

23 Bai, G., Song, Z., Geng, H. et al. (2017). Oxidized quasi-carbon nitride quantum dots inhibit ice growth. *Advanced Materials* 29 (28): 1606843.

24 Meister, K., DeVries, A.L., Bakker, H.J., and Drori, R. (2018). Antifreeze glycoproteins bind irreversibly to ice. *Journal of the American Chemical Society* 140 (30): 9365–9368.

25 Wu, J., Rong, Y., Wang, Z. et al. (2015). Isolation and characterisation of sericin antifreeze peptides and molecular dynamics modelling of their ice-binding interaction. *Food Chemistry* 174: 621–629.

26 Chen, X. and Wang, S.Y. (2019). Cryoprotective effect of antifreeze glycopeptide analogues obtained by nonenzymatic glycation on *Streptococcus thermophilus* and its possible action mechanism. *Food Chemistry* 288: 239–247.

27 Wang, F., Xiong, S., Li, X. et al. (2020). Cryoprotective effect of silver carp muscle hydrolysate on baker's yeast Saccharomyces cerevisiae and its underlying mechanism. *Food Science & Nutrition* 8 (1): 190–198.

28 He, Z., Zheng, L., Liu, Z. et al. (2017). Inhibition of heterogeneous ice nucleation by bioinspired coatings of polyampholytes. *ACS Applied Materials & Interfaces* 9 (35): 30092–30099.

29 Li, L., Wu, J.-H., Zhang, L. et al. (2017). Investigation of the physiochemical properties, cryoprotective activity and possible action mechanisms of sericin peptides derived from membrane separation. *LWT—Food Science and Technology* 77: 532–541.

30 Wang, S. and Damodaran, S. (2009). Ice-structuring peptides derived from bovine collagen. *Journal of Agricultural and Food Chemistry* 57 (12): 5501–5509.

31 Burkey, A.A., Riley, C.L., Wang, L.K. et al. (2018). Understanding poly(vinyl alcohol)-mediated ice recrystallization inhibition through ice adsorption measurement and pH effects. *Biomacromolecules* 19 (1): 248–255.

32 Du, L. and Betti, M. (2016). Chicken collagen hydrolysate cryoprotection of natural actomyosin: mechanism studies during freeze-thaw cycles and simulated digestion. *Food Chemistry* 211: 791–802.

33 Damodaran, S. and Wang, S. (2017). Ice crystal growth inhibition by peptides from fish gelatin hydrolysate. *Food Hydrocolloids* 70: 46–56.
34 Provesi, J.G., Valentim Neto, P.A., Arisi, A.C.M., and Amante, E.R. (2019). Extraction of antifreeze proteins from cold acclimated leaves of Drimys angustifolia and their application to star fruit (Averrhoa carambola) freezing. *Food Chemistry* 289: 65–73.
35 Ba, Y., Mao, Y., Galdino, L., and Günsen, Z. (2013). Effects of a type I antifreeze protein (AFP) on the melting of frozen AFP and AFP+solute aqueous solutions studied by NMR microimaging experiment. *Journal of Biological Physics* 39 (1): 131–144.
36 Moudrakovski, I.L., Ratcliffe, C.I., McLaurin, G.E. et al. (1999). Hydrate layers on ice particles and superheated ice: a 1H NMR microimaging study. *The Journal of Physical Chemistry A* 103 (26): 4969–4972.
37 Jiao, X., Cao, H., Fan, D. et al. (2019). Effects of fish oil incorporation on the gelling properties of silver carp surimi gel subjected to microwave heating combined with conduction heating treatment. *Food Hydrocolloids* 94: 164–173.
38 Jia, G., He, X., Nirasawa, S. et al. (2017). Effects of high-voltage electrostatic field on the freezing behavior and quality of pork tenderloin. *Journal of Food Engineering* 204: 18–26.
39 Kiani, H. and Sun, D.-W. (2011). Water crystallization and its importance to freezing of foods: a review. *Trends in Food Science & Technology* 22 (8): 407–426.
40 Ding, X., Zhang, H., Chen, H. et al. (2015). Extraction, purification and identification of antifreeze proteins from cold acclimated malting barley (*Hordeum vulgare L.*). *Food Chemistry* 175: 74–81.
41 Zhang, M., Li, F., Diao, X. et al. (2017). Moisture migration, microstructure damage and protein structure changes in porcine longissimus muscle as influenced by multiple freeze-thaw cycles. *Meat Science* 133: 10–18.
42 Nikoo, M., Benjakul, S., Ahmadi Gavlighi, H. et al. (2019). Hydrolysates from rainbow trout (Oncorhynchus mykiss) processing by-products: properties when added to fish mince with different freeze-thaw cycles. *Food Bioscience* 30: 100418.
43 Graham, L.A. and Davies, P.L. (2005). Glycine-rich antifreeze proteins from snow fleas. *Science* 310 (5747): 461.
44 Davies, P.L. (2014). Ice-binding proteins: a remarkable diversity of structures for stopping and starting ice growth. *Trends in Biochemical Sciences* 39 (11): 548–555.
45 DeVries, A.L. and Wohlschlag, D.E. (1969). Freezing resistance in some Antarctic fishes. *Science* 163 (3871): 1073–1075.
46 Hirano, Y., Nishimiya, Y., Matsumoto, S. et al. (2008). Hypothermic preservation effect on mammalian cells of type III antifreeze proteins from notched-fin eelpout. *Cryobiology* 57 (1): 46–51.
47 Wang, S., Zhao, J., Chen, L. et al. (2014). Preparation, isolation and hypothermia protection activity of antifreeze peptides from shark skin collagen. *LWT—Food Science and Technology* 55 (1): 210–217.
48 Wang, W., Chen, M., Wu, J., and Wang, S. (2015). Hypothermia protection effect of antifreeze peptides from pigskin collagen on freeze-dried Streptococcus thermophiles and its possible action mechanism. *LWT—Food Science and Technology* 63 (2): 878–885.
49 Zhang, Y., Liu, K., Li, K. et al. (2018). Fabrication of anti-icing surfaces by short α-helical peptides. *ACS Applied Materials & Interfaces* 10 (2): 1957–1962.

50 Liu, K., Wang, C., Ma, J. et al. (2016). Janus effect of antifreeze proteins on ice nucleation. *Proceedings of the National Academy of Sciences* 113 (51): 14739.

51 Yu, G., Yap, Y.R., Pollock, K., and Hubel, A. (2017). Characterizing intracellular ice formation of Lymphoblasts using low-temperature Raman spectroscopy. *Biophysical Journal* 112 (12): 2653–2663.

52 Raus, M., Kopecný, D., and Šebela, M. (2013). Program application for the prediction of results of protein digestion by proteolytic enzymes. *Chemické Listy* 107: 44–53.

53 Lahrichi, S.L., Affolter, M., Zolezzi, I.S., and Panchaud, A. (2013). Food Peptidomics: large scale analysis of small bioactive peptides — a pilot study. *Journal of Proteomics* 88: 83–91.

54 Siciliano, R.A., Mazzeo, M.F., Arena, S. et al. (2013). Mass spectrometry for the analysis of protein lactosylation in milk products. *Food Research International* 54 (1): 988–1000.

55 Damodaran, S. (2007). Inhibition of ice crystal growth in ice cream mix by gelatin hydrolysate. *Journal of Agricultural and Food Chemistry* 55 (26): 10918–10923.

56 Yang, H., Ma, C., Li, K. et al. (2016). Tuning ice nucleation with supercharged polypeptides. *Advanced Materials* 28 (25): 5008–5012.

57 Duman, J.G. and DeVries, A.L. (1972). Freezing behavior of aqueous solutions of glycoproteins from the blood of an Antarctic fish. *Cryobiology* 9 (5): 469–472.

58 Liu, Q., Liu, Y., Yu, J. et al. (2019). Identification and characterization of cryoprotective peptides extracted from silver carp (Hypophthalmichthys molitrix) hydrolysates. *International Journal of Food Properties* 22 (1): 1011–1023.

59 Nguyen, C.T., Yuan, M., Yu, J.S. et al. (2018). Isolation of ice structuring collagen peptide by ice affinity adsorption, its ice-binding mechanism and breadmaking performance in frozen dough. *Journal of Food Biochemistry* 42 (3): e12506.

60 Kun, H. and Mastai, Y. (2007). Activity of short segments of Type I antifreeze protein. *Peptide Science* 88: 807–814.

61 Sicheri, F. and Yang, D.S.C. (1995). Ice-binding structure and mechanism of an antifreeze protein from winter flounder. *Nature* 375: 427–431.

62 Kong, C.H.Z., Hamid, N., Liu, T., and Sarojini, V. (2016). Effect of antifreeze peptide pretreatment on ice crystal size, drip loss, texture, and volatile compounds of frozen carrots. *Journal of Agricultural and Food Chemistry* 64 (21): 4327–4335.

63 Knight, C.A. (2000). Adding to the antifreeze agenda. *Nature* 406 (6793): 249–251.

64 Raymond, J.A. and DeVries, A.L. (1977). Adsorption inhibition as a mechanism of freezing resistance in polar fishes. *Proceedings of the National Academy of Sciences* 74 (6): 2589–2593.

65 Fu, W., Wang, P., Chen, Y. et al. (2019). Preparation, primary structure and antifreeze activity of antifreeze peptides from *Scomberomorus niphonius* skin. *LWT Food Science and Technology* 101: 670–677.

66 Leygonie, C., Britz, T.J., and Hoffman, L.C. (2012). Impact of freezing and thawing on the quality of meat: review. *Meat Science* 91 (2): 93–98.

67 Ustun, S. and Turhan, S. (2015). Antifreeze proteins: characteristics, function, mechanism of action, sources and application to foods. *Journal of Food Processing and Preservation* 39: 3189–3197.

68 Soukoulis, C. and Fisk, I. (2016). Innovative ingredients and emerging Technologies for Controlling ice Recrystallization, texture, and structure stability in frozen dairy desserts: a review. *Critical Reviews in Food Science and Nutrition* 56 (15): 2543–2559.

69 Song, D.H., Kim, M., Jin, E.-S. et al. (2019). Cryoprotective effect of an antifreeze protein purified from Tenebrio molitor larvae on vegetables. *Food Hydrocolloids* 94: 585–591.

70 Feeney, R.E. and Yeh, Y. (1998). Antifreeze proteins: current status and possible food uses. *Trends in Food Science & Technology* 9 (3): 102–106.

71 Zhan, X., Sun, D.-W., Zhu, Z., and Wang, Q.-J. (2018). Improving the quality and safety of frozen muscle foods by emerging freezing technologies: a review. *Critical Reviews in Food Science and Nutrition* 58 (17): 2925–2938.

72 Li, Y. and Yu, J. (2015). Research Progress in structure-activity relationship of bioactive peptides. *Journal of Medicinal Food* 18 (2): 147–156.

73 Tao, H., Wang, P., Wu, F. et al. (2016). Particle size distribution of wheat starch granules in relation to baking properties of frozen dough. *Carbohydrate Polymers* 137: 147–153.

74 Tsolmonbaatar, A., Hashida, K., Sugimoto, Y. et al. (2016). Isolation of baker's yeast mutants with proline accumulation that showed enhanced tolerance to baking-associated stresses. *International Journal of Food Microbiology* 238: 233–240.

75 Ayati, S.-V., Hamdami, N., and Le-Bail, A. (2017). Frozen Sangak dough and bread properties: impact of pre-fermentation and freezing rate. *International Journal of Food Properties* 20 (4): 782–791.

76 Yu, W., Xu, D., Zhang, H. et al. (2020). Effect of pigskin gelatin on baking, structural and thermal properties of frozen dough: comprehensive studies on alteration of gluten network. *Food Hydrocolloids* 102: 105591.

77 Gong, S., Yang, D., Wu, Q. et al. (2019). Evaluation of the antifreeze effects and its related mechanism of sericin peptides on the frozen dough of steamed potato bread. *Journal of Food Processing and Preservation* 43 (8): e14053.1–e14053.8.

78 Zhang, C., Zhang, H., Wang, L., and Guo, X. (2008). Effect of carrot (Daucus carota) antifreeze proteins on texture properties of frozen dough and volatile compounds of crumb. *LWT—Food Science and Technology* 41 (6): 1029–1036.

79 Feng, X., Ng, V.K., Mikš-Krajnik, M., and Yang, H. (2017). Effects of fish gelatin and tea polyphenol coating on the spoilage and degradation of myofibril in fish fillet during cold storage. *Food and Bioprocess Technology* 10 (1): 89–102.

80 Banerjee, R. and Maheswarappa, N.B. (2019). Superchilling of muscle foods: potential alternative for chilling and freezing. *Critical Reviews in Food Science and Nutrition* 59 (8): 1256–1263.

81 Yeh, C.-M., Kao, B.-Y., and Peng, H.-J. (2009). Production of a recombinant type 1 antifreeze protein analogue by L. lactis and its applications on frozen meat and frozen dough. *Journal of Agricultural and Food Chemistry* 57 (14): 6216–6223.

82 Kittiphattanabawon, P., Benjakul, S., Visessanguan, W., and Shahidi, F. (2012). Cryoprotective effect of gelatin hydrolysate from blacktip shark skin on surimi subjected to different freeze-thaw cycles. *LWT—Food Science and Technology* 47 (2): 437–442.

83 Velickova, E., Tylewicz, U., Dalla Rosa, M. et al. (2013). Effect of vacuum infused cryoprotectants on the freezing tolerance of strawberry tissues. *LWT—Food Science and Technology* 52 (2): 146–150.

84 Upadhyay, R., Ghosal, D., and Mehra, A. (2012). Characterization of bread dough: rheological properties and microstructure. *Journal of Food Engineering* 109 (1): 104–113.

85 Kashyap, P., Kumar, S., and Singh, D. (2020). Performance of antifreeze protein HrCHI4 from Hippophae rhamnoides in improving the structure and freshness of green beans upon cryopreservation. *Food Chemistry* 320: 126599.

86 Alonso, J., Canet, W., and Rodriguez, T. (1997). Thermal and calcium pretreatment affects texture, Pectinesterase and Pectic substances of frozen sweet cherries. *Journal of Food Science* 62 (3): 511–515.

87 Regand, A. and Goff, H.D. (2002). Effect of biopolymers on structure and ice recrystallization in dynamically frozen ice cream model systems. *Journal of Dairy Science* 85 (11): 2722–2732.

88 Regand, A. and Goff, H.D. (2006). Ice recrystallization inhibition in ice cream as affected by ice structuring proteins from winter wheat grass. *Journal of Dairy Science* 89 (1): 49–57.

89 Ndoye, F.T. and Alvarez, G. (2015). Characterization of ice recrystallization in ice cream during storage using the focused beam reflectance measurement. *Journal of Food Engineering* 148: 24–34.

90 Cao, H., Zhao, Y., Zhu, Y.B. et al. (2016). Antifreeze and cryoprotective activities of ice-binding collagen peptides from pig skin. *Food Chemistry* 194: 1245–1253.

91 Wang, S., Agyare, K., and Damodaran, S. (2009). Optimisation of hydrolysis conditions and fractionation of peptide cryoprotectants from gelatin hydrolysate. *Food Chemistry* 115 (2): 620–630.

92 Reddy, K.B.P.K., Awasthi, S.P., Madhu, A.N., and Prapulla, S.G. (2009). Role of Cryoprotectants on the viability and functional properties of probiotic lactic acid Bacteria during freeze drying. *Food Biotechnology* 23 (3): 243–265.

93 Capicciotti, C.J., Kurach, J.D.R., Turner, T.R. et al. (2015). Small molecule ice recrystallization inhibitors enable freezing of human red blood cells with reduced glycerol concentrations. *Scientific Reports* 5 (1): 9692.

94 Hasan, M., Fayter, A.E.R., and Gibson, M.I. (2018). Ice recrystallization inhibiting polymers enable glycerol-free cryopreservation of microorganisms. *Biomacromolecules* 19 (8): 3371–3376.

95 Allison, S.D., Chang, B., Randolph, T.W., and Carpenter, J.F. (1999). Hydrogen bonding between sugar and protein is responsible for inhibition of dehydration-induced protein unfolding. *Archives of Biochemistry and Biophysics* 365 (2): 289–298.

4

Antioxidant Peptides

4.1 Sources

4.1.1 Marine Resources

Among numerous marine organisms, marine organisms have been widely recognized as abundant producers of bioactive chemical substances with promising medicinal prospects. With the continuous deepening of scientific research, peptides obtained from marine species have attracted increasing interest due to their extensive biological functions.

Marine organisms, with their vast diversity in species and a wide range of different living habits, serve as a rich source of bioactive nitrogen-containing compounds with diverse structures. Through persistent exploration, researchers have identified a large number of bioactive peptides in various marine protein resources, among which antioxidant peptides have attracted particular attention. Antarctic krill [1] (*Euphausia superba*), as an important source of marine protein, occupies a significant position in the field of bioactive peptide research. The proteolytic products of Antarctic krill exhibit high scavenging activities against 2,2-diphenyl-1-picrylhydrazyl (DPPH) radicals and hydroxyl radicals. Through a series of complex and meticulous experimental procedures, researchers have successfully purified and identified a variety of antioxidant peptides from the proteolytic products of Antarctic krill. These antioxidant peptides play an important role in living organisms, being able to effectively scavenge free radicals, reduce the damage of oxidative stress to cells and tissues, and possess great application potential [2].

In addition to Antarctic krill, marine organisms such as scallops, seaweeds, and sea cucumbers are also potential marine sources. The bioactive peptides contained in these marine organisms also possess unique structures and functions, demonstrating favorable effects in aspects such as anti-oxidation, immune regulation, and improvement of the intestinal microbiota [3–5].

The marine environment is a valuable yet underutilized treasure trove, harboring commercially viable natural products with antioxidant properties [6]. As shown in Figure 4.1, a series of marine animals, plants, and microorganisms all serve as sources of bioactive peptides [8]. Marine organisms have evolved unique metabolic patterns due to their harsh living environments—characterized by cold temperatures, darkness, and high pressure.

Food-Derived Bioactive Peptides: Green Preparation, Activity Mechanisms, and Health Applications, First Edition. Shao-Yun Wang, Xu Chen, and Xixi Cai.

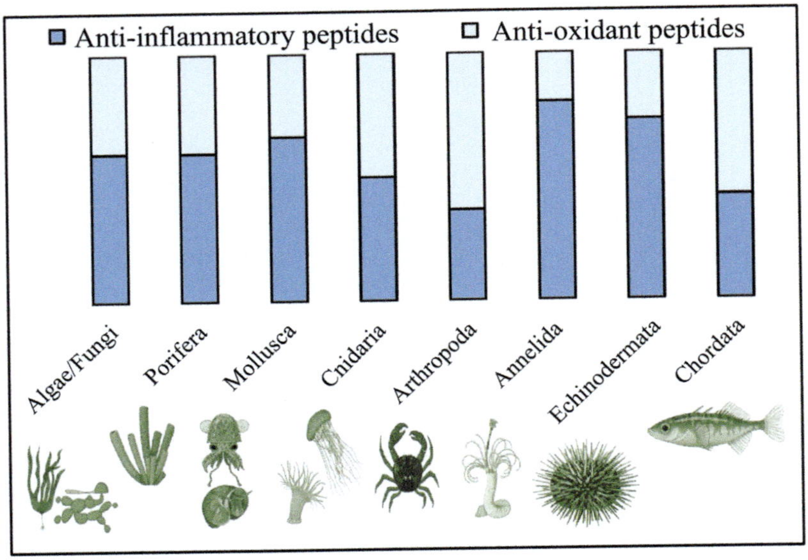

Figure 4.1 Distribution of anti-inflammatory and antioxidant peptides in marine organisms. The bar chart classifies marine organisms according to the proportion of bioactive peptides showing anti-inflammatory and antioxidant activities, including algae/fungi, porifera, mollusca, cnidaria, arthropoda, annelida, echinodermata, and chordata. *Source:* [7]/with permission of Elsevier

As a result, they have generated a large number of bioactive substances with novel structures and diverse functions. These substances not only possess antioxidant properties but also exhibit tremendous commercial potential in fields such as food, pharmaceuticals, and cosmetics [9].

4.1.2 Meat and By-Products

Meat proteins are considered to be an excellent source for obtaining antioxidant peptides. This is because meat proteins contain essential amino acids with high bioavailability, which are generally absent in plant proteins. For instance, methylhistidine and hydroxymethyllysine are such essential amino acids that are specifically found in meat proteins and play crucial roles in the formation and functionality of antioxidant peptides, making meat proteins a unique and valuable resource in the context of antioxidant peptide acquisition [10]. In addition, a large quantity of meat by-products, such as skins, bones, minced meat, blood, and horns, have no high-value applications in the meat industry [10]. However, they possess the potential for utilization as antioxidant hydrolysates and peptides. Table 4.1 presents the statistics of the antioxidant hydrolysates and peptides generated from meat proteins and by-products.

In the current era of continuous development in the fields of life sciences and food research, in recent years, researchers have made remarkable progress in the study of antioxidant peptides. Through the approach of enzymatic hydrolysis of poultry eggs, they have successfully obtained a variety of antioxidant peptides. The sources of these antioxidant

Table 4.1 Antioxidant hydrolysate and peptides generated from meat proteins and by-products.

Source	Antioxidant fractions	Parental protein	Reference
Fermented meat sauce	GYP	—	[11]
Pork hams	Water-soluble and salt-soluble protein hydrolysate	—	[12]
Spanish dry-cured ham	Elution zone between 220 and 345 mL	—	[13]
Spanish dry-cured ham	SAGNPN, GLAGA (2 most active peptides from 27 identified peptides)	Integrin alpha-3, Collagen, type VII, alpha 1	[14]
Spanish dry-cured ham	SNAAC (the most active peptide from 93 identified peptides)	myosin heavy chain 1,4	[15]
Jinhua ham	GKFNV	—	[16]
Xuanwei ham	DLEE	—	[17]
Duck breast meat	LQAEVEELRAALE	Myosin heavy chain	[18]
	IEDPFDQDDWGAWKK	Alpha-enolase	
Duck skin	HTVGCMPG	—	[19]
Duck skin	AVCGAAVAGT	—	[20]
Gelatin from duck skin	gelatin hydrolysate of two enzymes combination	—	[21]
Porcine myofibrillar protein	DSGVT	Actin	Saiga, A.; Tanabe, S.; [22]
	IEAEGE	Tropomyosin	
	DAQEKLE	Myosin heavy chain	
	EELDNALN	—	
	VPSIDDQEELM	—	
Chicken breast protein	Breast protein hydrolysate	—	[23]
Bovine brisket sarcoplasmic proteins	Potential peptides of Sarcoplasmic protein hydrolysates: EAWAEDVDLRVN	Pyruvate kinase,	[24]
	GGWQMEEADDWLR	Phosphorylase	
	GWQMEEADDWLR	Phosphoglycerate kinase 1	
	RIGEEYIADLDQLRKLL	Fructase-biphosphate aldolase	
	VFEWEAFAR	—	
	AIMENANVLAR	—	
	LAIMENANVLAR	—	
Porcine blood plasma	HNGN	—	[25]

(Continued)

Table 4.1 (Continued)

Source	Antioxidant fractions	Parental protein	Reference
Venison muscle	MQIFVKTLTG DLSDGEQGVL	—	[26]
Chicken breast meat	Hydrophilic fraction	—	[27]
Porcine muscle	DLYA, SLYA,VW	Actomyosin	[28]
Porcine skin collagen	QGAR	Collagen	[29]
Porcine plasma	Hydrolysates	Albumin and globulin	[30]
Porcine blood	Pepsin hydrolysate Papain hydrolysate	Plasma proteins	[31]
Porcine blood	Porcine hydrolysates	Hemoglobin	[32]□
Bovine skin gelatin bovine	GE-Hrp-GP-Hrp-GA-Hrp GP-Hrp-GP-Hrp-GP-Hrp-G GP-Hrp-GP-Hrp-GP-Hrp	Collagen	[33]

Source: Adapted from [10].

peptides are extremely abundant, mainly derived from the key components of poultry eggs, such as whole egg white, ovalbumin, ovotransferrin, lysozyme, and egg yolk proteins.

The composition of egg yolk proteins is complex, encompassing numerous different types of proteins, which provides a vast resource for the acquisition of antioxidant peptides. The antioxidant peptides obtained through enzymatic hydrolysis from egg yolk proteins possess diverse structures and biological activities.

The antioxidant peptides obtained by enzymatic hydrolysis of poultry eggs have several remarkable characteristics. They exhibit strong activity and can efficiently react with free radicals in the body, preventing the oxidative damage of free radicals to cells and biological macromolecules, thus effectively protecting the body from the harm of oxidative stress. Moreover, they have a small molecular mass, a feature that enables them to more easily penetrate biological membranes and exert their functions inside cells. Compared with macromolecular antioxidant substances, they have better bioavailability. At the same time, due to their small molecular mass, these antioxidant peptides also have the advantage of being easily absorbed. They can be rapidly absorbed into the bloodstream in the human digestive system and promptly exert their antioxidant functions, providing a powerful guarantee for human health.

Wang et al. have conducted a detailed collation and summary of the antioxidant peptides with determined amino acid sequences, as shown in Table 4.2 [34]. Understanding the amino acid sequences of these antioxidant peptides is like mastering their molecular "code". Based on this information, researchers can conduct in-depth studies on the relationship between their structures and functions, further optimize the preparation process of antioxidant

Table 4.2 Antioxidant peptides prepared by hydrolyzing eggs [34].

Source	Digestive enzyme	Amino acid sequence	Reference
Ostrich egg albumen	Pepsin	LTEQESGVPVMK	[35]
Chicken egg albumen	Papain	YLGAK	[36]
Chicken egg albumen	Papain	GGLEPINF	[36]
Ovalbumin	Pepsin	YAEERYPIL	[37]
Ovalbumin	Pepsin	SALAM	
Ovalbumin	Pepsin	YQIGL	
Ovalbumin	Pepsin	YRGGLEPIN	
Lysozyme	Papain/trypsin	NTDGSTDTYGILQINSR	[38]
Chicken egg yolk	Alkaline protease	LMSYMWSTSM	[39]
Chicken egg yolk	Alkaline protease	LELHKLRSSSHWFSRR	[39]

peptides, and even produce antioxidant peptides with specific functions through artificial synthesis or genetic engineering techniques, laying a solid foundation for the extensive application of antioxidant peptides in the fields of food, medicine, cosmetics, and others.

4.1.3 Plant Sources

In the complex ecosystem of nature, plants are constantly exposed to diverse and unpredictable environmental pressures. Whether facing intense solar radiation, heavy rainfall, or pathogen attacks, plants must adapt and respond effectively to survive. Throughout their evolutionary history, plants have developed a sophisticated and well-coordinated defense system to counteract these challenges. Upon exposure to external stressors, plants activate a series of defense responses, such as the synthesis of systemic response proteins, hypersensitive reactions, and the production of antimicrobial molecules, which collectively enable them to mitigate the adverse effects of environmental stressors and maintain physiological homeostasis. This intricate defense mechanism not only reflects their evolutionary adaptation to ecological pressures but also underscores the resilience and adaptability that have allowed plants to thrive in diverse environmental conditions [40].

In the life activities of plants, small molecules such as peptides, despite their minuscule size, play a vital role in various crucial aspects of plant growth, development, and reproduction. During the process of cell proliferation, peptides can regulate the expression of proteins related to the cell cycle, ensuring the orderly division and growth of cells. Regarding the growth of pollen tubes, specific peptides can guide the pollen tubes to accurately extend toward the ovules, thereby completing the fertilization process, which is of great significance for plant reproduction. In the process of organ senescence, peptides can participate in regulating the signal transduction of related hormones, influencing the rate of organ senescence [41].

As a special class of peptides, antioxidant peptides have received much attention in the field of plant research in recent years. Through unremitting efforts, researchers have successfully isolated and identified antioxidant peptides from various plant tissues.

Antioxidant peptides have been isolated and identified from various plants, including beans, nuts, seeds, grains, and leaves. For instance, antioxidant peptides have been found in Mucuna pruriens proteins, soybean protein isolates, defatted walnut meal proteins, pine nut meal proteins, hemp seed meal proteins, rapeseed proteins, soluble rice proteins, and sorghum kafirin hydrolysates [42–48]. Among the various organs of plants, legumes are distinguished by the relatively high number of antioxidant peptides identified compared to other plant parts. This phenomenon is primarily attributed to the fact that legumes typically contain a high protein content, which provides ample raw material for the formation of antioxidant peptides. Within the legume family, soybeans stand out as the most extensively studied species due to their widespread cultivation, exceptional protein content, and rich nutritional value. Consequently, research on soybean-derived antioxidant peptides has yielded the most fruitful outcomes, reflecting the significant attention this crop has received from researchers in the field of functional food and nutritional science. The unique combination of these factors has positioned soybeans as a model system for understanding the bioactive properties of legume-derived peptides and their potential applications in health promotion and disease prevention.

Leaves, with lower protein content, have received the least attention so far. However, the prospects for identifying antioxidant peptides from leaves remain promising, as demonstrated by recent reports of antioxidant peptides derived from mulberry leaves [48].

To valorize agricultural by-products, researchers have explored watermelon seeds [49], chia oil by-products [50], corn gluten meal [51], peanut meal [52], and perilla seed by-products [53] as sources of antioxidant peptides and hydrolysates.

Bioinformatics is increasingly playing a pivotal role in the field of plant-derived antioxidant peptide research. By harnessing computational modeling approaches, researchers can predict and analyze the enzymatic hydrolysis processes of plant proteins within a virtual environment. This not only streamlines the identification of bioactive peptides but also significantly enhances the efficiency of discovering antioxidant peptides of plant origin, thereby accelerating the translation of theoretical insights into practical applications in nutraceutical and functional food development.

For example, via computational simulation, researchers can predict the hydrolysis of ribulose-1,5-bisphosphate carboxylase/oxygenase (RuBisCO) by thermolysin [54]. Lin et al. proposed an efficient computational strategy (CAE-VD), which integrates deep learning models with virtual digestion, to enzymatically hydrolyze walnut protein for the preparation of antioxidant peptides with specific activities [55].

4.1.4 Dairy Products

Milk, as a precious gift from nature, represents a nutritionally complete and balanced food system. Its complex and intricate composition is characterized by the presence of proteins and essential amino acids, which play a crucial role in supporting biological functions. Proteins in milk exist in diverse forms and serve as the building blocks of life, providing an indispensable foundation for growth, development, and tissue repair in living organisms. Essential amino acids, which cannot be synthesized by the human body and must be obtained through dietary intake, further highlight the nutritional value of milk due to their adequate content. Beyond its fundamental nutritional components, milk is also recognized for its potential bioactive properties, most of which can be attributed to its protein content and the peptides derived from its composition. These bioactive properties underscore the functional significance of milk as a multifaceted resource for both nutrition and health [56].

Dairy-derived antioxidant peptides exhibit significant advantages compared to protein peptides from other sources. Their ubiquity spans the entire production and processing spectrum, from raw milk to various dairy products. Whether it is freshly harvested raw milk, liquid milk treated by pasteurization or ultra-high temperature (UHT) processing, or dairy products such as yogurt, powdered milk, and cheese manufactured through fermentation, concentration, and drying, all can serve as rich sources of dairy-derived antioxidant peptides. This broad availability ensures convenient acquisition and accessibility. Moreover, compared to antioxidant peptides derived from rare organisms or complex synthetic processes, dairy-derived peptides are relatively inexpensive, offering exceptional cost-effectiveness. Additionally, the unique compositional profile of milk enables non-peptidic components to interact with peptides, creating synergistic interactions that further enhance the antioxidant activity of dairy-derived peptides. These characteristics position dairy-derived antioxidant peptides as a highly promising resource for functional food and nutraceutical applications [57, 58].

Recent studies have demonstrated that Lactobacillus helveticus can hydrolyze bovine milk proteins during fermentation to produce bioactive peptides. Elfahri et al. confirmed through experiments that four strains of *Lactobacillus helveticus* (ASCC953, ASCC474, ASCC1188, and ASCC1315) fermented bovine milk at 37 °C for 12 hours, and the bioactive peptides produced exhibited a high DPPH free radical scavenging rate [59]. Kudoh et al. utilized *Lactobacillus delbrueckii* subsp. *bulgaricus* to ferment dairy products and obtained peptides with DPPH radical scavenging activity. They confirmed that these peptides were derived from β-casein [60].

4.2 Evaluation Method of Activity

The evaluation methods for antioxidant capacity have advanced rapidly in parallel with the design and development of antioxidants. Currently, the evaluation and mechanism investigation of the antioxidant capacity of peptides is primarily conducted through chemical methods, cellular models, and animal models. Giving the antioxidant activity of peptides arises from the synergistic effects of multiple properties, including free radical scavenging, transition metal chelation, intracellular reactive oxygen species (ROS) inactivation, and lipid peroxidation inhibition [61]. Therefore, recent studies have increasingly adopted integrated approaches combining chemical assays with cellular and animal-level testing to evaluate the antioxidant activity of peptides.

4.2.1 In vitro Evaluation

Antioxidant biomolecules acts as free radical scavengers; their antioxidant capacity is characterized by two primary mechanisms: single electron transfer (ET-based reactions) or hydrogen atom transfer (HAT-based reactions) [62]. The ET-based reactions involve redox processes where oxidants serve as both probes to monitor reactions and indicators of endpoints. Common ET-based assays include the DPPH radical scavenging ability, trolox equivalent antioxidant capacity (TEAC), and ferric reducing antioxidant power (FRAP) [63]. For HAT-based reactions, antioxidant activity of peptides is typically evaluated in peptides using assays such as oxygen radical absorbance capacity (ORAC), total radical trapping antioxidant parameter (TRAP), and lipid peroxidation inhibition tests (e.g., linoleic acid or low-density lipoprotein oxidation suppression) [62]. Notably, ET reactions occur more rapidly than HAT reactions and are diffusion-independent [64]. Beyond these two mechanisms, transient metal chelation is considered an indirect antioxidant strategy, as it inhibits Fenton reactions responsible for hydroxyl radical generation and subsequent radical chain propagation [62].

4.2.1.1 DPPH Radical Scavenging Assay

The DPPH (2,2-diphenyl-1-picrylhydrazyl) radical scavenging assay is one of the most widely used techniques for evaluating the antioxidant capacity of peptides. The DPPH radical contains an unpaired electron, exhibiting a characteristic purple color in alcoholic solutions with maximum absorption at 515 nm. As a stable free radical, DPPH can be quenched through both electron transfer and HAT mechanisms. Upon reaction with hydrogen-donating compounds, the radical is neutralized, resulting in color fading from purple to yellow [65]. The degree of radical scavenging is quantified spectrophotometrically by measuring the reduction in absorbance, with reference to standard antioxidants such as Trolox or BHT (2,6-di-tert-butyl-4-methylphenol). The peptides derived from crimson snapper scale presented a DPPH radical scavenging ability with the half maximal elimination concentrations (EC_{50}) of 1.85 mg mL^{-1} [66]. Ten antioxidant peptides isolated from protein hydrolysate of swim bladders from miiuy croaker, including FYKWP, FTGMD, GFEPY, YLPYA, FPPYERRQ, GFYAA, FSGLR, FPYLRH, VPDDD, and GIEWA, presented EC_{50} in scavenging DPPH radical in the range of 0.51–5.02 mg mL^{-1} [67]. In a recent study, the DPPH radical scavenging activities of the synthesized antioxidant peptides (VVFVDRL, VIYVVDLR, IYVVDLR, IYVFVR) based on the Alcalase-Hydrolyzed Soybean (*Glycine max* L.) Hydrolysate were characterized as 14.4–17.8 mM TE/mg peptide [61]. Although this method provides rapid and straightforward assessment, its sensitivity is subjected to multiple influencing factors, including solvent quality and volume, presence and concentration of hydrogen donors and metal ions, as well as the purity and freshness of DPPH reagent [68]. Moreover, this method demonstrates notable limitations in evaluating hydrophilic antioxidants, since DPPH is soluble exclusively in organic solvents and insoluble in aqueous media. Another inherent drawback lies in dual functionality of DPPH as both the oxidizing substrate and chromogenic probe, which inevitably introduces spectral interference [69]. Thus, the DPPH assay is typically employed in combination with other methods for comprehensive evaluation of the antioxidant activity of peptides.

4.2.1.2 ABTS Radical Scavenging Assay

The ABTS [2,2′-azino-bis (3-ethylbenzothiazoline-6-sulfonic acid)] radical scavenging assay provides a sensitive measurement of antioxidant capacity through the reduction of ABTS radicals to stable complexes. The antioxidant activity is measured exclusively by the ability of an antioxidant to act as a hydrogen or electron donor to neutralize preformed ABTS radicals. The ABTS is oxidized by oxidants such as potassium persulfate to generate a stable ABTS• + radical cation, which exhibits a characteristic blue-green color with maximum absorbance at 734 nm. The reduction of ABTS• + by antioxidants can be measured using Trolox as a reference standard to calculate TEAC. Yang [70] used the ATVY, an antioxidant peptide derived from black shark skin protein hydrolysate, to investigate the reaction kinetics and underlying mechanisms between the peptide and ABTS radicals. They found that the ATVY-mediated ABTS radical scavenging process consisted of two distinct phases: an initial rapid phase (within 30 seconds) followed by a slower phase persisting beyond 600 seconds. Notably, only partial ABTS radicals converted to ABTS, and the remaining portion transformed into a purple-colored compound exhibiting absorbance at 550 nm. To further elucidate the reaction mechanism, three variant peptides (GTVY, ATVG, and GTVG) were synthesized via amino acid substitution. Replacement of the C-terminal tyrosine (Tyr) with glycine (Gly) resulted in complete loss of both antioxidant activity and purple compound formation. Subsequently, the purple compound was identified as the derivation from ABTS molecules or their fragments through liquid chromatography-mass spectrometry analysis. Based on these findings, the researchers proposed the possible reaction mechanism between ATVY and ABTS: (i) the hydroxyl group on the phenol of ATVY donates a hydrogen to the ABTS radical, forming a semiquinone free radical; (ii) the ATVY losing a hydrogen reacts with the other ABTS• + to generate unstable polyphenol-derived adducts; (iii) these adducts undergo fragmentation, ultimately yielding the purple ABTS•-peptide complex. This ABTS radical scavenging assay offers several advantages, including operational simplicity, good reproducibility, and applicability to both hydrophilic and hydrophobic samples [68, 71]. However, variations in reaction rates between antioxidants and ABTS radicals may lead to different reaction times for complete assessment of antioxidant activities of different samples [72].

4.2.1.3 Hydroxyl Radical Scavenging Assay

Compared to DPPH and ABTS radicals, the hydroxyl radical (•OH) exhibits the highest reactivity. The hydroxyl radical possesses the highest 1-electron reduction potential among ROS [73]. Its extreme reactivity enables facile interactions with various biomolecules, including peptides, proteins, lipids, and DNA. Therefore, determining the scavenging capacity of antioxidant peptides against hydroxyl radicals serves as a crucial indicator for evaluating their antioxidant activity. Hydroxyl radicals can be generated through the interaction of superoxide anions ($O_2\bullet^-$) and hydrogen peroxide (H_2O_2) in the presence of transition metal ions (e.g., Cu^{2+} or Fe^{2+}) [65]. Zhao [67] found that 10 antioxidant peptides from the protein hydrolysate of swim bladders of miiuy croaker (*Miichthys miiuy*) all exhibited hydroxyl radical scavenging EC_{50} values below 5.0 mg/mL. Among them, SMP8 (FPYLRH) and SMP10 (GIEWA) showed the lowest hydroxyl radical scavenging EC_{50} values (0.68 and 0.71 mg/mL, respectively), which were close to that of ascorbic acid (0.52 mg/mL). Zheng [74] used papain, bromelain, neutral protease, and flavorzyme to extract antioxidant hydrolysates from abalone viscera via subcritical water-assisted enzymatic hydrolysis, and these hydrolysates demonstrated strong·OH scavenging ability (0.92–1.71 mg/mL). In another study, the synthetic peptide RQSHFANAQP derived from *Cicer arietinum* L. was found to have a hydroxyl radical scavenging IC50 of 3.92 μmol/mL [65].

4.2.1.4 Superoxide Anions Radical Scavenging Assay

Superoxide anions ($O_2\bullet^-$) can promote oxidation reactions to generate hydrogen peroxide (H_2O_2) and hydroxyl radicals ($\bullet OH$), which damage essential biomolecules and alter their biological functions [67]. The superoxide anion scavenging capacity of antioxidant peptides is primarily determined through two established methods: the pyrogallol autoxidation assay and the NADH-PMS-NBT system. For pyrogallol autoxidation method, pyrogallol undergoes spontaneous oxidation under alkaline conditions, generating superoxide anions $O_2\bullet^-$ and colored intermediates that exhibit a characteristic absorption peak at 320 nm. Antioxidant compounds capable of scavenging $O_2\bullet^-$ prevent the accumulation of these intermediates, allowing calculation of radical scavenging activity through monitoring absorbance changes [69]. Butylated hydroxyanisole (BHA) and ascorbic acid serve as standard references for this assay. The pyrogallol method is widely adopted in antioxidant research due to its high sensitivity and excellent reproducibility. The NADH-phenazine methosulfate-nitroblue tetrazolium chloride (NADH-PMS-NBT) system provides an alternative method for superoxide anion $O_2\bullet^-$ generation. The superoxide scavenging capacity of antioxidant peptides can be calculated by measuring absorbance at 560 nm. Zhao [67] used this method to evaluate the $O_2\bullet^-$ scavenging activity of 10 antioxidant peptides derived from the swim bladders of miiuy croaker (*Miichthys miiuy*). Among these, the peptides FPYLRH and GIEWA demonstrated the highest radical scavenging activity, with EC_{50} values of 0.34 and 0.30 mg/mL, respectively. Beyond the above assays using a spectrophotometer, superoxide anion radical scavenging capacity can also be determined using electron spin resonance (ESR) spectroscopy with appropriate spin-trapping agents. The trapping technique operates by detecting transitions of unpaired electrons in magnetic field, enabling identification of radical species. ESR spectroscopy provides a highly sensitive and direct analytical approach for monitoring reactive species at ambient temperature. The high sensitivity of ESR permits detection of low-concentration radicals in peptide samples [75]. In a study investigating the antioxidant activity of peptide fractions from whey protein hydrolysates, superoxide anion radicals ($O_2\bullet^-$) were generated using a riboflavin/EDTA system under 365 nm irradiation. In the presence of oxygen, riboflavin undergoes photooxidation to generate a flavin radical, while simultaneously producing superoxide anion radicals ($O_2\bullet^-$) through the following net reaction: Flavin (reduced) + $2O_2 \rightarrow$ Flavin (oxidized) + $O_2\bullet^- + 2H^+$. The superoxide anion radical scavenging capacity of antioxidant peptides can be determined by monitoring the corresponding decrease in ESR signal intensity.

4.2.1.5 Ferric-Reducing Antioxidant Power (FRAP)

FRAP is another commonly used technique for evaluating the antioxidant capacity of peptides, reflecting their electron transfer ability [71]. In this reaction, Fe^{3+}-TPTZ (2,4,6-tri [2-pyridyl]-1,3,5-triazine) is reduced by antioxidants to Fe^{2+}-TPTZ, a blue complex with maximum absorption at 593 nm [61]. The compounds such as reduced glutathione, ascorbic acid, α-tocopherol, or butylated hydroxytoluene can be used as standards. Lin [49] found that egg white protein hydrolysate fractions obtained through alkaline protease hydrolysis and ultrafiltration exhibited FRAP antioxidant activity. Among them, the lower molecular weight fraction-3 (<1 kDa) showed the highest Fe^{3+} reducing capacity (FRAP value: 0.382 mmol Fe^{2+}/mg sample), which was 2.81 and 1.92 times higher than that of the higher molecular weight Fraction-1 (<30 kDa) and Fraction-2 (1–30 kDa), respectively.

Harnedy [76] found that the antioxidant peptide Ser-Asp-Ile-Thr-Arg-Pro-Gly-Gly-Gln-Met, derived from an enzymatically hydrolyzed *Palmaria palmata* protein isolate, exhibited a high FRAP value, which was attributed to its Met residue. Research suggested that sulfur-containing amino acids, including Cys and Met, contribute positively to FRAP antioxidant activity because their thiol groups are readily oxidized by reactive species. It should be noted, however, that a high content of positively charged amino acids (particularly lysine) is considered unfavorable for the FRAP activity of hydrolysates and peptides, as it may weaken the interaction strength with Fe^{3+} [64, 77].

4.2.1.6 Oxygen Radical Absorption Capacity (ORAC) Method

Distinct from the aforementioned single electron transfer-based radical scavenging assays, the ORAC method measures the hydrogen-atom-donating capacity of antioxidants [61]. The ORAC assay is recognized as a more physiologically relevant method for in vitro antioxidant activity evaluation, as the free radicals it generates closely resemble those occurring in physiological processes [71]. The ORAC assay measures antioxidant activity by assessing the inhibition of peroxyl radical-mediated oxidation [68]. These radicals are produced through thermal decomposition of 2,2′-azobis(2-amidinopropane) dihydrochloride (AAPH) at 37 °C. The assay utilizes fluorescein, which exhibits weak fluorescence when it is oxidized. As peroxyl radicals oxidize the fluorescein probe, fluorescence intensity decreases progressively. When antioxidants are present, they scavenge the radicals, protecting fluorescein from oxidation and delaying fluorescence decay until all antioxidants are consumed. The assay quantifies antioxidant capacity using Trolox as a standard.

In summary, in vitro antioxidant capacity assays including ABTS, DPPH, hydroxyl radical scavenging capacity, and ORAC, etc. have been widely used in evaluation of the antioxidant activity of peptides. However, several critical issues must be considered when assessing peptide antioxidant capacity in practical food systems, including: (i) interference from food matrix coloration in absorbance measurements, (ii) influence of environmental pH, and (iii) differences in hydrophilic/lipophilic properties of antioxidants. Specifically, the presence of other food components with similar absorption characteristics to chromogenic reagents may significantly affect the determination of peptide antioxidant activity. For instance, anthocyanins exhibit absorption spectra overlapping with DPPH at 500–550 nm [68]. Furthermore, the intrinsic coloration of food matrices can significantly alter absorbance measurements, particularly affecting decolorization-based assays (e.g., ABTS and DPPH) more profoundly than color-developing methods such as FRAP and CUPRAC. System pH is another critical factor to antioxidant assessment. Different assays require specific pH conditions: FRAP functions under acidic conditions, whereas Folin–Ciocalteu and CUPRAC assays operate at neutral and alkaline pH, respectively [68]. Additionally, the distinction between hydrophilic and lipophilic antioxidants necessitates different analytical approaches. Although CUPRAC and ABTS methods can measure both hydrophilic and lipophilic antioxidants, FRAP and Folin–Ciocalteu are limited to hydrophilic compounds, and DPPH is exclusively suitable for hydrophobic systems. Finally, these analytical methods should be combined with complementary approaches to comprehensively evaluate the antioxidant capacity of bioactive peptides. The development of innovative assessment technologies, particularly user-friendly electrochemical biosensors, is urgently required to facilitate antioxidant research.

4.2.2 Cellular Antioxidant Effect

Although chemical assays facilitate rapid screening of antioxidant peptides from food hydrolysates, the underlying antioxidant mechanisms of peptides extend beyond the limited principles reflected in these chemical methods. For instance, antioxidant peptides exert their protective effects through modulating both harmful oxidative products and endogenous antioxidant levels within cellular systems. Compared to conventional chemical assays, cell-based models provide more physiologically relevant evaluations of antioxidant capacity by simulating in vivo conditions. Moreover, unlike costly and time-consuming animal models, the cell model represents a more practical approach for preliminary screening of antioxidant peptides. Consequently, the cellular antioxidant properties of food-derived bioactive peptides have attracted growing research interest worldwide.

Cell models are typically selected according to the target tissues where antioxidant peptides exert biological functions, thereby enabling more accurate evaluation of tissue-specific antioxidant activity and mechanistic pathways. Xiang [78] conducted a systematic review of decadal research (2010–2021) on antioxidant peptides derived from edible aquatic organisms, revealing that hepatic carcinoma cells (HepG2) and human Chang liver cell lines predominated as experimental models. They speculated that oxidative stress is thought to be closely linked to liver disease. Wang [79] reported that GPEGPMGLE, EGPFGPEG, and GFIGPTE derived from redlip croaker (*Pseudosciaena polyactis*) scale collagen hydrolysates could protect HepG2 cells against H_2O_2-induced oxidative stress by reducing ROS and malondialdehyde (MDA) levels, while enhancing the activities of endogenous antioxidant enzymes including superoxide dismutase (SOD), catalase (CAT), and glutathione peroxidase (GSH-Px). Pulsed electric field (PEF)-treated pine nut-derived antioxidant peptides demonstrated significant protective effects on H_2O_2-induced oxidative damage in HepG2 cells, including increased intracellular activities of total SOD, CAT, GSH-Px, and glutathione reductase (GSH-Rx), along with decreased MDA content.

PC12 cells, derived from transplantable rat pheochromocytoma, exhibit synaptic differentiation capacity and retain fundamental neuronal biological characteristics [78]. Thus, this cell is widely utilized to assess the neuroprotective effects of antioxidant peptides. Four shrimp protein hydrolysate-derived peptides (KMDDK, KMDDQ, QMDDK, and QMDDQ) demonstrated significant attenuation of ROS and lactate dehydrogenase (LDH) levels in scopolamine-treated PC12 cells, accompanied by upregulated SOD activity [80]. To exert biological effects in vivo, bioactive peptides must undergo gastrointestinal digestion (GID) and then reach their target sites in active form after absorption. The Caco-2 polarized monolayer system, which can display a well-developed apical brush border membrane expressing specific hydrolases, has been extensively employed to evaluate nutrient transport kinetics, including peptide absorption [81]. This model enables in vitro permeability assessments that correlate with oral bioavailability predictions for antioxidant peptides.

Wang [82] conducted an investigation to characterize antioxidant peptides exhibiting both digestive stability and intestinal absorption potential in silver carp muscle protein hydrolysates. Following simulated gastrointestinal digestion (SGID), the fractions <1 kDa of alcalase- and papain-digested hydrolysates demonstrated intestinal permeability rates of 9.21% and 11.45%, respectively, as determined by Caco-2 monolayer trans-cortical transport analysis. Subsequent LC–MS/MS analysis of the transported permeate identified the peptide fragment LVPVAVF as the most potent antioxidant, exhibiting significant radical scavenging capacity (DPPH·$EC_{50} = 0.65$ mg/mL) and ROS inhibition (27.23% at 50 μg/mL).

Human umbilical vein endothelial cells (HUVECs) serve as a well-established in vitro model for investigating the cardioprotective potential of antioxidant peptides [78]. In a study by Xu [83], the rapeseed-derived bioactive peptide WDHHAPQLR demonstrated significant cytoprotective effects against H_2O_2-induced oxidative damage in HUVECs by inhibiting cell apoptosis. In addition, the human cell models such as cervical adenocarcinoma (HeLa) and neuroblastoma (SH-SY5Y) cell, and various non-human cell models including primary rat hepatocytes, murine macrophage RAW 264.7 cells, and yeast strain Saccharomyces cerevisiae BY4741 were used to assess the antioxidant capacity of bioactive peptides [84]. The antioxidant effects exhibited by food-derived peptides at the cellular level are listed in Table 4.3.

H_2O_2 is the most commonly used chemical agent for inducing oxidative damage in cellular models. As a nonpolar molecule, H_2O_2 readily permeates cell membranes and undergoes Fenton reactions to generate ROS. These reactive intermediates subsequently induce oxidative damage to cellular components and can directly trigger apoptotic pathways. Furthermore, H_2O_2 offers practical experimental advantages, including chemical stability, cost-effectiveness, and dose-dependent effects, making it an indispensable tool for evaluating the antioxidant potential of bioactive peptides in cell-based assays. The muscle protein hydrolysate of Nile tilapia (*Oreochromis niloticus*) yielded three antioxidant peptides (SC, CH, and PGY) that demonstrated ROS-scavenging capacity and upregulated both catalase (CAT) and superoxide dismutase type 1 (SOD1) antioxidant enzymes in AAPH-induced HepG2 cells [87]. Under physiological conditions, ROS are effectively scavenged by endogenous antioxidant defense systems, including SOD and CAT. However, pathological conditions disrupt the redox homeostasis between ROS generation and elimination. Excessive ROS accumulation induces oxidative damage to critical cellular components, including DNA, proteins, and enzymes. Cellular protection against such damage is mediated through both enzymatic and non-enzymatic antioxidant systems [94]. Food-derived peptides could decrease the intracellular ROS and MDA levels [61, 85], enhance cell viability [79], and improve the activity of intracellular antioxidant enzymes, such as SOD, CAT, and GSH-Px [64, 85, 93]. Regulation of the intracellular glutathione levels, inhibition of membrane lipid peroxidation are also found in cellu;ar antioxidant effect of food-derived peptides [85, 93]. The fluorometric probe 2′,7′-dichlorodihydrofluorescein diacetate (DCFH-DA) has been widely employed to quantify intracellular ROS in studies of food-derived antioxidant peptides. Following cellular internalization, DCFH-DA is deacetylated by intracellular esterases to form non-fluorescent compound, which is subsequently oxidized by ROS to yield highly fluorescent dichlorofluorescein (DCF) [90]. Zhang [61] discovered five antioxidant peptides (VVFVDRL, VIYVVDLR, IYVVDLR, VVVDRL, and VIYVVDLR) in soybean (*Glycine max* L.) that effectively reduced ROS levels in H_2O_2-stressed Caco-2 cells. SOD catalyzed the conversion of superoxide radicals into oxygen and H_2O_2, which is subsequently decomposed by CAT into water [95]. Glutathione (GSH), as a free radical scavenger or an enzyme cofactor, played a key role in cellular defense against oxidative damage [96]. Furthermore, GSH-Px utilizes GSH to catalyze peroxide reduction [97]. MDA, a metabolite of cellular lipid peroxidation, serves as a reliable indicator of the extent of oxidative damage [79, 98]. The susceptibility of cell membranes to oxidative stress stems from their high content of polyunsaturated fatty acids (PUFAs), which are particularly vulnerable to free radical attack [99]. Through thiobarbituric acid reactive substances (TBARS) analysis, numerous studies have demonstrated the efficacy of

Table 4.3 Cellular antioxidant effects of food-derived peptides.

Cellular effect	Peptide	Source	Cell model	Reference
Decreased intracellular ROS levels	VVFVDRL VIYVVDLR IYVVDLR IYVFVR	Soybean (*Glycine max* L.)	H_2O_2-treated Caco-2 cell	[61]
	DDWENWAK	Seabass (*Lateolabrax japonicus*)	H_2O_2-treated Caco-2 cell	[85]
	KMDDK KMDDQ QMDDK QMDDQ	Shrimp muscle	H2O2-treated PC12 cells	[80]
	SSDAFFPFR	*Euphausia superba*	Scopolamine-treated PC12 cells	[86]
	GPEGPMGLE EGPFGPEG GFIGPTE	*Pseudosciaena polyactis* scale	H_2O_2-treated HepG2 cells	[79]
	SC CH PGY	*Oreochromis niloticus* muscle	AAPH-treated HepG2 cells	[87]
	AMVDAIAR	*Euphausia superba total* protein	H_2O_2-treated Chang liver cells	[88]
	CP_S (<1 kDa)	*Collagen from Acaudina molpadioides*	H_2O_2-treated RAW 264.7 cells	[89]
	VCSV CAAP	*Paralichthys olivaceus* muscle	AAPH-treated Vero cells	[90]
	GPA	Nile tilapia skin	H_2O_2-treated IPEC-J2 cells	[91]
	FWKVV FMPLH	Miiuy Croaker muscle	H_2O_2-treated HUVEC cells	[92]
Increased glutathione reductase activity	IYVVDLR, IYVFVR	Soybean (*Glycine max* L.)	H_2O_2-treated Caco-2 cell	[61]
	QDHCH	Pine nut proteins	H_2O_2-treated HepG2 cells	[93]
Increased catalase (CAT) activity	IYVVDLR, IYVFVR	Soybean (*Glycine max* L.)	H_2O_2-treated Caco-2 cell	[61]
	DDWENWAK	Seabass (*Lateolabrax japonicus*)	H_2O_2-treated Caco-2 cell	[85]
	GPEGPMGLE EGPFGPEG GFIGPTE	*Pseudosciaena polyactis* scale	H_2O_2-treated HepG2 cells	[79]

Table 4.3 (Continued)

Cellular effect	Peptide	Source	Cell model	Reference
	SC CH PGY	*Oreochromis niloticus* muscle	AAPH-treated HepG2 cells	[87]
	AMVDAIAR	*Euphausia superba total* protein	H_2O_2-treated Chang liver cells	[88]
	QDHCH	Pine nut proteins	H_2O_2-treated HepG2 cells	[93]
Increased glutathione peroxidase (GSH-Px) activity	GPEGPMGLEEG PFGPEGGFIGPTE	*Pseudosciaena polyactis* scale	H_2O_2-treated HepG2 cells	[79]
	AMVDAIAR	*Euphausia superba total* protein	H_2O_2-treated Chang liver cells	[88]
	CP_S (<1 kDa)	*Acaudina molpadioides* body wall	H_2O_2-treated RAW 264.7 cells	[89]
	FWKVV FMPLH	Miiuy Croaker muscle	H_2O_2-treated HUVEC cells	[92]
	QDHCH	Pine nut proteins	H_2O_2-treated HepG2 cells	[93]
Enhanced superoxide dismutase (SOD) activity	DDWENWAK	Seabass (*Lateolabrax japonicus*)	H_2O_2-treated Caco-2 cell	[85]
	GPEGPMGLEEG PFGPEGGFIGPTE	*Pseudosciaena polyactis* scale	H_2O_2-treated HepG2 cells	[79]
	KMDDK KMDDQ QMDDK QMDDQ	Shrimp muscle	H_2O_2-treated PC12 cells	[80]
	SSDAFFPFR	*Euphausia superba* total protein	Scopolamine-treated PC12 cells	[86]
	SC CH PGY	*Oreochromis niloticus* muscle	AAPH-treated HepG2 cells	[86]
	AMVDAIAR	*Euphausia superba total* protein	H_2O_2-treated Chang liver cells	[88]
	CP_S (<1 kDa)	*Acaudina molpadioides* body wall	H_2O_2-treated RAW 264.7 cells	[89]
	FWKVV FMPLH	Miiuy Croaker muscle	H_2O_2-treated HUVEC cells	[92]
	QDHCH	Pine nut proteins	H_2O_2-treated HepG2 cells	[93]

(Continued)

Table 4.3 (Continued)

Cellular effect	Peptide	Source	Cell model	Reference
Decreased malondialdehyde (MDA) level	DDWENWAK	Seabass (*Lateolabrax japonicus*)	H_2O_2-treated Caco-2 cell	[85]
	GPEGPMGLEEG PFGPEGGFIGPTE	*Pseudosciaena polyactis* scale	H_2O_2-treated HepG2 cells	[79]
	AMVDAIAR	*Euphausia superba total* protein	H_2O_2-treated Chang liver cells	[88]
	CP_S (<1 kDa)	*Acaudina molpadioides* body wall	H_2O_2-treated RAW 264.7 cells	[89]
	FWKVV FMPLH	Miiuy Croaker muscle	H_2O_2-treated HUVEC cells	[92]
	QDHCH	Pine nut proteins	H_2O_2-treated HepG2 cells	[93]
Decrease LDH content	DDWENWAK	Seabass (*Lateolabrax japonicus*)	H_2O_2-treated Caco-2 cell	[85]
	KMDDK KMDDQ QMDDK QMDDQ	Shrimp muscle	H_2O_2-treated PC12 cells	[80]

food-derived bioactive peptides in mitigating oxidant-induced lipid peroxidation across various cell models. Li Chen [85] reported five antioxidant peptides derived from seabass hydrolysate, with the peptide DDWENWAK demonstrating significant protection against H_2O_2-induced oxidative injury in Caco-2 intestinal epithelial cells. DDWENWAK exhibited no cytotoxicity within the effective concentration range, and attenuated H_2O_2-induced cell viability reduction as well as the accumulation of ROS and MDA. In addition, DDWENWAK significantly inhibited the release of LDH, a cytoplasmic enzyme marker whose extracellular accumulation directly correlates with H_2O_2-induced membrane permeability alterations in Caco-2 cells. Furthermore, the peptide significantly upregulated activities of antioxidant enzymes in oxidatively damaged Caco-2 cells, such as SOD and CAT.

4.2.3 In vivo Antioxidant Effect

Although animal models may not directly predict human responses, in vivo studies enable the expansion of knowledge regarding the biological effects of antioxidant peptides and hydrolysates [84]. Specifically, in vivo investigations allow the examination of biological activities that are difficult to explore using cellular models, such as lifespan-promoting effects and anti-fatigue properties. Currently, the in vivo antioxidant activity of peptides can be evaluated using invertebrate models (including *Drosophila melanogaster* and *Caenorhabditis elegans*) and vertebrate models (primarily zebrafish, rats, and mice), as tabulated in Table 4.4.

Table 4.4 In vivo antioxidant effects of food-derived peptides.

Animal model	Peptide/hydrolysate	Protein source	In vivo effect	Reference
Zebrafish	Hydrolysate	Hen egg-white lysozyme	• Reduced lipid peroxidation	[100]
Zebrafish under oxidative stress induced by Cu	LPLWPY	Crayfish shell	Liver: • Reduced MDA content • Upregulated expression of SOD1, SOD, GPx3, and GPx4b • Upregulated expression of *keap1* and *nrf2*	[101]
Lard-induced Drosophila melanogaster aging	Hydrolysate	Crimson snapper scale	• Prolonged lifespan • Decreased MDA, PCO content • Increased T-SOD, CAT content • Upregulated expression of SOD1, SOD2, CAT	[66]
Drosophila melanogaster under oxidative stress induced by H_2O_2, paraquat, and UV irradiation	Hydrolysate	Crimson snapper scale	• Prolonged lifespan • Decreased MDA, PCO content • Increased T-SOD, CAT, and GSH-Px content • Upregulated expression of SOD1, SOD2, CAT	[102]
Drosophila melanogaster under oxidative stress induced by H_2O_2	Hydrolysate	*Lateolabrax japonicus*	• Prolonged lifespan • Decreased MDA, PCO content • Increased T-SOD, CAT, and GSH-Px content • Upregulated expression of SOD1, SOD2, CAT • Downregulated expression of mTOR	[85]

(Continued)

Table 4.4 (Continued)

Animal model	Peptide/ hydrolysate	Protein source	In vivo effect	Reference
C. elegans	Hydrolysate	*Actinopyga miliaris*	• Prolonged lifespan even under stress conditions • Increased SOD, CAT, activities, and GSH content • Increased total antioxidant capacity • Decreased ROS level	[103]
Paraquat-induced oxidative stress C. elegans	ILGATIDNSK	*Decapterus maruadsi* total protein	• Increased survival rate • Decreased ROS level • Increased SOD and CAT activities	[104]
	AAVPSGASTG IYEALELRNPL LEAFGNAK	*Strongylocentrotus nudus* gonad	• Increased survival rate • Decreased ROS level • Increased SOD-3 activity • Activated DAF-16 signaling pathway	[105]
Nematode (*Caenorhabditis elegans*) under juglone-induced oxidative stress	FDPAL	Soybean	• Reduced intracellular ROS production • Increased SOD-3 expression • Increased survival rate	[106]
Healthy mice	Hydrolysate	Sesame cake	• Hippocampus tissue: • Reduced MDA level • Increased SOD and GPx activities	[107]
Healthy mice (after exercise)	<3 kDa hydrolysate fraction	Manchurian walnuts	Serum: • Reduced MDA level • Increased activities of SOD, GPx, and CAT	[108]

Ethanol-induced mice	MMPH	*Scomber japonicus* muscle	• Decreased xanthine oxidase activity • Increased SOD and CAT activities • Reduced the MDA level in liver tissues	[109]
High-fat diet and alcohol-induced rats	FCP	*Sphyrna mokarran* skin	• Increased SOD and CAT activities • Reduced the MDA level in liver tissues	[110]
Healthy mice injected with scopolamine to induce memory impairment	<3 kDa hydrolysate fraction	Manchurian walnuts	Brain tissue: Increased SOD and GPx activities	[111]
UV-irradiation-induced oxidative stress mice	HCP	*Hypophthalmichthys molitrix* skin	• Decreased MDA level • Suppressed protein carbonyl formation • Increased SOD and CAT activities • Normalized skin structure and collagen arrangement	[112]
Plasmodium berghei-infected mice	Cysteine-stabilized peptide fraction	Morinda lucida leaf	• Liver tissue and red blood cells: • Reduced MDA levels • Increased GSH level • Increased SOD, CAT, GR, and GPx activities	[113]
Diabetic mice	TCP	*Oreochromis niloticus* skin	• Decreased MDA level • Increased SOD and CAT activities	[114]

Drosophila melanogaster (fruit fly) is a well-established model organism extensively employed in nutritional and genetic research [115]. Functionally, key organ systems in *Drosophila*—including the trachea, Malpighian tubules, fat body, gut, and gonads—exhibit physiological parallels with mammals' organs [115, 116]. Furthermore, *Drosophila* shares similar life cycle features with humans, such as growth, reproduction, and senescence, including age-related declines in exercise and cognitive functions [117]. At the genetic level, *Drosophila* has homology with more than 70% of the known pathogenic genes in humans [118]. Additionally, *Drosophila* exhibits fundamental similarities with mammals in protein molecular basis, signal transduction pathways, synaptic transmission mechanisms, neural coding patterns, and the development and progression of neurological disorders. Metabolically generated ROS contribute to the accumulation of oxidative damage, ultimately accelerating aging and death [119]. Given these conserved biological and pathological mechanisms, *Drosophila* serves as an exceptional model for investigating oxidative stress and related aging processes. Chen [66] demonstrated that crimson snapper scale peptides (CSSPs) exhibit in vivo antioxidant effects in a lard-induced *Drosophila melanogaster* aging model. Dietary supplementation with CSSPs effectively reduced MDA and protein carbonylation (PCO) levels while increasing total superoxide dismutase (T-SOD) and catalase (CAT) activities in a dose-dependent manner. Furthermore, CSSPs upregulated the expression of antioxidant-related genes (SOD1, SOD2, and CAT) in *Drosophila*. In a subsequent study, Cai [102] further investigated the in vivo antioxidant capacity of CSSPs using *Drosophila* models subjected to oxidative stress induced by hydrogen peroxide (H_2O_2), paraquat, and UV irradiation. CSSPs supplementation significantly decreased oxidative markers (MDA and PCO) while enhancing the activities of T-SOD, CAT, and GSH-Px. Additionally, CSSPs upregulated the expression of key antioxidant genes. Notably, at a concentration of 6 mg/mL, CSSPs markedly suppressed the mTOR signaling pathway and activated autophagy in aged male *Drosophila*, with the inhibition of mTOR likely mediated by antioxidant properties of CSSPs. Consequently, CSSPs extended the lifespan of oxidative stress-exposed *Drosophila*.

The nematode *Caenorhabditis elegans* (*C. elegans*) serves as another widely used animal model for evaluating the in vivo antioxidant activity of peptides. With its ease of cultivation, rapid reproduction, and relatively short lifespan, *C. elegans* enables rapid and effective assessment of peptide antioxidant activity. Furthermore, this organism shares significant biochemical and genetic pathway homology with mammals, particularly humans [120]. When fed with Actinopyga miliaris hydrolysate (AMH), *C. elegans* exhibited increased SOD and CAT enzyme activities, elevated GSH levels, enhanced total antioxidant capacity, and reduced ROS levels. Notably, AMH supplementation not only improved these antioxidant parameters but also extended the nematodes' lifespan, even under thermal or oxidative stress conditions. Additionally, paraquat-induced oxidative stress models in *C. elegans* have been commonly employed to characterize in vivo antioxidant effects. Chen [104] reported that the antioxidant peptide ILGATIDNSK, purified from defatted round scad (*Decapterus maruadsi*) protein hydrolysate, significantly enhanced the survival rate of *C. elegans* under paraquat-induced oxidative stress. Furthermore, ILGATIDNSK supplementation reduced intracellular ROS accumulation while simultaneously increasing SOD and CAT activities in *C. elegans*. This simple model organism has proven valuable for screening antimicrobial peptides with in vivo antioxidant activity. Zhao [105] identified antioxidant peptides from purple sea urchin

(*Strongylocentrotus nudus*) gonad hydrolysates using a paraquat-induced *C. elegans* oxidative stress model. Through this approach, peptides SnP7 (AAVPSGASTGIYEALELR) and SnP10 (NPLLEAFGNAK) were shown to mitigate oxidative damage by decreasing ROS levels and modulating the expression of key stress-response proteins, including SOD-3 and HSP-16.2, in oxidatively stressed nematodes.

The zebrafish (*Danio rerio*) is an excellent animal model for biomedical research due to its cost-effectiveness, low maintenance requirements, and year-round reproductive capacity. Remarkably, zebrafish exhibit 70–80% genetic homology with humans. A key advantage of this model system lies in its externally developing transparent embryos, which complete organogenesis within the first 48 hours post-fertilization. Zebrafish development has been well characterized, and therefore, results from zebrafish are comparable to mammalian developmental studies. Furthermore, the well-characterized developmental stages of zebrafish allow direct comparison with mammalian developmental mechanisms. As a vertebrate model, zebrafish offer unique advantages for investigating behavioral and functional parameters relevant to human disease pathogenesis, as well as for potential clinical treatments screening. Therefore, zebrafish represent a valuable model organism for assessing the in vivo antioxidant properties of food-derived bioactive peptides. Carrillo [100] demonstrated that antioxidant peptides derived from hen egg-white lysozyme effectively inhibited lipid peroxidation in zebrafish larvae. In a more recent study, Li [101] investigated the crayfish shell-derived peptide LPLWPY in a copper-induced oxidative stress zebrafish model. Dietary supplementation with this peptide significantly ameliorated hepatic oxidative stress, as evidenced by upregulated expression of SOD1, SOD, GPx3, and GPx4b, along with reduced MDA content. Furthermore, LPLWPY enhanced cellular antioxidant capacity through increased T-SOD and T-AOC levels. The peptide also normalized lipid metabolism by decreasing triglyceride [121] content, increasing ACC activity, and upregulating the expression of *lpl*, *cpt2*, and *acox1* genes. These beneficial effects were mediated through modulation of the Nrf2 and PPARγ signaling pathways, ultimately reducing both oxidative stress and pathological lipid accumulation. While the models mentioned above effectively demonstrated in vivo antioxidant mechanisms, it exhibited limitations in evaluating tissue-specific antioxidant effects of peptides, as evidenced by complementary studies employing rodent (rat and mice) models [122].

Three primary mouse models are commonly employed to assess the antioxidant capacity of peptides: chemically induced, physically induced, and diet-induced oxidative stress models. Among these, ethanol-induced hepatic injury represents one of the most widely utilized approaches for establishing oxidative stress models. The liver synthesizes triglycerides from fatty acids and releases free fatty acids into the bloodstream. Chronic ethanol consumption leads to triglyceride accumulation in hepatocytes, resulting in hepatic steatosis. Concurrently, fatty acid accumulation in the liver impairs cholesterol conversion to bile salts, causing hypercholesterolemia. This condition enhances ROS generation, ultimately compromising the hepatic antioxidant defense system and leading to tissue damage and organ dysfunction. Bashir [109] established a 42-day ethanol-diet-induced liver injury model in mice. Administration of mackerel (*Scomber japonicus*) muscle protein hydrolysate (MMPH) significantly reduced triglyceride and cholesterol concentrations while decreasing lipid peroxidation levels. Additionally, MMPH treatment markedly enhanced SOD and CAT activities. Beyond ethanol-induced hepatotoxicity, high-fat diets constitute another major factor contributing to hepatic lipid accumulation. Vijayan [110] investigated the efficacy of fish collagen peptides (FCP) derived from great hammerhead shark (*Sphyrna mokarran*) skin in mitigating high-fat diet/alcohol-induced hyperlipidemia. Oral FCP supplementation increased SOD and CAT activities while reducing MDA levels in liver tissues of experimental rats. Physical induction methods, particularly radiation exposure, provide alternative approaches for establishing oxidative stress models. High antioxidant collagen peptides (HCP) from silver carp (*Hypophthalmichthys molitrix*) skin demonstrated protective effects against UV-induced cutaneous photoaging in mice [112]. Specifically, HCP supplementation significantly elevated SOD and CAT activities in both serum and skin tissues while reducing MDA content and inhibiting protein carbonyl formation.

Notably, food-derived antioxidant peptides have shown capacity to enhance endogenous antioxidant defenses across various disease models, including: diabetic mice [114], spontaneously hypertensive rats [46], plasmodium berghei-infected mice [113, 123], scopolamine-induced memory impairment models [111]. This protective effect against oxidative stress manifests across multiple tissue types. Existing research confirms the association between diabetes and oxidative stress, wherein excessive ROS accumulation impairs pancreatic β-cell function, diminishes insulin synthesis and secretion, and exacerbates insulin resistance. Alloxan solution injection serves as the common induction method for diabetic mouse models. High-dose tilapia skin collagen peptide (TCP) injection increased SOD and CAT levels by 23.0% and 59.2%, respectively, while reducing MDA by 39.1% in diabetic mice [114]. These values are approaching normal ranges, indicating the protective effect of TCP against pancreatic oxidative damage. Scopolamine, an anticholinergic drug, is frequently used to induce cognitive deficits in experimental animals. It produces dose-dependent effects on learning and memory in mice, establishing validated models of dementia [64]. Manchurian walnut (*Juglans mandshurica* Maxim.) derived antioxidant peptides were shown to enhance SOD and GPx activities in scopolamine-treated mice with memory impairment [111].

4.3 Mechanism Consideration

4.3.1 Mechanisms Underlying Cellular Antioxidant Effects

Some studies have investigated the cellular antioxidant mechanisms mediated by food-derived peptides, particularly those of animal origin [88]. These investigations employ purified peptide sequences, which have provided more definitive mechanistic insights. Oxidative stress ultimately induces cell death through both apoptotic and necrotic pathways. The Bcl-2 protein family plays a central role in regulating apoptotic signaling, making its members increasingly important targets for studying how antioxidant peptides inhibit oxidative stress-induced apoptosis. This family comprises two functionally distinct subgroups: (i) pro-apoptotic members (Bax, Bad, Bim, Bik, and Bcl-xs) that initiate or promote apoptotic signals, and (ii) anti-apoptotic members (Bcl-2, Bcl-xL, Mcl-1, and A1) that prevent effector caspase (e.g., caspase-3) activation, thereby blocking apoptotic signals. Xu [83] found that the rapeseed protein-derived peptide WDHHAPQLR effectively inhibited H_2O_2-induced apoptosis in HUVECs. To be specific, WDHHAPQLR downregulated

Bax and caspase-3 protein expression while upregulating anti-apoptotic Bcl-2. Since Bax mediates mitochondrial damage leading to caspase activation (including caspase-3), the suppression of Bax and mitochondrial-mediated apoptosis represents a key protective mechanism against H_2O_2-induced oxidative damage in HUVECs. Moreover, Bcl-2 upregulation enhances intracellular GSH levels, increases the $NAD^+/NADH$ ratio, and helps maintain redox homeostasis. Similarly, the octapeptide P2 (Ala-Met-Val-Asp-Ala-Ile-Ala-Ile-Ala-Arg) from krill pepsin hydrolysate exhibited cytoprotective effects in H_2O_2-treated Chang liver cells by decreasing Bax and caspase-3 expression while increasing Bcl-2 levels. Notably, P2 pretreatment also upregulated nuclear factor erythroid 2-related factor 2 (Nrf2) and heme oxygenase-1 (HO-1) expression via ERK pathway activation. Zhang [124] investigated the protective mechanisms of *Mytilus coruscus* hydrolysate fraction MCPH-T1 against H_2O_2-induced HUVEC damage. MCPH-T1 pretreatment. The results revealed that MCPH-T1 pretreatment attenuated apoptotic morphology by upregulating the anti-apoptotic Bax/Bcl-2 ratio while downregulating the expression of apoptotic effector proteins caspase-9 and caspase-3. Furthermore, MCPH-T1 treatment suppressed the pro-apoptotic factor NF-κB activation and inhibited caspase-3 activity. Notably, MCPH-T1 pretreatment activated the Nrf2 signaling pathway through Keap1 downregulation, thereby restoring redox homeostasis in endothelial cells under oxidative stress conditions. Deng [125] reported that the novel synthetic peptide CWHTH, derived from the natural peptide LWHTH in *Styela clava*, exhibited protective effects against H_2O_2– and AAPH-induced apoptosis in LO2 cells. This protection was mediated through upregulation of *p*PI3K/PI3K and *p*Akt/Akt expression, coupled with downregulation of *p*JNK/JNK, *p*-c-Jun/c-Jun, and cleaved caspase-9/3. In a separate study using IPEC-J2 cells, Zheng [91] showed that the tripeptide Gly-Pro-Ala, isolated from Nile tilapia skin gelatin hydrolysate, activated Nrf2/ARE transcriptional activity. The peptide dose-dependently enhanced mRNA expression of key antioxidant enzymes, including HO1, NQO1, and GCLM, thereby conferring cytoprotection against H_2O_2-induced oxidative damage.

4.3.2 Mechanisms Underlying *In Vivo* Antioxidant Effects

Current researches are increasingly focusing on elucidating the potential mechanisms underlying the in vivo antioxidant effects of bioactive peptides. However, antioxidant peptides derived from plant proteins may contain non-peptide phytochemicals such as phenolic compounds, saponins, and steroids [84]. These phytoconstituents may either directly participate in antioxidant regulation or interact synergistically with peptides to exert antioxidant effects in vivo. In contrast, peptides originating from animal proteins have been more extensively studied for their in vivo antioxidant activities. This section reviews the currently reported mechanisms on the in vivo antioxidant actions of food-derived peptides.

C. elegans serves as a prominent model organism for studying the in vivo antioxidant mechanism of food-derived peptides. Key transcription factors involved in stress resistance pathways, particularly DAF-16, SKN-1, and HSF-1, have been extensively investigated in antioxidant mechanism studies. DAF-16 regulates numerous downstream genes associated with stress resistance, including sod-3, which collectively influence organismal lifespan. For instance, antioxidant peptides from yak bone collagen have been shown to enhance antioxidant capacity and prolong C. *elegans* lifespan. Ma [106] demonstrated that the soybean-derived peptide FDPAL increased resistance to juglone-induced oxidative stress in wild-type C. *elegans*, while upregulating SOD-3::GFP expression under oxidative conditions. Similarly, Zhao [105] identified two antioxidant peptides (SnP7: AAVPSGASTGIYEALELR; SnP10: NPLLEAFGNAK) from purple sea urchin (*Strongylocentrotus nudus*) gonad hydrolysate that mitigated paraquat-induced oxidative stress in C. *elegans*. Using GR1352 transgenic models, quantitative analysis revealed enhanced nuclear localization of DAF-16::GFP following peptide treatment. Both peptides upregulated sod-3 expression, while only SnP10 increased hsp-16.2 transcription. More recently, Li [126] reported that millet bran-derived peptide (MBPE) extended C. *elegans* lifespan by reducing lipid oxidation, oxidative stress, and lipofuscin accumulation, with genetic analyses confirming upregulated expression of DAF-16 and SKN-1. In zebrafish models, Guru [127] characterized the cysteine/glycine-rich peptide WL15 (WHKNCFRCAKCGKSL), which attenuated H_2O_2-induced apoptosis in zebrafish larvae. To be specific, WL15 pretreatment suppressed activated caspase-3 expression while upregulating antioxidant enzymes, including glutathione S-transferase (GST), glutathione peroxide (GPx), and γ-glutamyl cysteine synthetase (GCS).

4.3.3 Structure–Activity Relationship

4.3.3.1 Effect of Molecular Weight

Researches indicated that food-derived antioxidant peptides typically consist of 2–20 amino acid residues with molecular weights ranging between 250 and 1800 Da [78, 84]. Lower molecular weight peptides generally demonstrate more potent antioxidant activity compared to their higher molecular weight counterparts. This enhanced efficacy is attributed to their superior capacity for target molecule interaction due to reduced steric hindrance [128]. In a comparative study, Wang [129] produced cottonseed peptides through enzymatic hydrolysis and microbial fermentation. The low-molecular-weight fraction CPH3 (<3 kDa) exhibited significantly stronger DPPH and hydroxyl radical scavenging capabilities than the higher molecular weight fractions CPH1 (>10 kDa) and CPH2 (3–10 kDa).

Chen [130] isolated three casein-derived fractions with distinct molecular weights: high (FH, >3 kDa), intermediate (FI, 1–3 kDa), and low (FL, <1 kDa), analyzing their antioxidant capacity changes during in vitro GID. Prior to simulated digestion, the FL fraction demonstrated superior ABTS and hydroxyl radical scavenging activity along with higher ORAC values. Although antioxidant capacity alterations during GID varied across fractions depending on bioactive peptide generation, the post-digestion FL-GID fraction maintained significantly higher antioxidant activity than both FH-GID and FI-GID fractions.

Smaller peptides demonstrate distinct advantages in cellular and in vivo systems owing to their enhanced absorption properties compared to larger peptides. Di- and tripeptides exhibit particularly efficient intestinal absorption, leading to superior bioavailability and greater potential for in vivo biological activity. Previous studies have confirmed that even relatively large peptides such as the 9-amino acid rapeseed-derived WDHHAPQLR (1,158.57 Da) can be absorbed intact through Caco-2 cell monolayers and detected in the serum of orally administered rats [83]. The transport mechanisms of peptides across epithelial barriers show significant molecular weight dependence, resulting in varying bioavailability. Wang [131] conducted a comprehensive investigation of this relationship using casein-derived peptides of different molecular weights in Caco-2 monolayers. The research revealed that larger peptide fractions (F1: 1,600–1,300 Da; F2: 1,000–500 Da) primarily underwent paracellular transport with bioavailability of 10.66% and 9.54% respectively, while the smaller fraction F3 (<500 Da) utilized PepT1-mediated transport, achieving 16.23% bioavailability. The study further demonstrated that amino acid sequence plays a crucial role in determining the absorption efficiency of di- and tripeptides. The antioxidant activity of peptides is significantly influenced by both their amino acid composition and sequence arrangement. Yu [132] observed this phenomenon when fractionating *Sepia esculenta* hydrolysate, where the lowest molecular weight fraction (F3) exhibited significantly weaker protective effects against paraquat-induced oxidative stress in C. *elegans* compared to higher molecular weight fractions (F1 and F2). This finding underscores the critical influence of amino acid composition and sequence on antioxidant activity, independent of molecular size considerations.

4.3.3.2 Effect of Amino Acid Composition and Sequence

The antioxidant capacity of peptides is critically determined by their amino acid composition and sequence. Generally, peptides exhibiting potent antioxidant activity contain a higher proportion of hydrophobic amino acids compared to hydrophilic ones, as these residues play a pivotal role in radical scavenging. The hydrophobic nature facilitates peptide interaction with lipid bilayers, enhancing cellular uptake and subsequent ROS neutralization. Furthermore, hydrophobic amino acids improve membrane affinity and accessibility to lipid-soluble ROS, thereby mitigating lipid peroxidation.

Matsui [133] reported that hydrophobic amino acids, including alanine (Ala), valine (Val), leucine (Leu), isoleucine (Ile), proline (Pro), phenylalanine, and methionine (Met), exhibited notable myoglobin protection rates (30%) against hydroxyl radicals, with leucine achieving 100% protection. Substantial hydrophobic group of methionine particularly enhances interaction with hydrophobic radical species. Ding [134] identified the radical-scavenging peptide SSGPPVPGPMGPMGPR from *Katsuwonus pelamis* bone hydrolysate. Through methionine-to-glycine substitution (SSGPPVPGPGGPMGPR), they confirmed the essential role of methionine in antioxidant activity, as evidenced by reduced DPPH radical scavenging capacity in the modified peptide. However, excessive hydrophobic residues can compromise peptide solubility and antioxidant efficacy.

Many polar amino acids contribute significantly to antioxidant activity through specific functional groups. For example, the thiol group in cysteine donates electrons to electrophilic species like ROS, neutralizing free radicals. Molecular docking studies by Agrawal [135] revealed that serine and threonine residues in *Eleusine coracana* peptides TSSSLNMAVRGGLTR and STTVGLGISMRSASVR form hydrogen bonds with DPPH and ABTS radicals, mediating antioxidant effects. Tyrosine represents another crucial antioxidant amino acid. Its phenolic hydroxyl group acts as a potent hydrogen donor, driving free radical scavenging. Liang [136] attributed the cytoprotective effects of *Moringa oleifera* seed peptide QY on oxidatively damaged Chang liver cells to its tyrosine residue. Using myoglobin assays, Matsui [133] evaluated amino acid antioxidant activity against various ROS and reactive nitrogen species (peroxynitrite), facilitating antioxidant peptide design. Tyrosine demonstrated strong activity against peroxyl radicals even at low concentrations. While tyrosine quantity (not position) primarily influenced peroxyl radical scavenging, residue placement critically determined peroxynitrite antioxidant activity, as shown by myoglobin protection rates: GYYG > GYG > YGGY. Shi [137] further identified cysteine, tyrosine, methionine, and tryptophan as the most effective inhibitors of linoleic acid oxidation among 20 amino acids. Notably, C-terminal tyrosine positioning enhanced peptide anti-linoleic acid radical capacity. Furthermore, acidic and basic amino acids play crucial roles in the antioxidant properties of peptides. The charged acidic residues aspartate [117] and glutamate (Glu) serve as effective electron donors and metal chelators, while positively charged basic amino acids like arginine (Arg) function as potent hydrogen donors [73]. Research indicates that the side chains of acidic amino acids (Glu and Asp) and their analogs (glutamine and asparagine) effectively protect myoglobin structure from peroxynitrite-induced damage [133]. Zhu [138] observed that simulated tryptic digestion dramatically enhanced the Fe^{2+}-chelating capacity of antioxidant peptides from Jinhua ham from 26.8% to 90.4%, attributed to increased exposure of acidic and basic amino acids whose carboxyl and amino side chains can bind Fe^{2+}. Casein hydrolysate fractions richer in basic amino acids reportedly exhibited higher TEAC and ORAC values than those with more acidic residues [139].

Aromatic amino acids contribute to antioxidant activity by donating protons to electron-deficient radicals [139]. The electron-dense aromatic rings in phenylalanine, tyrosine (Tyr), and tryptophan (Trp) residues enhance metal-chelating capacity, thereby reinforcing antioxidant potential [140]. Phe specifically acts as a direct radical scavenger by proton donation and resonance stabilization [73, 141], while also improving lipid solubility to inhibit peroxidation [141]. The peptide AFDEGPWPK demonstrated notable activity in ORAC test, primarily due to the proton-donating capacity of the imidazole N—H group in Trp [142].

Terminal amino acid composition represents another critical factor influencing antioxidant capacity in food-derived peptides. Hydrophobic residues at peptide termini (e.g., Lys at both the N-termini and C-termini of KTKK, Ile at both the N-termini and C-termini of IGSLI, and Leu at both the N-termini and C-termini of SKKGL) enhance antioxidant activity [143]. Bulky hydrophobic residues with low electronic or steric/hydrogen bonding properties (Trp, Tyr, Phe, Met, Leu, Ile) the third position of the amino acid next to the C-terminus also contribute significantly [143, 144]. Charged terminal residues can effectively quench radicals, as exemplified by the Fe^{2+}-chelating histidine at position C1 in APLEEPSSPH [145].

To investigate the antioxidant roles of Cys and Met residues in peptide SYPTECRMR, Lu [140] designed two analogs: SYPTECRCR and SYPTEMRMR. SYPTECRCR showed superior antioxidant activity to both SYPTECRMR and SYPTEMRMR, indicating that Cys contributes more effectively than Met to antioxidant enhancement. This phenomenon is caused by the larger steric hindrance of the methyl group of Met compared to thiol hydrogen of Cys, which impedes electron donation to electrophiles. These findings underscore how structural modifications critically influence antioxidant performance.

4.3.3.3 Effect of Secondary Structure

The antioxidant capacity of peptides is significantly influenced by their secondary structure, with most bioactive peptides exhibiting chiral properties and asymmetric conformations such as α-helices, β-sheets, and β-turns. Structural modifications can enhance antioxidant activity by exposing previously concealed reactive sites that facilitate ROS scavenging. Zhang [146] demonstrated that pulsed electric field (PEF) treatment improved the DPPH and ABTS radical scavenging capacity of shrimp-derived antioxidant peptide

QMDDQ. Circular dichroism spectroscopy revealed that PEF induced a conformational transition from β-sheet to random coil, which reduced steric hindrance and consequently enhanced radical scavenging efficiency. Complementary studies using molecular dynamics simulations have further elucidated structure–activity relationships. Ding [134] found that the secondary structure of SSGPPVPGPMGPMGPR critically determines its antioxidant potential. In the folded conformation, the S–CH3 group in the Met residue, which is on top of a peptide clump and has been experimentally observed to be effective, is exposed to the environment, while another buried Met residue remains sterically hindered from interacting with dissociated radicals.

4.4 Security Assessment

Most bioactive peptides are generally considered safe due to the use of food-grade proteases, which minimize toxicity associated with chemical usage. However, the employment of novel processes, cultures, and unconventional proteases in the production of certain bioactive peptides raises safety concerns, such as potential allergenic risks to consumers. These peptides are often the first such products consumed by individuals seeking specific health benefits [147].

The protein extraction, pre-hydrolysis, fermentation, or proteolysis steps in bioactive peptide production significantly increase the likelihood of safety risks. Toxic substances or proteinaceous materials may be generated during these stages. During protein extraction, amino acid (AA) racemization and isopeptide bond formation can occur, and the resulting bioactive peptides may undergo sequence modifications, leading to increased safety risks [148]. In addition to protein compounds, polyphenols, carbohydrates, and lipids can increase the risk of reactions such as the Maillard reaction and oxidative toxicity, as well as lead to the loss of bioactivity [147].

Studies have shown that specific amino acid sequences such as Lys-Lys-Leu-Leu, Phe-Lys-Lys, Lys-Trp-Lys, Leu-Lys-Leu, and Cys-Tyr-Cys-Arg, as well as residues like Cys, Asn, His, and Pro, exhibit toxicity [149].ToxinPred serves as a time-saving, cost-effective, and reliable safety tool for identifying and eliminating unique peptide sequences that may be toxic, thereby ensuring the safety of bioactive peptides (MDP) used in the development of functional foods [150]. Some proteins are natural allergens, notorious for mediating IgE-mediated allergic reactions. Some peptides inherit this allergenicity from their parent proteins based on resistance to proteases used in their production. However, allergenic peptides can still function, so robust allergenicity studies are needed for peptide-based products.

The protein extraction, pre-hydrolysis, fermentation, or proteolysis steps in bioactive peptide production can significantly increase the likelihood of safety risks, as toxic substances or proteinaceous materials may be generated during these stages.

In general, further rigorous safety assessments, including toxicological studies and clinical trials, are essential to establish the safety of antioxidant peptides and ensure their regulatory approval for food and pharmaceutical applications. This process typically involves multiple aspects, such as toxicological evaluation, allergenicity assessment, and studies on bioavailability and metabolism, to gain a comprehensive understanding of the potential risks and safety of the peptides.

4.4.1 Toxicological Assessment

The potential toxicity of peptides under different doses and durations is investigated through acute, subchronic, and chronic toxicity studies. For instance, acute toxicity studies typically use mouse or rat models to evaluate the LD50 (median lethal dose) of peptides, thereby determining their safety range.

4.4.2 Immunogenicity Assessment

Peptides may possess antigenicity, which can trigger allergic reactions. The production of anti-drug antibodies can lead to reduced drug efficacy, altered clearance rates, and other safety issues, including allergic reactions. Common evaluation methods include skin prick tests, serum IgE antibody detection, and in vitro allergenic activity assays to assess potential allergy risks.

4.4.3 Genotoxicity Studies

Some peptides may exhibit potential genotoxicity at high concentrations or under specific conditions, leading to DNA damage, mutations, or chromosomal abnormalities, which could increase the risk of cancer or other genetic diseases. These studies are conducted through in vitro and in vivo experiments. In vitro experiments include the use of bacteria (e.g., Ames test) or mammalian cells (e.g., micronucleus test, comet assay) to detect DNA damage or mutations caused by peptides. In vivo experiments involve animal models (e.g., mice) to evaluate the genotoxicity of peptides within living organisms, typically focusing on dosage and exposure time. Most approved peptide drugs have shown negative results in genotoxicity studies.

4.4.4 Bioavailability and Metabolism Studies

Bioavailability and metabolism studies of antioxidant peptides are crucial for evaluating their absorption, distribution, metabolism, and excretion (ADME) properties. These studies are significant for understanding peptide efficacy, safety, and clinical applications.

Bioavailability refers to the ability of peptides to enter systemic circulation and exert biological activity, typically involving three stages: absorption, distribution, and metabolism. Peptide absorption primarily depends on transmembrane transport across intestinal epithelial cells, which may involve carrier-mediated active transport or passive diffusion. Testing often combines in vitro models (e.g., Caco-2 cell model to simulate intestinal absorption) with in vivo models (e.g., measuring peptide concentration in the blood of animal models such as mice or rats).

Metabolism is the process by which peptides are broken down or transformed in the body, primarily occurring in the liver, intestines, and kidneys. This includes proteolytic degradation by peptidases (e.g., trypsin, enterokinase) into amino acids or small fragments, oxidative metabolism under oxidative stress conditions, and conjugation reactions with glucuronic acid or sulfate to form inactive metabolites. In vitro metabolism can be simulated using liver microsomes or recombinant enzymes, while in vivo metabolism experiments track the distribution and transformation of peptides and their metabolites in animal models.

Peptides and their metabolites are primarily excreted via the kidneys, intestines, or bile. Studies typically focus on toxicology and dose dependency of metabolites. The safety of metabolites requires assessment of their toxicity and potential genotoxicity. High doses of peptides may increase the metabolic burden and lead to toxic reactions, necessitating dose-dependent studies.

4.5 Application of Antioxidant Peptides

Antioxidant peptides, as bioactive molecules with stable antioxidant properties, exhibit a broad spectrum of sources, diverse functionalities, low molecular weight, and are readily absorbed and digested. They demonstrate high biological safety and significant potential in delaying aging, preventing chronic diseases, and improving overall health, offering additional opportunities for industries such as food, medicine, and cosmetics. This section provides an overview of research examples of antioxidant peptides and discusses their application value in a brief manner.

4.5.1 Applications in the Food Sector

Oxidative damage to food reduces its nutritional value and leads to food degradation. Particularly, oxidative rancidity in high-fat foods generates undesirable odors, significantly reducing food quality and negatively impacting human health. It is increasingly recognized that robust evidence links chain oxidation reactions of fats to cardiovascular diseases [151]. In addition, fresh fruits and vegetables are particularly susceptible to browning due to oxidative reactions caused by mechanical damage sustained during transport or storage [152]. This process results in the loss of key nutrients such as vitamins, carotenoids, anthocyanins, and other bioactive compounds. In the food industry, methods such as cooling, light avoidance, or modified atmosphere packaging are implemented at various stages, including raw material handling, processing, and storage, to prevent oxidation and spoilage of food. Antioxidants are also added to reduce oxidation damage, enhance food stability, and extend shelf-life duration. Currently, most of the antioxidants commonly used in food are synthetic, such as butylated hydroxyanisole [153], butylated hydroxytoluene [154], propyl gallate (PG), tertiary butylhydroquinone (TBHQ) [155], ascorbic palmitate (AP), and others. Although these synthetic antioxidants [156] can effectively improve food oxidation and spoilage, they pose potential health risks to humans. Therefore, natural antioxidants that can inhibit food oxidation have garnered significant attention.

Recent studies have demonstrated the presence of antioxidant peptides derived from natural sources that effectively inhibit lipid oxidation and prevent food rancidity in meat products, which plays a crucial role in maintaining the freshness and nutritional value of food. Hirose et al. evaluated proteins from various sources and found that soy protein peptides not only scavenge free radicals but are also involved in inhibiting lipid oxidation in meat products [157]. Furthermore, Elias et al. demonstrated that specific Tyr and Met residues in β-lactoglobulin and its hydrolysates were able to scavenge free radical species, thereby ameliorating lipid oxidation in food products. These findings further highlight the antioxidant activity of antioxidant peptides [158]. Peptides isolated from Caragana ambigua seeds can be used as potent antioxidants in edible oils and emulsions, which inhibit the oxidation of linolenic acid and slow down the autoxidation of walnut oil [159]. Preparation of cottonseed protein peptides by Wang et al. exhibited both free radical scavenging and Fe^{2+} chelating activities. These peptides are characterized by their enrichment with acidic, basic, aromatic, and hydrophobic amino acids, which contribute to the high antioxidant capacity [129]. Further analysis from the hydrolysates identified 19 peptides, which enhance their free radical binding capacity through specific hydrogen bonding, π-π interactions, and van der Waals interactions with Tyr334, Arg380, Arg415, Ser508, Tyr572, Ser602, Tyr525, Gln530, and Ser555 residues. These findings underscore the potential of these peptides to protect food quality and improve human health. Zheng et al. found that mung bean protein hydrolysate and its lipophilic peptides exhibited excellent DPPH radical scavenging activity, ABTS radical scavenging activity, and metal ion chelating activity [160]. Moreover, mung bean peptides significantly enhanced the oxidative stability of linoleic acid, sunflower oil, and sunflower oil-in-water emulsions. In addition to plant-derived antioxidant peptides, animal-derived antioxidant peptides have also been shown to inhibit lipid oxidation and improve food flavor. Shen et al. reported that the addition of silver carp protein hydrolysate to surimi effectively attenuated myofibrillar fibronectin carbonyl formation, reduced free sulfhydryl groups, and slowed the rate of lipid peroxide formation (MDAs) as well as the alteration of flavor compounds in food products [161].

In addition, natural antioxidant peptides are also utilized as food preservatives to extend the shelf life of foods. Their exceptionally high safety profile positions them as promising green and safe alternatives to conventional synthetic preservatives, which include nitrite [162], sulphates [163], benzoic acid and its salt [164], sorbic acid and its salts [121], dehydroacetic acid and its sodium salt [165], and synthetic flavorings and colors [166]. Jridi [167] et al. added 0.5 mg/g of cuttlefish skin gelatin hydrolysate, which not only slowed fat oxidation and ensured sausage flavor preservation but also extended shelf life to 10 days. Qiu et al. investigated the antioxidant effect of fish fillets using silver carp antioxidant peptide (SAP) solution, and the results showed that SAP significantly reduced free fatty acids and inhibited the increase of TBARS content during storage [168]. Consequently, SAP emerges as an effective natural antioxidant to extend the storage time of frozen fish fillets. In addition, antioxidant peptides are versatile and can be utilized as freshness-preserving agents in various forms, such as coatings or films, enhancing the freshness and shelf life of packaging materials and food products. Abdelhedi et al. successfully developed films based on zebra hemiramphus viscera proteins with excellent tensile strength and low elongation at break [169]. Furthermore, the incorporation of zebra-haulmus peptides significantly enhanced the antioxidant properties of these films, demonstrating their potential for innovative applications in the food industry, particularly in bioactive packaging materials.

In addition, antioxidant peptides have the advantage of being easily absorbed by the intestinal tract and entering the circulatory system with a relatively intact structure, fully leveraging their antioxidant functions and making them suitable for the development of functional foods. Dai et al. investigated the bioaccessibility of rapeseed soluble proteins (RSPs) by simulating their digestion in the gastrointestinal tract. During the gastric phase of their study, RSPs exhibited an increase in hydrolysis from 2.6% to 8.2% accompanied by a decrease in surface hydrophobicity from 62.8% to 59.1%. Further digestion in the intestinal phase reduced both hydrolysis and surface hydrophobicity to 18.2% and 50.5%, respectively [170]. The ability of RSPs to scavenge DPPH and hydroxyl radicals showed a slight increase after pepsin digestion and a more significant enhancement after trypsin digestion. Additionally, the intestinal uptake of RSPs reached its peak at 28.6% when administered at a concentration of 5 mg/mL after 60 minutes of incubation. These findings underscore the dynamic interactions of RSPs in the gastrointestinal tract and their potential role in mitigating oxidative stress and enhancing nutrient absorption Another study simulating the bioavailability of soy protein peptides (SPH) in the gastrointestinal tract demonstrated that SPH exhibited DPPH radical scavenging activity (IC_{50} = 4.22 mg/mL), $ABTS^+$ radical scavenging activity (IC_{50} = 2.93 mg/mL), and metal ion chelating activities (IC_{50} = 0.67 mg/mL) [171]. Additionally, SPH significantly suppressed intracellular ROS in Caco-2 cells. After simulated digestion, the antioxidant activity of SPH was enhanced, though its $ABTS^+$ free radical scavenging activity decreased. Post-translocation through epithelial transport, the permeate retained residual antioxidant activity. Therefore, SPH contains bioavailable antioxidant peptides and represents a promising source for functional food ingredients. In an evaluation of faba bean protein, Samaei investigated the properties of low molecular weight peptides (PHs) generated through hydrolysis, which exhibited enhanced solubility, foaming capacity, and stability under both acidic and neutral pH conditions [172]. Moreover, all PHs demonstrated enhanced antioxidant activity, including free radical scavenging and Fe^{2+} chelating ability. Peptidomic analysis identified homology sequences with known antioxidant peptides. In addition, apple juice spiked with broad bean hydrolysate underwent organoleptic evaluation, and no significant difference in flavor was observed at a concentration of 3% (w/v).

4.5.2 Applications in the Pharmaceutical Field

Oxidation is a physiological mechanism required for cell survival and is an essential basis for life activities. During normal life activities, the body produces corresponding ROS. Reactive oxygen includes both free radicals forms (such as O_2^-, -OH, etc.) and non-free radicals (such as O_3, H_2O_2, $^1O_2$, etc.). Under normal circumstances, the ROS in the body exist in a balance between production and elimination, with low concentrations playing important roles in immune responses, signal transduction, cell proliferation, differentiation, and apoptosis. When cells are exposed to harmful factors or in a pathological state, the active oxygen in the body increases substantially. Additionally, with the increasing operation time, ROS accumulate in the organism. When there is an imbalance between the generation and removal of ROS, the body is placed in a state of oxidative stress, leading to oxidative stress-induced damage to cells and tissues, thus promoting the occurrence of cardiovascular disease, diabetes, cancer, and neurodegenerative diseases, among others. There is evidence demonstrating that the formation of ROS is implicated in the pathogenesis of multiple diseases.

Antioxidants effectively counteract oxidative stress by employing a multifaceted approach. They achieve this through mechanisms such as the direct removal of ROS, enzymatic processes that break down these species, metal ion chelation to neutralize harmful metals, regulation of cellular signaling pathways, protection of cellular membranes, and modulation of inflammatory responses. These actions collectively safeguard the health of cells and organisms, underscoring the medicinal potential of antioxidants in preventing and treating conditions like cardiovascular disease, diabetes, cancer, and neurodegenerative disorders. Antioxidant peptides, with their small molecular weight, are widely sourced, highly safe, and easily absorbed, making them effective in preventing and treating various diseases. They neutralize free radicals, inhibit lipid oxidation, and protect cellular structures, while also exerting effects on immune regulation, anti-inflammatory responses, anti-fatigue, and anti-cancer processes. Their mechanisms of action, unique properties, sources, safety, and clinical applications are key areas for further exploration to enhance understanding and utilize their benefits effectively. In recent years, antioxidant peptides from various sources with anti-inflammatory, anti-cancer, and anti-fatigue effects have shown potential application value in the field of medicine and have become a research hotspot both domestically and internationally.

4.5.2.1 Anti-Inflammatory and Liver-Protecting Drugs

Zhao's discovery of the bioactive peptide RAP-8 represents a significant advancement in the understanding and treatment of non-alcoholic steatohepatitis (NASH) and liver fibrosis [173]. The study has shown that RAP-8 significantly reduces hepatic steatosis and improves lipid metabolism to attenuate hepatitis, fibrosis and liver injury. Moreover, the administration of RAP-8 reduced apoptosis in carbon tetrachloride (CCl4)-induced liver injury, which is mediated by oxidative stress. Therefore, RAP-8 could serve as a potential therapeutic candidate for the treatment of NASH and related metabolic diseases. Poor drinking habits not only lead to great addiction but also cause different health risks to the human body, especially the most direct damage to liver function [174]. Yu et al. extracted corn germ protein peptide (CGMAP) from the industrial by-products of corn germ oil and investigated their liver-protective effects against chronic alcoholic liver injury. [175]. The results showed that an oligopeptide (APF4) rich in glutamic acid, alanine, leucine, and proline from CGMAP significantly increased liver transaminase [164], alanine transaminase [176], MDA, and triglyceride [121] levels during alcohol administration, while effectively lowering SOD, catalase [177] activity, and glutathione [178] levels. Therefore, APF4 can exert a great protection against chronic alcohol-induced liver injury by enhancing the body's antioxidant capacity. In another study, an antioxidant peptide (His-Thr-Val-Gln-Cys-Met-Phe-Gln) was obtained from duck skin by-products, with scavenging activities for hydroxyl, DPPH, alkyl, and superoxide radicals measured at 32.6, 22.7, 55.1, and 49.8 μg/mL, respectively [179]. In evaluating the protective effects of this peptide against 3.5% alcohol-induced damage in normal hepatocytes, it was found that the peptide inhibits the production of ROS and alleviates alcohol-induced hepatocyte cell death. In a study on the hepatoprotective effects of krill protein hydrolysate on ethanol-induced liver injury in mice, Park et al. observed that the peptide increased the body weight and survival rate of the mice, reduced serum levels of glutamic oxaloacetic transaminase and glutamic pyruvic transaminase, and inhibited ethanol-induced apoptosis in the liver [180]. In addition, Zhu et al. isolated and identified an antioxidant peptide from Spurgeon's ginseng and investigated its benefits in an acute ethanol-induced male C57BL/6J mouse model. Spiny ginseng reduced hepatomegaly, hepatitis, and lipid accumulation in mice with acute alcoholic liver injury, and increased antioxidant enzyme activity [181]. The underlying mechanism of action may involve the activation of the Nrf2/HO^{-1} pathway and the blocking of NF-κB nuclear translocation, thereby alleviating oxidative stress and inflammation.

4.5.2.2 Adjuvant Cancer Treatment Drugs

The tumor microenvironment is increasingly characterized by an accumulation of ROS, which is believed to promote tumor growth, metastasis, and recurrence [182]. It has been shown that ROS play different roles at different stages of cancer, and that ROS-induced nucleic acid/mitochondrial DNA damage and metabolism are critical for cancer development and promotion [183]. In the advanced stages of cancer progression, tumor cells generate high levels of intracellular antioxidants, such as NADPH and glutathione [178], through the pentose phosphate pathway to evade apoptosis and buffer the excessive production of ROS and associated oxidative damage within the tumor [184]. Current antioxidants such as Vitamin C, Vitamin E, and β-carotene, which are used to reduce ROS production and oxidative damage in cancer tissues, have unsatisfactory effects. This may be because they target ROS during apoptosis rather than during ROS-induced proliferation [183]. Therefore, natural antioxidant peptides have emerged as promising ingredients for cancer therapy or cancer prevention drugs. A variety of antioxidant peptides have been shown to inhibit cancer cell proliferation, as shown in Table 4.5. The application of antioxidant peptides to cancer prevention and its

Table 4.5 Performance of antioxidant peptides in inhibiting cancer cell proliferation.

Antioxidant peptide name	Cancer cell name	Proliferation inhibition EC_{50} value	Reference
Oyster antioxidant peptide	Human colon cancer HT-29	90.3 μg/mL	[185]
Chickpea Protein Antioxidant Peptide	Breast cancer cell MCF-7	2.4 μmol/mL	[186]
	Breast cancer cell MDA-MB-231	1.5 μmol/mL	
Blood Clam Muscle Protein Peptide	Human prostate cancer cell PC-3	2.0 mg/mL	[187]
	Human prostate cancer cell DU-145	2.8 mg/mL	
	Human lung cancer cell H-1299	3.3 mg/mL	
	Cervical cancer cells NIH3T3	2.5 mg/mL	
Soy peptide	Cervical cancer cells HeLa	16.2 mg/mL	[188]
	Cervical cancer cell CasKi	14.3 mg/mL	
	Breast cancer cell MDA-MB-231	15.2 mg/mL	
Silk protein peptide	Lung cancer cell A549	9.9 mg/mL	[189]
	Lung cancer cell H460	9.1 mg/mL	
Fava bean peptide	Human breast cancer cell MCF-7	0.9 mg/mL	[190]
	Human prostate cancer cell PC3	0.5 mg/mL	
	Human hepatocellular carcinoma cell HepG2	7.6 mg/mL	

therapeutic drug development can not only reduce oxidative stress damage in patients but also inhibit the proliferation of cancer cells.

4.5.2.3 Anti-Fatigue Drug

According to the "free radical theory", high-intensity or exhaustive exercise leads to an increase in the metabolism of free radicals, disrupting the balance of the body's oxidative and antioxidant systems, generating a large amount of ROS, accelerating the oxidation of proteins and lipids, and hastening the fatigue of the organism [191], which is usually manifested as a decrease in blood glucose, a large amount of depletion of hepatic glycogen, and hypoxia of the organism, lactic acid accumulation, increased urea nitrogen content, and decreased creatine kinase activity. Therefore, the anti-fatigue efficacy of antioxidant peptides is generally determined by measuring these biochemical indices and exercise endurance tests. Biochemical indicators typically include several categories: (i) energy substances, such as liver glycogen, myo-glycogen, blood glucose, phosphocreatine; (ii) metabolic regulatory substances, hormones, enzymes; (iii) metabolites, such as lactate and urea nitrogen in serum. Exercise endurance tests measure the time of continuous exercise to exhaustion; there are animal swimming, pole climbing test, etc., which are the most direct and objective indicators to reflect the body's fatigue resistance.

In recent years, studies have shown that antioxidant peptides such as whey peptide, soybean peptide, fish protein peptide, seaweed peptide, shrimp and crab crustacean hydrolyzed peptide have good anti-fatigue effects. Moreover, antioxidant peptides are easily absorbed, provide nitrogen source for the body, eliminate the accumulation of fatigue substitutes, reduce lactic acid as well as reduce the content of serum urea nitrogen levels during exercise to alleviate the exercise-induced fatigue. The mechanism of action of antioxidant peptides is to enhance the body's relief by scavenging free radicals and promoting the synergistic effect between antioxidant substances, which has a potential role in the treatment of chronic fatigue syndrome.

You et al. found that protein peptides extracted from loach prolonged exhaustive swimming time by 20–28% in mice, increased blood glucose levels (28–42%) and liver glycogen levels (2.3–3.0 times), and reduced lactate (10.9–27.5%) and blood urea nitrogen (8.6–17.5%) levels [192]. It also improved endogenous cellular antioxidant enzymes in mice by increasing the activity of SOD, CAT, and GSH-Px. Sun et al. extracted egg white protein peptides (EWPs2) and evaluated their antioxidant and anti-fatigue effects. EWPs2 increased blood glucose levels (28.4–42.2%), decreased muscle glycogen (6.4–10.6%) and liver glycogen (10.7–23.8%), and significantly reduced lactate levels in muscles and urea nitrogen levels in the blood. The results suggest that EWPs can be used as safe and natural antioxidants and anti-fatigue agents [193]. Zhao et al. investigated the biological activities of peptide fragments from protein hydrolysates of Tibetan yellow flower fish [194], from which swim bladder peptide fractions (SBP-III-3) were isolated, and the swim-to-exhaustion time of mice in the experimental group of SBP-III-3 was prolonged by 57.9%–107.5% compared with the control group, and it could be increased by 107.5%. SBP-III-3 mice in the experimental group swam to exhaustion 57.9% to 107.5% longer than the control group and increased muscle glucose (9.4%–115.2%) and hepatic glycogen (35.7%–157.3%) levels, and decreased blood urea nitrogen (16.4%–22.4%), lactic acid (LA) (13.9%–20.1%), and MDA (28.0%–53.6%) levels. SBP-III-3 also enhances LDH activity to clear excess LA, thereby slowing the progression of fatigue. In another study, Ren et al. evaluated and characterized tilapia skin collagen-derived antioxidant hydrolysate (TSCP) and showed that intragastric administration of TSCP at a dose of 2.5 mg/(10 g d) significantly prolonged exhaustion swimming time in loaded mice by mechanisms that included a reduction in serum lactate, serum urea nitrogen, and creatine kinase activity, which resulted in an improvement in liver and muscle glycogen levels and endogenous SOD activity [195].

4.5.2.4 Drugs for Other Diseases

Antioxidant peptides have also widespread applications in the treatment and prevention of chronic diseases such as diabetes and cardiovascular diseases. Chen et al. isolated an antioxidant peptide (LSGYGP) from gelatin hydrolysate of tilapia skin, which possessed the ability to inhibit angiotensin-I-converting enzyme (ACE) and antihypertensive effects [196]. Rezvankhah et al. found that lentil peptides showed promising bioactive properties in terms of antioxidant activity and ACE inhibitory activity and ACE inhibition, with excellent antioxidant, antihypertensive, and antidiabetic characteristics [197]. In addition, antioxidant peptides also play a role in the relief of neuropathic pain. Vijayanand et al. isolated the peptide KC14 from carp and found that it has protective effects against H_2O_2-induced developmental abnormalities [198]. Furthermore, it restores the activity of antioxidant enzymes, reduces neuronal damage, and decreases the levels of lipid peroxidation products (such as malondialdehyde) and nitric oxide induced by H_2O_2. This suggests that KC14 has effective antioxidant activity and neuroprotective ability against oxidative stress, providing a novel therapeutic agent for combating neurodegenerative diseases caused by oxidative stress.

4.5.3 Applications for Skincare

Against the backdrop of heightened uncertainty, obtaining emotional fulfillment has become a new consumer imperative. Beauty and skincare have become an important part of life. The consumers' heightened concerns regarding the safety of skincare product ingredients have grown increasingly stringent. Anti-aging, green, and safe have become new trends in beauty products in recent years. Antioxidant peptides can effectively fight against free radicals, slow down skin aging, and maintain skin health, showing great potential for application in the field of skin care products.

4.5.3.1 Anti-Aging Effect

When exposed to ultraviolet (UV) radiation for a long period of time, the skin will show obvious photo-aging characteristics, including aging, wrinkles, dryness, hyperpigmentation, and skin cancer. Under UV irradiation, the skin can undergo photochemical reactions, leading to the generation of significant amounts of free radicals. Free radicals are highly reactive molecules that attack lipids, proteins, and DNA in cells, leading to cell damage, apoptosis, and even gene mutations. Therefore, promptly removing or inhibiting radiation-induced free radicals plays a role in slowing down skin aging and protecting skin luster. Numerous studies have shown that bioactive peptides have the potential effects to protect the skin from radiation damage, photoaging, and other skin harms.

Rosena et al. utilized chlorpromazine (CPZ), a representative phototoxic drug, to investigate the protective effects of silk extracts derived from three different silks of Thai silkworms against drug-induced phototoxicity [199]. In vitro models, including the A431 human epidermal cell line and RhE human epidermal cells, were employed to assess the photoprotective properties of these silk extracts. Silk extract significantly increased the viability of A431 cells after CPZ exposure and ultraviolet (UVA) irradiation, with a possible protective mechanism being the alleviation of intracellular oxidative stress by increasing intracellular glutathione levels. Similarly, silk extract showed protective effects against CPZ-induced UVA phototoxicity in an RhE model, improving tissue viability and inhibiting the release of interleukin-1α. Kong et al. investigated the protective functions and mechanisms of three peptides from skipjack tuna cardiac arteries against skin photo-aging using a HaCaT cell model of UVB-irradiation [200]. The results showed that skipjack tuna peptides could protect HaCaT cells from apoptosis by restoring the mitochondrial membrane potential (MMP), increase the downstream SOD activity by up-regulating the expression levels of Nrf2, NQO1, and HO^{-1}, scavenge reactive ROS, and decrease the level of intracellular MDA. Feng et al. found that the frog peptide could directly and rapidly scavenge intracellular ROS of HaCaT cells under UVB irradiation and rapidly eliminate artificial free radicals $ABTS^{+}$ to alleviate UVB-induced DNA damage, cell apoptosis, and inflammation, thus inhibiting UVB-induced skin photoaging [201]. In addition to scavenging or inhibiting free radicals, antioxidant peptides can also improve collagen production and achieve anti-aging effects. Yang purified an antioxidant peptide (SHP2) from the hydrolysate of hippocampus abdominalis alkaline protease, which could reduce the intracellular level of ROS in UVB-stimulated HaCaT cells to attenuate apoptosis and significantly enhance cell death. SHP2 reduced intracellular ROS levels in UVB-stimulated HaCaT cells to mitigate apoptosis and significantly enhanced cell viability [202]. Furthermore, SHP2 effectively inhibits ROS, improves collagen synthesis, and reduces cell death by inhibiting ROS generation and lipid peroxidation, thereby inhibiting photoaging. Therefore, SHP2 has antioxidant properties and modulates skin photoaging activity and has potential for use in cosmetic development.

4.5.3.2 Restoration and Conservation

Skin burns, ulcers, and acne wounds are also common forms of skin damage. After skin burns, oxidative stress ensues, and excessive ROS can lead to re-injury of the scalded wound, forming edema, inflammatory response, and organ damage, and even producing shock in severe cases [203]. Excessive oxidative stress and inflammatory response will promote each other, which will cause a vicious circle leading to wound necrosis, hindering wound healing. It has shown that antioxidant peptides have a restorative effect on skin burns, ulcers, and wound infections, in addition to UV radiation-induced skin damage.

Zhang et al. isolated and identified an active peptide (livisin) from the skin secretions of Odorrana livida and showed that livisin was effective in stimulating keratinocyte migration [204]. Using a mouse wound model, livisin was shown to accelerate skin wound healing by promoting epithelialization and collagen deposition. Yu et al. isolated supermolecular peptide self-assembled materials (CAPs) from the Pacific oyster (Crassostrea gigas) and demonstrated their hemostatic effects, free radical scavenging activity (112.74%), and the ability to promote HaCaT cell proliferation (127.61%) [205]. Furthermore, in vivo experiments revealed that CAPs exhibit anti-inflammatory effects, promote fibroblast proliferation, and facilitate blood vessel reconstruction, thereby accelerating wound healing and skin regeneration. Pozzolini et al. studied collagen extracts (MCHs) from sponges and showed that MCHs stimulated cell growth with a ROS scavenging activity of up to 60%. In vitro wound healing assay, MCHs significantly increased the rate of cell proliferation and inhibited UV-induced cell death [206].

4.5.3.3 Whitening and Moisturizing Effect

The role of melanin and melanocytes is crucial in the study of skin whitening and pigmentation. By regulating melanocyte activity and melanin synthesis, uneven skin tone and pigmentation can be effectively improved. Tyrosine is the precursor substance for melanin synthesis. The enzyme tyrosinase catalyzes the oxidation of tyrosine to dopa (DOPA), which is then further converted to dopaquinone and ultimately to melanin [207]. Many ingredients in whitening products aim to inhibit the activity of tyrosinase, reducing melanin synthesis to achieve a whitening effect. Therefore, inhibitors targeting tyrosinase (e.g., hydroquinone, tretinoin, etc.) have been widely used in skin whitening care. Among them, there are numerous research reports on the antioxidant, whitening, and moisturizing properties aspects of the actives.

Yingchutrakul et al. identified and investigated a novel multifunctional peptide from Ganoderma lucidum by means of a targeted proteomics approach and measured the antioxidant capacity of the peptide using DPPH, ABTS, and FRAP assays [208]. The results showed that the antioxidant activity of the Ganoderma lucidum peptide was comparable to that of antioxidants such as ascorbic acid, gallic acid, $FeSO_4$, and others. In Ding's study, novel peptides with antioxidant, lipoxygenase (LOX), xanthine oxidase (XOD), and tyrosinase activities were found after simulating gastrointestinal digestion of processed antler blood (VB) [209]. It was shown that the hydrolysates of VB significantly increased their antioxidant, anti-LOX, anti-XOD, and anti-tyrosinase activities and remained active under in vitro gastrointestinal digestion. Kriangkrai et al. obtained protein hydrolysates from shiitake mushrooms and shiitake mushroom roe by pepsin hydrolysis and ultrafiltration, yielding peptides with molecular weights below 6.5 kDa [210]. These low molecular weight peptides had significant tyrosinase inhibitory activity. The peptides can interact with tyrosinase through hydrogen bonding at specific residues in DPPH and ABTS assays, showing potent tyrosinase inhibition without cytotoxic effects, providing a promising solution to overpigmentation caused by excessive tyrosinase activity due to stress or UV exposure.

With the accelerated pace of people's lives, irregular diet, insufficient sleep, life stress, bad habits, and many other factors lead to many people being in a state of sub-health. Therefore, health products with the effect of relieving fatigue, delaying the aging of the organism, and enhancing the immunity of the organism are getting more and more attention from people. As a bioactive peptide of natural origin, food-borne antioxidant peptide has the characteristics of high safety, small molecular mass, easy absorption, strong activity, etc. It is excellent in the absorption rate and biological function and thus has potential development value. With the development of life science and biological mechanism of action, antioxidant peptide structure–effect relationship will be clearer, and antioxidant peptides will be widely applied in the food industry, cosmetics, and pharmaceuticals. Antioxidant peptides have garnered significant attention as potential drugs or functional factors in health supplements, emerging as a hot topic in food research and high-tech industries and presenting promising prospects and advantages in the pharmaceutical field.

4.6 Limitation and Challenges

Antioxidant peptides represent a novel class of antioxidants that exhibit significant efficacy in aging prevention, anti-aging effects, and cardiovascular disease prevention. However, their practical application is hindered by numerous limitations and challenges.

4.6.1 Preparation Method

Biologically active peptides are abundant in natural sources, including animal, plant, and marine organisms, and play diverse physiological roles to maintain normal metabolic functions. To date, numerous biologically active peptides have been isolated from animal [211–213], plant [178, 214, 215], microbial [154, 216], and marine [217–219] sources. Nevertheless, their low abundance in biological systems and the lack of advanced, efficient techniques for their isolation, purification, and identification pose significant obstacles to the development of antioxidant peptides. Achieving high efficiency in the production of antioxidants is a prerequisite for their large-scale synthesis and application. Chemical synthesis and recombinant protein production methods bypass the complex isolation and purification processes. However, these methods still face challenges such as low yields, high economic consumption, and instability in the final products, which limit their practical utility. SCAP1 is a pentapeptide isolated from the hydrolysate of oyster (Saccostrea cucullata), which has antioxidant and anticancer properties. Sabana [220] synthesized SCAP1 through the solid-phase peptide synthesis (SPPS) method with Fmoc strategy on 2-chlorotrityl chloride resin. 1H-NMR spectra of the peptide revealed proton signals consistent to the structure of SCAP1. During the synthesis, the coupling between Ala and Asn was not straightforward, which led to a low yield of the final product. The use of Trt-based Asn was not a suitable option, as it could cause aggregation during the synthesis, eventually leading to a poor yield of the final product. SCAP1 was obtained as a white solid with a percentage yield of 8.28%. Using solid-phase methods for the synthesis of SCAP1 analogs, the yield achieved by Rani is the highest, reaching up to 57.8% [221]. The replacement of Asn with other amino acid residues (Thr, Pro, Tyr, Trp, and Phe) resulted in a decrease in the IC50 values of the peptides. Notably, however, the replacement of the Lys residue with Val marginally increased the activity. Except for chemical synthesis methods, Sousa [222] et al. demonstrated that recombinant peptides produced by fixed enzyme technology exhibit strong antioxidant potential. Gene engineering synthesis methods allow for the design of specific DNA sequences to guide protein expression in microorganisms, enabling the production of complex biologics. Xiang [223] et al. proposed a novel genetic engineering approach to produce active ginseng peptides through multicopy tandem insertion (5 and 15 times). The recombinant ginseng peptides were successfully produced from engineered Bacillus subtilis with an increasing yield from 356.55 to 2,900 mg/L as the repeats multiplied. Additionally, an oxidative stress-induced aging model caused by H_2O_2 was established to evaluate whether the recombinant ginseng peptides, without enzymatic hydrolysis into individual peptides, also have positive effects on antiaging. The results demonstrated that all two kinds of recombinant ginseng peptides could also delay cellular aging through various mechanisms, such as inhibiting cell cycle arrest, suppressing the expression of pro-inflammatory factors, and enhancing cellular antioxidant capacity. The advantages of the gene synthesis method lie in the low cost, ease of culturing, and ability to efficiently produce target peptide segments using cheap and easily cultured microorganisms. However, this method also faces challenges such as the instability of multi-peptide and low expression yields.

4.6.2 Structural and Properties

In the preparation of drugs and food products, it is essential to consider the inherent properties of biologically active peptides, such as their active peptides, potential susceptibility to allergens, safety, and stability. Proteins are produced during the fermentation and hydrolysis processes, and they are characterized by the production of a group of structurally diverse, highly unpleasant, and hydrophilic oligopeptides. Deng [224] et al. have shown that bitter peptides derived from the hydrolysates of pepsin-digested fish protein can increase lymphocyte and lysozyme activity and improve fish survival in aquaculture. Bitterness is a major problem limiting the uses of protein hydrolysates in foods; the approaches for debittering of protein include physicochemical and biological methods. Among those methods, masking of bitter tastes, encapsulation, Maillard reaction, and enzymatic hydrolysis have been used in the industry-scale production, while selective separation, enzymatic deamidation, and plastein reaction have not been scaled up to the industrial level. Jiang [225] et al. further enhanced the antioxidant ability of enzymatic milk through lactic acid bacteria (LAB) fermentation and reduced the bitterness of the milk protein. To overcome the bitterness problem of proteolysate, xylitol, sucrose, α-cyclodextrin, maltodextrin, and combinations of these were tested systematically as bitter-masking agents of an enzyme-treated soy protein in an aqueous model and in a bread model in the study of Bertelsen. Results in water, xylitol, sucrose, and maltodextrin reduced bitterness significantly, whereas α-cyclodextrin did not. Clearer bitter-masking effects were seen in the aqueous model compared with the bread model. Plastein reactions [226] originally utilized for the debittering of protein hydrolysates, are a process wherein the concentrated peptide mixtures or protein hydrolysates are catalyzed by proteases to form elastic, gelatinous, and water-insoluble products. Sharma [227] et al. explored the impacts of plastein reaction on bitterness and the physicochemical and antioxidant properties of salmon frame hydrolysate. The bitterness of hydrolysate was markedly debittered via plastein reaction under optimal conditions. Plastein generally had a lighter color and still possessed antioxidant activity. Additionally, the plastein reaction can be used to both mask the bitterness of proteins and enhance their antioxidant activity. In Xu's study [228], fish scale collagen hydrolysate (FH) was used as the raw material for plastein reaction. Various enzymes, including papain, alcalase, flavourzyme, and the combination of alcalase and flavourzyme, were employed for this purpose. The plastein reaction significantly improved the thermal stability, chemical antioxidant activity, and capacity to scavenge cellular ROS of FH.

During the process of protein hydrolysis, low-molecular-weight peptides are typically generated, which are often allergens, or allergens are transferred from the parent protein level to the newly formed peptides during hydrolysis. This limitation has hindered the application of antioxidant peptides in the nutritional drug industry. The application of peanut in foods and dietary supplements is limited due to the fear of peanut allergy [229]. Yu [230] et al. indicated that hydrolysis of peanut flour (12% lipid) up to 4 hour with Alcalase lowered the immunoreactivity of both soluble and insoluble portions of peanut flour, and generated peanut flour hydrolysate (PFH) with good in vitro antioxidant and ACE-inhibitory activities. Importantly, the fraction smaller than 5 kDa did not bind IgE of peanut-allergic patients but exhibited higher ACE-inhibitory activity than the crude PFH. Huo et al. reduced the beany flavor, and the sensitization of soymilk and improved the digestibility of soymilk through LAB fermentation, making it easy to be accepted by consumers. The fermented soymilk showed higher free amino acids (FAA) content, peptide content, and stronger antioxidant activity than soymilk after digestion.

Even if the issues of flavor and safety are overcome, peptide substances are easily decomposed under conditions of high heat, high pH, or nutrition, resulting in poor stability. Peptide-based drugs may have several drawbacks, including their inability to penetrate cell membranes, limited bioavailability, and non-specific interactions with plasma proteins [231]. In contrast to proteins, endogenous peptides do not possess the additional structural stability conferred by secondary or tertiary structures. Thus, peptides can potentially lose their secondary structure, leading to a decline in the mechanism behind some activities. This problem is particularly evident in the food and pharmaceutical industries, thereby affecting their application effectiveness.

Preparation and purification processes for antioxidant peptides are complex, and their structural properties are unstable, which poses challenges in their application. Additionally, the lack of clinical trials, unclear mechanisms of antioxidant agents, low acceptance in the market, and incomplete standardized regulations further hinder their practical use. The future study of antioxidant peptides may focus on the following key areas to enhance their effectiveness and applications: (i) Diversify the species sources of antioxidant peptides, optimize the extraction, identification, and production processes, and establish a more comprehensive antioxidant peptide information database based on proteomics and artificial intelligence algorithms. Additionally, optimize the structure of antioxidant peptides and design them to address issues such as low activity, toxicity, and instability. (ii) The mechanisms of action of antioxidant peptides remain controversial, and further investigation is required to establish their bioavailability, persistence in vivo, and specificity in vivo before converting these findings to clinical applications. Incorporate human subject testing into the design of such studies. (iii) To expand the application scope of antioxidant peptides, currently, the application of antioxidant peptides is limited, primarily used in drugs. In future studies, it is necessary to expand their application, and they have widespread application prospects in food, pharmaceuticals, cosmetics, and agriculture. Additionally, corresponding laws, regulations, and management standards should also be established and improved progressively.

References

1 RiBang, W., CuiLing, W., Dan, L. et al. (2015). Overview of antioxidant peptides derived from marine resources: the sources, characteristic, purification, and evaluation methods. *Applied Biochemistry and Biotechnology* 176 (7): 1815–1833.

2 ShuangYi, Z., GuoXu, Z., ShiKun, S. et al. (2021). Purification, identification, activity evaluation, and stability of antioxidant peptides from Alcalase hydrolysate of Antarctic krill (Euphausia superba) proteins. *Marine Drugs* 19 (6): 347–347.

3 Li, X.-R., Chi, C.-F., Li, L., and Wang, B. (2017). Purification and identification of antioxidant peptides from protein hydrolysate of scalloped hammerhead (Sphyrna lewini) cartilage. *Marine Drugs* 15 (3): 61.

4 Zhou, X., Wang, C., and Jiang, A. (2012). Antioxidant peptides isolated from sea cucumber Stichopus Japonicus. *European Food Research And Technology* 234 (3): 441–447.

5 Bousquet, M.S., Ratnayake, R., Pope, J.L. et al. (2020). Seaweed natural products modify the host inflammatory response via Nrf2 signaling and alter colon microbiota composition and gene expression. *Free Radical Biology and Medicine* 146: 306–323.

6 Todorka, V., Nelly, G., Anna, S., and Dilyana, G. (2022). Recent Progress in antioxidant active substances from marine biota. *Antioxidants* 11 (3): 439–439.

7 Jegani, K.T., Balde, A., and Nazeer, R.A. (2025). A review on anti-inflammatory and antioxidant peptides derived from marine organisms: mechanism of action and therapeutic applications. *Food Bioscience* 63: 105745.

8 Susmita, G., Tanmay, S., Siddhartha, P. et al. (2022). Novel bioactive compounds from marine sources as a tool for functional Food development&#13. *Frontiers in Marine Science* 9: 832957.

9 Vaibhav, M. and Akanksha, S. (2018). Marine peptides act as novel chemotherapeutic agent. *Journal of Microbiology & Experimentation* 6 (6): 267–270.

10 Liu, R., Xing, L., Fu, Q. et al. (2016). A review of antioxidant peptides derived from meat muscle and by-products. *Antioxidants* 5 (3): 32.

11 Ohata, M., Uchida, S., Zhou, L., and Arihara, K. (2016). Antioxidant activity of fermented meat sauce and isolation of an associated antioxidant peptide. *Food Chemistry* 194: 1034–1039.

12 Park, S.Y. and Chin, K.B. (2011). Antioxidant activities of pepsin hydrolysates of water- and salt-soluble protein extracted from pork hams. *International Journal of Food Science & Technology* 46 (2): 229–235.

13 Escudero, E., Aristoy, M.-C., Nishimura, H. et al. (2012). Antihypertensive effect and antioxidant activity of peptide fractions extracted from Spanish dry-cured ham. *Meat Science* 91 (3): 306–311.

14 Escudero, E., Mora, L., Fraser, P.D. et al. (2013). Identification of novel antioxidant peptides generated in Spanish dry-cured ham. *Food Chemistry* 138 (2–3): 1282–1288.

15 Mora, L., Escudero, E., Fraser, P.D. et al. (2014). Proteomic identification of antioxidant peptides from 400 to 2500Da generated in Spanish dry-cured ham contained in a size-exclusion chromatography fraction. *Food Research International* 56: 68–76.

16 Chao-Zhi, Z., Wan-Gang, Z., Guang-Hong, Z. et al. (2013). Isolation and identification of antioxidant peptides from Jinhua ham. *Journal of Agricultural and Food Chemistry* 61 (6): 1265–1271.

17 Xing, L.-j., Hu, Y.-y., Hu, H.-y. et al. (2016). Purification and identification of antioxidative peptides from dry-cured Xuanwei ham. *Food Chemistry* 194: 951–958.

18 Lu-Sha, W., Ji-Chao, H., Yu-Lian, C. et al. (2015). Identification and characterization of antioxidant peptides from enzymatic hydrolysates of duck meat. *Journal of Agricultural and Food Chemistry* 63 (13): 3437–3444.

19 Lee, S.-J., Kim, Y.-S., Hwang, J.-W. et al. (2012). Purification and characterization of a novel antioxidative peptide from duck skin by-products that protects liver against oxidative damage. *Food Research International* 49 (1): 285–295.

20 Lee, S.-J., Kim, E.-K., Hwang, J.-W. et al. (2010). Purification and characterisation of an antioxidative peptide from enzymatic hydrolysates of duck processing by-products. *Food Chemistry* 123 (2): 216–220.

21 Lee, S.-J., Kim, K.H., Kim, Y.-S. et al. (2012). Biological activity from the gelatin hydrolysates of duck skin by-products. *Process Biochemistry* 47 (7): 1150–1154.

22 Saiga, A., Tanabe, S., and Nishimura, T. (2003). Antioxidant activity of peptides obtained from porcine Myofibrillar proteins by protease treatment. *Journal of Agricultural and Food Chemistry* 51 (12): 3661–3667.

23 Sun, Y., Pan, D., Guo, Y., and Li, J. (2012). Purification of chicken breast protein hydrolysate and analysis of its antioxidant activity. *Food and Chemical Toxicology* 50 (10): 3397–3404.

24 Di Bernardini, R., Mullen, A.M., Bolton, D. et al. (2012). Assessment of the angiotensin-I-converting enzyme (ACE-I) inhibitory and antioxidant activities of hydrolysates of bovine brisket sarcoplasmic proteins produced by papain and characterisation of associated bioactive peptidic fractions. *Meat Science* 90 (1): 226–235.

25 Liu, Q., Kong, B., Xiong, Y.L., and Xia, X. (2010). Antioxidant activity and functional properties of porcine plasma protein hydrolysate as influenced by the degree of hydrolysis. *Food Chemistry* 118 (2): 403–410.

26 Kim, E.-K., Lee, S.-J., Jeon, B.-T. et al. (2009). Purification and characterisation of antioxidative peptides from enzymatic hydrolysates of venison protein. *Food Chemistry* 114 (4): 1365–1370.

27 Sacchetti, G., Di Mattia, C., Pittia, P., and Martino, G. (2008). Application of a radical scavenging activity test to measure the total antioxidant activity of poultry meat. *Meat Science* 80 (4): 1081–1085.

28 Arihara, K. (2006). Strategies for designing novel functional meat products. *Meat Science* 74 (1): 219–229.

29 Li, B., Chen, F., Wang, X. et al. (2007). Isolation and identification of antioxidative peptides from porcine collagen hydrolysate by consecutive chromatography and electrospray ionization–mass spectrometry. *Food Chemistry* 102 (4): 1135–1143.

30 Wang, J.-Z., Zhang, H., Zhang, M. et al. (2008). Antioxidant activity of hydrolysates and peptide fractions of porcine plasma albumin and globulin. *Journal of Food Biochemistry* 32 (6): 693–707.

31 Xu, X., Cao, R., He, L., and Yang, N. (2009). Antioxidant activity of hydrolysates derived from porcine plasma. *Journal of the Science of Food and Agriculture* 89 (11): 1897–1903.

32 Chang, C.-Y., Wu, K.-C., and Chiang, S.-H. (2007). Antioxidant properties and protein compositions of porcine haemoglobin hydrolysates. *Food Chemistry* 100 (4): 1537–1543.

33 Kim, S.K., Kim, Y.T., Byun, H.G. et al. (2000). Purification and characterization of antioxidative peptides from bovine skin. *BMB Reports* 34 (3): 219–224.

34 Wang, J.J. (2014). Advances in research on egg-derived antioxidant peptides. *Science and Technology of Food Industry* 35 (02): 358–363.

35 Tanzadehpanah, H., Asoodeh, A., and Chamani, J. (2012). An antioxidant peptide derived from ostrich (Struthio camelus) egg white protein hydrolysates. *Food Research International* 49 (1): 105–111.

36 Chen, C., Yu-Jie, C., Ming-Yang, Z., and Lei, L. (2012). Purification and identification of antioxidant peptides from egg white protein hydrolysate. *Amino Acids* 43 (1): 457–466.

37 Dávalos, A., Miguel, M., Bartolome, B., and Lopez-Fandino, R. (2004). Antioxidant activity of peptides derived from egg white proteins by enzymatic hydrolysis. *Journal of Food Protection* 67 (9): 1939–1944.

38 Memarpoor-Yazdi, M., Asoodeh, A., and Chamani, J. (2012). A novel antioxidant and antimicrobial peptide from hen egg white lysozyme hydrolysates. *Journal of Functional Foods* 4 (1): 278–286.

39 Park, P.-J., Jung, W.-K., Nam, K.-S. et al. (2001). Purification and characterization of antioxidative peptides from protein hydrolysate of lecithin-free egg yolk. *Journal of The American Oil Chemists' Society* 78 (6): 651–656.

40 Tiffin, P. and Moeller, D.A. (2006). Molecular evolution of plant immune system genes. *Trends in Genetics* 22 (12): 662–670.

41 Thomas, R., Markus, S., Daniel, B. et al. (2019). Plant cell packs: a scalable platform for recombinant protein production and metabolic engineering. *Plant Biotechnology Journal* 17 (8): 1560–1566.

42 Chalé, F.G.H., Ruiz, J.C.R., Fernández, J.J.A. et al. (2014). ACE inhibitory, hypotensive and antioxidant peptide fractions from Mucuna pruriens proteins. *Process Biochemistry* 49 (10): 1691–1698.

43 Ma, H., Liu, R., Zhao, Z. et al. (2017). A novel peptide from soybean protein isolate significantly enhances resistance of the organism under oxidative stress. *PLoS One* 11 (7): e0159938.

44 Gu, M., Chen, H.-P., Zhao, M.-M. et al. (2015). Identification of antioxidant peptides released from defatted walnut (Juglans Sigillata Dode) meal proteins with pancreatin. *LWT—Food Science and Technology* 60 (1): 213–220.

45 Yang, R., Li, X., Lin, S. et al. (2017). Identification of novel peptides from 3 to 10kDa pine nut (Pinus koraiensis) meal protein, with an exploration of the relationship between their antioxidant activities and secondary structure. *Food Chemistry* 219: 311–320.

46 Girgih, A.T., Alashi, A.M., He, R. et al. (2014). A novel hemp seed meal protein hydrolysate reduces oxidative stress factors in spontaneously hypertensive rats. *Nutrients* 6 (12): 5652–5666.

47 Bok-Kyung, H., Yooheon, P., Hyeon-Son, C., and Joo, S.H. (2016). Hepatoprotective effects of soluble rice protein in primary hepatocytes and in mice. *Journal of the Science of Food and Agriculture* 96 (2): 685–694.

48 Sun, C., Tang, X., Ren, Y. et al. (2019). Novel Antioxidant Peptides Purified from Mulberry (Morus atropurpurea Roxb.) Leaf Protein Hydrolysates with Hemolysis Inhibition Ability and Cellular Antioxidant Activity. *Journal of Agricultural and Food Chemistry* 67 (27): 7650–7659.

49 Wen, C., Zhang, J., Zhang, H. et al. (2019). Effects of divergent ultrasound pretreatment on the structure of watermelon seed protein and the antioxidant activity of its hydrolysates. *Food Chemistry* 299: 125165.

50 Coelho, M.S., de Araujo Aquino, S., Latorres, J.M., and de las Mercedes Salas-Mellado, M. (2019). In vitro and in vivo antioxidant capacity of chia protein hydrolysates and peptides. *Food Hydrocolloids* 91: 19–25.

51 Wang, L., Ding, L., Wang, Y. et al. (2015). Isolation and characterisation of in vitro and cellular free radical scavenging peptides from corn peptide fractions. *Molecules* 20 (2): 3221–3237.

52 Lin, Z., Guowan, S., Jiaoyan, R. et al. (2012). Isolation and characterization of an oxygen radical absorbance activity peptide from defatted peanut meal hydrolysate and its antioxidant properties. *Journal of Agricultural and Food Chemistry* 60 (21): 5431–5437.

53 Juanjuan, Y., Lei, H., Tiantian, C. et al. (2018). Purification and identification of two novel antioxidant peptides from perilla (*Perilla frutescens* L. Britton) seed protein hydrolysates. *PloS one* 13 (7): e0200021.

54 Je, J.-Y., Cho, Y.-S., Gong, M., and Udenigwe, C.C. (2015). Dipeptide Phe-Cys derived from in silico thermolysin-hydrolysed RuBisCO large subunit suppresses oxidative stress in cultured human hepatocytes. *Food Chemistry* 171: 287–291.

55 Lin, L., Xiao, D., Song, W., and Lu, W. (2025). A breakthrough computational strategy for efficient enzymatic digestion of walnut protein to prepare antioxidant peptides. *Food Chemistry* 476: 143311.

56 Fitzgerald, R.J. and Meisel, H. (2003). *Milk Protein Hydrolysates and Bioactive Peptides.* Springer US.

57 Pihlanto, A. (2006). Antioxidative peptides derived from milk proteins. *International Dairy Journal* 16 (11): 1306–1314.

58 Power, O., Jakeman, P., and FitzGerald, R.J. (2013). Antioxidative peptides: enzymatic production, in vitro and in vivo antioxidant activity and potential applications of milk-derived antioxidative peptides. *Amino Acids* 44 (3): 797–820.

59 Elfahri, K.R., Vasiljevic, T., Yeager, T., and Donkor, O.N. (2016). Anti-colon cancer and antioxidant activities of bovine skim milk fermented by selected lactobacillus helveticus strains. *Journal of Dairy Science* 99 (1): 31–40.

60 Kudoh, Y., Matsuda, S., Igoshi, K., and Oki, T. (2001). Antioxidative peptide from Milk fermented with lactobacillus delbrueckii subsp. bulgaricus IFO13953. *Nippon Shokuhin Kagaku Kogaku Kaishi* 48 (1): 44–50.

61 Zhang, Q., Tong, X., Li, Y. et al. (2019). Purification and characterization of antioxidant peptides from Alcalase-hydrolyzed soybean (Glycine max L.) hydrolysate and their Cytoprotective effects in human intestinal Caco-2 cells. *Journal of Agricultural and Food Chemistry* 67 (20): 5772–5781.

62 Canabady-Rochelle, L.L.S., Harscoat-Schiavo, C., Kessler, V. et al. (2015). Determination of reducing power and metal chelating ability of antioxidant peptides: revisited methods. *Food Chemistry* 183: 129–135.

63 Ohashi, Y., Onuma, R., Naganuma, T. et al. (2015). Antioxidant properties of tripeptides revealed by a comparison of six different assays. *Food Science and Technology Research* 21 (5): 695–704.

64 Park, S.Y., Fernando, I.P.S., Han, E.J. et al. (2019). In vivo Hepatoprotective effects of a peptide fraction from krill protein hydrolysates against alcohol-induced oxidative damage. *Marine Drugs* 17 (12).

65 Xue, Z., Wen, H., Zhai, L. et al. (2015). Antioxidant activity and anti-proliferative effect of a bioactive peptide from chickpea (Cicer arietinum L.). *Food Research International* 77: 75–81.

66 Chen, S., Yang, Q., Chen, X. et al. (2020). Bioactive peptides derived from crimson snapper and in vivo anti-aging effects on fat diet-induced high fat Drosophila melanogaster. *Food & Function* 11 (1): 524–533.

67 Zhao, W.-H., Luo, Q.-B., Pan, X. et al. (2018). Preparation, identification, and activity evaluation of ten antioxidant peptides from protein hydrolysate of swim bladders of miiuy croaker (Miichthys miiuy). *Journal of Functional Foods* 47: 503–511.

68 Mardani, M., Badakne, K., Farmani, J., and Aluko, R.E. (2023). Antioxidant peptides: overview of production, properties, and applications in food systems. *Comprehensive Reviews in Food Science and Food Safety* 22 (1): 46–106.

69 Tang, X., He, Z., Dai, Y. et al. (2010). Peptide fractionation and free radical scavenging activity of Zein hydrolysate. *Journal of Agricultural and Food Chemistry* 58 (1): 587–593.

70 Yang, Q., Cai, X., Yan, A. et al. (2020). A specific antioxidant peptide: its properties in controlling oxidation and possible action mechanism. *Food Chemistry* 327: 126984.

71 Yu, Y., Ma, S., Han, Y. et al. (2025). A novel antioxidant peptide from soybean meal alleviates H_2O_2-induced oxidative damage via the Keap1-Nrf2-HO-1 pathway. *Food Research International* 206: 116084.

72 Munteanu, I.G. and Apetrei, C. (2022). A review on electrochemical sensors and biosensors used in assessing antioxidant activity. *Antioxidants* 11 (3): 584.

73 Wang, L., Ma, M., Yu, Z., and Du, S.-k. (2021). Preparation and identification of antioxidant peptides from cottonseed proteins. *Food Chemistry* 352: 129399.

74 Zheng, P., Hao, G., Weng, W., and Ren, H. (2019). Antioxidant activities of hydrolysates from abalone viscera using subcritical water-assisted enzymatic hydrolysis. *Food and Bioprocess Technology* 12 (6): 910–918.

75 Peng, X., Xiong, Y.L., and Kong, B. (2009). Antioxidant activity of peptide fractions from whey protein hydrolysates as measured by electron spin resonance. *Food Chemistry* 113 (1): 196–201.

76 Harnedy, P.A., O'Keeffe, M.B., and FitzGerald, R.J. (2017). Fractionation and identification of antioxidant peptides from an enzymatically hydrolysed Palmaria palmata protein isolate. *Food Research International* 100: 416–422.

77 Phongthai, S. and Rawdkuen, S. (2020). Fractionation and characterization of antioxidant peptides from rice bran protein hydrolysates stimulated by in vitro gastrointestinal digestion. *Cereal Chemistry* 97 (2): 316–325.

78 Xiang, Z., Xue, Q., Gao, P. et al. (2023). Antioxidant peptides from edible aquatic animals: preparation method, mechanism of action, and structure-activity relationships. *Food Chemistry* 404: 134701.

79 Wang, W.-Y., Zhao, Y.-Q., Zhao, G.-X. et al. (2020). Antioxidant peptides from collagen hydrolysate of Redlip croaker (Pseudosciaena polyactis) scales: preparation, characterization, and Cytoprotective effects on H_2O_2-damaged HepG2 cells. *Marine Drugs* 18 (3): 156.

80 Wu, D., Sun, N., Ding, J. et al. (2019). Evaluation and structure-activity relationship analysis of antioxidant shrimp peptides. *Food & Function* 10 (9): 5605–5615.

81 Amigo-Benavent, M., Clemente, A., Caira, S. et al. (2014). Use of phytochemomics to evaluate the bioavailability and bioactivity of antioxidant peptides of soybean -conglycinin. *Electrophoresis* 35 (11): 1582–1589.

82 Wang, K., Han, L., Hong, H. et al. (2021). Purification and identification of novel antioxidant peptides from silver carp muscle hydrolysate after simulated gastrointestinal digestion and transepithelial transport. *Food Chemistry* 342: 128275.

83 Xu, F., Zhang, J., Wang, Z. et al. (2018). Absorption and metabolism of peptide WDHHAPQLR derived from rapeseed protein and inhibition of HUVEC apoptosis under oxidative stress. *Journal of Agricultural and Food Chemistry* 66 (20): 5178–5189.

84 Wong, F.-C., Xiao, J., Wang, S. et al. (2020). Advances on the antioxidant peptides from edible plant sources. *Trends in Food Science & Technology* 99: 44–57.

85 Li, C., Li, L., Cheng, J. et al. (2024). Anti-fatigue effect of Lateolabrax japonicus peptides in mice and the underlying action mechanism via in vitro and in vivo assays. *Food Bioscience* 58: 103763.

86 Zheng, J., Li, M., Wang, C. et al. (2021). A novel nonapeptide SSDAFFPFR from Antarctic krill exerts a protective effect on PC12 cells through the BCL-XL/Bax/Caspase-3/p53 signaling pathway. *Food Bioscience* 43: 101345.

87 Zhang, X., Noisa, P., and Yongsawatdigul, J. (2021). Identification and characterization of tilapia antioxidant peptides that protect AAPH-induced HepG2 cell oxidative stress. *Journal of Functional Foods* 86: 104662.

88 Fernando, I.P.S., Park, S.Y., Han, E.J. et al. (2020). Isolation of an antioxidant peptide from krill protein hydrolysates as a novel agent with potential hepatoprotective effects. *Journal of Functional Foods* 67: 103889.

89 Li, Y., Li, J., Lin, S.-J. et al. (2019). Preparation of antioxidant peptide by microwave-assisted hydrolysis of collagen and its protective effect against H_2O_2-induced damage of RAW264.7 cells. *Marine Drugs* 17 (11): 642.

90 Ko, J.-Y., Lee, J.-H., Samarakoon, K. et al. (2013). Purification and determination of two novel antioxidant peptides from flounder fish (Paralichthys olivaceus) using digestive proteases. *Food and Chemical Toxicology* 52: 113–120.

91 Zheng, L., Yu, H., Wei, H. et al. (2018). Antioxidative peptides of hydrolysate prepared from fish skin gelatin using ginger protease activate antioxidant response element-mediated gene transcription in IPEC-J2 cells. *Journal of Functional Foods* 51: 104–112.

92 Wang, Y.-Z., Wang, Y.-M., Pan, X. et al. (2021). Antioxidant mechanisms of the oligopeptides (FWKVV and FMPLH) from muscle hydrolysate of Miiuy croaker against oxidative damage of HUVECs. *Oxidative Medicine and Cellular Longevity* 2021: 9987844.

93 Liang, R., Zhang, Z., and Lin, S. (2017). Effects of pulsed electric field on intracellular antioxidant activity and antioxidant enzyme regulating capacities of pine nut (Pinus koraiensis) peptide QDHCH in HepG2 cells. *Food Chemistry* 237: 793–802.

94 Chen, J., Yan, Y., Zhang, L. et al. (2021). Purification of novel antioxidant peptides from myofibrillar protein hydrolysate of chicken breast and their antioxidant potential in chemical and H_2O_2-stressed cell systems. *Food & Function* 12 (11): 4897–4908.

95 Islam, M.N., Rauf, A., Fahad, F.I. et al. (2022). Superoxide dismutase: an updated review on its health benefits and industrial applications. *Critical Reviews in Food Science and Nutrition* 62 (26): 7282–7300.

96 Wang, L., Ding, L., Yu, Z. et al. (2016). Intracellular ROS scavenging and antioxidant enzyme regulating capacities of corn gluten meal-derived antioxidant peptides in HepG2 cells. *Food Research International* 90: 33–41.

97 Jin, J.-E., Ahn, C.-B., and Je, J.-Y. (2018). Purification and characterization of antioxidant peptides from enzymatically hydrolyzed ark shell (*Scapharca subcrenata*). *Process Biochemistry* 72: 170–176.

98 Hu, X.-M., Wang, Y.-M., Zhao, Y.-Q. et al. (2020). Antioxidant peptides from the protein hydrolysate of monkfish (*Lophius litulon*) muscle: purification, identification, and cytoprotective function on HepG2 cells damage by H_2O_2. *Marine Drugs* 18 (3): 153.

99 Zhu, Y., Wang, K., Jia, X. et al. (2024). Antioxidant peptides, the guardian of life from oxidative stress. *Medicinal Research Reviews* 44 (1): 275–364.

100 Carrillo, W., Gomez-Ruiz, J.A., Miralles, B. et al. (2016). Identification of antioxidant peptides of hen egg-white lysozyme and evaluation of inhibition of lipid peroxidation and cytotoxicity in the zebrafish model. *European Food Research and Technology* 242 (10): 1777–1785.

101 Li, H., Zhang, Y., Wang, J. et al. (2024). Crayfish shell peptide mediated Nrf2-PPARγ pathway to improve cu-induced oxidative stress and lipid metabolism disorders in zebrafish. *Aquaculture Reports* 38: 102336.

102 Cai, X., Chen, S., Liang, J. et al. (2021). Protective effects of crimson snapper scales peptides against oxidative stress on Drosophila melanogaster and the action mechanism. *Food and Chemical Toxicology* 148: 111965.

103 Chen, K., Shi, L., Ren, Z., and Weng, W. (2023). Antioxidant characteristics of hydrolysate from low-value sea cucumber: in vitro and in vivo activities of *Caenorhabditis elegans*. *Food Chemistry: X* 19: 100836.

104 Chen, H., Wang, S., Zhou, A. et al. (2020). A novel antioxidant peptide purified from defatted round scad (Decapterus maruadsi) protein hydrolysate extends lifespan in Caenorhabditis elegans. *Journal of Functional Foods* 68: 103907.

105 Zhao, S., Cheng, Q., Peng, Q. et al. (2018). Antioxidant peptides derived from the hydrolyzate of purple sea urchin (Strongylocentrotus nudus) gonad alleviate oxidative stress in Caenorhabditis elegans. *Journal of Functional Foods* 48: 594–604.

106 Ma, H., Liu, R., Zhao, Z. et al. (2016). A novel peptide from soybean protein isolate significantly enhances resistance of the organism under oxidative stress. *PLoS One* 11 (7): e0159938.

107 Shu, Z., Liu, L., Geng, P. et al. (2019). Sesame cake hydrolysates improved spatial learning and memory of mice. *Food Bioscience* 31: 100440.

108 Fang, L., Ren, D., Cui, L. et al. (2018). Antifatigue, antioxidant and immunoregulatory effects of peptides hydrolyzed from Manchurian walnut (*Juglans mandshurica maxim.*) on mice. *Grain & oil Science and Technology* 1 (1): 44–52.

109 Bashir, K.M.I., Mohibbullah, M., An, J.H. et al. (2018). In vivo antioxidant activity of mackerel (Scomber japonicus) muscle protein hydrolysate. *PeerJ* 6: e6181.

110 Vijayan, D.K., Perumcherry Raman, S., Dara, P.K. et al. (2022). In vivo anti-lipidemic and antioxidant potential of collagen peptides obtained from great hammerhead shark skin waste. *Journal of Food Science and Technology-Mysore* 59 (3): 1140–1151.

111 Ren, D., Zhao, F., Liu, C. et al. (2018). Antioxidant hydrolyzed peptides from Manchurian walnut (*Juglans mandshurica maxim.*) attenuate scopolamine-induced memory impairment in mice. *Journal of The Science of Food and Agriculture* 98 (13): 5142–5152.

112 Zhang, L., Zheng, Y., Cheng, X. et al. (2017). The anti-photoaging effect of antioxidant collagen peptides from silver carp (Hypophthalmichthys molitrix) skin is preferable to tea polyphenols and casein peptides. *Food & Function* 8 (4): 1698–1707.

113 Adebayo, J.O., Adewole, K.E., and Krettli, A.U. (2017). Cysteine-stabilised peptide extract of Morinda lucida (Benth) leaf exhibits antimalarial activity and augments antioxidant defense system in P. Berghei-infected mice. *Journal of Ethnopharmacology* 207: 118–128.

114 Zhang, R., Chen, J., Jiang, X. et al. (2016). Antioxidant and hypoglycaemic effects of tilapia skin collagen peptide in mice. *International Journal of Food Science and Technology* 51 (10): 2157–2163.

115 Staats, S., Lueersen, K., Wagner, A.E., and Rimbach, G. (2018). Drosophila melanogaster as a versatile model organism in Food and nutrition research. *Journal of Agricultural and Food Chemistry* 66 (15): 3737–3753.

116 Leao, M.B., Goncalves, D.F., Miranda, G.M. et al. (2019). Toxicological evaluation of the herbicide palace® in Drosophila melanogaster. *Journal of Toxicology and Environmental Health-Part A-Current Issues* 82 (22): 1172–1185.

117 He, Y. and Jasper, H. (2014). Studying aging in Drosophila. *Methods* 68 (1): 129–133.

118 Minois, N. (2006). How should we assess the impact of genetic changes on ageing in model species? *Ageing Research Reviews* 5 (1): 52–59.

119 Qiu, W., Chen, X., Tian, Y. et al. (2020). Protection against oxidative stress and anti-aging effect in Drosophila of royal jelly-collagen peptide. *Food and Chemical Toxicology* 135: 110881.

120 Shen, P., Yue, Y., Zheng, J., and Park, Y. (2018). Caenorhabditis elegans: A convenient in vivo model for assessing the impact of Food bioactive compounds on obesity, aging, and Alzheimer's disease [M]//Doyle M P, Klaenhammer T R. *Annual Review of Food Science and Technology* 9: 1–22.

121 Jung, R., Cojocel, C., Muller, W. et al. (1992). Evaluation of the genotoxic potential of SORBIC acid and potassium sorbate. *Food and Chemical Toxicology* 30 (1): 1–7.

122 Abeer, M.M., Trajkovic, S., and Brayden, D.J. (2021). Measuring the oral bioavailability of protein hydrolysates derived from food sources: A critical review of current bioassays. *Biomedicine & Pharmacotherapy* 144: 112275.

123 Adewole, K.E. and Adebayo, J.O. (2017). Antioxidant defense system induced by cysteine-stabilized peptide fraction of aqueous extract of &ITMorinda lucida&IT leaf in selected tissues of &ITPlasmodium&IT &ITberghei&IT-infected mice. *Journal of Integrative Medicine* 15 (5): 388–397.

124 Zhang, Z., Jiang, S., Zeng, Y. et al. (2020). Antioxidant peptides from Mytilus Coruscus on H_2O_2-induced human umbilical vein endothelial cell stress. *Food Bioscience* 38: 100762.

125 Deng, X., Mai, R., Zhang, C. et al. (2021). Synthesis and pharmacological evaluation of a novel synthetic peptide CWHTH based on the Styela clava-derived natural peptide LWHTH with improved antioxidant, hepatoprotective and angiotensin converting enzyme inhibitory activities. *International Journal of Pharmaceutics* 605: 120852.

126 Li, C., Xu, W., Zhang, X. et al. (2023). Antioxidant peptides derived from millet bran promote longevity and stress resistance in Caenorhabditis elegans. *Plant Foods for Human Nutrition* 78 (4): 790–795.

127 Guru, A., Lite, C., Freddy, A.J. et al. (2021). Intracellular ROS scavenging and antioxidant regulation of WL15 from cysteine and glycine-rich protein 2 demonstrated in zebrafish in vivo model. *Developmental and Comparative Immunology* 114: 103863.

128 Wen, C., Zhang, J., Feng, Y. et al. (2020). Purification and identification of novel antioxidant peptides from watermelon seed protein hydrolysates and their cytoprotective effects on H_2O_2-induced oxidative stress. *Food Chemistry* 327: 127059.

129 Wang, L.Y., Ma, M.T., Yu, Z.P., and Du, S.K. (2021). Preparation and identification of antioxidant peptides from cottonseed proteins. *Food Chemistry* 352: 129399.

130 Chen, M. and Li, B. (2012). The effect of molecular weights on the survivability of casein-derived antioxidant peptides after the simulated gastrointestinal digestion. *Innovative Food Science & Emerging Technologies* 16: 341–348.

131 Wang, B. and Li, B. (2017). Effect of molecular weight on the transepithelial transport and peptidase degradation of casein-derived peptides by using Caco-2 cell model. *Food Chemistry* 218: 1–8.

132 Yu, X., Su, Q., Shen, T. et al. (2020). Antioxidant peptides from Sepia esculenta Hydrolyzate attenuate oxidative stress and fat accumulation in Caenorhabditis elegans. *Marine Drugs* 18 (10): 490.

133 Matsui, R., Honda, R., Kanome, M. et al. (2018). Designing antioxidant peptides based on the antioxidant properties of the amino acid side-chains. *Food Chemistry* 245: 750–755.

134 Ding, D., Du, B., Zhang, C. et al. (2019). Isolation and identification of an antioxidant collagen peptide from skipjack tuna (Katsuwonus pelamis) bone. *RSC Advances* 9 (46): 27032–27041.

135 Agrawal, H., Joshi, R., and Gupta, M. (2019). Purification, identification and characterization of two novel antioxidant peptides from finger millet (Eleusine coracana) protein hydrolysate. *Food Research International* 120: 697–707.

136 Liang, L.-l., Cai, S.-y., Gao, M. et al. (2020). Purification of antioxidant peptides of *Moringa oleifera* seeds and their protective effects on H_2O_2 oxidative damaged Chang liver cells. *Journal of Functional Foods* 64.

137 Shi, C., Liu, M., Zhao, H. et al. (2022). A novel insight into screening for antioxidant peptides from hazelnut protein: based on the properties of amino acid residues. *Antioxidants* 11 (1): 127.

138 Zhu, C.-Z., Zhang, W.-G., Kang, Z.-L. et al. (2014). Stability of an antioxidant peptide extracted from Jinhua ham. *Meat Science* 96 (2): 783–789.

139 Wang, C., Wang, B., and Li, B. (2016). Bioavailability of peptides from casein hydrolysate in vitro: amino acid compositions of peptides affect the antioxidant efficacy and resistance to intestinal peptidases. *Food Research International* 81: 188–196.

140 Lu, X., Zhang, L., Sun, Q. et al. (2019). Extraction, identification and structure-activity relationship of antioxidant peptides from sesame (Sesamum indicum L.) protein hydrolysate. *Food Research International* 116: 707–716.

141 Zhang, J., Zhang, H., Wang, L. et al. (2009). Antioxidant activities of the rice endosperm protein hydrolysate: identification of the active peptide. *European Food Research and Technology* 229 (4): 709–719.

142 Ren, L.-k., Yang, Y., Fan, J. et al. (2023). Novel antioxidant peptide from broken rice resist H_2O_2-induced premature senescence in 2BS cells via PI3K/Akt and JNK/Bax signaling pathway. *Food Bioscience* 52: 102376.

143 Huang, Y., Ruan, G., Qin, Z. et al. (2017). Antioxidant activity measurement and potential antioxidant peptides exploration from hydrolysates of novel continuous microwave-assisted enzymolysis of the Scomberomorus niphonius protein. *Food Chemistry* 223: 89–95.

144 Li, Y.-W. and Li, B. (2013). Characterization of structure-antioxidant activity relationship of peptides in free radical systems using QSAR models: key sequence positions and their amino acid properties. *Journal of Theoretical Biology* 318: 29–43.

145 Zhang, J., Du, H., Zhang, G. et al. (2020). Identification and characterization of novel antioxidant peptides from crucian carp (*Carassius auratus*) cooking juice released in simulated gastrointestinal digestion by UPLC-MS/MS and in silico analysis. *Journal of Chromatography B-Analytical Technologies in the Biomedical and Life Sciences* 1136: 121893.

146 Zhang, S., Qi, L., Li, D. et al. (2021). The regulatory mechanism of pulsed electric field (PEF) targeting at C-terminal glutamine of shrimp antioxidant peptide QMDDQ based on MD simulation. *LWT—Food Science and Technology* 141: 110930.

147 Ashaolu, T.J., Le, T.-D., and Suttikhana, I. (2023). An updated review of the biological activities, production and safety of meat-derived peptides. *International Journal of Food Science & Technology* 58 (4): 1712–1719.

148 Li, T., Shi, C., Zhou, C. et al. (2020). Purification and characterization of novel antioxidant peptides from duck breast protein hydrolysates. *LWT—Food Science and Technology* 125: 109215.

149 Chaudhary, K., Kumar, R., Singh, S. et al. (2016). A web server and mobile app for computing hemolytic potency of peptides. *Scientific Reports* 6 (1): 22843.

150 Gupta, S., Kapoor, P., Chaudhary, K. et al. (2015). Peptide toxicity prediction. *Methods in Molecular Biology* 1268: 143–157.

151 Wang, Z.M., Wu, Z.Y., Tu, J.C., and Xu, B.C. (2023). Muscle food and human health: a systematic review from the perspective of external and internal oxidation. *Trends in Food Science & Technology* 138: 85–99.

152 Li, Z. and Thomas, C. (2014). Quantitative evaluation of mechanical damage to fresh fruits. *Trends in Food Science & Technology* 35 (2): 138–150.

153 Zhang, X.J., Diao, M.N., and Zhang, Y.F. (2023). A review of the occurrence, metabolites and health risks of butylated hydroxyanisole (BHA). *Journal of the Science of Food and Agriculture* 103 (13): 6150–6166.

154 Jodhani, K.A., Bethsheba, B., Sujit, D. et al. (2022). Antioxidative, antimicrobial and anti-inflammatory activities and release of ultra-filtered antioxidative and antimicrobial peptides during fermentation of sheep milk: in-vitro, in-silico and molecular interaction studies. *Food Bioscience* 47: 101666.

155 Wang, W.Y., Xiong, P., Zhang, H. et al. (2021). Analysis, occurrence, toxicity and environmental health risks of synthetic phenolic antioxidants: a review. *Environmental Research* 201: 111531.

156 Xu, X.Q., Liu, A.M., Hu, S.Y. et al. (2021). Synthetic phenolic antioxidants: metabolism, hazards and mechanism of action. *Food Chemistry* 353: 129488.

157 Hirose, A. and Miyashita, K. (1999). Inhibitory effect of proteins and their hydolysates on the oxidation of triacylglycerols containing docosahexaenoic acids in emulsion. *Journal of the Japanese Society for Food Science and Technology* 46 (12): 799–805.

158 Elias, R.J., Bridgewater, J.D., Vachet, R.W. et al. (2006). Antioxidant mechanisms of enzymatic hydrolysates of β-lactoglobulin in food lipid dispersions. *Journal of Agricultural and Food Chemistry* 54 (25): 9565–9572.

159 Jie, Y., Zhao, H., Sun, X. et al. (2019). Isolation of antioxidative peptide from the protein hydrolysate of Caragana ambigua seeds and its mechanism for retarding lipid auto-oxidation. *Journal of the Science of Food and Agriculture* 99 (6): 3078–3085.

160 Zheng, Z.J., Wang, M., Li, J.X. et al. (2020). Comparative assessment of physicochemical and antioxidative properties of mung bean protein hydrolysates. *RSC Advances* 10 (5): 2634–2645.

161 Shen, X., Li, T., Li, X. et al. (2022). Dual cryoprotective and antioxidant effects of silver carp (*Hypophthalmichthys molitrix*) protein hydrolysates on unwashed surimi stored at conventional and ultra-low frozen temperatures. *LWT—Food Science and Technology* 153: 112563.

162 Karwowska, M. and Kononiuk, A. (2020). Nitrates/nitrites in Food—risk for Nitrosative stress and benefits. *Antioxidants* 9 (3): 241.

163 Florin, T.H.J., Neale, G., Goretski, S., and Cummings, J.H. (1993). The sulfate content of foods and beverages. *Journal of Food Composition and Analysis* 6 (2): 140–151.

164 Esimbekova, E.N., Asanova, A.A., Deeva, A.A., and Kratasyuk, V.A. (2017). Inhibition effect of food preservatives on endoproteinases. *Food Chemistry* 235: 294–297.

165 Chen, X., Hao, F., Zhang, M. et al. (2021). Sex metabolic differences and effects on blood coagulation among rats exposed to sodium dehydroacetate. *Frontiers in Pharmacology* 12: 727084.

166 Nunes, R.D.M., Sales, I.M.S., Silva, S.I.O. et al. (2017). Antiproliferative and genotoxic effects of nature identical and artificial synthetic food additives of aroma and flavor. *Brazilian Journal of Biology* 77 (1): 150–154.

167 Jridi, M., Lassoued, I., Nasri, R. et al. (2014). Characterization and potential use of cuttlefish skin gelatin hydrolysates prepared by different microbial proteases. *BioMed Research International* 2014: 461728.

168 Qiu, X., Chen, S., and Dong, S. (2014). Effects of silver carp antioxidant peptide on the lipid oxidation of Sierra fish fillets (*Scomberomorus niphonius*) during frozen storage. *Journal of Food Biochemistry* 38 (2): 167–174.

169 Abdelhedi, O., Jridi, M., Nasri, R. et al. (2019). Rheological and structural properties of skin gelatin: potential use as an active fish coating agent. *Food Hydrocolloids* 87: 331–341.

170 Dai, C.H., Zhang, W.W., He, R.H. et al. (2017). Protein breakdown and release of antioxidant peptides during simulated gastrointestinal digestion and the absorption by everted intestinal sac of rapeseed proteins. *LWT-Food Science and Technology* 86: 424–429.

171 Zhang, Q.Z., Tong, X.H., Qi, B.K. et al. (2018). Changes in antioxidant activity of Alcalase-hydrolyzed soybean hydrolysate under simulated gastrointestinal digestion and transepithelial transport. *Journal of Functional Foods* 42: 298–305.

172 Samaei, S.P., Ghorbani, M., Tagliazucchi, D. et al. (2020). Functional, nutritional, antioxidant, sensory properties and comparative peptidomic profile of faba bean (L.) seed protein hydrolysates and fortified apple juice. *Food Chemistry* 330: 127120.

173 Zhao, Q., Xu, H., Hong, S. et al. (2019). Rapeseed protein-derived antioxidant peptide RAP ameliorates nonalcoholic steatohepatitis and related metabolic disorders in mice. *Molecular Pharmaceutics* 16 (1): 371–381.

174 Hendriks, H.F.J. (2020). Alcohol and human health: what is the evidence? *Annual Review of Food Science and Technology* 11 (1): 1–21.

175 Yu, Y.L., Wang, L.J., Wang, Y. et al. (2017). Hepatoprotective effect of albumin peptides from corn germ meal on chronic alcohol-induced liver injury in mice. *Journal of Food Science* 82 (12): 2997–3004.

176 Vijayanand, M., Issac, P.K., Velayutham, M. et al. (2024). Exploring the neuroprotective potential of KC14 peptide from Cyprinus carpio against oxidative stress-induced neurodegeneration by regulating antioxidant mechanism. *Molecular Biology Reports* 51 (1): 990.

177 Plácido, A., do Pais do Amaral, C., Teixeira, C. et al. (2022). Neuroprotective effects on microglia and insights into the structure–activity relationship of an antioxidant peptide isolated from *Pelophylax perezi*. *Journal of Cellular and Molecular Medicine* 26: 2793–2807.

178 Chen, Y., Liang, Y., He, H. et al. (2023). Optimisation process of walnut protein hydrolysed as an antioxidant candidate. *Journal of Food Measurement and Characterization* 17: 5845–5855.

179 Lee, S.J., Kim, Y.S., Hwang, J.W. et al. (2012). Purification and characterization of a novel antioxidative peptide from duck skin by-products that protects liver against oxidative damage. *Food Research International* 49 (1): 285–295.

180 Park, S.Y., Fernando, I.P.S., Han, E.J. et al. (2019). In vivo Hepatoprotective effects of a peptide fraction from krill protein hydrolysates against alcohol-induced oxidative damage. *Marine Drugs* 17 (12): 690.

181 Zhu, Q.L., Zhuo, H.L., Yang, L.M. et al. (2022). A peptide HEPFYGNEGALR from alleviates acute alcoholic liver injury by enhancing antioxidant response in male C57BL/6J mice. *Molecules* 27 (18): 5839.

182 Malla, R., Surepalli, N., Farran, B. et al. (2021). Reactive oxygen species (ROS): critical roles in breast tumor microenvironment. *Critical Reviews in Oncology/Hematology* 160: 21–38.

183 Assi, M. (2017). The differential role of reactive oxygen species in early and late stages of cancer. *American Journal of Physiology-Regulatory Integrative and Comparative Physiology* 313 (6): R646–R653.

184 Wu, C.L., Huang, A.C., Yang, J.S. et al. (2011). Benzyl Isothiocyanate (BITC) and Phenethyl Isothiocyanate (PEITC)-mediated generation of reactive oxygen species causes cell cycle arrest and induces apoptosis via activation of Caspase-3, mitochondria dysfunction and nitric oxide (NO) in human osteogenic sarcoma U-2 OS cells. *Journal of Orthopaedic Research* 29 (8): 1199–1209.

185 Umayaparvathi, S., Arumugam, M., Meenakshi, S. et al. (2013). Purification and characterization of antioxidant peptides from oyster (*Saccostrea cucullata*) hydrolysate and the anticancer activity of hydrolysate on human colon cancer cell lines. *International Journal of Peptide Research and Therapeutics* 20 (2): 231–243.

186 Xue, Z.H., Wen, H.C., Zhai, L.J. et al. (2015). Antioxidant activity and anti-proliferative effect of a bioactive peptide from chickpea (*L.*). *Food Research International* 77: 75–81.

187 Chi, C.F., Hu, F.Y., Wang, B. et al. (2015). Antioxidant and anticancer peptides from the protein hydrolysate of blood clam muscle. *Journal of Functional Foods* 15: 301–313.

188 Marcela, G.M., Eva, R.G., del Carmen, R.R.M., and Rosalva, M.E. (2016). Evaluation of the antioxidant and Antiproliferative effects of three peptide fractions of germinated soybeans on breast and cervical Cancer cell lines. *Plant Foods for Human Nutrition* 71 (4): 368–374.

189 Wang, M.-s., Du, Y.-b., Huang, H.-m. et al. (2019). Silk fibroin peptide suppresses proliferation and induces apoptosis and cell cycle arrest in human lung cancer cells. *Acta Pharmacologica Sinica* 40 (4): 522–529.

190 Kuerban, A., Al-Ghafari, A.B., ALGhamadi, S.A. et al. (2020). Potential antiglycation, antioxidant and antiproliferative activities of peptides. *Journal of Food Measurement and Characterization* 14 (4): 2155–2162.

191 Sjödin, B., Westing, Y.H., and Apple, F.S. (1990). Biochemical mechanisms for oxygen free radical formation during exercise. *Sports Medicine* 10 (4): 236–254.

192 You, L., Zhao, M., Regenstein, J.M., and Ren, J. (2011). In vitro antioxidant activity and in vivo anti-fatigue effect of loach (Misgurnus anguillicaudatus) peptides prepared by papain digestion. *Food Chemistry* 124 (1): 188–194.

193 Sun, S.G., Niu, H.H., Yang, T. et al. (2014). Antioxidant and anti-fatigue activities of egg white peptides prepared by pepsin digestion. *Journal of The Science of Food and Agriculture* 94 (15): 3195–3200.

194 Zhao, Y.Q., Zeng, L., Yang, Z.S. et al. (2016). Anti-fatigue effect by peptide fraction from protein hydrolysate of Croceine croaker swim bladder through inhibiting the oxidative reactions including DNA damage. *Marine Drugs* 14 (12): 221.

195 Ren, Y., Wu, H., Chi, Y.L. et al. (2020). Structural characterization, erythrocyte protection, and antifatigue effect of antioxidant collagen peptides from tilapia (L.) skin. *Food & Function* 11 (11): 10149–10160.

196 Chen, J., Ryu, B., Zhang, Y. et al. (2020). Comparison of an angiotensin-I-converting enzyme inhibitory peptide from tilapia (Oreochromis niloticus) with captopril: inhibition kinetics, in vivo effect, simulated gastrointestinal digestion and a molecular docking study. *Journal of the Science of Food and Agriculture* 100 (1): 315–324.

197 Rezvankhah, A., Yarmand, M.S., Ghanbarzadeh, B., and Mirzaee, H. (2023). Development of lentil peptides with potent antioxidant, antihypertensive, and antidiabetic activities along with umami taste. *Food Science & Nutrition* 11 (6): 2974–2989.

198 Vijayanand, M., Issac, P.K., Velayutham, M. et al. (2024). Exploring the neuroprotective potential of KC14 peptide from against oxidative stress-induced neurodegeneration by regulating antioxidant mechanism. *Molecular Biology Reports* 51 (1): 990.

199 Rosena, A., Koobkokkruad, T., Eaknai, W. et al. (2018). Protective effect of Thai silk extracts on drug-induced phototoxicity in human epidermal A431 cells and a reconstructed human epidermis model. *Journal of Photochemistry and Photobiology B-Biology* 188: 50–59.

200 Kong, J., Hu, X.M., Cai, W.W. et al. (2023). Bioactive peptides from skipjack tuna cardiac arterial bulbs (II): protective function on UVB-irradiated HaCaT cells through antioxidant and anti-apoptotic mechanisms. *Marine Drugs* 21 (2): 105.

201 Feng, G.Z., Wei, L., Che, H.L. et al. (2022). A frog peptide ameliorates skin photoaging through scavenging reactive oxygen species. *Frontiers in Pharmacology* 12: 761011.

202 Yang, F., Yang, Y., Xiao, D. et al. (2024). Anti-Photoaging effects of antioxidant peptide from seahorse (Hippocampus abdominalis) in in vivo and in vitro models. *Marine Drugs* 22 (10): 471.

203 Parihar, A., Parihar, M.S., Milner, S., and Bhat, S. (2008). Oxidative stress and anti-oxidative mobilization in burn injury. *Burns* 34 (1): 6–17.

204 Zhan, X., Wang, D., Wang, H. et al. (2023). Revitalizing skin repair: unveiling the healing power of Livisin, a natural peptide calcium mimetic. *Toxins* 16 (1): 21–38.

205 Yu, D.Y., Cui, S.H., Chen, L.Q. et al. (2023). Marine-derived bioactive peptides self-assembled multifunctional materials: antioxidant and wound healing. *Antioxidants* 12 (6): 1190.

206 Pozzolini, M., Millo, E., Oliveri, C. et al. (2018). Elicited ROS scavenging activity, Photoprotective, and wound-healing properties of collagen-derived peptides from the marine sponge. *Marine Drugs* 16 (12): 465.

207 Li, Y.J., Liu, J.J., Wang, Y.J. et al. (2015). Mass spectrometric and spectrophotometric analyses reveal an alternative structure and a new formation mechanism for melanin. *Analytical Chemistry* 87 (15): 7958–7963.

208 Yingchutrakul, Y., Krobthong, S., Choowongkomon, K. et al. (2022). Discovery of a multifunctional Octapeptide from Lingzhi with antioxidant and Tyrosinase inhibitory activity. *Pharmaceuticals* 15 (6): 684.

209 Ding, C.B., Hao, M.Q., Ma, S. et al. (2022). Identification of peptides with antioxidant, anti-lipoxygenase, anti-xanthine oxidase and anti-tyrosinase activities from velvet antler blood. *LWT—Food Science and Technology* 168: 113889.

210 Kriangkrai, W., Kantasa, T., Sagasae, W. et al. (2024). Discovery of superior bioactive peptides of two edible mushrooms protein hydrolysate in biological activities: tyrosinase inhibitory and antioxidant activity. *Food Science and Biotechnology* 33 (13): 3105–3117.

211 Tian, X., Zhang, Z., Zhao, Y. et al. (2024). Isolation and characterization of antioxidant peptides from dairy cow (*Bos taurus*) placenta and their antioxidant activities. *Antioxidants* 13: 913.

212 Jin, Y., Zhou, P., Zhu, C. et al. (2024). Preparation of antioxidant peptides from chicken bone proteins and the influence of their compositional characteristics on antioxidant activity. *Foods* 13: 4171.

213 Fukada, Y., Mizutani, S., Nomura, S. et al. (2016). Antioxidant activities of a peptide derived from chicken dark meat. *Journal of Food Science and Technology* 53: 2476–2481.

214 Luo, X., Fei, Y., Xu, Q. et al. (2020). Isolation and identification of antioxidant peptides from tartary buckwheat albumin (Fagopyrum tataricum Gaertn.) and their antioxidant activities. *Journal of Food Science* 85 (3): 611–617.

215 Torres-Fuentes, C., del Mar Contreras, M., Recio, I. et al. (2015). Identification and characterization of antioxidant peptides from chickpea protein hydrolysates. *Food Chemistry* 180: 194–202.

216 Xiang, Z., Xue, Q., Gao, P. et al. (2022). Antioxidant peptides from edible aquatic animals: preparation method, mechanism of action, and structure-activity relationships. *Food Chemistry* 404: 134701.

217 Wu, J., Wang, X., Qiao, K. et al. (2025). Screening and antioxidant activities evaluation of peptides from abalone (*Haliotis discus hannai Ino*). *Food Science & Nutrition* 13: e70028.

218 Qian, B., Zhao, X., Yang, Y., and Tian, C. (2020). Antioxidant and anti-inflammatory peptide fraction from oyster soft tissue by enzymatic hydrolysis. *Food Science & Nutrition* 8: 3947–3956.

219 Huang, P., Miao, J., Li, J. et al. (2023). Novel antioxidant peptides from pearl shell meat hydrolysate and their antioxidant activity mechanism. *Molecules* 28: 864.

220 Sabana, I., Naufal, M., Wiani, I. et al. (2019). Synthesis of antioxidant peptide SCAP1 (Leu-ala-Asn-ala-Lys). *Egyptian Journal of Chemistry* 63 (3): 921–926.

221 Maharani, R., Hidayat, A.T., Sabana, I.R. et al. (2021). Synthesis, antioxidant activity, and structure–activity relationship of SCAP1 analogues. *International Journal of Peptide Research and Therapeutics* 27: 17–23.

222 Sousa, N.F.C., Santos, M.P.F., Barbosa, R.P. et al. (2024). Pepsin immobilization on activated carbon and functionalized with glutaraldehyde and genipin for the synthesis of antioxidant peptides of goat casein. *Food Research International 186: 114161.*

223 Xiang, Q., Qu, L., Lei, H. et al. (2024). Expression of multicopy tandem recombinant ginseng Hexapeptide in Bacillus subtilis and the evaluation of antiaging activity. *Journal of Agricultural and Food Chemistry* 72: 7266–7278.

224 Deng, S., Choto Lutema, P., Gwekwe, B. et al. (2019). Bitter peptides increase engulf of phagocytes in vitro and inhibit oxidation of myofibrillar protein in peeled shrimp (*Litopenaeus vannamei*) during chilled storage. *Aquaculture Reports* 15: 100234.

225 Jiang, Y., Zhang, L., Jin, Y. et al. (2024). *Lactiplantibacillus plantarum* for the preparation of fermented low-bitter enzymatic skim milk with antioxidant ability. *Foods* 13: 3828.

226 Sun, X., Acquah, C., Gazme, B. et al. (2020). Mechanisms of plastein formation influence the IgE-binding activity of egg white protein hydrolysates after simulated static digestion. *Food Chemistry* 345: 128783.

227 Idowu, A.T. and Benjakul, S. (2019). Bitterness of fish protein hydrolysate and its debittering prospects. *Journal of Food Biochemistry* 43: e12978.

228 Xu, C., Cai, C., Liu, T. et al. (2024). Enhancing the antioxidant activity of fish scale collagen hydrolysates through Plastein reaction. *Food and Bioprocess Technology* 17: 3693–3703.

229 Yu, J., Mikiashvili, N., Bonku, R., and Smith, I.N. (2021). Allergenicity, antioxidant activity and ACE-inhibitory activity of protease hydrolyzed peanut flour. *Food Chemistry* 360: 129992.

230 Huo, C., Yang, X., and Li, L. (2023). Non-beany flavor soymilk fermented by lactic acid bacteria: characterization, stability, antioxidant capacity and in vitro digestion. *Food Chemistry: X* 17: 100578.

231 Jegani, K.T., Balde, A., and Nazeer, R.A. (2024). A review on anti-inflammatory and antioxidant peptides derived from marine organisms: mechanism of action and therapeutic applications. *Food Bioscience* 63: 105745.

5

Antimicrobial Peptides

The global pandemic of coronavirus disease (COVID-19) has underscored the critical need for multidisciplinary antimicrobial strategies, a demand further intensified by the antibiotic resistance (AMR) crisis—a silent epidemic causing 1.27 million deaths annually worldwide [1]. This dual crisis of viral pandemics and bacterial resistance demands innovative solutions, particularly as simultaneous resistance to ≥3 antibiotic classes in pathogens like methicillin-resistant *Staphylococcus aureus* (MRSA) and carbapenem-resistant *Klebsiella pneumoniae* renders traditional antibiotics obsolete.

Antimicrobial peptides (AMPs) represent a conserved evolutionary defense mechanism across biological kingdoms, characterized by amphipathic α-helices with cationic surface charges that facilitate membrane permeabilization. Their bactericidal efficacy against G^- bacteria—exemplified by *Salmonella enterica* and *Escherichia coli*—stems from synergistic interactions with lipopolysaccharide (LPS) outer membranes [2]. Notably, their multifactorial killing mechanisms (including pore formation, metabolic interference, and immune modulation) substantially reduce the likelihood of resistance development compared to conventional antibiotics.

The golden age of antibiotic discovery (1940s–1960s) established β-lactam antibiotics as cornerstones of modern medicine through targeted inhibition of bacterial cell wall biosynthesis. However, microbial-host co-evolutionary adaptations in G^- bacteria have disrupted this therapeutic paradigm, which exploits LPS outer membrane modifications and efflux pump overexpression to evade host defenses. Such evolutionary strategies have precipitated a therapeutic impasse, with multidrug-resistant (MDR) infections complicating 28% of sepsis cases and 45% of pneumonia-related deaths in hospitalized patients.

AMPs emerge as a paradigm-shifting alternative, exhibiting three-dimensional antimicrobial efficacy: Membrane disruption via amphipathic α-helical structures targeting lipid bilayers. Metabolic interference through intracellular enzyme engagement. Immune modulation by activating pattern recognition receptors. These mechanisms collectively reduce resistance development by >80% compared to conventional antibiotics. Notably, AMPs demonstrate enhanced stability in food systems (pH 2.5 ~ 7.5; temperature range: 4 ~ 80°C) with lipid solubility coefficients (logD) ranging from −0.3 to 2.1, enabling their application

Food-Derived Bioactive Peptides: Green Preparation, Activity Mechanisms, and Health Applications, First Edition. Shao-Yun Wang, Xu Chen, and Xixi Cai.

in functional food preservation while maintaining Generally Recognized as Safe (GRAS) compliance and low cytotoxicity ($CC_{50} > 100\,\mu M$).

This chapter provides a systematic panorama of cationic AMPs in addressing the dual crisis of AMR and viral pandemics. By synthesizing state-of-the-art research on source diversity, activity evaluation, mechanistic innovation, safety profiling, functional applications, and limitation management, we demonstrate that AMPs represent a transformative solution with proven efficacy and a controllable risk profile.

5.1 Sources

As an age-old host defense mechanism against pathogen intrusion, AMPs are extensively dispersed throughout both prokaryotic and eukaryotic organisms. This is due to the following reasons: (i) AMP production is an innate defense strategy that requires considerably less time and energy than antibody synthesis through adaptive immunity; (ii) their relatively low molecular weight enables AMPs to reach target sites more rapidly than immunoglobulins; and (iii) some eukaryotic organisms lack lymphocyte-mediated immune systems. For example, insects primarily depend on the secretion of diverse antimicrobial agents to combat invading microbes [3]. Since the first AMP was found in the chrysalis of the American silkworm, numerous peptides have subsequently been reported across a wide range of organisms [3]. Ever since the initial AMP was detected in the American silkworm chrysalis, a vast number of AMPs have been extensively discovered in diverse organisms. These include microbial-derived AMPs (such as those from bacteria and archaea), plant-derived AMPs, and animal-derived AMPs, which can be acquired through fermentation or enzymatic hydrolysis [4].

5.1.1 Plant-Derived Antimicrobial Peptides

Plants are exposed to various pathogens during different seasons and must rely on self-protection to survive. However, plants do not possess the specific immune system found in mammals. During the growth process, exposure to pathogenic microorganisms makes the nonspecific immune system particularly vital, with AMPs serving as a key group of molecules in this defense response [5]. According to variations in amino acid composition and secondary structural features, plant AMPs are categorized into nine types, including plant defensins, thionins, lipid transfer proteins, hevein-like peptides, knottins, glycine-rich peptides, hairpin AMPs, cyclic peptides, and dermaseptins, as shown in Table 5.1.

5.1.1.1 Thionins

Thionins were first discovered in wheat flour by Balls and co-workers and are obtained through extraction with petroleum ether. These crucial peptides are present in both monocotyledons and dicotyledons, typically consisting of 45–48 amino acid residues and having a molecular mass of around 5 kDa. Structurally, they feature a β-sheet configuration enriched with lysine, arginine and cysteine residues [6]. Thionins are sorted

Table 5.1 Classification of Plant-derived AMPs.

Plant-derived AMPs	Activities	Examples
Defensins	Antifungal	NaD1, PhD1, Rs-AFP1
Thionins	Antibacterial, antifungal	Tu-AMP1, Cp-thionin II
Lipid transfer proteins	Anti-parasitic, antibacterial, antifungal	Cc-LTP1, LTP110
Knottin peptides	Antiviral	PAFP-S, Mj-AMP2
α-Hairpinin family peptides	Antibacterial	MBP-1, EcAMP1
Cyclotides	Antibacterial, antifungal, antiviral	Kalata B1, Cliotide 20
Snakins	Antibacterial, antifungal	Snakin-1, Snaking-Z

into two main classes, namely α/β-thionins and γ-thionins. Beyond their antimicrobial properties, these peptides play roles in plant defense, storage, seed maturation, and protein packaging. Table 5.1 provides a comprehensive overview of the origins, functions, and other details of various plant AMPs.

A notable thionin, CaThi, was isolated from jalapeño (*Capsicum annuum*) fruit, exhibiting both antifungal and antibacterial activities. Interestingly, CaThi damaged the membranes of six *Candida* species, but nuclear damage and reactive oxygen species (ROS) burst were observed only in *Candida tropicalis* [7]. Furthermore, CaThi exhibited a synergistic interaction with the azole antifungal fluconazole, increasing susceptibility in *Fusarium solani* and suppressing growth of all six tested *Candida* species. Additional research revealed that CaThi triggers apoptosis in *Candida tropicalis* via mechanisms that involve caspase activation and disruption of pH balance [8].

Two additional thionins, NsW1 and NsW2, obtained from *Nigella sativa*, exhibited strong antibacterial effects against *Bacillus subtilis* and *Staphylococcus aureus* by enhancing membrane permeability and modifying cellular ultrastructure. In a recent study analyzing transcriptome data from 1,000 plants, Höng and colleagues identified 133 novel thionin sequences [9]. A key finding was that some of these sequences did not align with established thionin families, implying the presence of a novel class with a distinctive three-dimensional structure, as evidenced by unusual cysteine patterns.

5.1.1.2 Plant Defensins

Plant defensins represent a large and diverse superfamily of AMPs ubiquitously found across all plant species, characterized by highly conserved structural frameworks. These peptides are typically non-toxic to human and plant cells while demonstrating strong antifungal properties [10]. They are predominantly localized in various plant tissues, including leaves, flowers, fruits and other organs. Genomic analyses, such as those conducted on *Arabidopsis thaliana*, have identified defensins as one of the most prominent and extensive families of membrane-soluble pathogen-associated molecular patterns [11]. Generally, these peptides are composed of approximately 45–54 amino acid residues and are categorized as cysteine-rich peptides (CRPs) carrying a net positive charge, with an average molecular mass of about 5 kDa. Their structural stability is conferred by eight cysteine residues establishing four disulfide linkages, which stabilize the α-helical segments and triple-stranded β-sheets, conferring resistance to proteolysis and stability under harsh temperature and pH conditions.

Certain defensins, such as HsAFP1, demonstrate broad-spectrum fungicidal activity through diverse mechanisms, including via perturbation of the fungal cell cycle [12]. For instance, RsAFP2, a radish (*Raphanus sativus*)-derived defensin, specifically binds to fungal glucosylceramide sphingolipids while showing no affinity for plant or human counterparts. Moreover, RsAFP2 has been shown to provide prophylactic protection against candidiasis in murine models, highlighting its therapeutic potential. This is particularly significant given the resilience of *Candida albicans* biofilms to conventional antifungals and immune responses [13]. Another example is Dm-AMP1, a defensin from *Dahlia merckii*, which exhibited an IC50 value of 15 μg/mL against *Bacillus subtilis*, as reported by Osborn et al. Similarly, PaDef, an avocado (*Persea americana var. drymifolia*)-origin defensin, displayed robust activity against drug-resistant strains of *Staphylococcus aureus* and *Escherichia coli*. Although significant progress has been made, the exact mechanisms by which plant defensins exert their antimicrobial effects remain largely unknown [14].

5.1.1.3 Snakins

Most AMPs possess two to four disulfide linkages, yet the snakin group within the PAMP family is notable for having 12 cysteine residues, which form the highest number of disulfide linkages. Snakins generally consist of approximately 63 amino acid residues, contain six disulfide linkages, and are characterized by a high proportion of cysteine residues (19% Cys) [15]. The first snakin peptide to be found, Snakin-1, was extracted from potato tubers (*Solanum tuberosum*) and shares a certain similar sequence with snake venom proteins. Snakins can be expressed either constitutively or induced in various plant organs in response to biotic or abiotic stress [16]. For instance, Snakin-Z, a cationic antimicrobial peptide found in jujube fruit (*Ziziphus jujuba*), exhibits activity against *Staphylococcus aureus* while demonstrating excellent compatibility with blood cells [17]. Interestingly, the Snakin-1 gene from *Solanum tuberosum* is not markedly involved in tissue development and shows no response to biotic or abiotic stressors, whereas the Snakin-2 gene elicits a distinct response upon pathogenic infection. These peptides are notable for their high cysteine content, which allows the establishment of six disulfide linkages, providing structural stability. Snakins are generally considered inducible plant defense molecules due to their activity toward a broad spectrum of G^- and G^+ bacterial strains.

The prospective use of snakins in managing plant diseases has attracted considerable interest. Numerous studies have focused on developing transgenic plants that overexpress snakin genes to enhance disease resistance. For instance, potato plants overexpressing StSN1 and StSN2 have been demonstrated to exhibit enhanced resistance to blackleg disease [18]. While experimental evidence suggests that the cationic nature of snakins facilitates binding to negatively charged pathogen membranes, resulting in membrane permeabilization, how they precisely exert these effects is still unknown.

5.1.1.4 Hevein-Like Peptides

Hevein-like peptides (HLPs) are alkaline peptides initially identified in the latex of the rubber tree (*Hevea brasiliensis*), typically comprising 29 ~ 45 amino acid residues [19]. These peptides feature a structural arrangement of three antiparallel β-strands and a short α-helix, which is reinforced by three to five disulfide linkages (Figure 5.1). *In vitro* experiments have demonstrated the potent antifungal activity of these peptides. Due to their chitin-binding motif, hevein serves as an excellent model for investigating carbohydrate–peptide interactions. Research has shown that hydrogen bonds and van der Waals forces are key mediators in these interactions [20]. HLPs share functional

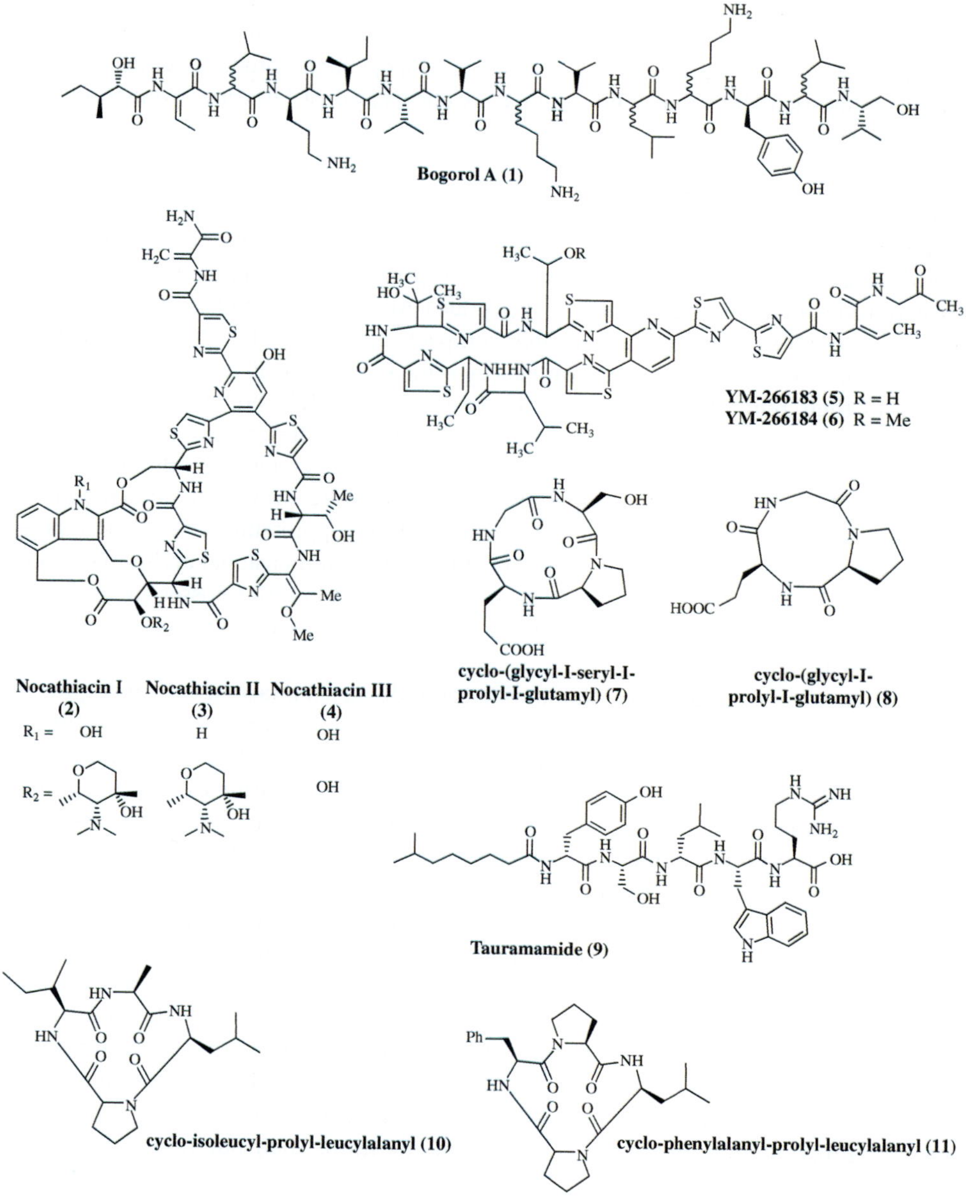

Figure 5.1 Structures of NRPs from bacteria (1–11). *Source:* Adapted from [53].

similarities with hevein, as they inhibit the growth of fungi possessing chitin in their cell walls and shield vegetation against diverse fungal pathogens. Their antifungal activity is mediated by two main mechanisms, namely electrostatic interactions that compromise fungal cytomembranes and inhibition of chitin biosynthesis.A unique 10-cysteine peptide derived from *Triticum kiharae* wheat seeds exhibits chitinase activity, effectively inhibiting *Fusarium solani* and *Fusarium oxysporum*. In the medicinal herb group *Panax* (ginseng), a distinct family of HLPs, known as ginsentides, has been identified [21]. These cysteine- and glycine-rich peptides possess a ring-like conformation and demonstrate exceptional thermal stability. Current computational analyses on gymnosperms have revealed that HLPs are widespread in this group. For instance, *Ginkgo biloba* produces ginkgotides, a class of proline-rich HLPs that exhibit inhibitory effects against *Fusarium oxysporum* and *Aspergillus niger*.

5.1.1.5 Knottin-Type Peptides

Plant knottins constitute a large family of small peptides, typically comprising around 30 amino acids, and include members such as the trypsin inhibitors, alpha-amylase inhibitors, carboxypeptidase inhibitors and cyclic peptides. Among pathogen-associated molecular patterns (PAMPs), knottins are notably compact [22]. Their characteristic structure features a trio of disulfide linkages, forming a cysteine knot motif (Figure 5.1), which contributes to their remarkable functional diversity. The disulfide bonds in knottins are arranged in a conserved pattern, with linkages between CysI–CysIV, CysII–CysV and CysIII–CysVI, creating stable cystine junctions. However, the cysteine motifs vary among different knottin subfamilies. This superfamily displays a wide array of functional properties, such as insect-killing, microbial inhibition and HIV-suppressing effects. These peptides rely on their disulfide linkages to preserve their three-dimensional architecture and stability under chemical and enzymatic stress. Additionally, knottins possess membranolytic properties and are amphipathic, enabling them to interact with microbial membranes and induce antibacterial effects [23].

Knottins are considered ideal candidates for peptide-based pharmaceutical engineering due to several key attributes: (i) their small size, which facilitates chemical synthesis; (ii) exceptional stability under thermal, chemical and proteolytic conditions; and (iii) the ability to maintain functional sequences despite variations in loop regions. These features make them highly versatile for therapeutic applications. Knottins, also described as "versatile peptides", can engage a broad spectrum of molecular targets, including acid-sensing ion channels, potassium and sodium conduits, together with cytomembranes. Their antimicrobial efficacy extends to insects, along with bacterial species, fungal organisms and viral pathogens. The amphipathic nature of knottins, characterized by cationic and hydrophobic regions, facilitates their binding to membranes and the exertion of their biological effects. These molecules have the capacity to switch between hydrophobic contacts and ionic interactions, thereby improving their membrane affinity and biological activity. However, their high membrane interaction also results in significant cytotoxicity to human cells, as they do not selectively target microbial membranes. This lack of specificity limits their potential therapeutic applications.

5.1.2 Animal-Derived AMPs

Among aquatic vertebrates, α-helical amphipathic peptides represent the predominant class of AMPs. The paradigmatic example, pardaxin—initially identified in marine vertebrates—comprises 33 amino acid residues and adopts a pore-forming α-helical conformation [24]. These evolutionarily conserved molecules serve as essential effectors of innate immunity across animal taxa, exerting cytolytic and membrane-disruptive activities against invading pathogens. Their strategic localization in epithelial barriers and bodily fluids—including mucosal surfaces, integumentary secretions, and circulatory systems—reflects their frontline defensive role. Comparative analyses by Rajanbabu [25] cataloged AMP diversity spanning multiple vertebrate lineages, with notable representatives including cathelicidin, hepcidin (liver-expressed antimicrobial peptide), dicentracin, epinecidine, and phylogenetically conserved families such as β-defensins, histone-derived peptides, and the piscidin group—a taxonomically restricted AMP class originally described in teleosts. Current databases document 122 distinct AMP variants across aquatic vertebrates, systematically classified into five structural families (Table 5.2).

5.1.2.1 Piscidin-Like Peptides

This evolutionarily conserved AMP group, exemplified by piscidin and pleurocidin, shares structural homology with α-helical antimicrobial effectors found in arthropods and amphibians. Comparative sequence analyses reveal substantial diversification across vertebrate lineages, with pleurocidin, a 25-residue peptide first isolated from *Pseudopleuronectes americanus* (winter flounder) integument, serving as the archetypal member [26]. Subsequent discoveries identified structural analogs, including misgurin in *Misgurnus anguillicaudatus* (Oriental weatherfish) and piscidins in Morone chrysops × Morone saxatilis (hybrid striped bass) mast cells [27, 28]. These molecules exhibit functional convergence with magainins from Xenopus laevis (African clawed frog), showing comprehensive inhibitory effects on prokaryotic and eukaryotic pathogens [29]. Notably, cecropin isoforms from *Hyalophora cecropia* (cecropia moth) show therapeutic potential against piscine viral pathogens [30]. Recent discoveries continue expanding this pharmacopeia, with novel

Table 5.2 Classification of animal-derived AMPs.

Animal-derived AMPs	Activities	Examples
Piscidins	Antibacterial, antiviral, anti-protozoa, antifungal	Pleurocidin, Misgurin, Cecropin B
Cathelicidins	Antibacterial, antifungal	Cathelicidin, Cathelicidin-1, Cathelicidin-2
Histone-derived AMPs	Anti-parasitic, Antibacterial, antifungal	Histones H1, H2A, H2B and H6
Hepcidins	Antibacterial, anti-tumor	TP-LEAP, HAMP1, HAMP2
Defensins	Antibacterial, antiviral	α-defensins, β-defensins, θ-defensins

analogs like dicentracin from *Dicentrarchus labrax* (European sea bass) demonstrating immunomodulatory properties [31].

5.1.2.2 Cathelicidins

These 12 kDa polypeptides, characterized by conserved cathelin domains, represent ancient components of vertebrate immunity. The prototypical form in basal vertebrates was identified in *Myxine glutinosa* (Atlantic hagfish), revealing deep evolutionary conservation [32]. Tissue-specific expression patterns vary considerably across species, with *Oncorhynchus mykiss* (rainbow trout) exhibiting infection-responsive regulation. Functional diversification is evident in antimicrobial spectra—*Gadus morhua* (Atlantic cod) cathelicidins selectively target G^- pathogens, while M. glutinosa variants display broader activity. Developmental expression patterns underscore their primordial immune functions.

5.1.2.3 Histone-Derived Antimicrobial Effectors

This unconventional AMP class originates from the proteolytic processing of nuclear histones. The pioneering example, buforin I from *Bufo gargarizans* (Asian toad), demonstrated histone H2A-derived antimicrobial activity [33]. Beyond nucleic acid packaging, histones in aquatic vertebrates serve dual roles in pathogen defense and immune signaling. Their broad tissue distribution and pleiotropic functions, from microbial membrane disruption to apoptotic regulation, highlight evolutionary repurposing of conserved proteins.

5.1.2.4 Hepcidin Family

Characterized initially as hepatic iron regulators, these cysteine-rich β-sheet peptides exhibit pan-vertebrate conservation [34]. In teleost models like *Megalobrama amblycephala* (blunt snout bream), hepcidin expression initiates during embryogenesis [35]. Multi-tissue expression profiles in species such as *Sparus aurata* (gilthead seabream) reveal immune cell-specific localization [36]. Functional studies confirm dual roles in microbial defense and iron homeostasis through ferroportin regulation.

5.1.2.5 Defensin Superfamily

Cysteine-stabilized β-sheet peptides represent ancient antimicrobial weapons across metazoans. Phylogenetic analyses position β-defensins as the predominant vertebrate subclass, with *Danio rerio* (zebrafish) expressing multiple isoforms [37]. Their multifunctional nature spans direct microbial killing, immune cell recruitment, and viral neutralization. Expression dynamics in *Cynoglossus semilaevis* (tongue sole) exemplify pathogen-responsive regulation across epithelial barriers.

5.1.3 Microbial-Derived Antimicrobial Peptides

Microorganisms exhibit a sophisticated capacity to biosynthesize structurally and functionally diverse peptide-based metabolites, which are sorted into ribosomally synthesized peptides (RPs, Table 5.3) and nonribosomal peptides (NRPs, Table 5.4). Since the discovery of penicillin, the first NRP, microorganisms have been the primary source of novel AMPs to combat MDR strains [38]. These molecules play pivotal roles in mediating microbial interactions within polymicrobial ecosystems, ranging from competitive antagonism to cooperative

Table 5.3 Classification of RPs.

Animal-derived AMPs	Activities	Examples
Nucleotide peptides	Antibacterial	Microcin C
Siderophore peptides	Antibacterial	Microcins E492, Microcins M, Microcins H47
Lasso peptides	Antibacterial	Microcin J25
Circular bacteriocins	Antibacterial	enterocin AS-48
TOMMs-linear azol(in) e-containing peptides (LAPs)	Antibacterial	listeriolysin S
Lantibiotics	Antibacterial	Nisin

Table 5.4 Classification of NRPs.

Animal-derived AMPs	Activities	Examples
Polymyxins	Antibacterial	Polymyxin B, Polymyxin E
Bogorols	Antibacterial, anti-tumor	Bogorol A, Bogorol B-JX, Bogorols I
Paenipeptins	Antibacterial	Paenipeptin A, Paenipeptin B, Paenipeptin C
Ulleungmycins	Antibacterial	Ulleungmycin A, Ulleungmycin B
Malacidins	Antibacterial	Malacidin A, Malacidin B
Odilorhabdins	Antibacterial	Odilorhabdin A, Odilorhabdin B, Odilorhabdin C

signaling. RiPPs originate from ribosomally translated precursor peptides that are extensively processed through posttranslational modifications (PTMs), including dehydration, cyclization, and glycosylation, orchestrated by dedicated enzymatic machineries. In contrast, NRPs are assembled by modular nonribosomal peptide synthetases (NRPSs), multifunctional enzymatic complexes that operate independently of ribosomal translation. NRPSs employ both proteinogenic and noncanonical amino acid substrates, facilitating the incorporation of D-amino acids, lipid moieties, and heterocyclic elements into their scaffolds. Such structural flexibility drives their broad biofunctional repertoire, including antimicrobial defense, metal acquisition, and quorum-sensing modulation. Notably, while NRPs like vancomycin and polymyxin B dominate clinical antibiotic arsenals, RPs-derived bacteriocins exhibit narrower phylogenetic targeting, primarily inhibiting closely related microbial competitors. Both classes of peptides exemplify evolutionary adaptations that enhance microbial fitness in ecological niches, with emerging implications for biotechnological applications and holobiont dynamics.

5.1.3.1 RPs

Nucleotide Peptides The nucleotide peptide microcin C (McC) originates from selected *Escherichia coli* lineages, generated in the form of a heptapeptide lacking a leader sequence, which functions as an intermediate requiring two-stage enzymatic maturation for functional activation. In the producer cell, initial PTMs generate an inactive intermediate characterized by a formylated N-terminus covalently linked to a 5′-phosphoribosyl adenosine moiety through a noncanonical phosphoramidate bond [39]. This prodrug-like conformation is actively exported via ABC transporters to extracellular environments.

Upon entry into susceptible Escherichia or related enterobacteria, the compound undergoes sequential proteolytic processing by bacterial aminopeptidases (e.g., PepN/PepA) that cleave the peptide backbone at two specific sites [40]. The final bioactive entity, a nonhydrolyzable aspartyl-adenylate analog (Asp-AMS), acts as a transition-state mimic that competitively inhibits aspartyl-tRNA synthetase (AspRS) with subnanomolar affinity (Ki ~0.2 nM). This inhibition disrupts tRNA aminoacylation, abruptly terminating protein synthesis at the elongation phase.

Critically, the dual activation mechanism ensures target-specific toxicity: the initial PTMs prevent premature activity in producer cells, while the pathogen's proteolytic machinery converts the inert precursor into the lethal inhibitor. Structural studies reveal that Asp-AMS binds AspRS with 10^3-fold higher affinity than the native substrate aspartic acid, explaining its potent translational blockade effect.

Siderophore Peptides The family of siderophore peptides is typified by microcins E492, M, and H47. *Klebsiella pneumoniae* RYC492 produces microcin E492 (MccE492), a ribosomally synthesized peptide initially characterized as an unmodified 84-residue molecule with selective activity against *Escherichia* and *Salmonella* species [41]. Its dual mechanism involves (i) membrane pore formation through β-barrel structural motifs that disrupt proton gradients and (ii) competitive inhibition of the ManYZ subunits within the mannose permease complex, blocking carbohydrate uptake. Under iron-limited conditions, MccE492 undergoes a PTM at Ser84, where a linear catecholate siderophore moiety—derived from the nonribosomal enterobactin pathway—is covalently attached via a C-glucosylated ester bond. This PTM-enhanced variant (MccE492m) exhibits a 10-fold increase in potency and broader antimicrobial spectrum encompassing Klebsiella and Enterobacter species, while the unmodified form represents an incompletely processed intermediate [41].

The biosynthetic pathway requires four enzymes (MceC/D/I/J) coordinating two precursors: ribosomal precursor MceA and enterobactin. Enterobactin is first modified by MceC (C-glucosyltransferase) and MceD (esterase) to generate a linear, glucose-decorated derivative, which is subsequently conjugated to MceA's C-terminal serine through an ester linkage. Notably, MceC and MceD exhibit functional homology to *Salmonella* iroB/iroD enzymes involved in salmochelin biosynthesis, highlighting evolutionary conservation of siderophore hijacking strategies. The mature siderophore-microcin is exported via an ABC transporter after leader peptide cleavage, with its siderophore moiety binding ferric ions to facilitate receptor-mediated uptake through Escherichia catecholate receptors FepA, Cir and Fiu.

Lasso Peptides Lasso peptides are biosynthesized by members of *Enterobacteriaceae* (Proteobacteria), *Actinobacteria* and select *Firmicutes*, defined by their unique rotaxane topology resembling a cowboy's lasso. This architecture comprises a macrocyclic ring generated by an isopeptide linkage that joins the N-terminal α-amino group and the Asp/Glu side-chain carboxyl, with the C-terminal tail traversing the ring. The kinetically trapped conformation is stabilized by steric hindrance from bulky residues (e.g., Trp/Tyr), disulfide bridges, or a combination thereof [42]. Microcin J25, the archetypal lasso peptide originating from *Escherichia coli* AY25, exemplifies this RiPP family with its narrow-spectrum activity directed toward *Salmonella* and *Escherichia* [43]. The three-dimensional conformation of this peptide, elucidated during the early years of the 21st century, revealed a 21-residue scaffold resistant to thermal and proteolytic degradation.

The biosynthesis of microcin J25 involves a four-gene plasmid-borne operon (*mcjABCD*). The precursor peptide McjA undergoes ATP-dependent enzymatic processing by McjB (a leader peptidase) and McjC (an asparagine synthetase homolog), which collectively form a lasso synthetase complex to establish the threaded topology *in vitro*. The mature peptide is exported via the ABC transporter McjD, which confers self-immunity by coupling secretion with producer cell protection [44].

Microcin J25 employs a dual-membrane translocation strategy: it hijacks the FhuA/TonB-ExbB-ExbD siderophore uptake system for outer membrane penetration and utilizes the inner membrane transporter SbmA. Subsequent inhibition of RNA polymerase occurs through direct binding to the β' subunit, blocking transcription initiation. Notably, while many lasso peptides lack broad antibacterial activity, their antimicrobial potency correlates strongly with receptor-mediated uptake efficiency, as evidenced by RNA polymerase inhibition in sensitive strains.

Circular Bacteriocins Circular bacteriocins are defined by their unique N-to-C terminal covalent cyclization, the sole PTM of ribosomal precursors. The gene clusters governing circular bacteriocins invariably encode a membrane-integrated cyclase, postulated to direct both the topological ligation and proteolytic cleavage of the precursor peptide. To date, a total of 14 circular bacteriocins have undergone structural elucidation, with enterocin AS-48 derived from *Enterococcus faecalis* (recovered from dietary matrices and hospital isolates) serving as the archetype. These peptides adopt a conserved compact tertiary structure characterized by four to five α-helical elements that assemble into either a bundle-like configuration or a saposin-inspired fold, stabilized by hydrophobic core interactions and electrostatic surface patches. The antimicrobial mechanism involves membrane-targeted disruption through pH-dependent oligomerization [45]. Enterocin AS-48 dimerizes at physiological pH (7.0 ~ 7.5), forming amphipathic pores that induce ion efflux, collapse of proton motive force, and rapid cell death. While most circular bacteriocins exhibit receptor-independent membrane permeabilization, garvicin ML demonstrates a dual concentration-dependent mode: at sublethal concentrations, it competitively inhibits the maltose ABC transporter via structural mimicry of substrate-binding domains, whereas higher concentrations trigger nonspecific membrane poration. This functional plasticity correlates with the amphipathic molecular surface—cationic regions (Lys/Arg clusters near the ligation site) mediate initial electrostatic adherence to anionic phospholipid bilayers, while hydrophobic helices (Phe/Trp/Val residues) drive membrane insertion [46].

TOMMs-linear Azol(in)e-containing Peptides (LAPs) TOMMs, also termed linear azole-containing peptides (LAPs), are ribosomally synthesized antimicrobials characterized by heterocyclic thiazole/oxazole motifs originating from the enzymatic conversion of cysteine, serine, and threonine residues during dehydration-cyclization [47]. While the term "microcin" traditionally denotes G^- bacterial products, TOMMs encompass both G^+ and G^- variants, unified by their conserved B17 synthetase-like modification machinery. The nomenclature originates from the prototype microcin B17 (*Escherichia coli*-derived), the first RiPP with *in vitro* reconstituted biosynthesis, where a tripartite enzyme complex (McbBCD) converts precursor residues into azoline intermediates, subsequently oxidized to azoles [48].

The TOMM family includes pathogen-derived virulence factors (e.g., *Streptococcus pyogenes* streptolysin S, *Listeria monocytogenes* listeriolysin S, *Clostridium botulinum* clostridiolysin S) and ecological modulators from nonpathogens (*Klebsiella pneumoniae* klebsazolicin, *Bacillus amyloliquefaciens* FZB42 plantazolicin, *Rhizobium leguminosarum* trifolitoxin). Mechanistically, microcin B17 inhibits DNA gyrase by mimicking the enzyme's ATP-binding pocket, while *Rhizobium* sp. Pop5 phazolicin and *Klebsiella klebsazolicin* target ribosomal subunits, blocking translation elongation. However, functional ambiguity persists for many TOMMs, particularly those from pathogens like *Staphylococcus aureus* stapholysin S, whose structural elusiveness hampers mechanistic studies.

Lantibiotics Lantibiotics are a class of bacteriocins characterized through the incorporation of Lan and/or Melan ring structures. Lanthipeptide refers to a class encompassing lantibiotics as well as related non-antimicrobial molecules. (Me)Lan residues arise from leader peptide-mediated dehydration of Ser and Thr, producing 2,3-dehydroalanine (Dha) and 2,3-dehydrobutyrine (Dhb), which are subsequently converted into Lan or MeLan via Michael-type addition of the Cys thiol to the β-carbon of Dha or Dhb (Figure 5.1). These cyclic architectures confer structural rigidity and enhanced protease resistance. Lanthipeptides are divided into four groups according to the biosynthetic enzymes responsible [49]. This review concentrates exclusively on types I and II, as the characterized representatives of types III and IV exhibit no antimicrobial activity. Nevertheless, it is important to emphasize that this does not imply that all type I and II lanthipeptides predominantly act as bacteriocins. In certain instances, the functional properties of these lanthipeptides remain uncharacterized. In class I lantibiotics, a LanB dehydratase mediates a transesterification between glutamyl-tRNAGlu and Ser/Thr, followed by β-elimination to yield the dehydro residues. Thereafter, a zinc-dependent LanC cyclase facilitates the nucleophilic attack of the Cys thiol on Dha/Dhb. In class II lantibiotics, the dehydration and cyclization reactions are both mediated by a bifunctional synthetase, commonly called LanM, which contains an N-terminal dehydratase domain and a C-terminal LanC-like cyclase domain. LanM catalyzes the phosphorylation of Ser/Thr residues using ATP and Mg^{2+}, followed by elimination of the resulting phosphate ester to yield Dha/Dhb. Previous reviews have thoroughly examined lanthipeptide biosynthesis and the dynamic classification framework reflecting their biosynthetic enzymes. Beyond Dha, Dhb, and (Me)Lan, numerous additional atypical residues have been identified in lanthipeptides, such as D-Ala, D-aminobutyrate, and S-[(Z)-2-aminovinyl]-D-cysteine (AviCys). A recent review comprehensively covers the enzymatic processes underlying these PTMs [50]. The spatial arrangement of (Me)Lan was initially considered to be solely DL-(Me)Lan. The earliest LL-(Me)Lan motifs were detected in cytolysin in 2013 and later reported in carnolysin [51].

Extensive structure–activity relationship (SAR) analyses have been performed on lantibiotics to assess how their structural features influence biological activity. Structurally, lantibiotics are divided into three categories—AI, AII, and B—representing linear (e.g., nisin), hybrid linear-globular (e.g., bovicin HJ50), and globular (e.g., mersacidin) conformations, respectively. Nuclear magnetic resonance analyses of type AI lantibiotics, including nisin, epilancin 15X, and microbisporicin, revealed a common elongated conformation within this subgroup. Nisin was additionally structurally resolved in complex with its receptor, lipid II, a peptidoglycan biosynthesis precursor. The nisin-lipid II complex structure revealed that the N-terminal segment of nisin interacts with the pyrophosphate group of lipid II. This interaction inhibits cell wall biosynthesis, while the C-terminal portion facilitates pore formation, causing efflux of key metabolites and disruption of membrane potential. Conversely, microbisporicin, currently in the late preclinical stage as an antibiotic targeting multi-drug-resistant bacteria, appears to incorporate the lipid II binding motifs seen in both nisin and mersacidin. Mersacidin, together with actagardine and cinnamycin, represents a type B lantibiotic featuring compact and globular three-dimensional structures. In contrast to nisin, mersacidin associates with the disaccharide portion of lipid II through its C-terminal domain and does not induce pore formation. Lipid II has been recognized as the primary the main binding target of the majority of lantibiotics, whereas cinnamycin has been reported to target a membrane phospholipid, phosphatidylethanolamine. In type AII lantibiotics, bovicin HJ50 represents a typical solution structure. The structure displayed a random-coil N-terminal region and a flexible globular C-terminal domain. It harbors a disulfide bond, which is infrequently observed in lantibiotics. Bovicin HJ50 is suggested to interact with lipid II via its N-terminal ring A. The C-terminal segment subsequently associates with the membrane, while the disulfide-containing region forms a hairpin-like structure, promoting membrane insertion and pore formation.

Two-component lantibiotics, defined by their synergistic antimicrobial activities, form a distinctive subgroup within this peptide family. Lacticin 3,147, the best-characterized representative, consists of a type B α-peptide (LtnA1) and a type A1 β-peptide (LtnA2). The α-peptide interacts with lipid II through its C-terminal residues and subsequently recruits the β-peptide in a 2:2:1 α:β:lipid II stoichiometry, leading to pore formation. A comparable mechanism has been proposed for Halα and Halβ of haloduracin. A representative 3D structure of this class is lichenicidin VK21, composed of Lchα and Lchβ, where Lchα contains a mersacidin-like lipid II-binding motif and Lchβ harbors a central hydrophobic α-helix that may facilitate bilayer interaction. Lantibiotics are of considerable interest owing to their remarkable structural diversity and potent activities against G^+ bacteria. Among them, nisin is the most extensively studied and has been employed as a natural preservative for nearly half a century, while several others are currently undergoing clinical trials as promising antimicrobial agents.

5.1.3.2 NRPs

NRPs, generated by modular NRPSs, comprise a highly heterogeneous family of microbial bioactive compounds endowed with multifunctional pharmacological effects [52]. These compounds are characterized by non-canonical amino acid incorporation, cyclic/linear hybrid architectures, and extensive post-assembly modifications. Below is a functional classification of NRPs with expanded mechanistic insights: (i) Antimicrobials: Vancomycin: a glycopeptide compound originating from Amycolatopsis orientalis, which interferes with cell wall formation via recognition of D-Ala-D-Ala units in peptidoglycan precursors. Vancomycin: A glycopeptide antibiotic produced by *Amycolatopsis orientalis*, which inhibits bacterial cell wall synthesis by binding to D-Ala-D-Ala termini of peptidoglycan precursors. Bleomycin: A glycopeptide from *Streptomyces verticillus* that induces DNA strand breaks via metal-dependent oxidative cleavage. Polymyxin B: A cyclic lipopeptide isolated from *Paenibacillus polymyxa*, targets G^- bacterial membranes through electrostatic interactions. (ii) Antineoplastic Agents: Bleomycin: Clinically used for treating squamous cell carcinomas and lymphomas. Romidepsin: A bicyclic depsipeptide derived from *Chromobacterium violaceum*, inhibits histone deacetylases (HDACs) to induce tumor cell apoptosis. (iii) Immunosuppressants: Cyclosporine A: A cyclic undecapeptide produced by *Tolypocladium inflatum*, blocks calcineurin-mediated T-cell activation by forming complexes with cyclophilins. (iv) Biosurfactants: Surfactin: A lipoheptapeptide from *Bacillus subtilis* with exceptional surface tension reduction (critical micelle concentration: 10–20 μM). (v) Feed Additives: Bacitracin: A branched cyclic peptide from *Bacillus licheniformis*, enhances livestock growth by inhibiting G^+ pathogens in animal guts. (vi) Agrobiological Agents: Fengycin and Iturin: Lipopeptides from *Bacillus amyloliquefaciens* suppress phytopathogens like *Fusarium oxysporum* through membrane disruption and biofilm inhibition. (vii) Specialized Therapeutics: Daptomycin: A cyclic lipopeptide isolated from *Streptomyces roseosporus*, which displays bactericidal activity against MRSA in a calcium-dependent manner. Enniatin B: A cyclohexadepsipeptide from *Fusarium* spp., exhibits ionophoric activity and synergizes with azole antifungals [53].

Biosynthetic origins of NRPs are predominantly found in bacteria such as *Bacillus* spp. (surfactin), *Streptomyces* spp. (daptomycin) and *Pseudomonas* spp. (pyoverdine). Fungi: *Aspergillus* spp. (gliotoxin) and *Penicillium* spp. (penicillin precursor peptides). Cyanobacteria: *Microcystis aeruginosa* (microcystins) and *Nostoc* spp. (nostocyclopeptides). Advances in genome mining tools (e.g., antiSMASH, NRPminer) have accelerated the discovery of cryptic NRPS gene clusters, enabling structural diversification through combinatorial biosynthesis [53].

Bacterial-Derived NRPs Bogorol A (Figure 5.1), a cationic lipopeptide derived from *Bacillus laterosporus* PNG276 (Papua New Guinea), exhibits marked bactericidal efficacy against MRSA (MIC 2 μg/mL) and VRE (MIC 10 μg/mL). Nocathiacins I–III (Figure 5.1), thiazolyl peptide natural products isolated from *Nocardia* sp., show broad-spectrum potency against MDR pathogens (MIC 0.0005–0.06 μg/mL) and demonstrate notable therapeutic efficacy in murine *Staphylococcus aureus* infection models (PD_{50} 0.6–0.8 mg/kg/d) [54]. Marine sponge-associated *Bacillus cereus* produces cyclic thiopeptides YM-266183 (5) and YM-266184 (6), featuring thiazole-pyridine scaffolds with potent anti-staphylococcal (MIC 0.013–0.2 μg/mL) and anti-enterococcal (MIC 0.006–0.05 μg/mL) activity [55]. The sponge-derived *Ruegeria* sp. (*Suberites domuncula*, Gulf of Naples) yields cyclic peptides 7–8 (MIC 25–50 μg/mL against *Bacillus subtilis*) [56], while *Bacillus laterosporus* PNG276 synthesizes the lipopeptide Tauramamide (MIC 0.1 μg/mL against *Enterococcus* spp.) [57]. Cyclic tetrapeptides 10–11 from seaweed-associated *Pseudomonas* sp. (*Diginea* sp., Okinawa) inhibit marine bacterial growth through membrane disruption. *Photobacterium* sp. MBIC06485 produces depsipeptides Unnarmicins A and C (Figure 5.2), selectively targeting *Pseudovibrio* spp. The thiopeptide TP-1161 from *Nocardiopsis* sp. (MIC 0.25–4 μg/mL against G^+) [58] contains a rare aminoacetone moiety, with its biosynthetic gene cluster identified via degenerate PCR. Marine-derived *Photobacterium halotolerans* synthesizes Solonamides A–B (Figure 5.2), inhibiting *Staphylococcus aureus* virulence gene expression.

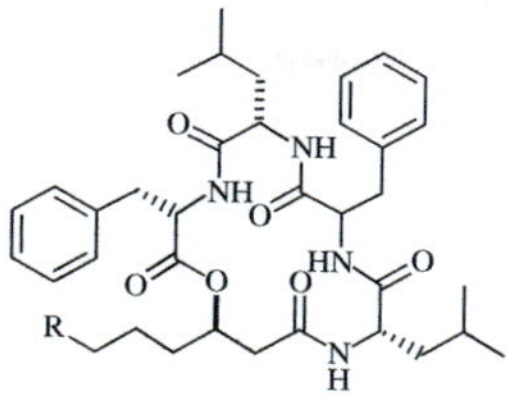

Unnarmicin A (12) (R) = H
Unnarmicin C (13) (R) = Et

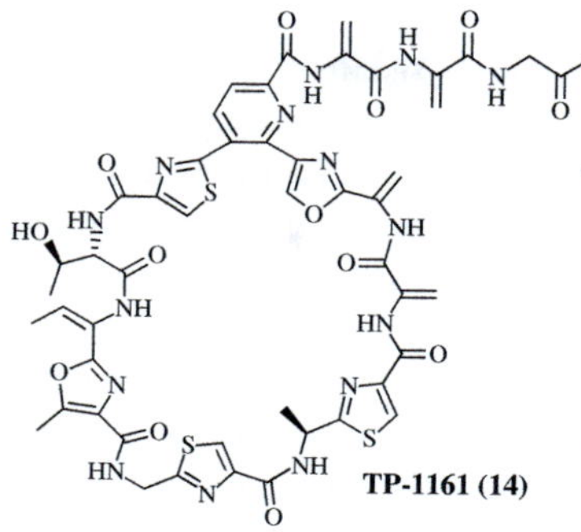

TP-1161 (14)

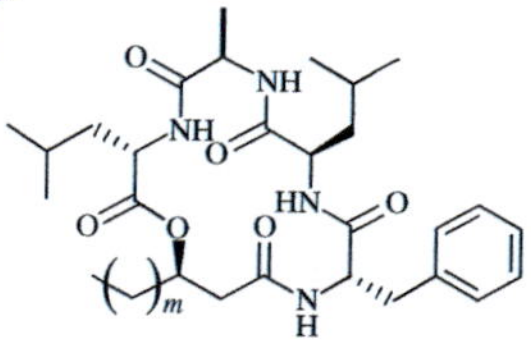

Solonamide A (15) (m) = 2
Solonamide B (16) (m) = 4

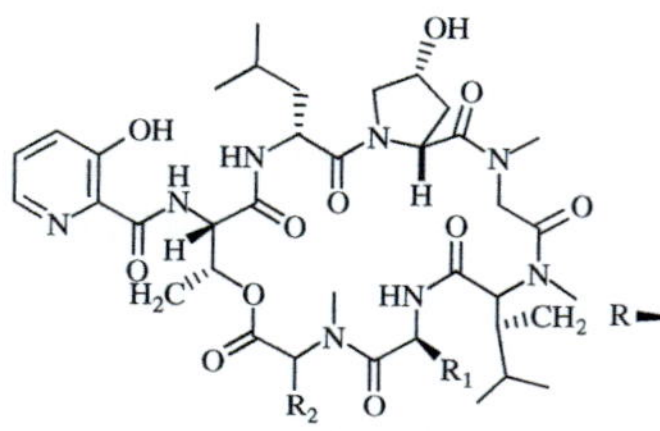

Fijimycin A (17) (R_1) = CH_3 (R_2) = Phe
Fijimycin B (18) (R_1) = CH_3, (R_2) = *i*Bu
Fijimycin C (19) (R_1) = OH, (R_2) = Phe

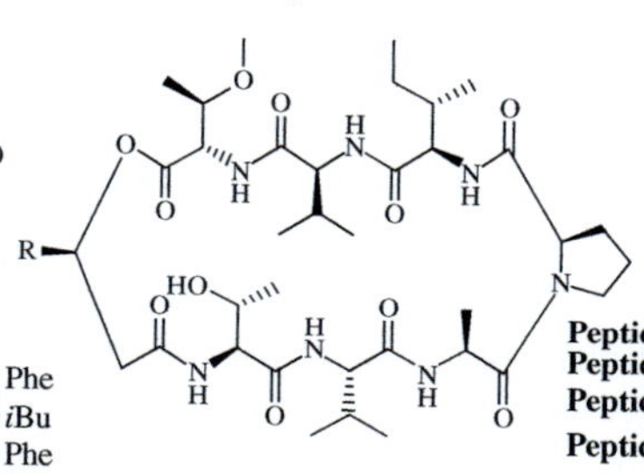

Peptidolipin B (20) (R) = $nC_{23}H_{47}$
Peptidolipin C (21) (R) = $nC_{25}H_{51}$
Peptidolipin D (22) (R) = $nC_{27}H_{55}$
Peptidolipin E (23) (R) =
Peptidolipin F (24) (R) =
m+n = 21

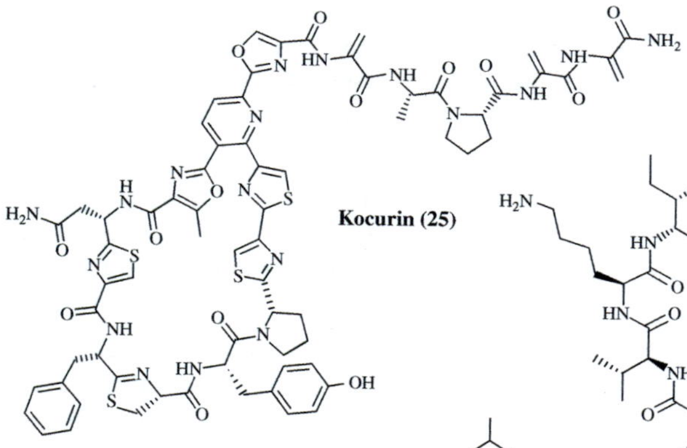

Kocurin (25)

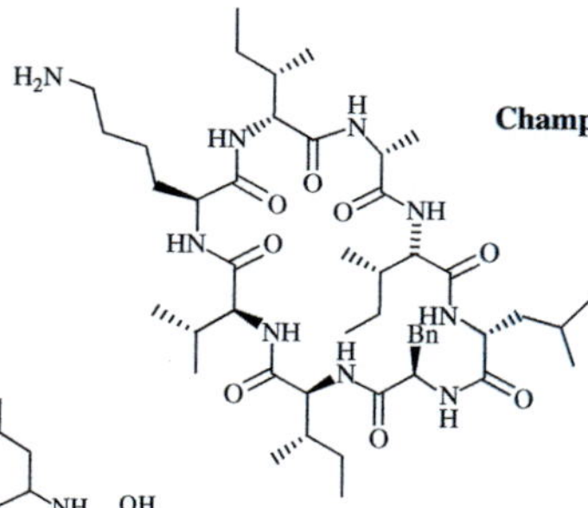

Champacyclin (26)

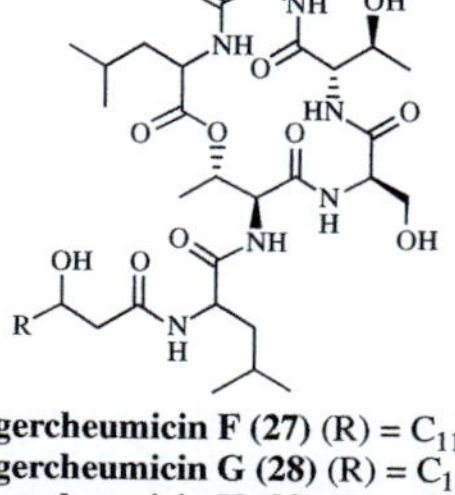

Ngercheumicin F (27) (R) = $C_{11}H_{21}$
Ngercheumicin G (28) (R) = $C_{11}H_{23}$, saturated
Ngercheumicin H (29) (R) = $C_{13}H_{25}$
Ngercheumicin I (30) (R) = $C_{13}H_{27}$, saturated

Figure 5.2 Structures of NRPs from bacteria (12–30). *Source:* Adapted from [53].

Fijian marine sediment *Streptomyces* spp. produce Fijimycins A–C (MIC100 4–16 μg/mL against MRSA) [59], while ascidian-associated *Nocardia* sp. yields Peptidolipins B—F with anti-MRSA/MSSA activity. *Kocuria palustris* biosynthesizes the thiazolyl peptide Kocurin (MIC 0.25 μg/mL against MRSA) [60]. Champacyclin from Baltic Sea *Streptomyces champavatii* inhibits *Erwinia amylovora*, and Ngercheumicins F—I from *Pseudomonas halotolerans* disrupt *Staphylococcus aureus* quorum sensing.

Fungal-Derived NRPs Bioactivity-guided fractionation of an unidentifiable marine sponge-associated fungal extract resulted in the isolation of two novel cyclodepsipeptides, Guangomides A and B (Figure 5.3). Both compounds exhibited modest antibacterial effects toward *Staphylococcus epidermidis* and *Enterococcus durans*, each with an MIC of 100 μg/mL, potentially via membrane permeability alteration [61], potentially via membrane permeability alteration. The marine-derived fungus *Aspergillus fumigatus* biosynthesized 11-O-methylpseurotin A (Figure 5.3), a polyketide-nonribosomal peptide hybrid (PKS/NRPS), which selectively inhibits *Saccharomyces cerevisiae* ΔHof1 strains through interference with cell wall integrity pathways. Cocultivation of the sponge-derived fungus *Emericella* sp. and actinomycete *Salinispora arenicola* induced production of Emericellamides A and B (Figure 5.3), displaying anti-MRSA activity (MIC 3.8–6.0 μM) [62] via putative inhibition of penicillin-binding proteins. *Scopulariopsis brevicaulis* (obtained from the Mediterranean sponge *Tethya aurantium*) produced Scopularides A and B (Figure 5.3), cyclodepsipeptides with weak G^+ antibacterial activity, suggesting ecological roles in sponge microbiome defense. The sediment-derived fungus *Alternaria* sp. (Masan Bay, South Korea) synthesized Alternaramide (42) (Figure 5.3), a pentacyclic depsipeptide demonstrating zone of inhibition (ZOI) values of 8 mm (*Bacillus subtilis*) and 13 mm (*Staphylococcus aureus*), indicative of quorum sensing modulation. Marine sponge-associated *Trichoderma* sp. yielded Trichoderins A, A1, and B (Figure 5.3), aminolipopeptides with potent activity against *Mycobacterium smegmatis*, *Mycobacterium bovis* BCG, and *Mycobacterium tuberculosis* H37Rv (MIC 0.02–2.0 μg/mL) [63] under both aerobic and hypoxic conditions, likely targeting mycobacterial membrane energetics. *Aspergillus* sp. AF119 produced Unguisin E (46) (Figure 5.3), a γ-aminobutyric acid-containing cyclic heptapeptide, while the halophilic fungus *Aspergillus sclerotiorum* PT06-1 (Putian Sea Salt Field, China) biosynthesized: Sclerotides A and B (Figure 5.3): Photointerconvertible cyclodepsipeptides with anthranilic acid and dehydroamino acid residues, showing moderate activity against *Candida albicans* (MIC 15–30 μM) and *Pseudomonas aeruginosa* (ZOI 8 mm) [64]. Sclerotiotides A–K (49–59) (Figure 5.3): Cyclic tripeptides with selective anti-*Candida* activity (MIC 3.8–30 μM) [65], highlighting structural conservation in aspochracin-type metabolites.

Cyanobacteria-Derived AMPs The marine cyanobacterium *Lyngbya confervoides* biosynthesizes lobocyclamide B (Figure 5.4), a cyclododecapeptide featuring 5 β-hydroxy-α-amino acid residues arranged in a saddle-shaped macrocyclic conformation. This compound exhibits excellent antifungal potency toward fluconazole-resistant *Candida albicans* (MIC_{90} 4.5 μM) through disruption of fungal membrane ergosterol biosynthesis. Its absolute stereochemistry was resolved via chiral HPLC analysis of Marfey's derivatives (FDAA-L/D-amino acid adducts) [66], confirming non-canonical D-configurations at

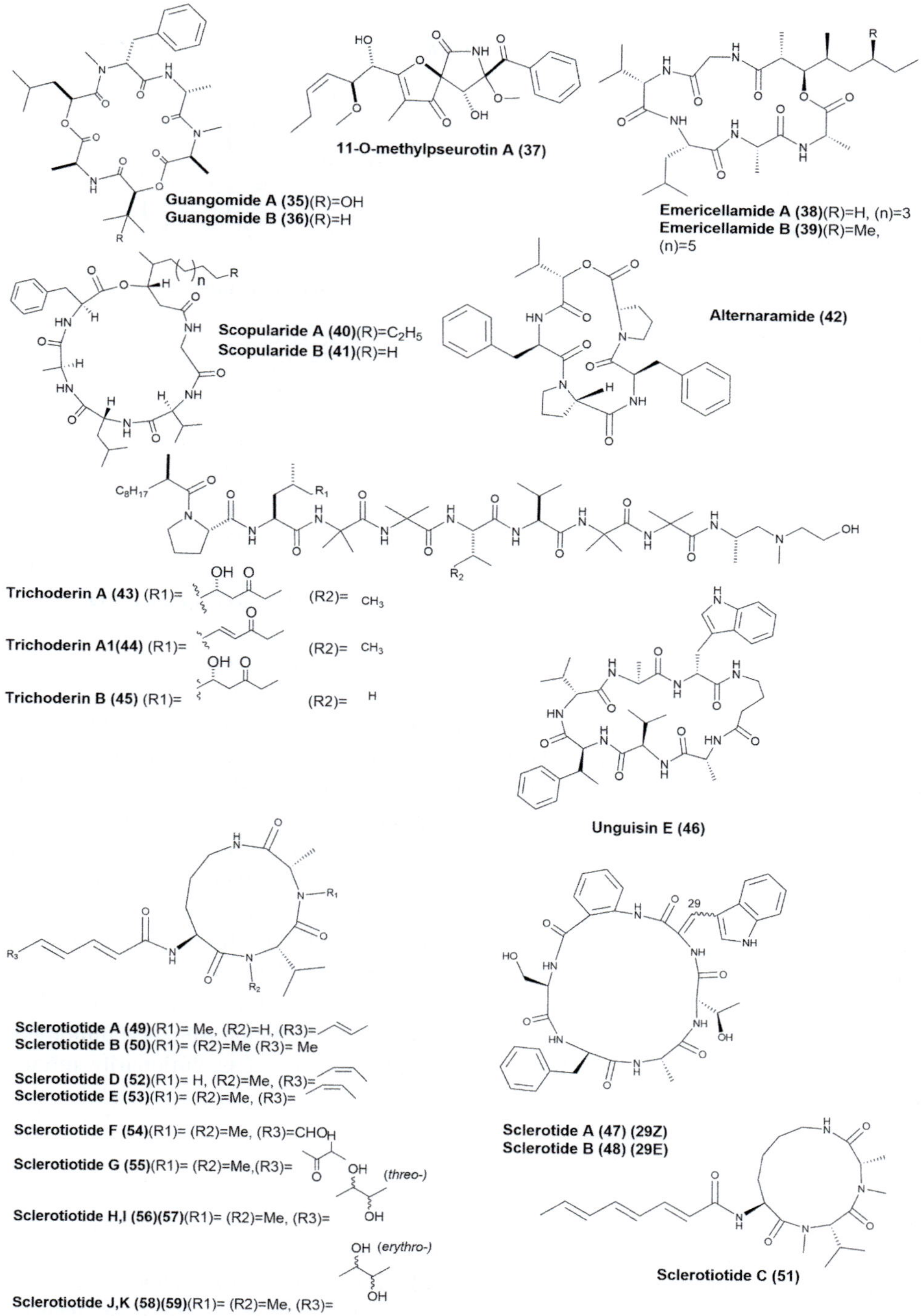

Figure 5.3 Structures of NRPs from fungi (35–59). *Source:* Adapted from [53].

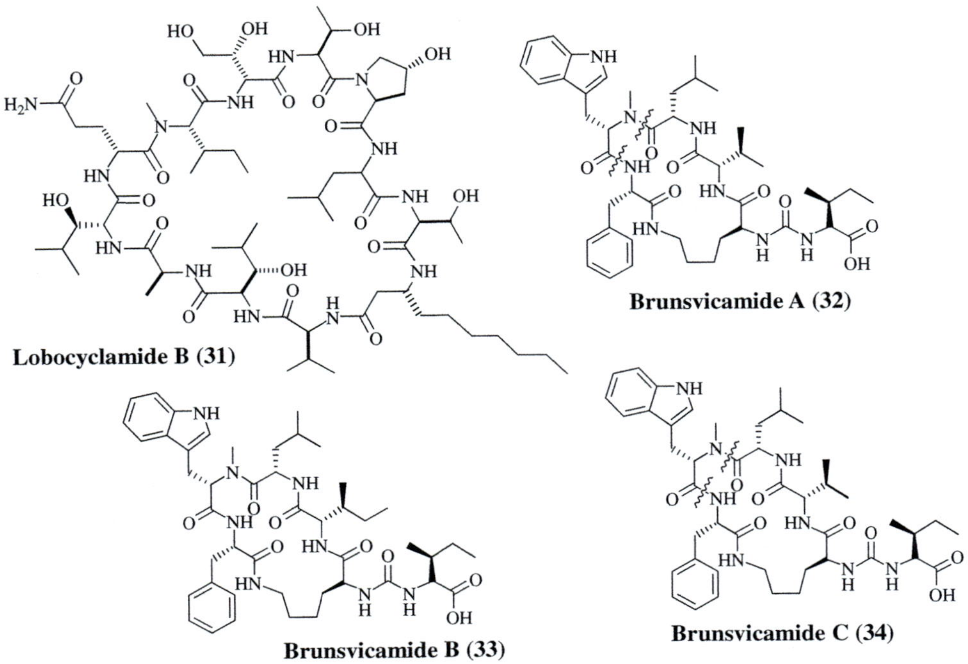

Figure 5.4 Structures of NRPs from cyanobacteria (31–34). *Source:* Adapted from [53].

residues 3 and 7. Brunsvicamides A–C (Figure 5.4), isolated from the freshwater cyanobacterium *Tychonema* sp. (Lake Brunsvic, Germany), represent a novel class of hexacyclic peptides with modified tryptophan derivatives. Notably, Brunsvicamide C features an N-methylated N'-formylkynurenine group, which is a rare post-translational modification in cyanopeptides. The compound Brunsvicamide B targets *Mycobacterium tuberculosis* protein tyrosine phosphatase B (MptpB) with an IC50 of 7.3 μM, a virulence factor critical for host cell manipulation [67], a virulence factor critical for host cell manipulation. Kinetic studies reveal non-competitive inhibition through binding to an allosteric pocket adjacent to the catalytic site. Structural-activity analyses indicate the β-hydroxy-α-amino acid motifs in lobocyclamide B enhance membrane permeability, while brunsvicamide B's kynurenine derivatives mediate target-specific phosphatase inhibition. These findings underscore cyanobacteria as prolific sources of evolutionarily refined enzyme inhibitors and antimicrobial scaffolds.

5.2 Evaluation Method of Activity

At present, research on AMPs mainly focuses on extraction, isolation, purification, and expression. However, ensuring the activity of AMPs during their isolation and purification is crucial. Therefore, each step of antimicrobial peptide isolation should track and measure its activity; otherwise, purification loses its significance. This section mainly discusses the methods for measuring and evaluating the *in vitro* and *in vivo* antibacterial activity and cytotoxicity (hemolytic activity) of AMPs.

5.2.1 *In Vitro* Antimicrobial Activity

AMPs are well documented for their wide-ranging antibacterial effects, capable of inhibiting or killing various G^- bacteria (commonly *Acinetobacter baumannii, Salmonella enterica,* and *Pseudomonas aeruginosa*) and G^+ counterparts (commonly *Clostridium botulinum, Bacillus cereus,* and *Streptococcus agalactiae*). However, the inhibitory potency exhibited by different AMPs differs greatly, and thus, the methods for measuring and evaluating *in vitro* antibacterial activity also differ accordingly. Numerous studies have shown that AMPs have broad-spectrum antibacterial activity, capable of inhibiting or killing various G^- bacteria (commonly *Escherichia coli, Vibrio parahaemolyticus,* and *Pseudomonas aeruginosa*) and G^+ bacteria (commonly *Staphylococcus aureus, Bacillus cereus,* and *Listeria monocytogenes*). However, the antibacterial activity of different AMPs varies greatly, and thus, the methods for measuring and evaluating *in vitro* antibacterial activity also differ accordingly. The primary qualitative methods include agar diffusion methods such as the Oxford cup method, the plate hole-punch method, and the paper disk method (K—B method), all of which visually reflect antimicrobial activity through the size of the inhibition zones. Quantitative methods mainly include turbidimetric methods and viable cell count methods.

5.2.1.1 Agar Diffusion Method

Principle of the agar diffusion method: On one hand, the added bacteria start to grow and multiply under suitable conditions; on the other hand, the antimicrobial peptide solution diffuses across the surface, where the peptide concentration decreases with increasing distance from the center. At the lowest inhibitory concentration, a zone forms in which bacteria cannot grow, creating a clear ring known as the inhibition zone [68]. The larger the inhibition zone, the better the antimicrobial peptide's effect and the stronger its antibacterial activity.

Oxford Cup Method The Oxford cup method, also known as the tube plate method, is a commonly used method for determining antibiotic potency both domestically and internationally [69]. Specific method: Take sterilized microbial culture media (e.g., LB agar, PDA medium) and perform mixed inoculation on the plates (bacteria as bacterial suspension, fungi as spore suspension, adjusted to a target density of 1×10^7 CFU/mL). Specific method: Take sterilized microbial culture media (e.g., LB agar, PDA medium) and perform mixed inoculation on the plates (bacteria as bacterial suspension, fungi as spore suspension, added to a final concentration of 1×10^7 CFU/mL). After solidification, place an Oxford cup (a small cylindrical tube with an inner diameter of 6 mm, outer diameter of 8 mm, and height of 10 mm) vertically on the surface of the medium, ensuring no gaps between the cup and the medium, then add 200 µL of antimicrobial peptide solution into the cup. Each group has three replicates, and after incubating for a certain period, the inhibition zone diameter is determined by the cross-measurement technique.

Plate Hole-Punch Method The procedure of the agar well diffusion method is like that of the disk diffusion method. First, perform a bacterial inoculation by pouring the agar plate, following the same steps as the disk diffusion method. After the agar plate solidifies, punch wells at the designated positions (the inner diameters of the puncher are 3, 4, 5 mm, etc.), remove the agar plugs, and inject the appropriate volume of antimicrobial peptide solution into the wells. After incubating for a certain period, measure the diameter of the inhibition zones.

Filter Paper Method The filter paper is folded into eight layers and punched into 6-mm circular discs, which are then autoclaved at 121°C and dried for later use. The mixed bacterial culture is poured into a flat plate, and the operation steps are the same as those for the tube disk method. After the medium solidifies, the prepared filter paper disk is positioned at the midpoint of the agar surface, and 100 μL of antimicrobial peptide solution of a certain concentration is slowly dripped onto the paper disk. After incubating in the incubator for a set period, the inhibition zone diameter is determined by the crossover measurement technique.

Evaluation of the Antibacterial Effect of AMPs by Three Agar Diffusion Methods: The antibacterial effect of AMPs in the agar diffusion method is influenced by their diffusion in the medium, and the concentration of the peptides decreases after diffusion, making it difficult to control. Therefore, it is generally only used for qualitative analysis. Numerous studies have shown that, in terms of clarity and shape of the inhibition zone, the punch method is the most effective. This may be because the large contact area between the AMP and the agar in the holes allows for more uniform and thorough diffusion into the medium, interacting with the target bacteria, resulting in a well-defined and clear inhibition zone. However, in the Oxford cup method, it is difficult to control the force when placing the Oxford cup, which may cause the cup to slide on the surface of the medium or lead to uneven diffusion of the antimicrobial peptide. The inhibition zone produced by the filter paper disk method is relatively regular, but since the drug-laden filter paper is fully exposed to the air, some crystallization of the antimicrobial peptide may occur, resulting in lower sensitivity.

5.2.1.2 Turbidimetric Assay Method

First, prepare the bacterial solution as follows: Under sterile conditions, centrifuge the freshly activated bacterial solution to remove the medium, then dilute it with sterile medium to a concentration of 10^7 CFU for later use. The diluted bacterial solution is mixed with an antimicrobial peptide solution of a specific concentration, placed on a shaker, and cultured with agitation for 24 hours. Subsequently, absorbance is measured, and the inhibition rate is calculated based on the absorbance data.

Inhibition rate (%) = (OD600 nm control group−OD600 nm experimental group) / OD600 nm control group × 100%

When the concentration of AMPs is below the optimal range, they do not exhibit significant inhibitory effects on pathogenic microorganisms. Within the optimal range, the inhibition rate increases significantly as the dosage of AMPs, and thereafter, the inhibition rate elevates slowly with further concentration increases. The antimicrobial efficacy measured by the turbidimetric assay method is higher than that measured by the agar diffusion method, which may be related to the permeability of AMPs in agar as the peptides have limited penetration in agar, but in a liquid state, they can interact more fully with target bacteria. The turbidimetric assay is developed according to the concept that the turbidity of the bacterial suspension is correlated with its OD value. Compared to the traditional spectrophotometric measurement of OD values, this method offers the advantages of speed, accuracy and high throughput. It not only reduces random errors caused by the sedimentation of the bacterial suspension, but the microplate has 96 wells, allowing multiple samples to be measured in a short time. However, this method has certain limitations, such as the heterogeneity and instability of the liquid buffer system and slight variations in the control of termination time, which can all affect the measurement results, leading to poor reproducibility.

5.2.1.3 Viable Cell Count Method

In short, 5 mL of bacterial culture at 10^7 CFU/mL is added into sterile shaking tubes, with one tube as the experimental tube and the other as the control tube. A certain concentration of AMPs is added to the experimental tube, while the same volume of sterile water is added to the control tube. Each group has three replicates. The tubes are placed in an incubator and shaken for one hour before performing serial dilutions. 50 μL of each dilution is plated on a solid medium. After 24 hours of incubation, colony counts for each plate are observed and recorded. Bacterial counts in the original solution are determined according to the dilution factor and plated volume. The inhibition rate is determined by the equation below:

Inhibition rate = (Number of bacteria in control tube−Number of bacteria in experimental tube)/Number of bacteria in control tube × 100%.

As the dosage of AMPs increases, the inhibition rate first increases significantly and then slowly. Compared with the turbidimetric method, the viable cell count method is more sensitive in measuring the antibacterial activity of AMPs, showing inhibitory effects even at lower concentrations. This is because, during the test, AMPs interact with target bacteria without the addition of a culture medium, and the longer the interaction time, the greater the inhibition rate.

5.2.1.4 Microdilution Method

Dissolve the AMPs in sterile distilled water, then perform serial dilutions using liquid medium. Afterward, add 50 μL to a 96-well plate. The overnight bacterial culture is diluted to 10^8 CFU/mL and used as the inoculum. In a microplate, 50 μL of bacterial inoculum is dispensed into each well, achieving a resulting bacterial concentration of 10^7 CFU/mL. The 11th row of the microplate, without bacterial inoculum or AMPs, serves as the blank control, while the 12th row with bacterial inoculum and medium is the negative control. The results are observed after incubating for 16–18 hours. In the standard microdilution method, after confirming bacterial growth in the control wells without AMPs, the drug concentration in the wells without visible bacterial growth is defined as MIC [70].

The criterion for determining growth inhibition is the absence of turbidity or precipitation in the culture medium as observed by the naked eye. The standard microdilution method can quantitatively reflect the biological activity of AMPs. The disk diffusion method is not ideal for testing AMPs. This is because AMPs have a more complex molecular structure and larger molecular weight compared to antibiotics, and their biological activity requires a specific spatial structure. The agar plate, due to its mesh-like structure, does not allow AMPs to diffuse quickly, resulting in an unclear inhibition zone. Sometimes bacterial colonies may appear within the inhibition zone, leading to erroneous conclusions. Therefore, compared to the disk diffusion and agar diffusion methods, the standard microdilution method overcomes the problem of slow diffusion caused by the complex molecular structure of AMPs. It enables AMPs to engage directly with the test bacteria, leading to rapid inhibition results.

5.2.2 *In Vitro* Antimicrobial Activity

The evaluation of AMPs' activity is currently focused mainly on the *in vitro* experimental stage, and these experiments play a key role in elucidating the antimicrobial mechanisms of AMPs and screening promising candidate molecules. However, further *in vivo* studies are essential to translate AMPs from laboratory research to clinical application. *In vivo* studies can provide direct evidence of the action of AMPs within living organisms, which is critical for assessing their potential and safety as therapeutic agents.

5.2.2.1 Principles of Model Construction

When designing bacterial infection animal models, a series of principles must be followed to ensure the scientific validity and practicality of the model. First, the replicated animal model should simulate spontaneous bacterial infections in humans as closely as possible to enhance the clinical relevance of the model. Second, the repeatability of the model is crucial, meaning that the model's establishment should be stable, reproducible, and even standardized. Third, the reliability of the model must be ensured, as it should accurately reflect the pathology and symptoms of infection, and a sufficient sample size is required to ensure the model's statistical significance. Additionally, biosafety is another key factor, ensuring that the model does not lead to the spread of infection, does not pose a threat to the health of personnel, and is easy to control. Finally, feasibility and cost-effectiveness are also very important, and the methods used should be as simple and economical as possible. In experimental animal management, factors such as environmental control, daily husbandry, and animal experiments should all be considered animal welfare. At the same time, animal experiments should follow the "3R" principles: reducing the number of animals used, finding alternative methods, and optimizing animal testing protocols. Currently, the primary animal systems employed for *in vivo* studies of AMPs are rodents, especially mice. Mice are widely used because of their compact body, rapid growth and breeding, ease of husbandry, established quality control standards, diverse breeds and strains, sensitivity to various pathogens, quick response, well-researched genetic background, minimal individual variation, and precise, reliable experimental results. Moreover, different strains, ages, weights and sexes of mice can be selected according to experimental requirements, increasing the model's flexibility. When establishing a bacterial infection model in mice, the most critical issue is choosing the appropriate inoculation route and dose. The inoculation typically varies depending on the mouse's weight, the bacterial strain, and the inoculation route. Before conducting *in vivo* studies of AMPs activity, the appropriate inoculation route and bacterial dose must be determined. Commonly used inoculation routes cover peritoneal, intravenous, subcutaneous, nasal, and topical wound administrations, and other specialized methods. Identification methods for bacterial infection mouse models generally include peripheral blood leukocyte counts, histological examination of tissues including liver and spleen, or bacterial culture at specific time points. Additionally, numerous studies agree that for lethal models, the successful establishment of the model can be defined by the complete death of the mice in the model group within a specific timeframe. These identification methods help ensure the accuracy and reliability of the model, providing a solid foundation for *in vivo* research on AMPs.

5.2.2.2 Inoculation Routes

When establishing a bacterial infection model in mice, the most critical issue is inoculating pathogenic bacteria. The inoculum typically varies based on the weight of the mice, the bacterial strain, and the inoculation route. Before conducting *in vivo* activity studies of AMPs, it is essential to determine the appropriate inoculation route and bacterial dose. Common inoculation routes include intraperitoneal injection, intravenous injection, subcutaneous injection, nasal inoculation, wound smear inoculation and other special methods.

5.2.2.3 Model Identification

The identification methods for bacterial infection models in mice generally include peripheral blood white blood cell counts, pathological sections of organs such as the liver and spleen, or bacterial culture at specific times. Additionally, numerous studies agree that in the case of lethal models, they can be identified by the death of all mice in the model group within a specific time.

5.2.2.4 Common Mouse Models of Bacterial Infection and Experimental Methods

Acute Peritonitis Model Bacterial peritonitis is a common complication of abdominal surgery, trauma, or abdominal cavity infections, which can lead to multi-system failure and ultimately death. There are three methods to establish an acute peritonitis model in mice: *Escherichia coli*-induced acute bacterial peritonitis, acetic acid-induced experimental peritonitis and *Enterococcus*-related peritonitis. The first method is commonly used for *in vivo* efficacy studies of AMPs. Additionally, models can be established using other pathogenic bacteria, such as MRSA and *Pseudomonas aeruginosa* [71]. The main experimental methods to evaluate the efficacy of AMPs in mouse peritonitis models include intraperitoneal inoculation followed by intraperitoneal drug administration, and intraperitoneal inoculation followed by tail vein drug administration.

Bacteremia Model Bacteremia refers to a condition where external bacteria enter the bloodstream through the body surface or infection sites, proliferate in the bloodstream, and spread throughout the body via circulation, eventually causing systemic infection and leading to death due to various symptoms. The bacteremia model in mice is also a commonly used model for *in vivo* studies of AMPs, and the method of intraperitoneal inoculation followed by intraperitoneal drug administration is widely used. The main experimental method for evaluating the efficacy of AMPs in the mouse bacteremia model is intraperitoneal inoculation followed by intraperitoneal drug administration. The bacteremia model typically reflects the short-term inhibitory effect of AMPs on bacteria. The bacterial count in the animal's blood serves as an indicator to assess the degree of systemic infection and reflects the strength of AMPs' therapeutic effect. It is also one of the commonly used methods for studying the dose–response relationship of AMPs *in vivo*.

Wound Infection Model Trauma can result in various types of wounds, such as lacerations, cuts, puncture wounds, and penetrating wounds, leading to different types of infections. The severity of the infection largely relies on the size, depth and degree of contamination of the wound. In severe cases, systemic poisoning symptoms like high fever and coma may occur. Bacterial infection of wounds is relatively common in clinical practice, and studying the effect of AMPs on wound infection models can not only provide a basis for clinical research but also expand the *in vivo* application of AMPs. The wound infection model in mice is usually established on the back of the mice by cutting, scraping, or burning to create a wound, followed by inoculation with bacteria at the wound site to form an infection model. AMPs are then applied in different ways to assess their antimicrobial activity. This model is used for studying local effects, and *in vivo* efficacy research; it is relatively intuitive. The antimicrobial activity *in vivo* can be evaluated through dual methods: observing wound healing and counting the changes in bacterial count at the wound site. Additionally, this model is suitable for investigating AMP formulations and combination therapies *in vivo* and can also allow assessment of AMPs in promoting wound healing. However, this model has slightly lower stability and reproducibility, and it has higher requirements for environmental conditions and procedures, which limits its use to some extent.

Organ Infection Model Organ bacterial infections are also relatively common in clinical practice and can be roughly classified into respiratory infections, surgical site infections, urinary system infections, and genital infections. AMPs can also be used in studies of organ bacterial infection models. The inoculation methods for organ infection models are generally more specialized, and the administration routes are diverse. However, before application, assessment is required to determine whether the therapeutic concentration of AMPs has any toxic impact on the target organ. Only after ensuring no toxicity can the experiments proceed. In addition to assessing the changes in bacterial counts in the organ, the efficacy of AMPs in this model can also be evaluated more intuitively by making pathological sections of the test organ and observing the changes in tissue cells.

Other Models *In vivo* studies of AMPs have also been conducted in other mouse models, such as the obstructive jaundice model. Moreover, *in vivo* efficacy studies of AMPs are not limited to rodents but have also been extended to other animal models, such as the avian coccidiosis model and the *Mannheimia haemolytica* infection model in adult sheep.

5.3 Mechanism Consideration of AMPs

5.3.1 Structure–Mechanism Relationship of AMPs

The chemical and physical characteristics of AMPs are associated with their antibacterial efficacies, including amino acid composition, peptide length, presence of positively charged residues, lipid composition, hydrophobicity, net charge, and spatial helicity [72]. Notably, AMPs can exhibit distinct subcellular localizations despite sharing similar structural features. For instance, the P11A mutant of buforin II is restricted to membrane localization, whereas the wild-type buforin II translocates across the membrane into the cytoplasm [73]. Moreover, it has been observed that the same peptides can exert varying modes of action across different organisms. Many membrane-active peptides are naturally disordered in aqueous environments but adopt defined conformations when associating with lipid bilayers. For example, human histatin 5 adopts α-helical conformations exclusively in non-aqueous solvents.

5.3.1.1 Constituents of AMPs

The amino acid composition and the configuration of side chains profoundly influence the antimicrobial potency and target specificity of AMPs. Based on their amino acid content and relative abundance, AMPs are commonly categorized as proline-rich, arginine-rich, cysteine-rich, or glycine-rich peptides. Peptides dominated by a single amino acid type generally lack a well-defined three-dimensional structure [74]. Representative peptides comprise abeaecin from honeybees, bovine indolicidin, human histatins and hymenoptaecins isolated from various insects [75]. The position of particular amino acids also modulates the antimicrobial potency and target specificity of AMPs. For example, studies have indicated that residue 11 plays a critical role in determining the antimicrobial potency of oncocin derivatives. Substitution with Pro11 or Thr11 leads to higher MICs against *Escherichia coli* compared to derivatives containing Arg11 [76]. Research has shown that the initial 35 N-terminal residues of Bac 7 are sufficient to retain the complete antibacterial activity of the full-length native peptide. The initial 16 residues at the N-terminus constitute the minimal sequence required for effective inhibition of bacterial growth, with the deletion of even a single amino acid abolishing its activity entirely. In contrast, the C-terminal region is crucial for the activity of apidaecins, whereas the N-terminal region is comparatively nonessential. Acetylation at the C-14 hydroxyl group of virginiamycin M1 causes the molecule to rotate roughly 90° compared to its orientation in the non-acetylated form. In peptides containing proline residues, the proline can adopt a wedge-like conformation that facilitates insertion into the lipid bilayer. Replacement of this residue results in reduced membrane-binding capacity.

5.3.1.2 Molecular Length of AMPs

Shorter membrane-active AMPs, such as aurein 1.2 and citropin 1.1, are thought to disrupt membranes via micellization, whereas longer peptides like caerin 1.1 may span greater distances and bend to stabilize transmembrane pores [77]. The antifungal efficacy of synthetic peptides composed of different repeats of lysine and tryptophan (KWn-NH2) has been found to be dependent on peptide length. An increase in peptide length was associated with enhanced antifungal activity, while cytotoxic effects on a human keratinocyte line appeared only in the case of the longest peptide, KW5.

5.3.1.3 Charges of AMPs

Peptide biophysical traits are tightly linked to antimicrobial function, with the high net positive charge and amphipathicity of AMPs enabling strong electrostatic and hydrophobic engagement with cytomembrane surfaces. The amphipathic profile of many AMPs arises from their pronounced membrane affinity, indicating that such interactions extend beyond α-helical peptide structures [78]. Both pH and ionic strength markedly influence the antibacterial effectiveness of AMPs. Under mildly acidic conditions, antimicrobial potency is generally elevated due to enhanced protonation of residues such as histidine, aspartate and glutamate. The pore-forming capacity of cationic amphipathic helices has been linked to the charges at their termini. Specifically, peptides carrying an N-terminal Lys residue display higher pore-forming activity, whereas a C-terminal positive charge reduces membrane activity relative to C-terminal Ile-Ala. In addition, many AMPs are C-terminally amidated, such as the cysteine-rich peptide PpCrAMP identified in Echinodermata. Certain AMPs exhibit diminished activity as a result of self-association or aggregation within the lipopolysaccharide-rich outer membrane, thereby preventing effective translocation and leading to functional loss. Interestingly, when combined with a cationic/hydrophobic LPS-binding β-boomerang motif (WKRKRF), these peptides regain strong bactericidal effects. Furthermore, the lethality of AMPs is strongly associated with their capacity to trigger membrane depolarization [79]. Comprehensive investigations indicate that the presence of cationic residues in the LPS-binding sequence underpins bactericidal activity and promotes membrane permeabilization, thereby underscoring the mechanistic significance of positive charge in this process.

5.3.1.4 Hydrophobicity of AMPs

The ratio of hydrophobicity to cationicity, which is also known as the amphiphilic balance, holds great significance for AMPs. It exerts an influence on both the antibacterial activity and selectivity of AMPs. The antibacterial potency and selectivity of AMPs are strongly influenced by their amphiphilic conformation, enabling the alignment of charged and nonpolar residues to optimize binding and disruption of bacterial membranes. This conformation also contributes to the difficulty in the development of drug resistance by bacteria.

5.3.1.5 Secondary Structure of AMPs

The optimal secondary structure of AMPs is defined by the spatial arrangement of hydrophobic and cationic residues into distinct domains, a feature termed amphipathic design. The most common structural motifs include α-helices, relaxed coils, and antiparallel β-sheets. Peptides with amphipathic α-helices generally display higher antimicrobial activity than those with less well-defined secondary structures, and an increase in α-helical content is linked to enhanced potency. Typically, small AMPs contain at least 22 residues for α-helical configurations or eight residues for β-sheet structures [80]. These findings indicate that α-helical peptides require at least 22 residues to penetrate a lipid bilayer. Linear AMPs can form an amphipathic α-helical conformation upon interacting with membranes composed of both zwitterionic and anionic vesicles. Moreover, β-sheet structures provide a framework for the spatial organization of cationic and hydrophobic residues [81]. Disulfide bonds in β-sheet-containing AMPs stabilize the peptide conformation and facilitate their passage across cell membranes.

5.3.1.6 Curvature of AMPs

The curvature of AMPs is influenced by the amphipathic arrangement of their amino acid sequences, allowing them to adjust conformation according to membrane thickness. For example, previous research indicates that membrane composition significantly affects the activity of the human dermcidin channel, with thicker, cholesterol-rich membranes exhibiting lower conductance than thinner membranes [82]. The curvature of α-helices is considered a valuable descriptor for assessing AMP activity and mode of action. Specifically, linear amphipathic helices favor barrel-stave pore formation, while curved amphipathic peptides are better suited for toroidal pore structures. Many investigations have shown that the inherent bending of membranes influences the antimicrobial performance of peptides.

5.3.2 Mode of Actions of AMPs

5.3.2.1 Targeting Cell Wall Biosynthesis

The bacterial envelope, a rigid exterior, confers structural integrity and mechanical strength, shielding cells against external environmental challenges. Severe cell wall disruption leads to lysis and cell death, making peptidoglycan biosynthesis a prime antibiotic target. Clinically used antibiotics, such as penicillin and vancomycin, exploit this vulnerability, exhibiting potent activity against $Gram^+$ bacteria but limited efficacy against G^- bacteria due to differences in peptidoglycan content (90% in G^+ vs. 5 ~ 20% in G^-).

In the cytoplasm, peptidoglycan synthesis starts with UDP-N-acetylglucosamine (GlcNAc), which is enzymatically converted to UDP-MurNAc through sequential actions of MurA (UDP-GlcNAc enolpyruvyl transferase) and MurB (UDP-MurNAc dehydrogenase). The enzymes MurC, MurD, MurE and MurF sequentially append amino acids to UDP-MurNAc, ultimately forming the UDP-N-acetylmuramyl pentapeptide [83]. Lipid I and lipid II are synthesized by MraY and MurG, respectively, and the subsequent addition of a glycine cross-bridge by FemXAB proteins produces lipid III (Figure 5.5). The flippase MurJ mediates the translocation of lipid II and lipid III to the periplasm, where peptidoglycan assembly and cross-linking are carried out by glycosyltransferases and transpeptidases. Cell wall biosynthesis involves multiple antibiotic targets: Penicillin-binding proteins (PBPs), structurally like GTases/TPases, are inhibited by β-lactams (e.g., penicillins, cephalosporins and carbapenems). Lipid II, an extracellular precursor, is targeted by AMPs like vancomycin, nisin, and plectasin. A variety of novel membrane-active peptides (MAPs), such as malacidins, mannopeptimycins, hypeptin, teixobactin, cilagicin, friulimicins, glycinocins, humimycin A, muraymycins and corbomycin, have broadened the spectrum of druggable bacterial targets [84].

Inhibition of Cell Wall Synthase The penicillin-binding proteins (PBPs) represent the primary enzymatic targets in cell wall synthesis, with penicillins and cephalosporins serving as classic inhibitors. Another critical target is peptidoglycan transferase (PGT), inhibited by the phosphoglycolipid antibiotic moenomycin. A group of uridyl peptide antibiotics—including pacidamycins, mureidomycins, and napsamycins—was initially noted for remarkable broad-spectrum activity, yet their mechanism of action was not fully understood until muraymycins were discovered and MraY's structure was resolved (Figure 5.6) [86].

Muraymycins are nucleoside peptide antibiotics from *Streptomyces* species that show strong activity against G^- and G^+ bacteria, including *Escherichia coli* and *Mycobacterium tuberculosis*. To date, 19 structural variants have been identified, with muraymycin A1 exhibiting therapeutic potential in a murine model of *Staphylococcus aureus* infection ($ED_{50} = 1.1\,mg/kg$) [87].

As a key membrane-bound catalyst, MraY initiates peptidoglycan assembly by forming lipid I and UMP from phospho-MurNAc-pentapeptide and undecaprenyl phosphate [88]. Challenges in large-scale production and stabilization have hindered detailed characterization of MraY. A breakthrough came when Lee's group determined the crystal structure of MraYAA,

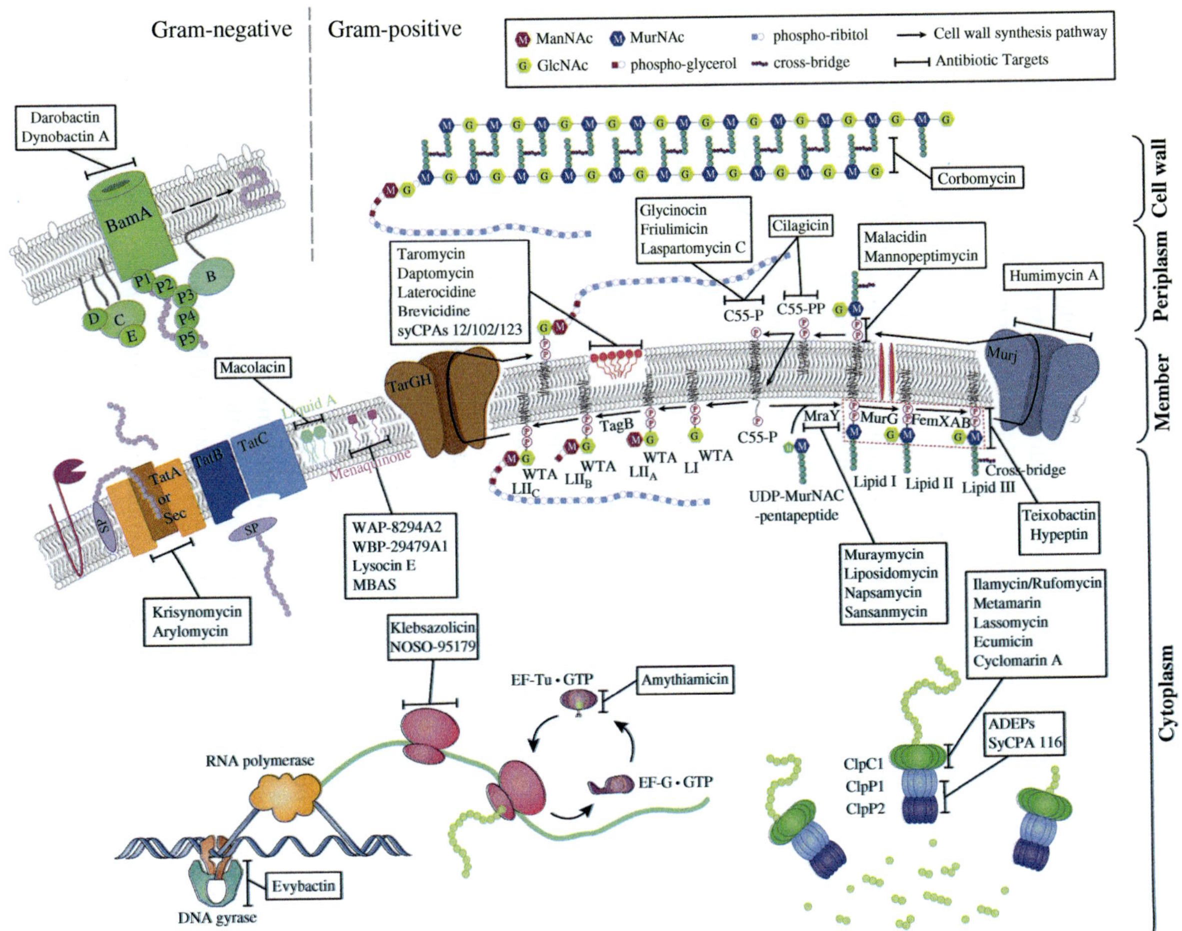

Figure 5.5 Schematic showing sites of action of AMPs mentioned in this work. *Source:* [85]/Royal Society of Chemistry.

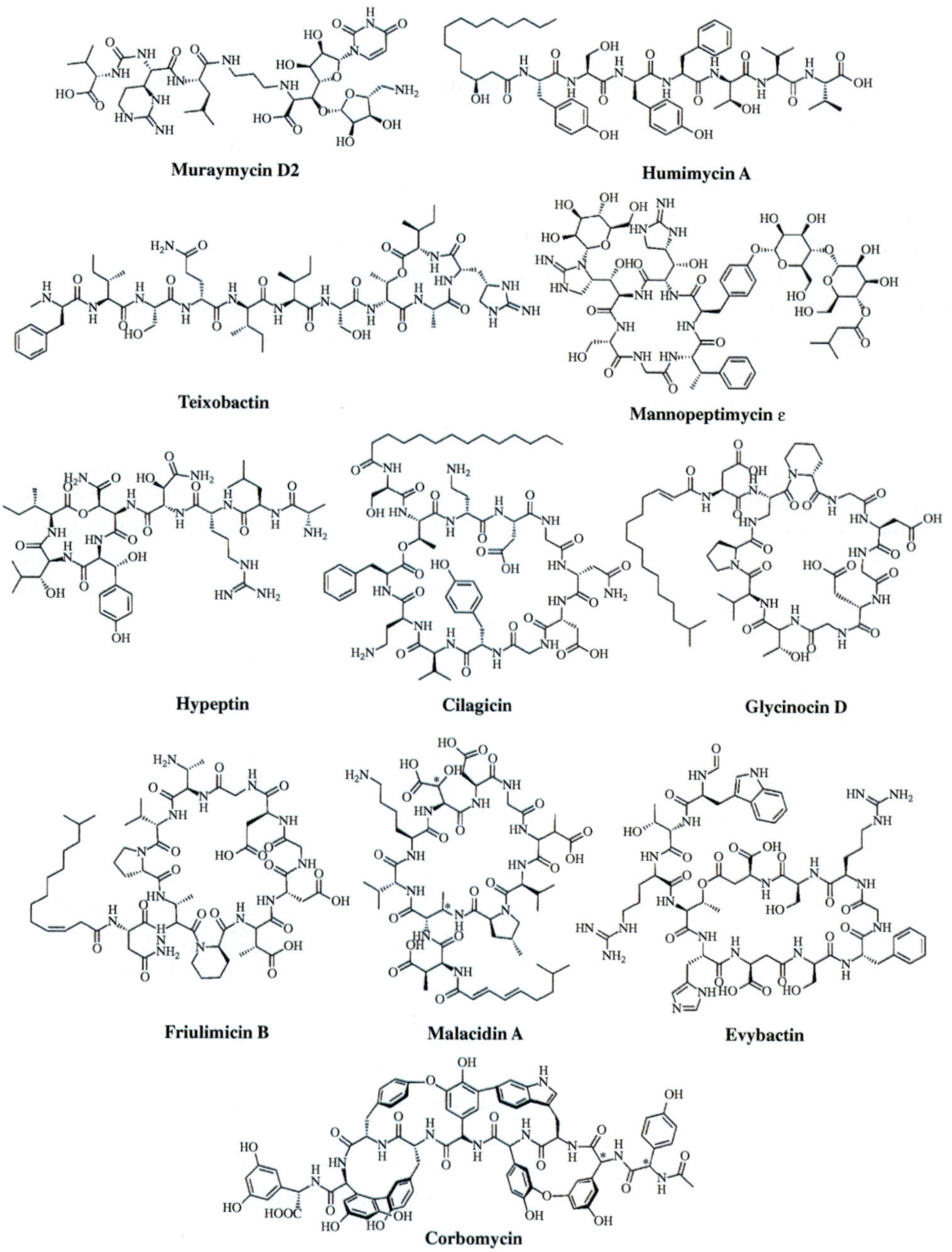

Figure 5.6 Representative antimicrobial peptides inhibiting cell wall formation with DNA gyrase inhibitor evybactin. Ambiguous chiral centers are denoted by asterisk. *Source:* [85]/Royal Society of Chemistry.

an Aquifex aeolicus homolog expressed in *Escherichia coli*, thereby unveiling the complete architecture of MraY [89]. MraYAA consists of 10 transmembrane helices and five cytoplasmic loops, assembling as a homodimer. A conserved hydrophilic cleft, enriched in 34 residues and the catalytic DDD motifs, serves as the putative binding site for UM5A, with D265 interacting with Mg^{2+}. Although substrate-bound structures remain unresolved, TM9b and loop E are thought to recognize the glycosyl portion of UM5A, while D265 and Mg^{2+} engage its pyrophosphate moiety.

The co-crystal structure of muraymycin D2 (MD2) bound to MraYAA represents the first structural determination of a nucleoside inhibitor complex of MraY, offering critical insights into the inhibitory mechanism of this group of microbial antimicrobial peptides (MAPs). Upon MD2 binding, the active site undergoes conformational rearrangements that generate a nucleoside-binding pocket together with a peptide-binding site, thereby improving ligand affinity. MD2 occupies the nucleoside-binding pocket in a manner resembling a dual-headed plug fitting into a socket. Specifically, the uridine moiety forms π–π stacking or hydrogen-bond interactions with Phe262, Asn255, Asp196, Gly194, and Lys70, whereas the 5′-aminoribose establishes hydrogen bonds with Thr75, Asp193, Asn190, and Gly264. Moreover, the peptide side chain contributes additional hydrogen-bonding contacts with Ala321, Gln305 and His325 in the peptide-binding region, further reinforcing the overall interaction between MD2 and $MraY_{AA}$ [90]. Notably, MD2 bypasses interactions with the catalytic DDD motif and Mg^{2+}, implying that its binding site partially overlaps with UM5A but follows a distinct mode, independent of the pyrophosphate and glycosyl groups essential for UM5A.

Subsequent investigations have determined the co-crystal structures of several nucleoside antibiotics targeting MraY, including tunicamycin, capuramycin, carbacaprazamycin, and 3′-hydroxymureidomycin A. Except for 3′-hydroxymureidomycin A, which carries a 4′,5′-enamide linker instead of a 5′ group, these compounds consistently contain a uridine core and a 5′-aminoribose moiety. Additionally, aside from 3′-hydroxymureidomycin A, their side chains are composed of varied aliphatic groups rather than peptide chains. Comparative analysis underscores the essential role of the uridine unit, while 5′-aminoribose primarily contributes to spatial orientation rather than absolute binding necessity. The pocket near uridine is sufficiently adaptable to fit different peptide chains from the mureidomycin family, yet the presence of 5′-aminoribose typically enhances both binding strength and selectivity. For example, tunicamycin lacks 5′-aminoribose and, besides acting on MraY, can also engage the human enzyme dolichyl-phosphate N-acetylglucosamine phosphotransferase (GPT/GlcNAc-1-P-transferase), which may result in unintended off-target effects.

Beyond MraY, the discovery of humimycins has enabled the identification of MurJ, the lipid II flippase, as a novel antibacterial target for small-molecule agents, expanding the scope of nucleoside antibiotic research [91]. Humimycins A and B belong to a class of linear heptapeptides with N-terminal acylation by (3S)-hydroxytetradecanoic acid, representing a novel antimicrobial framework (Figure 5.3). The corresponding gene clusters for humimycins A and B are encoded in Rhodococcus equi and Rhodococcus erythropolis, both strains being derived from the human microbiome [92]. Humimycins exhibit a broad spectrum of strong antibacterial activity against members of the *Firmicutes phylum*, notably targeting Staphylococcus and Streptococcus species, including *Staphylococcus aureus*

(MIC, 8 mg/mL) and *Streptococcus pneumoniae* (MIC, 4 mg/mL), and also display activity against certain bacteria within the *Actinomycetes phylum* [93]. Moreover, Humimycin A shows outstanding biocompatibility *in vivo*, tolerable at doses above 50 mg/kg in murine experiments. Although its activity as a single agent in mice was modest, combination therapy with dicloxacillin substantially improved survival outcomes in a peritonitis-sepsis model. Mechanistic studies suggest that humimycins target the lipid II flippase MurJ, embedded in the bacterial inner membrane and responsible for moving lipid II from the cytoplasmic to the outer membrane, a crucial step for peptidoglycan synthesis. As a result, MurJ inhibition blocks cell wall formation. Additionally, humimycins have been found to enhance the efficacy of β-lactam antibiotics, effectively controlling multiple MRSA and VRE clinical isolates without eliciting resistance. Humimycin A can resensitize β-lactam-resistant strains, likely because both drugs act on the same essential pathway, peptidoglycan biosynthesis. Currently, humimycins represent the only small-molecule inhibitors of MurJ reported to date, aside from the viral protein LysM, which has also been shown to interfere with MurJ. While the full mechanism of humimycins remains under investigation, this groundbreaking discovery paves the way for the design of novel antibiotics [94].

5.3.2.2 Targeting Precursors of Peptidoglycan Biosynthesis

Glycopeptide antibiotics (GPAs), with vancomycin as a representative, inhibit bacterial growth by targeting lipid II, a vital and highly conserved peptidoglycan intermediate required for cell wall assembly in G^+ bacteria [95]. This group of GPAs, such as balhimycin, chloroeremomycin, ristocetin A, and teicoplanin, acts by binding to the D-Ala–D-Ala terminal of lipid II. This reversible association blocks both glycosyl transfer and peptide cross-linking, thereby inhibiting the extension of peptidoglycan chains and impairing bacterial cell wall formation. Vancomycin, the most widely employed GPA, has been used to treat infections resistant to penicillins or cephalosporins, earning its status as a critical final-line therapeutic agent. Nevertheless, the rapid appearance and hospital spread of vancomycin-resistant Enterococcus (VRE) endanger this important therapeutic option [96]. Recent investigations reveal that VRE generally carries a resistance gene cluster comprising 5–7 genes, which gives rise to nine distinct genotypes: VanA, VanB, VanC, VanD, VanE, VanG, VanL, VanM and VanN. Of these, the VanA, VanB, VanD and VanM genotypes produce ligases that can partially swap the D-Ala–D-Ala moiety at the lipid II terminus with D-Ala–D-Lac, while VanC, VanE, VanG, VanL and VanN encode ligases that replace the same terminal sequence with D-Ala–D-Ser [97]. When modifications occur at the lipid II terminal binding site for vancomycin, the interaction between vancomycin and its target is significantly weakened, enabling bacteria to develop resistance.

Besides the vancomycin-type GPAs, cyclic lipopeptides such as laspartomycins and amphomycins act on a different class of cell wall synthesis precursors, namely undecylenyl phosphate (C55—P) or undecaprenyl pyrophosphate (C55-PP) [98]. C55—P is a distinct biomolecule composed of an 11-isoprenoid unit and typically functions as a general lipid carrier together with its pyrophosphate derivative, C55-PP [99]. C55—P plays a critical role in numerous bacterial biosynthetic processes. During peptidoglycan assembly, it serves as a transmembrane carrier that anchors intermediates on the cytoplasmic membrane and assists their transport across the membrane to the extracellular environment. Upon completion of the transglycosylation and transpeptidation of lipid II units, C55—P is recycled in its pyrophosphate form, dephosphorylated and repositioned by UPP phosphatases, and restored to the C55—P state to participate in subsequent biosynthetic cycles [100]. Laspartomycin C belongs to the class of calcium-dependent antibiotics (CDAs), whose antimicrobial efficacy is strongly influenced by the presence of calcium ions [101]. Structural analysis via co-crystallization of laspartomycin C with C10—P (geranyl-phosphate, a water-soluble mimic of C55—P) demonstrated that laspartomycin C coordinates with both C10—P and calcium ions to form a stoichiometric 1:1:2 complex [101]. The macrocyclic framework of laspartomycin C adopts an amphiphilic fold, resulting in a saddle-shaped conformation in solution, where polar residues occupy the upper surface and hydrophobic residues the lower, with an aliphatic tail protruding from the bottom. The phosphate moiety of C10—P is accommodated within the cyclic backbone of laspartomycin C via hydrogen bonding to polar residues and coordination with two calcium ions. One calcium ion associates with the conserved Asp-X-Asp-Gly motif of CDAs, facilitated by a water molecule, and the overall binding is further reinforced by hydrophobic interactions between the C10—P fatty chain extending from the lower plane and the fatty chain of laspartomycin C [101]. Simultaneously, driven primarily by intermolecular hydrogen interactions connecting the D-Dap_9 macrocyclic amide and the Asp^7 residue's carboxylic moiety of an adjacent lipopeptide, the complex assembles into a more stable dimer, fully sequestering C10—P from the surrounding solution. The structural framework of amphomycins is highly similar to that of laspartomycin C, and their co-crystallization with C10—P exhibits a comparable configuration.

Recently, numerous newly identified AMPs targeting precursors of bacterial cell wall synthesis have attracted significant scientific interest. Representative examples include mannopeptimycins, teixobactin, hypeptin, cilagicin, friulimicins, glycinocins B—D and malacidins. Mannopeptimycins a–3, a group of cyclic hexapeptides isolated from *Streptomyces hygroscopicus* LL-AC98, exhibit potent inhibitory effects against various G^- bacteria, such as *Bacillus subtilis*, *Staphylococcus aureus*, *Staphylococcus hemolyticus*, and *Micrococcus luteus*. These compounds also demonstrate remarkable *in vivo* activity, with the ED_{50} of mannopeptimycin 3 against *Staphylococcus aureus* determined to be 0.59 mg/kg in a mouse acute lethal infection model [102]. Mannopeptimycins belong to the glycopeptide antibiotic (GPA) family and also act on lipid II, yet they employ a mechanism distinct from that of vancomycin. These peptides do not interfere with vancomycin's binding to lipid II, indicating that their functional site is located elsewhere and not at the D-Ala–D-Ala terminus [103]. Additionally, studies indicate that mannopeptimycins interact non-specifically with lipophosphatidic acid, a property that may promote the swift localization of these antibiotics on the bacterial surface [104]. The unique targeting pattern of mannopeptimycins allows them to effectively eliminate VRE, attracting significant interest among chemists focused on molecule design. Despite this, the precise way in which mannopeptimycins interact with lipid II remains unclear, highlighting a critical question that demands prompt investigation.

Teixobactin, an uncommon cyclic undecapeptide containing the rare residues enduracididine and methylated phenylalanine, was obtained from a strain of Eleutheria terrae through in situ cultivation methods [105]. This particular strain cannot be grown using standard laboratory techniques. Teixobactin exhibits remarkable antibacterial activity against G^+ bacteria, with MIC values below 0.5 mg/mL for *Enterococcus faecium* and *Enterococcus faecalis*, and MICs of 20 ng/mL and 5 ng/mL against *Bacillus anthracis* and

Clostridium difficile, respectively [105]. Moreover, teixobactin demonstrates remarkable *in vivo* efficacy, with a PD_{50} of 0.2 mg/kg in an MRSA-induced septicemia protection model, which is over tenfold more effective than vancomycin used as a control.

Hypeptin, a cyclic peptide antibiotic generated by *Lysobacter* sp. K5869 through in situ cultivation, exhibits strong inhibitory activity against diverse G^+ pathogens [106]. It shows MIC values of 0.0625, 0.125, 0.0625, 0.25, 0.5, 2 and 0.25 mg/mL for *Bacillus subtilis*, *Staphylococcus simulans*, *Staphylococcus aureus*, MRSA, vancomycin-intermediate *Staphylococcus aureus* (VISA), VRE, and *Mycobacterium bovis*, respectively [106]. Both teixobactin and hypeptin belong to new classes of cyclic depsipeptides capable of targeting multiple cell wall biosynthesis intermediates, such as C55-PP and lipid I, II and III. Their novel mode of action relies on specific interactions with the pyrophosphate group and the initial sugar attached to the lipid carriers, allowing them to engage several targets simultaneously. By binding to lipid II and III, these compounds concurrently block peptidoglycan formation and wall teichoic acid (WTA) synthesis, with the inhibition of WTA triggering uncontrolled autolysis in bacterial cells. Furthermore, elevated levels of teixobactin inhibit the YbjG-mediated dephosphorylation process, thereby interfering with the conversion of C55-PP to C55—P and blocking the downstream formation of lipids I, II and III [106]. Recent studies have shown that teixobactin interacting with lipid II on the extracellular side of the membrane rapidly promotes polymerization, forming teixobactin-lipid II fibrils that cause membrane disruption and enhance bacterial killing [107]. Although the exact binding site of hypeptin on its target remains unconfirmed experimentally, it is plausible that it involves the pyrophosphate group, similar to teixobactin. The mechanisms of action of mannopeptimycins, teixobactin and hypeptin all bypass the conventional D-Ala–D-Ala site, making them promising therapeutic options against clinically vancomycin-resistant strains.

Glycinocins B–D, friulimicins, and malacidins are recently identified calcium-dependent antibiotics (CDAs). In the presence of calcium ions, Glycinocins B–D demonstrate potent antibacterial effects against G^+ bacteria such as *Bacillus subtilis* and *Staphylococcus aureus* [108]. Friulimicin B displays low MIC values of 0.078 mg/mL against *Staphylococcus simulans* and *Bacillus subtilis*, comparable to those of daptomycin. Malacidin A exerts strong antibacterial activity against G^+ bacteria, including various multidrug-resistant MRSA strains (MIC 0.2–0.8 mg/mL) and VRE strains (MIC 0.8 ~ 2.0 mg/mL) [109]. When evaluated *in vivo*, Malacidin A effectively eradicated MRSA from infected wounds in a rat model of cutaneous wound infection [109]. Cilagicin, a novel cyclic lipopeptide derived from the BGC of *Paenibacillus* mucilaginosus, exhibits potent broad-spectrum antibacterial activity (MIC 0.125–2 mg/mL) against G^+ bacteria and shows particularly strong efficacy against multiple clinically resistant MRSA and VRE strains (MIC 0.125–0.5 mg/mL) [110]. Additionally, a semi-synthetic derivative, cilagicin BP, showed superior clearance of *Staphylococcus aureus* and *Streptococcus pyogenes* compared with vancomycin in a mouse neutropenic thigh infection model. Glycinocins and friulimicins harbor the conserved Asp-X-Asp-Gly calcium-binding motif and act on C55—P similarly to laspartomycin C. By contrast, malacidins target lipid II rather than C55—P in a calcium-dependent manner, demonstrating potent activity against VRE, indicating a binding mechanism distinct from vancomycin. Despite lacking the Asp-X-Asp-Gly motif, malacidins possess a structurally analogous module, Asp-X-Asp-(β-OH-Asp), and their interaction with lipid II likely involves the pyrophosphate moiety. Cilagicin represents a unique cyclic lipopeptide that binds both C55—P and C55-PP simultaneously without calcium, differing from previous reports in which bacitracin and amphomycin interact only with C55-PP or classical CDAs. This highlights a novel interaction pattern for cilagicin, warranting further mechanistic studies. Furthermore, a recent bioinformatic investigation based on cilagicin identified two additional MAPs, paenilagicin and virgilagicin, capable of dual targeting C55—P and C55-PP, as well as one antibiotic, bacilagicin, selectively targeting C55—P [111]. Growing evidence suggests that antimicrobial agents capable of targeting multiple cellular sites are less likely to induce resistance in clinical environments; therefore, agents such as cilagicin, paenilagicin, virgilagicin, teixobactin and hypeptin may offer enhanced potential for therapeutic development.

5.3.2.3 Blockage of Peptidoglycan Remodeling

Considerable efforts have been devoted by researchers to exploring the biosynthesis of peptidoglycan in the quest for novel antimicrobial targets, while relatively few studies have delved into the hydrolysis of peptidoglycan, a process equally vital for bacterial survival. During their growth, maturation, and division, bacteria constantly change in morphology and size, which necessitates the continuous remodeling of peptidoglycan. Hence, controlled peptidoglycan breakdown is a crucial biochemical process for bacterial growth, and any disturbance of this process can rapidly lead to bacterial cell lysis [112].

Corbomycin (Figure 5.6), a novel glycopeptide antibiotic (GPA), is structurally like the known antibiotic complestatin. This correlation arises from the parallel alignment of a phylogenetic tree constructed from glycopeptide biosynthetic gene clusters and associated resistance genes. Corbomycin primarily exhibits inhibitory activity against G^+ bacteria, with MICs ranging from 0.5 to 4 mg/mL and efficacy comparable to that of vancomycin, especially against MRSA and daptomycin-resistant *Staphylococcus aureus* (MICs of 0.5–2 mg/mL). Importantly, corbomycin also shows activity against VRE and vancomycin-intermediate *Staphylococcus aureus* (VISA), indicating a mechanism different from conventional GPAs. Its *in vivo* performance is noteworthy as well. In a mouse neutropenic superficial skin infection model, corbomycin cleared MRSA at levels comparable to the control treatment with fusidic acid [113].

Complestatin, an antimicrobial agent identified in the previous century, was once thought to act by suppressing bacterial fatty acid synthesis. Recent research, however, indicates that both complestatin and corbomycin act as inhibitors of peptidoglycan remodeling. This remodeling process in bacteria depends on autolysins, a naturally extensive and redundant family of enzymes [114]. In general, no individual autolysin is indispensable for bacterial survival. Corbomycin and complestatin cleverly inhibit a broad spectrum of autolytic enzymes by directly binding to the peptidoglycan substrate. Since both compounds display activity against various G^+ bacteria, their peptidoglycan interaction is likely directed at a conserved structural motif, allowing them to retain strong affinity across different bacterial species [115]. To date, the precise binding sites of corbomycin and complestatin on peptidoglycan remain unreported, necessitating further research.

5.3.3 Inhibition of DNA Gyrase

During bacterial proliferation and cell cycle progression, genome duplication and RNA synthesis are essential biochemical processes. These processes require the cleavage and religation of DNA strands as well as chromatin supercoiling, primarily regulated by topoisomerases. Quinolone antibiotics, including nalidixic acid, ciprofloxacin, fleroxacin, levofloxacin, moxifloxacin, and gemifloxacin, mainly target bacterial topoisomerase II (DNA gyrase) and topoisomerase IV, disrupting chromatin topology and inhibiting DNA synthesis, thereby causing bacterial cell lethality. Although topoisomerase II and IV share structural and functional features, their sensitivity to quinolones differs among bacterial species. In G^- bacteria, topoisomerase IV is the primary target, whereas in G^+ bacteria, topoisomerase II is the main target and topoisomerase IV plays a secondary role [116]. Moreover, multiple studies have demonstrated that quinolone antibiotics interact with their targets during the dynamic cleavage of DNA double strands catalyzed by topoisomerases, resulting in the formation of quinolone–topoisomerase–DNA complexes. These complexes ultimately cause the DNA replication machinery to stall at obstructed replication forks [117]. Quinolones have been widely employed in treating clinical infections, although their effectiveness is generally restricted to Enterobacteriaceae and a limited range of G^+ bacteria, including *Staphylococcus aureus*. Prolonged administration of quinolones has facilitated the development of resistant strains.

Evybactin, a distinctive cyclic depsipeptide synthesized by *Pseudomonas noeneputensis* DSM 25462, contains an uncommon N-methylhistidine (MeHis) residue and an N-terminal formyl modification [118]. Evybactin displays selective inhibition toward *Mycobacterium tuberculosis* (MIC = 0.25 mg/mL) and *Escherichia coli* (MIC = 8 mg/mL), showing negligible activity against other bacterial species. In a septicemia model, evybactin's effectiveness against *Escherichia coli* was comparable to that of gentamicin, without causing significant toxicity at 100 mg/kg [118]. The discovery of evybactin has expanded the range of antibacterial agents that inhibit topoisomerases. This antibiotic displays selective activity against Mycobacterium tuberculosis H37Rv, offering a promising approach for addressing clinical treatment difficulties. Structurally distinct from quinolones, evybactin targets DNA gyrase (topoisomerase II) but binds at a completely different site, an allosteric pocket located far from the moxifloxacin-binding region. Its antimicrobial effect relies heavily on ATP, reflecting a mechanism divergent from conventional quinolones. The high specificity of evybactin toward Mycobacterium tuberculosis H37Rv is attributed to unique bacterial transport proteins. Cellular uptake is mediated by the BacA transporter, a rare ABC-type "multi-solute transporter" primarily involved in vitamin B12 translocation. Among G^+ bacteria, *Mycobacterium tuberculosis* H37Rv is one of the few species expressing BacA. In contrast, in Enterobacteriaceae such as *Escherichia coli* K-12, although SbmA is highly homologous to BacA, its activity is largely offset by multidrug efflux systems [118]. Evybactin, therefore, exhibits a selective antimicrobial effect specifically against *Mycobacterium tuberculosis*.

5.3.4 Suppression of Protein Synthesis and Breakdown

Proteins represent one of the most crucial biological macromolecules and serve as the primary bearers and executors of life activities. Within normal cells, the processes of protein expression, modification, functionalization, and degradation are tightly regulated to ensure the orderly progression of life activities. Considering the indispensable role that proteins play in life activities, bacteria have evolved a variety of antibiotics that interfere with protein translation, post-modification, and degradation during their evolutionary process.

5.3.4.1 Interrupting Protein Translation

In both eukaryotic and prokaryotic organisms, the synthesis of nascent polypeptides is carried out by ribosomes. The bacterial ribosome consists of 30S and 50S subunits and coordinates all four phases of polypeptide production, including initiation, chain elongation, termination, and ribosome recycling [119]. Initially, the bacterial ribosome assembles into the full 70S complex, comprising the large (50S) and small (30S) subunits, the mRNA start codon, and the initiator tRNA. This assembly can be blocked by antibiotics: aminoglycosides such as kasugamycin and pactamycin act on the 30S subunit, whereas orthomycins including avilamycin and evernimicin target the 50S subunit [119]. During the elongation phase of protein synthesis, four sequential processes occur: delivery of aminoacyl-tRNA (aa-tRNA) facilitated by the elongation factor Tu (EF-Tu)-GTP complex, formation of peptide bonds catalyzed by the ribosome, mRNA translocation mediated by the elongation factor G (EF-G)-GTP complex (translation), and extension of the nascent polypeptide from the 50S subunit. Each of these steps is a potential antibiotic target. Specifically, aa-tRNA delivery is inhibited by pulvomycin, streptomycin, tetracyclines and glycylcyclines; peptide-bond formation is blocked by blasticidin S, clindamycin, puromycin, streptogramin A and sparsomycin; translation can be disrupted by capreomycin, viomycin and aminoglycosides such as hygromycin B, neomycin, and paromomycin; and the elongation of the nascent chain is targeted by erythromycin, streptogramin B, and telithromycin [119]. Termination release factors, RF1 and RF2, identify stop codons and facilitate the cleavage of the bond between the peptide and tRNA, thereby liberating the nascent polypeptide from the ribosome into the cytoplasm—this process is referred to as termination. Subsequently, the translation machinery disassembles, and the 70S ribosome dissociates into its 30S and 50S subunits, a process known as recycling. The termination stage is susceptible to inhibition by peptidyl-transferase inhibitors such as blasticidin S, chloramphenicol, puromycin, and sparsomycin. In contrast, the recycling phase has very few known inhibitors, with fusidic acid being one of the rare examples. Among recently identified AMPs, klebsazolicin and odilorhabdins specifically target bacterial ribosomes, while amythiamicins act as novel EF-Tu inhibitors (Figure 5.7) [120].

Klebsazolicin (KLB) is a thiazole/oxazole-rich linear peptide produced by *Klebsiella pneumoniae* subsp. *ozaenae* and is distinguished by an amidine heterocycle at its N-terminal end (Figure 5.7) [121]. The compound exhibits a narrow antibacterial profile, specifically inhibiting *Escherichia coli*, *Klebsiella pneumoniae*, and *Yersinia pseudotuberculosis*. NOSO-95A-C (Figure 5.7) belongs to the newly identified family of linear decapeptides known as odilorhabdins (ODLs). It is isolated from Xenorhabdus nematophila K102 and demonstrates potent antibacterial efficacy toward G^+ bacteria and G^- ones. Meanwhile, NOSO-95179, a semi-synthetic derivative of NOSO-95A-C, also shows good safety and efficacy. NOSO-95179 reduced the CFU (colony forming unit) value of *Klebsiella pneumoniae* by three orders of magnitude in the mouse sepsis infection model and the lung infection model at doses of 25 and 100 mg/kg, respectively [121].

Klebsazolicin (KLB) acts on the 70S ribosome, impeding peptide elongation in a manner comparable to macrolides. Structural studies showed that KLB binds at the ribosomal exit tunnel, sterically blocking nascent chains beyond three residues and defining its principal interactions with 23S rRNA. The binding site partially overlaps with erythromycin but KLB's larger size enhances the blockade. Unlike erythromycin, KLB does not interfere with

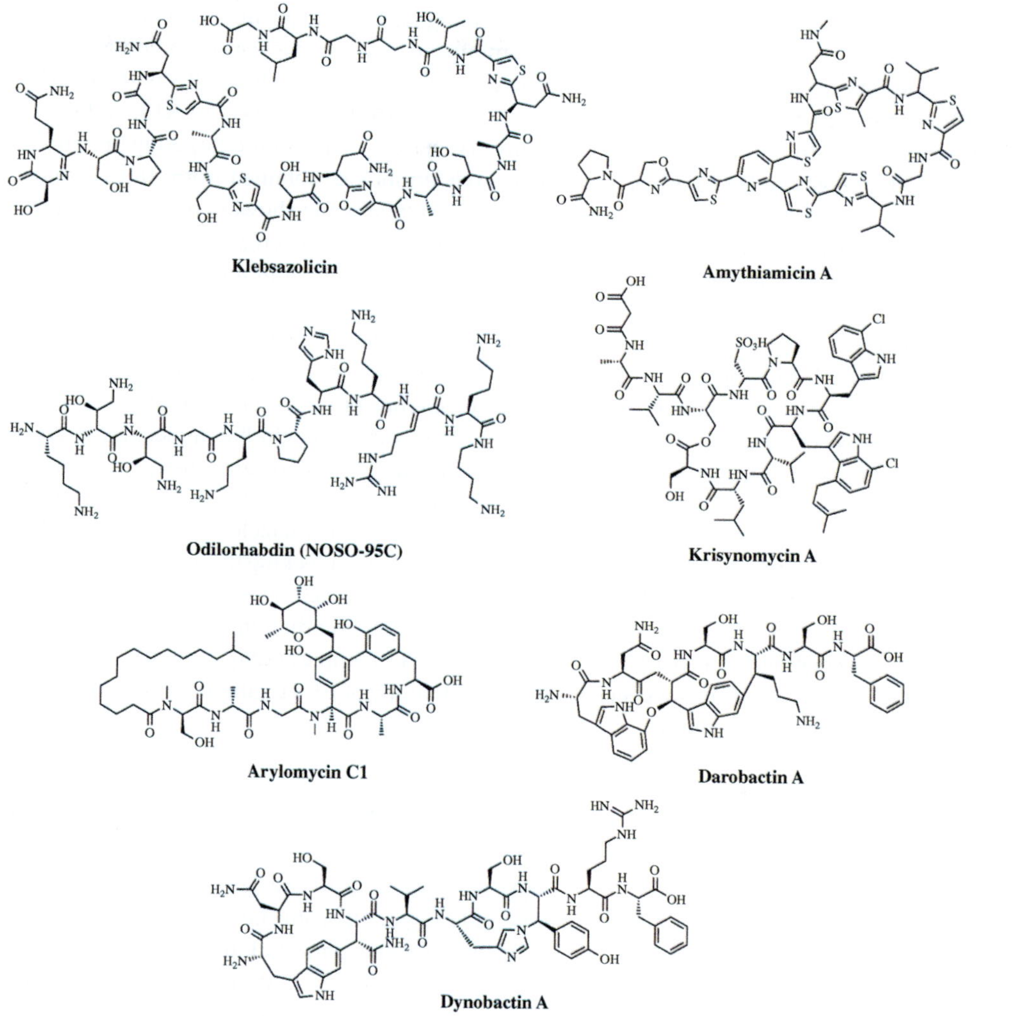

Figure 5.7 Representative structures of AMPs that interrupt protein translation and post-modification. *Source:* [85]/Royal Society of Chemistry.

aminoacyl-tRNAs at the A- and P-sites, leaving the peptidyl transferase center active and permitting minimal extension. An amidine ring further stabilizes interactions with conserved rRNA motifs. Functionally, the N-terminal residues mediate ribosome binding, whereas the C-terminal region is required for cellular translocation without affecting target recognition [121].

In contrast to KLB, which attaches to the 50S subunit, ODLs engage the 70S ribosome near the 30S subunit [120]. Co-crystal structures indicate that ODLs localize close to the anticodon loop of A-site tRNA, where the N-terminal lysine establishes a hydrogen bond with the non-bridging phosphate oxygen at C32. Persistent interaction with tRNA reduces translation fidelity, promoting miscoding. Tight binding may also block the transfer of A-site tRNA to the

P-site, halting elongation. Although the ODL binding region is adjacent to that of negamycin and overlaps spatially with tetracycline, paromomycin, and hygromycin B, its contacts with the ribosome remain distinct [120]. This novel binding mode endows ODLs with superior affinity for their target relative to negamycin. In summary, the discovery of KLB and ODLs diversifies the structural classes and mechanistic repertoire of ribosome-targeting inhibitors.

Amythiamicins A–D (Figure 5.7) are thiopeptide antibiotics obtained from *Amycolatopsis* sp. MI48 1-42F4 [122]. They selectively inhibit G^+ bacteria while showing negligible activity against G^- bacteria or fungi, with amythiamicin A displaying the strongest potency. The MICs for *Staphylococcus aureus*, MRSA, *Micrococcus luteus*, *Bacillus anthracis*, *Bacillus subtilis*, *Bacillus cereus*, and *Clostridium bovis* are 0.1, 0.2, 0.2, 0.78, 0.1, 0.2, 0.1, and 0.78 mg/mL, respectively [123]. These molecules inhibit EF-Tu and are structurally analogous to GE2270A [124]. Although EF-Tu inhibition by amythiamicins was reported previously, their binding interactions remained unclear due to the absence of co-crystal structures. Later, co-crystallization of GE2270A with EF-Tu revealed how amythiamicin-like molecules interact with their target. GE2270A occupies the aa-tRNA site on EF-Tu, stabilizing the nonproductive EF-Tu–GTP complex and preventing aa-tRNA binding. Its binding pocket lies between EF-Tu domains D1 and D2, with the main chain contacting D2 and the thiazole ring interacting with D1, thereby widening the interdomain interface and sterically blocking the 3′ end of aa-tRNA. Pulvomycin, another EF-Tu inhibitor, engages domains D1–D3, overlapping partially with GE2270A only via its terminal glycosyl moiety. This distinct binding mode not only prevents aa-tRNA association but also destabilizes the EF-Tu–GTP complex [125]. Overall, the elucidation of GE2270A's mechanism deepens our understanding of EF-Tu inhibitors and guides the design of enhanced analogues.

5.3.4.2 Disruption of Protein Post-Translational Modifications

Newly synthesized peptide chains emerging from the ribosome usually do not possess full biological function and must undergo post-translational alterations to attain functionality. This process is known as PTM [126]. PTMs generally comprise covalent chemical modifications and the removal of non-functional peptide segments from the nascent chain. These representative modifications include glycosylation, phosphorylation, acetylation, succinylation, pupylation, and ubiquitin-like conjugation in bacteria. Modifications play essential roles in bacterial protein biosynthesis and trafficking, nitrogen metabolism, cell cycle regulation, dormancy, sporulation, germination, and other physiological processes [126]. The latter process, excision of non-functional segments, primarily involves cleaving signal peptides, intronic sequences, and other dispensable portions of the nascent chain that do not contribute to the mature protein's activity. This modification plays a pivotal role in ensuring correct protein maturation and functionality. Although PTMs are fundamental to bacterial physiology, antibiotics directly interfering with these processes remain exceedingly uncommon. For instance, tunicamycin can suppress bacterial N-glycosylation by targeting GlcNAc-1-P-transferase [127]. Tunicamycin not only acts on N-glycosylation but also targets the enzyme MraY, making it uncertain whether blocking N-glycosylation alone can directly account for bacterial cell death. Compared with PTMs driven by chemical modifications, recent discoveries of AMPs, including arylomycin and krisnomycin A, highlight that enzymatic processing of superfluous peptide segments offers more attractive antibacterial targets, with signal peptidase I (SPase I) being a representative example. SPase

I, a membrane-integrated endopeptidase, plays a central role in the bacterial secretion system by removing signal peptide sequences from precursor proteins. Inhibition of this enzyme results in secretory proteins retaining their signal peptides, which subsequently accumulate in the membrane compartment and eventually lead to programmed cell death. As an essential transmembrane protein indispensable for bacterial survival, SPase I has been a longstanding focus of investigation; nevertheless, highly effective inhibitors with clinical potential are still largely lacking [128].

The arylomycins family, comprising subgroups A1–A5, B1–B7, C1–C8, and D (Figure 5.7), was first isolated from *Streptomyces* sp. Tu 6,075, *Streptomyces* sp. PTA-3546, and *Actinoplanes ferrugineus* MA7383, and exhibited wide-ranging activity against G^+ bacteria. To enhance therapeutic potential, a semisynthetic analog designated G0775 was generated, which demonstrated exceptional antibacterial efficacy. In a neutropenic mouse thigh infection model, G0775 achieved robust bactericidal outcomes, reducing bacterial loads by about 2 $\log_{10}$ when administered at 1 mg/kg for *Escherichia coli* and 5 mg/kg for *Klebsiella pneumoniae* [129]. Additionally, M131, a further semisynthetic derivative of arylomycins, has demonstrated synergistic antibacterial activity with imipenem in murine models. In both the MRSA COL deep-thigh infection model and the systemic infection model, neither M131 nor imipenem alone produced a notable effect, whereas co-administration at doses of 10 mg/kg and 20 mg/kg, respectively, resulted in a significant therapeutic benefit [130]. Arylomycins were the first reported inhibitors of SPase I, exhibiting broad-spectrum bactericidal activity, which could be further potentiated by derivatives such as M131 and G0775. Co-crystallization of arylomycins A2 with *Escherichia coli* SPase I (SPase D2–75) revealed that the macrocyclic portion of A2 is fully accommodated within the SPase I substrate-binding pocket, while the aliphatic tail chains extend outward, with the C-terminal carboxylate forming hydrogen bonds with adjacent residues Ser88, Ser90, and Lys145. Compared to the A series, arylomycins C series contain an additional glycosyl moiety. Co-crystallization of BAL4850C, a C-series derivative, with SPase I indicated that its glycosyl side chain projects outside the binding pocket and does not interact with the enzyme, suggesting that C-series compounds retain similar binding and antibacterial properties as the A series, while the sugar group enhances aqueous solubility. Naturally occurring arylomycins mainly target G^+ bacteria and are largely ineffective against G^- bacteria, as the active site of SPase I in G^- bacteria resides in the periplasm, which limits access by natural arylomycins. Furthermore, mutations in SPase I (LepB) in some G^- pathogens reduce arylomycin affinity. However, G0775, optimized from arylomycins, acquires activity against SPase I from G^- bacteria. Co-crystallization studies with SPase LepB demonstrated that G0775's amide macrocycle engages the target similarly to natural arylomycins, with the exception that the C-terminal aminoacetonitrile forms a covalent amidine linkage with the catalytic Lys146, firmly anchoring G0775 in the binding pocket. Additionally, lipophilic modifications of residues outside the macrocycle enhance G0775's capacity to penetrate the outer membrane of G^- bacteria [129]. In conclusion, a series of studies on arylomycins and SPase I have aided in the design of broad-spectrum antibiotics with a novel mechanism.

Krisynomycin A (Figure 5.7), a cyclic decapeptide derived from *Streptomyces fradiae* MA7310, has structural counterparts, Krisynomycins B and C, obtained from *Streptomyces canus* CA 091830. Krisynomycin A demonstrates moderate activity against MRSA, with an MIC of 16–32 mg/mL, whereas Krisynomycins B and C are largely inactive. As a cyclic depsipeptide, Krisynomycin A alone exhibits limited affinity for SPase I, but it can function as a potentiator, restoring MRSA sensitivity to imipenem [131]. It has been proposed that the activity of Krisynomycin A may stem from its interference with signal peptide-mediated secretion of β-lactam hydrolase, thereby compromising MRSA's resistance to imipenem. Nevertheless, β-lactam compounds are known to suppress *Escherichia coli* SPase I in a time- and pH-dependent fashion. Accordingly, co-crystallization of the morpholino-β-sultam derivative (BAL0019193) with arylomycin A2 in a ternary SPase I complex suggests an alternative mechanism. Structural studies showed that BAL0019193 is positioned adjacent to the binding region of arylomycin A2 within the SPase I active site, where it forms non-covalent interactions with residues Asn277, Ala279, Glu307, Lys145, Ser88, and Ser90, including direct hydrogen bonds with the tyrosine side chain of arylomycin A2. This stabilizes the ternary complex more than any binary complex and effectively blocks the SPase I catalytic site. Consequently, the potentiation of imipenem by Krisynomycin A likely arises from the formation of this stabilized ternary complex, further inhibiting SPase I function and inducing bacterial death. Of course, both mechanisms may operate concurrently. In conclusion, utilizing SPase I inhibitors to enhance β-lactam efficacy represents a viable therapeutic approach, though the detailed molecular mechanism underlying this synergy remains to be elucidated.

Beyond SPase I, BamA has recently been recognized as a novel focal point for antimicrobial design. Together with its partners BamB–E, it constitutes the β-barrel assembly machinery (BAM) located in the outer membrane of G^- bacteria, which is essential for the biogenesis of β-barrel outer membrane proteins (OMPs) [132]. Under standard conditions, BamA binds the β-signal at the C-terminal end of nascent OMPs, subsequently driving their folding and insertion into the outer membrane bilayer. These OMPs, exposed at the bacterial cell surface in organisms such as *Escherichia coli*, are indispensable for nutrient exchange, environmental signaling, and host colonization.

Darobactin A (DAR A), a newly discovered bicyclic heptapeptide produced by Photorhabdus khanii HGB1456 (Figure 5.7), demonstrates pronounced bactericidal activity against MDR pathogens, including *Klebsiella pneumoniae*, *Pseudomonas aeruginosa*, *Escherichia coli*, and *Salmonella typhimurium*, with MIC values reported at 2–4 mg/mL. In mouse infection models, DAR A was able to clear systemic infections caused by polymyxin-resistant *P. aeruginosa*, carbapenemase-producing *Klebsiella pneumoniae*, and polymyxin-resistant *Escherichia coli*. Effective treatment was observed with a single 50 mg/kg administration or three sequential doses of 25 mg/kg at 6-hours intervals, with no evident toxicity [133].

Dynobactin A (DYN A), obtained by genome mining of biosynthetic gene clusters in Photorhabdus strains (Figure 5.7), likewise displays potent G^- antibacterial efficacy together with favorable safety *in vivo*. Both DAR A and DYN A act on BamA, and darobactin was the first naturally occurring compound shown to inhibit this protein. Co-crystallographic analysis revealed that DAR A attaches to the lateral gate region of BamA, where its backbone interacts via hydrogen bonding with N422, G424, F426, N427, F428, and K808 on strand β1, as well as the amino groups of I430, G807, and L780. This occupancy blocks the recognition site for OMP β-signals and disables BamA activity [134].

Although the binding positions of DAR A and DYN A largely overlap, DYN A engages in additional nonpolar interactions. Besides hydrogen bonds with G424, F426, N427, F428, and I430 on β1, its tryptophan residues form hydrophobic contacts with K808, P782, and

M777 on β16. These interactions strengthen the binding, giving DYN A nearly tenfold greater affinity for BamA compared with DAR A.

5.3.4.3 Dysregulation of Protein Degradation

The protein degradation machinery is indispensable for both prokaryotic and eukaryotic cells. It removes misfolded, aggregated, inactive, or excess proteins to sustain cellular or organismal metabolic balance. Therefore, efficient operation of the proteolytic network is crucial for protein quality surveillance and homeostasis. In bacteria, apart from lysosomes, the AAA+ ATPase–protease assembly (AAA+: ATPases associated with diverse cellular activities) serves as another central proteolytic pathway. This complex generally consists of two serine proteases, ClpP and ClpQ, together with an AAA+ ATPase. Members of the AAA+ protein family fall into two categories: those with a single ATP-binding site, such as ClpM, ClpN, ClpX, and ClpY, and those with two ATP-binding motifs, including ClpA, ClpB, ClpC, ClpD, ClpE, and ClpL [135]. Among them, ClpY has only been reported to form the ClpYQ complex with ClpQ, which predominantly occurs in G^- bacteria like *Escherichia coli*. ClpP, however, interacts with multiple AAA+ ATPases, such as ClpA, ClpC, ClpE, and ClpX, with the ClpCP system being typical in G^+ bacteria, including *Staphylococcus aureus* and *Mycobacterium tuberculosis*. Thus far, all microbial antimicrobial peptides (MAPs) identified as acting on the AAA+ ATPase–protease complex target the ClpC1/ClpP1/ClpP2 assembly. Acyldepsipeptides (ADEPs), discovered decades ago, bind directly to ClpP, whereas another cyclic depsipeptide, cyclomarin A, attaches to the N-terminal domain (NTD) of ClpC1. In the past two decades, most newly characterized MAPs targeting this complex have been specific for ClpC1, including ilamycins/rufomycins, metamarin, lassomycin, and ecumicin, while SyCPA 116 is the only exception, sharing the same target as ADEPs.

Ilamycins and rufomycins are cyclic heptapeptides produced by Streptomyces (Figure 5.8). Metamarin, a structural analog of rufomycins, was obtained from soil metagenomic sequencing, whereas SyCPA 116 was identified through the biosynthetic gene cluster of Collimonas fungivorans using a similar genomics-based strategy (Figure 5.8) [136]. Lassomycin is a ribosomally synthesized cyclic polypeptide derived from the soil actinomycete *Lentzea kentuckyensis* IS009804 (Figure 5.8) [137], and ecumicin is a cyclic tridecapeptide isolated from Nonomuraea sp. MJM5123 (Figure 5.8). These compounds display strong and selective activity against *Mycobacterium tuberculosis*, although some rufomycins also affect other G^+ bacteria.

Functionally, Ilamycins/rufomycins, lassomycin, and ecumicin inhibit the proteolytic activity of the ClpC1/ClpP1/ClpP2 complex by binding to ClpC1. Docking studies indicate that lassomycin targets an acidic region of ClpC1-NTD, with Gln17 playing a pivotal role. Co-crystallization experiments show that ecumicin binds ClpC1-NTD as a pseudo-dimer in a 2:1 ratio, distinct from cyclomarin A, which interacts with ClpC1-NTD in a 1:1 ratio and ClpC2-NTD in a 1:2 ratio. By contrast, rufomycin I occupies a unique site on ClpC1-NTD, separate from cyclomarin A and ecumicin, and its epoxy side chain opens to form a covalent bond with Met1, representing a rare covalent mechanism.

These differences in binding modes result in varied effects on the ClpC1/ClpP1/ClpP2 complex. Cyclomarin A enhances ClpC1 activity, stimulating ClpP hydrolysis and causing uncontrolled protein degradation. Although ecumicin and lassomycin also activate ClpC1,

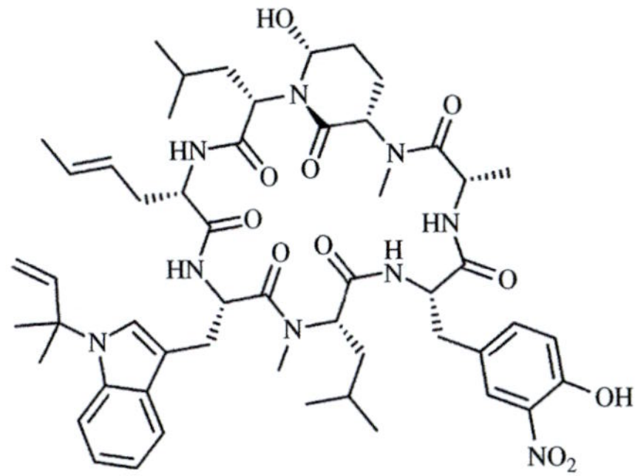

Ilamycins E1

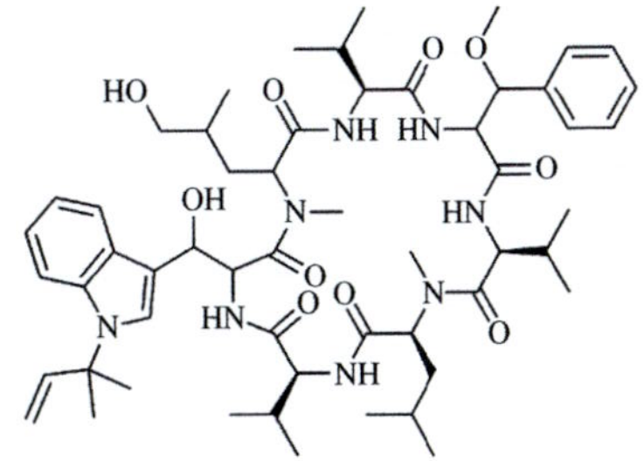

Metamarin

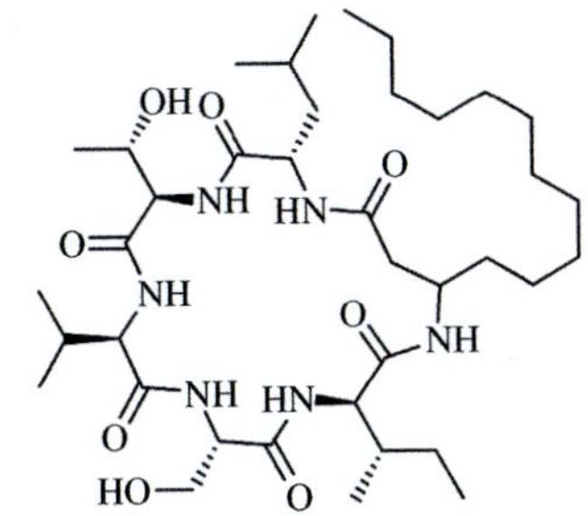

SyCPA 116 (collimosyn)

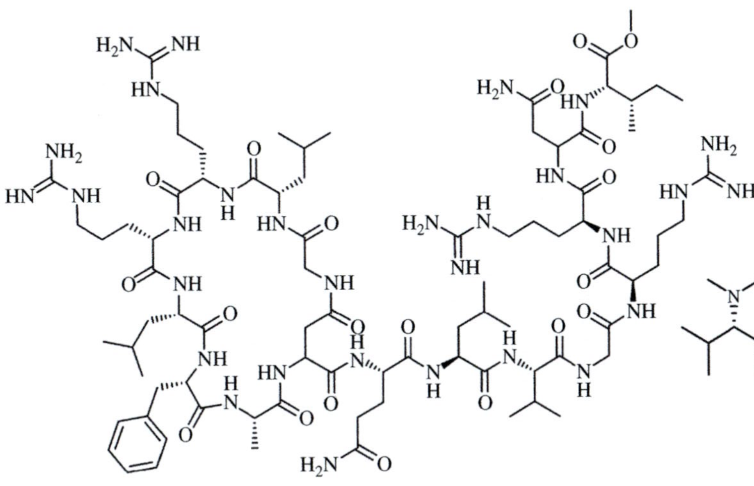

Lassomycin

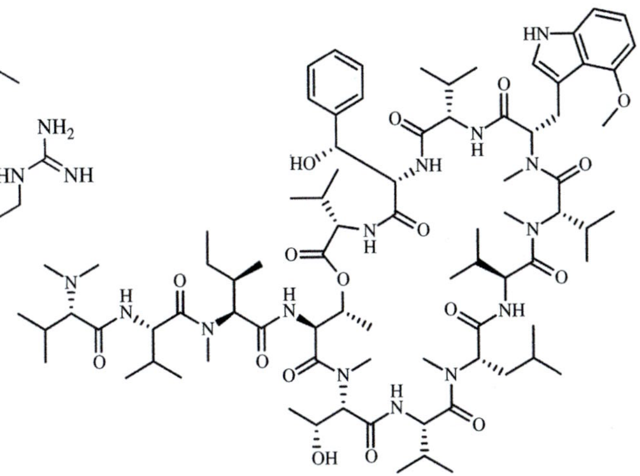

Ecumicin

Figure 5.8 Representative structures of AMPs that disrupt protein degradation. *Source:* [85]/Royal Society of Chemistry.

they ultimately inhibit protein hydrolysis, likely by obstructing substrate entry into ClpP via ClpC1. Rufomycin I, however, suppresses ClpC1 ATPase activity, thereby inactivating the ATP-dependent ClpC1/ClpP1/ClpP2 complex. Metamarin, as a cyclomarin A analog, exhibits comparable effects on the complex, although structural confirmation is lacking due to absent co-crystallization data. SyCPA 116 represents a rare case of direct ClpP inhibition, but whether its mechanism differs from ADEPs remains to be elucidated.

5.3.5 Destabilization of Cell Membranes

The semi-permeable cytoplasmic membrane, located immediately beneath the cell wall, plays a critical role in controlling the transport of molecules, which is essential for bacterial viability. Marked differences exist in both the structure and lipid composition of membranes between G^+ and G^- bacteria. In G^- bacteria, the membrane system comprises an inner membrane and an outer membrane, with LPS, a G^--specific lipid, forming a predominant constituent of the outer membrane. G^+ bacteria, in contrast, feature a thick, rigid cell wall but lack an outer membrane. From a compositional standpoint, the cytoplasmic membrane of G^+ bacteria resembles the inner membrane of G^- bacteria, containing phosphatidylglycerol, phosphatidylethanolamine, cardiolipin and menaquinone (vitamin K2), although the relative proportions vary among species [138]. Membrane-targeting antibiotics interact with these lipids, causing abnormal clustering and pore formation, which either compromise membrane integrity or induce depolarization, ultimately impairing membrane function and leading to rapid bacterial death.

Recent investigations have revealed that daptomycin specifically interacts with phosphatidylglycerol in bacterial membranes, rapidly concentrating on the membrane surface [139]. This localized accumulation exerts a directional disruptive force, causing lipid depletion and the creation of circular pores, ultimately resulting in bacterial death. Polymyxin, via its positively charged residues, establishes electrostatic interactions with the negatively charged LPS in the membrane, allowing its fatty acyl chains to insert into the lipid bilayer. As polymyxin accumulates, it triggers widespread membrane disruption, leading to bacterial lethality. Although lipid A enhances polymyxin binding, it is not strictly required for membrane insertion. Additional studies have shown that polymyxin promotes the assembly of hexagonal LPS crystal structures, which decrease membrane thickness and increase rigidity. These alterations compromise the mechanical stability of the membrane, making it prone to rupture under external stress and resulting in cell death. This work further clarifies how polymyxin mediates membrane damage upon binding. Furthermore, the identification of novel MAPs has broadened our perspective on bacterial membranes as antimicrobial targets. Examples include lysocin E and WBP-29479A1, along with structurally diverse agents with established mechanisms, such as brevicidine, laterocidine, and SyCPAs 12, 102, and 123. Collectively, these findings provide valuable guidance for designing more effective and safer antibiotics [85].

Brevicidine, laterocidine, and SyCPAs 12, 102, and 123 were discovered via genomic mining of BGCs from *Brevibacillus laterosporus* DSM 25, *Dickeya dadantii*, *Rhodococcus opacus* and *Rhodococcus jostii* (Figure 5.9). Brevicidine and laterocidine display comparable broad-spectrum activity against G^- bacteria. In a neutropenic mouse thigh infection model with *Escherichia coli*, these two compounds showed notable antibacterial effects at doses of 30

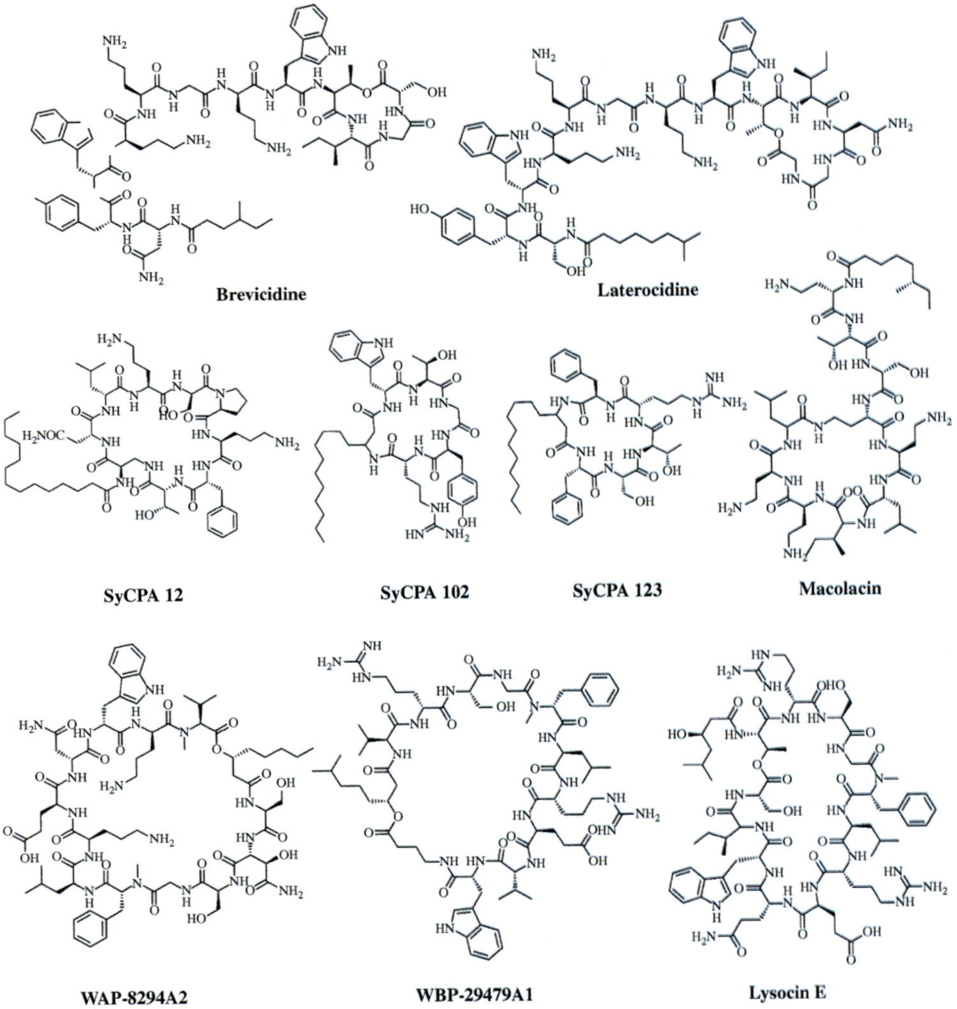

Figure 5.9 Representative architectures of AMPs acting through membrane disruption. *Source:* [85]/Royal Society of Chemistry.

and 15 mg/kg, respectively [140]. In contrast, all SyCPAs demonstrate inhibitory effects against G^+ bacteria and certain G^- strains.

Structurally, brevicidine and laterocidine are cationic NRPs with high similarity, exhibiting nearly identical antimicrobial profiles and membrane-targeting abilities. At elevated concentrations, brevicidine non-specifically enhances membrane permeability. At lower doses, it penetrates the bacterial inner membrane through interactions with LPS, targeting phosphatidylglycerol and cardiolipin, thereby disturbing the proton motive force. Such disruption leads to bacterial metabolic dysfunctions, including reduced ATP synthesis, impaired NADH dehydrogenation, and suppressed protein production, accompanied by accumulation of intracellular reactive oxygen species (ROS). Furthermore, combining brevicidine with erythromycin intensifies ATP inhibition and ROS buildup, improving the bactericidal effect against *Acinetobacter baumannii*. SyCPAs 12, 102, and 123 are cationic cyclic peptides capable of disrupting bacterial membranes. Notably, unlike most cationic peptides that primarily inhibit G^- bacteria but are less effective against G^+ bacteria, these SyCPAs show stronger activity against G^+ strains [141]. This implies that their antimicrobial mechanisms may involve more than membrane disruption, potentially including other pathways that have yet to be identified, which warrants further experimental investigation.

Macolacin (Figure 5.9) was discovered from the BGC of *Paenibacillus xylanexedens*, and it exhibits potent antibacterial activity toward G^+ bacteria, particularly high efficacy against clinical isolates resistant to colistin. In addition, its derivative, biphenyl-macolacin, exhibited remarkable antibacterial potency against highly colistin-resistant *Acinetobacter baumannii* (strain SM1536-mcr-1) and extensively drug-resistant *A. baumannii* (strain 0301) in a neutropenic mouse thigh infection model at 20 mg/kg, outperforming colistin by nearly three orders of magnitude at the same dose. Mechanistic studies indicate that macolacin shares the binding site with colistin, exerting bactericidal activity by interacting with lipid A and disrupting membrane integrity. The phosphoethanolamine (PEtN) transferase encoded by the mcr-1 resistance gene modifies lipid A through the addition of PEtN, thereby decreasing the affinity of colistin. Nevertheless, owing to its distinct structural characteristics, macolacin preserves its capacity to bind modified lipid A [142]. The exact molecular basis enabling macolacin to maintain this binding capacity has yet to be elucidated.

WAP-8294As constitute a group of cyclic lipopeptides discovered from Lysobacter enzymogenes WAP-8294, encompassing six analogues. Among them, WAP-8294A2 (Figure 5.9), also named lotilibcin, exhibits broad antimicrobial potency against G^+ pathogens, with particularly strong inhibition of methicillin-resistant *Staphylococcus aureus* (MRSA, MIC, 2 mg/mL). The compound also shows remarkable *in vivo* performance, as evidenced in a murine systemic infection model where its ED50 was determined to be 0.38 mg/kg, nearly an order of magnitude more effective than vancomycin. Other related members, WBP-29479A1 and lysocin E, were purified from *Lysobacter antibioticus* ATCC 29479 and *Lysobacter enzymogenes* RH2180-5, respectively (Figure 5.9). Synthetic-bioinformatic AMPs (syn-BAMPs), collectively termed MBAs, were generated through genome mining combined with total synthesis (Figure 5.10), and all display pronounced activity against G^+ bacteria. Among them, lysocin E provided strong therapeutic benefit in an MRSA murine systemic infection model, achieving an ED50 of 0.5 mg/kg, while intraperitoneal injection at 400 mg/kg produced no observable mortality. Furthermore, MBA3 and MBA6 successfully cured MRSA-induced murine peritonitis at minimal doses of 10 and 30 mg/kg, respectively [143].

From a mechanistic perspective, this group of MAPs binds to menaquinone in a Ca^{2+}-dependent fashion. Menaquinone, a vital component of the cytoplasmic membrane in G^+, G^-, cyanobacterial, and mycoplasmal species, participates both in electron transport and active translocation processes. To date, no other AMPs have been shown to target menaquinone within bacterial membranes. All known menaquinone-binding MAPs have confirmed their reliance on menaquinone for antimicrobial action through rigorous biochemical assays. Specifically, for lysocin E and MBAs, experimental evidence indicates that menaquinone binding leads to depolarization of the cell membrane, ultimately resulting in cell lysis.

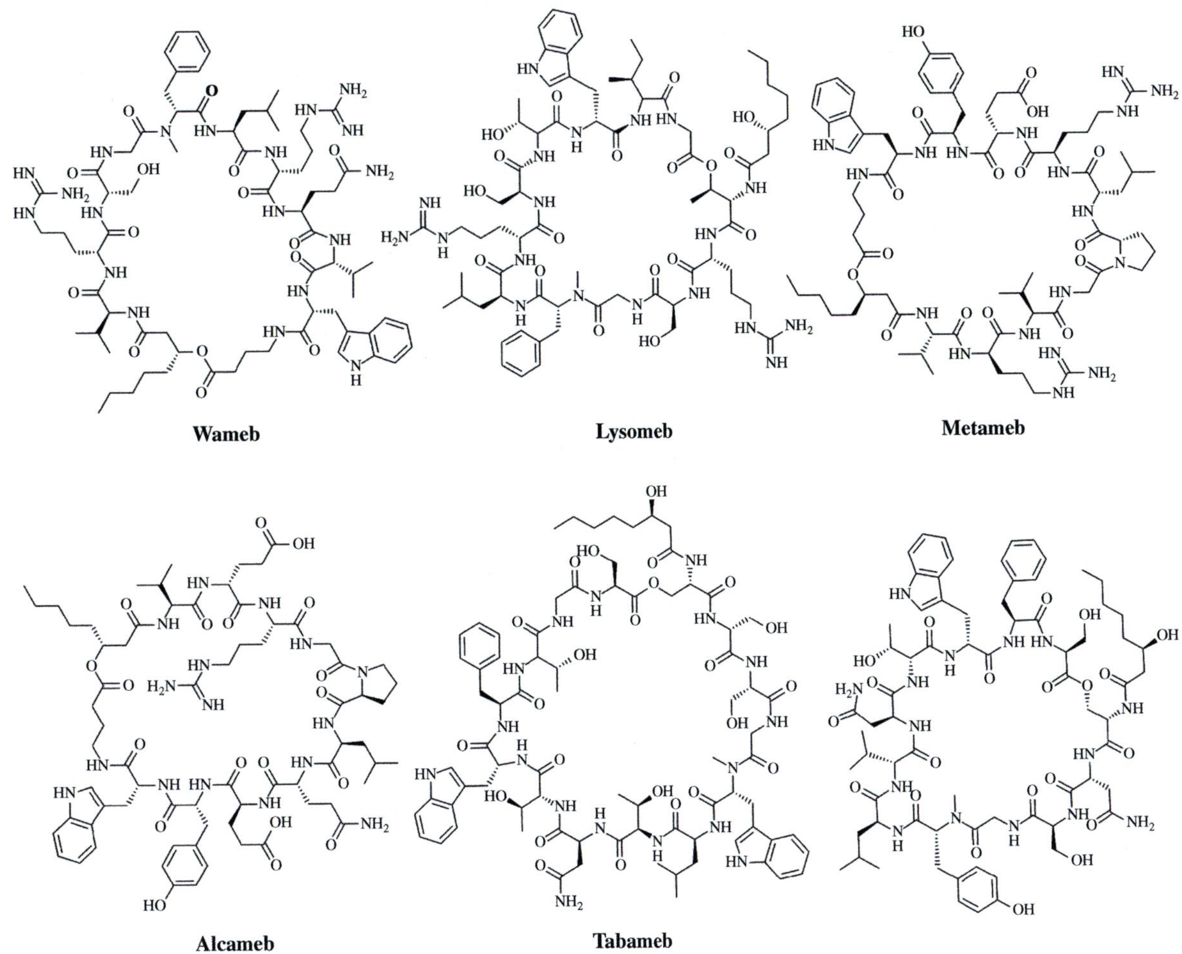

Figure 5.10 Structures of AMPs. *Source:* [85]/Royal Society of Chemistry.

Nevertheless, the precise cascade by which menaquinone engagement provokes membrane depolarization and disintegration remains unresolved. Interestingly, additional findings suggest that lysocin E is also capable of binding lipid II, thereby suppressing cell wall biosynthesis, which implies that lysocin E may function as a multi-target antimicrobial agent. In conclusion, although menaquinone-binding MAPs represent a new class of membrane-acting antibiotics, many aspects of their molecular mechanism remain to be fully elucidated.

5.4 Security Assessment

AMPs represent a versatile class of molecules with extensive antimicrobial spectra and have garnered considerable interest for their strong activity against MDR pathogens. Nevertheless, issues related to biosafety and cytotoxicity remain major challenges, making safety evaluation a critical aspect in their research, development, and clinical translation.

5.4.1 Cytotoxicity of AMPs

The cytotoxicity of AMPs is typically evaluated through *in vitro* experiments. It reflects the ability of AMPs to damage cells at specific concentrations, including changes in cell membrane integrity, cellular metabolic activity, and cell morphology. Usually, the hemolytic activity of AMPs is tested using their effect on normal human red blood cells. The minimum dosage of the antimicrobial peptide that does not cause hemolysis, known as the minimum hemolytic concentration, is used to quantify and compare the hemolytic activity of AMPs. The toxicity of AMPs to normal cells is assessed using their cytotoxicity to human embryonic stem cells, and the concentration of the antimicrobial peptide that causes 50% cell death, known as the half-maximal lethal concentration, is used to quantify and compare the cytotoxicity of AMPs. The toxicity trend of AMP analogs to embryonic stem cells is consistent with their hemolytic activity. Therefore, their hemolytic activity is generally used to measure the cytotoxicity of AMPs to normal human cells, and the results are expressed as the minimum hemolytic concentration.

5.4.2 *In Vivo* Toxicity of AMPs

Assessment of the biosafety of AMPs requires *in vivo* toxicity studies, which serve as an indispensable step in determining their overall safety profile. Through experiments in animal models, it is possible to thoroughly assess the potential hazards posed by AMPs in living organisms. Typically, the method of determining the median lethal dose (LD50) is used to assess the acute toxicity of AMPs *in vivo*, providing an initial indication of their toxicity to experimental animals. Furthermore, chronic toxicity studies are conducted to evaluate the effects of repeated dosing on animal organ function and tissue structure, as well as to identify potential chronic toxic reactions. In studies of *in vivo* toxicity, the pharmacokinetic behavior of AMPs is closely examined, encompassing their uptake, tissue distribution, metabolic processing, and clearance within the organism. This information is critical for understanding the pharmacokinetic properties of AMPs and predicting their behavior in humans. Additionally, histopathological analysis through observation of tissue slices from the major organs of experimental animals can help detect whether AMPs cause cell damage, inflammation, or other pathological changes.

Overall, *in vivo* toxicity testing provides essential safety data for the clinical application of AMPs, ensuring that their potential harm to the host is minimized while maintaining their effective antimicrobial properties.

5.5 Application

5.5.1 Food Preservation

Microbial contamination remains a primary contributor to food spoilage and foodborne illnesses, necessitating the use of preservatives to inhibit microbial proliferation. However, the extensive application of synthetic preservatives has contributed to the development of bacterial strains resistant to conventional antimicrobial agents, diminishing their efficacy. In response, consumer demand has shifted toward minimally processed, naturally derived food products [144]. This trend underscores the importance of developing novel, safe, and effective antimicrobial alternatives that can preserve food quality without altering its physicochemical properties. Among these alternatives, AMPs exhibit significant potential as food preservatives.

AMPs offer a viable substitute for traditional preservatives, capable of suppressing spoilage and pathogenic microorganisms either through direct addition to food products or via incorporation into protective packaging layers. For instance, Gao et al. employed Sephadex column chromatography, pre-HPLC, and LC–MS/MS to isolate and characterize three AMPs from Laba garlic, designated as F3–3-a, F3–3-b, and F3–3-c. Notably, F3–3-c demonstrated potent antimicrobial activity, effectively inhibiting *Escherichia coli* and *Staphylococcus aureus* growth by disrupting cytomembrane integrity at an MIC of 100 μmol/L. These findings suggest that F3–3-c holds promise as an antibiotic alternative or functional food ingredient [145].

Additionally, polylysine, which exists in α- and ε-forms (α-PL and ε-PL), has been investigated for its antimicrobial potential. Park et al. evaluated a composite antimicrobial formulation comprising ε-PL, rosemary extract, camellia extract, and citric acid, demonstrating enhanced inhibitory effects against multiple foodborne pathogens while maintaining environmental sustainability [146].

Future applications of AMPs may extend to dairy, meat, canned vegetables, fruit juices, and alcoholic beverages, where they could serve as natural preservatives to enhance shelf life and medicinal value. Although AMPs are predominantly utilized in pharmaceuticals and agriculture, their role in food preservation warrants further exploration, given their natural origin and broad-spectrum antimicrobial properties. A schematic representation of the multifunctional applications of AMPs is illustrated in Figure 5.11.

5.5.2 Bioactive Food Ingredients

AMPs exhibit notable health-promoting effects through their immunoregulatory and inflammation-modulating properties, rendering them important bioactive ingredients for health-oriented food products and dietary supplements. Beyond their primary

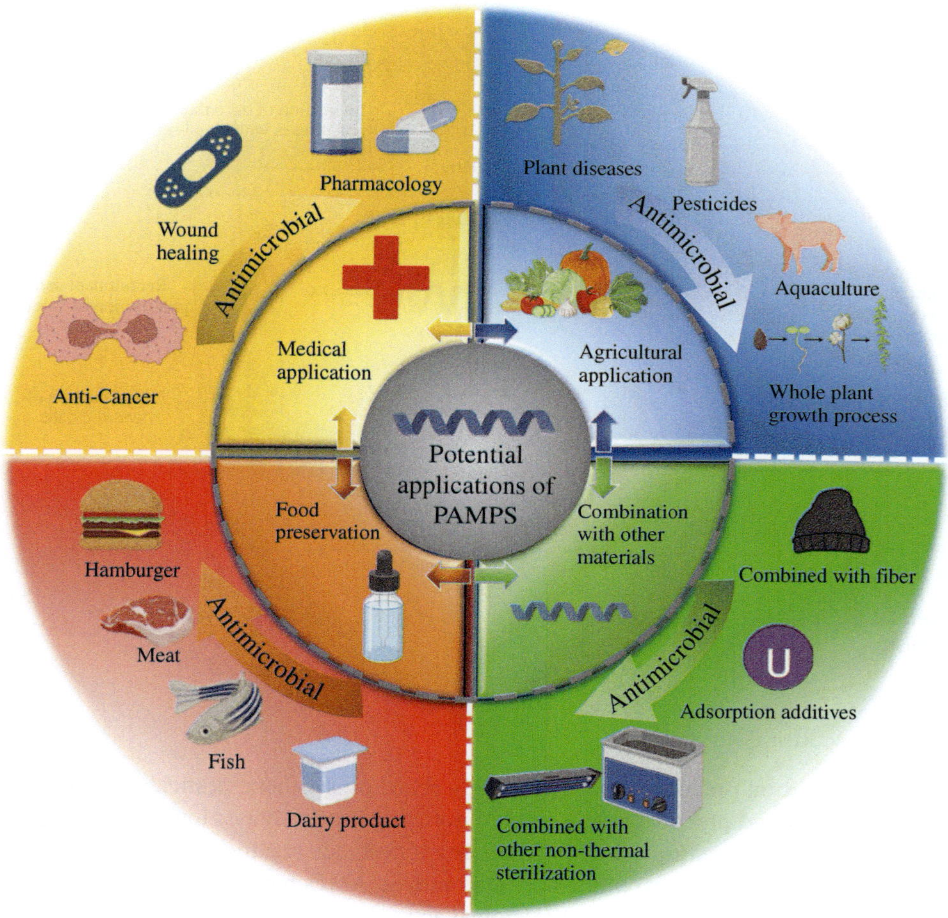

Figure 5.11 Integrated application of AMPs in different fields. *Source:* [6]/Royal Society of Chemistry.

antimicrobial activity, these peptides impact the body systemically through the regulation of cellular signaling pathways, thus influencing host defense mechanisms against various diseases. Their immunomodulatory capacity encompasses multiple mechanisms: activation and differentiation of immune cells, suppression of pro-inflammatory cytokine production, stimulation of anti-inflammatory mediators, promotion of wound healing with reduced scarring, and facilitation of angiogenesis. These multifunctional attributes stem from their ability to simultaneously target diverse biological pathways through synergistic interactions, highlighting their potential for therapeutic development (Figure 5.12) [147]. The physiological regulatory function of anti-inflammatory peptides is mediated through the precise modulation of cytokine networks and inflammatory mediators. Given the established association between chronic mild inflammation and the development of metabolic disorders such as cardiovascular disease and type 2 diabetes, AMPs present a safer alternative to conventional anti-inflammatory pharmaceuticals owing to their natural origin and favorable safety profile.

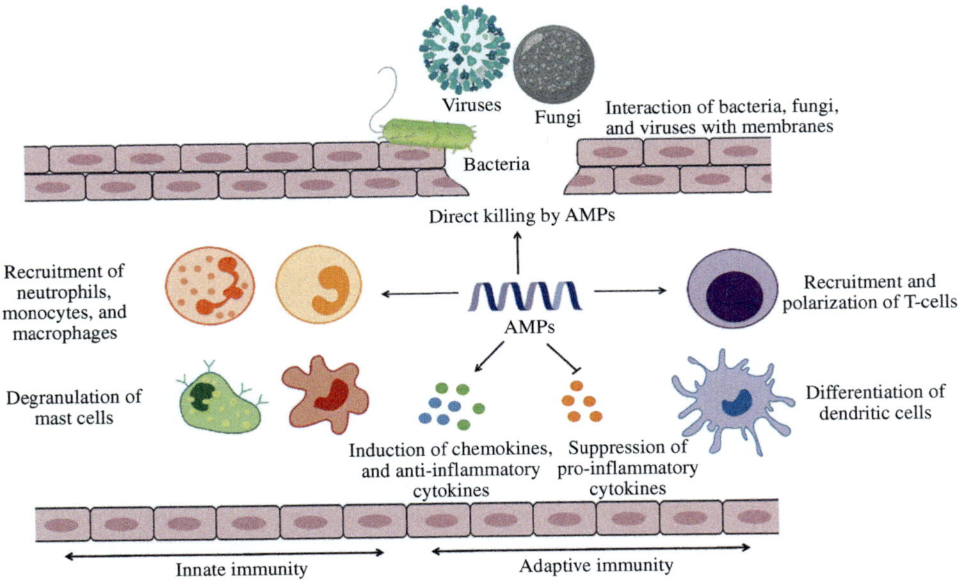

Figure 5.12 The Diverse Immunoregulatory Activities of AMPs. *Source:* [6]/Royal Society of Chemistry.

5.5.3 Agricultural Applications

Modern agricultural practices have increasingly relied on synthetic pesticides to manage pest infestations and phytopathogens while improving productivity. However, mounting evidence reveals the hazardous effects of persistent pesticide residues on ecosystems and human health, driving the search for safer alternatives. AMPs have emerged as promising biopesticides due to their targeted antimicrobial activity and environmental compatibility. These bioactive compounds can be administered through foliar application or seed coating to prevent microbial infections, offering a sustainable approach to crop protection. Research has elucidated the dual functionality of AMPs in plant systems, serving not only as defense molecules but also as regulators of plant physiology throughout developmental stages—from germination and root establishment to flowering and seed maturation [148]. Experimental evidence demonstrates their practical applications in disease management; for instance, citrus-AMP effectively controlled *Xanthomonas citri* with minimal phytotoxicity, while novel peptide formulations outperformed conventional fungicides against *Ramularia areola* in cotton. Transgenic approaches have further enhanced crop resistance, as demonstrated by *Arabidopsis thaliana* expressing barely LTP genes showing improved defense against multiple pathogens. Notably, synergistic effects occur when multiple defense genes are co-expressed, with rice plants containing both Rs-AFP2 and Dm-AMP1 defensins exhibiting superior antifungal activity compared to single-gene transformations. Beyond pathogen control, AMPs contribute to agronomic improvement by stimulating plant growth parameters, suggesting their potential as multifunctional agents that simultaneously enhance crop yield and ensure sustainable food production.

5.5.4 Animal Feed Additives

AMPs are increasingly being incorporated into animal feed formulations as sustainable substitutes for conventional antibiotics, serving the dual purposes of growth promotion and disease prevention in livestock production systems. These bioactive compounds enhance digestive efficiency and gastrointestinal health by modulating gut microbiota composition while demonstrating multiple beneficial effects. Their growth-promoting properties primarily stem from two mechanisms: direct antimicrobial action against enteric pathogens and immunoregulatory functions that improve nutrient assimilation. Compared to synthetic antibiotics, AMPs offer distinct advantages, including minimal toxicity, negligible residual accumulation, and reduced risk of inducing microbial resistance in farm animals [149]. Experimental evidence from Tang et al. [150] demonstrated significant improvements in weaned piglets supplemented with AMPs, showing a 21% increase in average daily weight gain compared to controls. The treatment group exhibited enhanced intestinal morphology (elongated villi structure), improved immune parameters (elevated immunoglobulin levels), and superior antioxidant capacity. Microbiological analysis revealed favorable shifts in gut microbial populations, characterized by increased *Lactobacillus* and Bifidobacterium counts alongside decreased *Escherichia coli* colonization, correlating with reduced incidence of diarrheal episodes. These findings collectively underscore the promise of AMPs in serving as versatile dietary supplements that simultaneously enhance animal performance and welfare.

5.5.5 Medical Applications

5.5.5.1 Antimicrobial Agent

Antimicrobial agents significantly contributed to modern medicine by effectively suppressing life-threatening microbial infections. Nevertheless, the rise of drug-resistant microbial strains has created substantial challenges for antibiotic development, necessitating urgent exploration of novel therapeutic alternatives. AMPs represent a promising solution because they act through a distinct mode of action, which operates independently of traditional antibiotic pathways. In contrast to conventional drugs directed at particular bacterial structures, AMPs exhibit broad-spectrum activity through nonspecific interactions with microbial membranes, making them particularly effective against drug-resistant strains. AMPs have exhibited significant antibacterial activity against G^+ and G^- bacteria via multiple mechanisms, including membrane permeabilization and disruption of vital intracellular functions, such as DNA/RNA synthesis, protein production, and cell wall assembly [151]. Beyond their antimicrobial properties, certain AMPs exhibit selective cytotoxicity against malignant cells while sparing healthy tissues, positioning them as potential anticancer therapeutics. For instance, cycloviolacin O_2 (CyO2), a cyclotide obtained from *Viola odorata*, induces necrosis in lymphoma cells through membrane disruption [152]. The clinical potential of AMPs is further evidenced by ongoing trials investigating their applications in diverse medical conditions, including systemic infections (sepsis), inflammatory disorders (dermatitis, periodontitis), and localized infections (otitis, dental caries), as summarized in Table 5.5. These developments highlight the versatility of AMPs as next-generation therapeutic agents capable of addressing multiple clinical challenges.

Table 5.5 AMPs in commercialization.

AMPs	Stage of development	Application
TAPS-18	Early phase I	Periodontitis
Liver-enriched antimicrobial peptide 2	Early phase I	Type 2 diabetes and obesity
DPK-060	Phase I	Atopic dermatitis
	Phase II	Acute otitis externa
C16G2	Phase II	Dental caries
LL-37	Phase II	Melanoma
Omiganan	Phase II	Seborrheic dermatitis, atopic dermatitis, acne vulgaris, condylomata acuminata, and papulopustular rosacea
PXL01	Phase II	Surgical adhesions
MBI 226	Phase III	Bacteremia, sepsis, and fungemia

5.5.5.2 Wound Healing

The healing of cutaneous wounds in humans and animals represents a critical biological process, where delayed or impaired tissue regeneration can lead to functional impairment. Contemporary wound care strategies emphasize the importance of advanced dressings that facilitate tissue repair while minimizing complications associated with conventional materials such as gauze, which may impede recovery through mechanical trauma during frequent dressing changes. A fundamental requirement for effective wound management involves preventing microbial colonization, making AMPs particularly valuable due to their combined microbiocidal properties and biocompatibility. These attributes position AMPs as innovative therapeutic agents capable of simultaneously reducing infection risks and accelerating tissue regeneration processes.

Among clinically relevant AMPs, the cathelicidin-derived peptide LL-37 has been recognized as a highly promising candidate. This 37-residue polypeptide, generated through proteolytic processing of hCAP18 precursor protein, demonstrates distinct spatial localization patterns—showing significant accumulation at acute wound margins while being conspicuously absent in chronic wounds. Subsequent *in vitro* and *in vivo* investigations have substantiated LL-37's therapeutic potential in promoting epithelialization and granulation tissue formation [153]. Beyond their direct antimicrobial effects, AMPs offer strategic advantages in addressing the growing challenge of AMR. Particularly problematic are biofilm-associated infections, where microbial communities exhibit dramatically enhanced (up to 1,000-fold) resistance compared to their planktonic counterparts [154]. This has spurred the development of synthetic anti-biofilm peptides (SAAPs), exemplified by SAAP-148, which has demonstrated both safety and efficacy against MDR pathogens in preclinical wound infection models, including established biofilms.

5.5.5.3 Other Applications

Beyond their biomedical applications, AMPs demonstrate significant potential in environmental remediation and resource recovery. A notable application involves their use as specialized adsorbents for uranium extraction from marine environments. Li and colleagues [155] developed an innovative composite nanofiber system combining poly (ethylene glycol) methyl ether-b-poly(l-arginine) (PPLA) with amidoxime-functionalized polyacrylonitrile (AOP) for efficient seawater uranium recovery. In textile engineering, AMP integration addresses critical limitations of conventional polyamide (PA) fabrics, which typically exhibit poor hydrophilicity and lack antimicrobial properties. Zhang et al. [156] demonstrated that polyline modification can simultaneously enhance the wettability and microbial resistance of PA-based textiles, including fibers, yarns, and knitted fabrics. These emerging applications highlight the versatility of AMPs as functional biomaterials. Ongoing research into peptide-based antimicrobial materials promises to unlock novel applications that combine the intrinsic antimicrobial properties of AMPs with advanced material science, potentially leading to transformative technological developments with broad societal benefits.

5.6 Limitations and Challenges

Over the past several decades, researchers have isolated and characterized hundreds of AMPs, explored their antimicrobial mechanisms, and clarified certain processes responsible for their pathogen-inhibiting effects. Despite these advances, significant gaps remain in understanding how AMPs can be developed as potent microbicidal agents for clinical, agricultural, food industry, and other commercial purposes (Figure 5.13). (i) Mechanistic insights: The precise molecular mechanisms remain largely unresolved. While AMPs are known to interact with pathogen membranes and cell wall components,

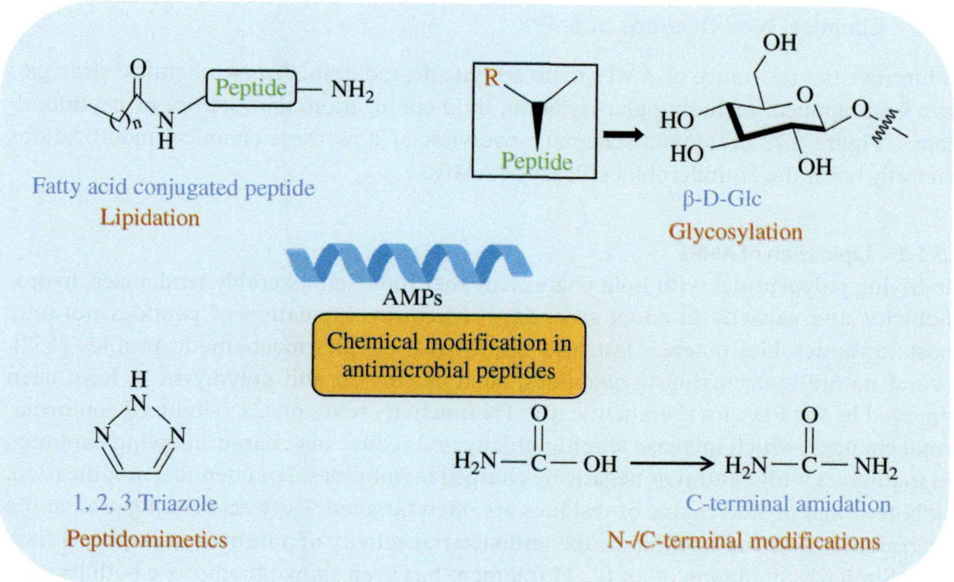

Figure 5.13 General mechanism of different chemical modifications to improve the efficacy and antimicrobial activity of AMPs. *Source:* [6]/Royal Society of Chemistry.

causing membrane disruption and leakage of intracellular contents, the detailed pathways governing these effects are still unclear. Questions remain regarding how AMPs penetrate cells, modulate intracellular signaling, and identify specific targets. This lack of mechanistic knowledge has hindered a deeper understanding of PAMP function. (ii) Synergistic interactions: Certain AMPs are upregulated in plants or animals following pathogen challenge, collectively contributing to immune defense. Optimizing these synergistic effects is critical for maximizing their practical utility. (iii) Ex vivo applications: Applying AMPs outside of living organisms encounters challenges, including rapid degradation, limited stability, and stringent conditions required for maintaining bioactivity. (iv) Stability and safety concerns: Key issues restrict therapeutic use, including poor oral bioavailability, susceptibility to proteolytic enzymes (e.g., trypsin and pepsin), instability under gastric pH, and potential cytotoxicity in mammalian cells, which may limit their broader application. (v) Cost limitations: Low natural abundance, difficulty in purification, and high costs of chemical synthesis impede widespread implementation. (vi) Pathogen specificity: AMPs exhibit variable activity against different microbes, restricting their use against particular bacterial or fungal strains. (vii) Delivery challenges: Effective application may require tailored delivery systems or carriers, particularly in contexts such as food preservation. (viii) Regulatory hurdles: Approval for AMP use in food or other sectors often involves complex, time-consuming, and expensive regulatory procedures. (xi) Resistance potential: As with other antimicrobial agents, the development of microbial resistance could limit the long-term effectiveness of AMPs. To mitigate these shortcomings and improve AMP effectiveness, researchers have explored multiple strategies, such as chemical modifications and varied delivery methods.

5.6.1 Chemical Modifications of AMPs

To improve the resistance of AMPs to proteolytic degradation, diverse chemical strategies have been applied, including glycosylation, lipid conjugation, and the use of peptidomimetics. Figure 5.13 presents a schematic overview of how these chemical modifications generally boost the antimicrobial efficacy of AMPs.

5.6.1.1 Lipidation of AMPs

Modifying polypeptides with lipid chains can alter their self-assembly tendencies, hydrophobicity, and capacity to adopt secondary structures. Lipidation of peptides not only boosts antimicrobial potency but also improves their pharmacokinetic profiles [157]. Several naturally occurring lipopeptides, such as colistin and polymyxin B, have been approved by the FDA for therapeutic use. Their activity relies on Ca^{2+}-induced conformational changes, which increase amphipathicity and reduce net charge, allowing daptomycin to interact with neutral or negatively charged membranes. For chemical modification, the N-terminal or side chains of residues are often targeted. Fatty acid conjugation at the N-terminal has served to enhance the antibacterial activity of a human cathepsin G fragment. Similarly, lipidation of an LL-37 fragment has been shown to improve both its antibiofilm and antibacterial functions [158].

5.6.1.2 Glycosylation of AMPs

Glycosylation, as a form of post-translational modification, is essential for expanding protein structural diversity. Specific amino acid residues form covalent bonds with carbohydrate moieties, which contribute to correct protein folding, enhance host immune recognition, and participate in various physiological functions. The major types—N-glycosylation, O-linked glycosylation, C-glycosylation and S-glycosylation—are defined according to the particular amino acid targeted by the sugar attachment. For AMPs, introducing glycosylation increases their solubility in water-based systems and significantly improves their stability against proteolytic enzymes.

5.6.1.3 Peptidomimetics

The effectiveness of peptide therapeutics is frequently limited by restricted bioavailability, rapid degradation, and suboptimal target specificity. Peptidomimetics, synthetic analogues that replicate the structural and functional features of natural peptides, have emerged as a promising strategy. These compounds are generally constructed on chemically engineered backbones, including extended or modified chains, which preserve the native two- and three-dimensional architecture while conferring resistance to enzymatic hydrolysis [159]. The effectiveness of peptide therapeutics is often constrained due to poor systemic availability, rapid degradation, and suboptimal target specificity. In comparison with conventional peptides, peptidomimetic molecules exhibit highly selective antimicrobial activity against pathogenic organisms, with minimal cytotoxic effects on mammalian cells and tissues. Overall, the rational design of peptidomimetics not only augments peptide stability and absorption but also substantially improves their therapeutic specificity and functional resilience.

5.6.2 Delivery Systems for AMPs

Another effective strategy for enhancing efficacy and stability, as well as *in vivo* half-life of AMPs, employs advanced delivery systems. These platforms encompass self-assembling carriers, polymer-based matrices, metallic and inorganic nanoparticles, as well as liposome-based formulations. The incorporation of AMPs into nanotechnology-driven systems, particularly nanoparticle (NP) conjugates, represents a rapidly advancing area. Such nanoformulations can mitigate cytotoxicity toward host tissues while simultaneously enhancing the bioactivity and proteolytic stability of AMPs. Owing to their nanoscale dimensions, low inherent toxicity, and precise targeting ability, AMP-loaded nanocapsules offer promising potential for next-generation antimicrobial delivery.

5.6.2.1 Inorganic and Metallic Nanoparticles for AMPs

Inorganic carriers have been applied for AMP delivery because they can effectively inhibit peptide aggregation, protect against enzymatic degradation, and preserve structural integrity. Commonly employed inorganic materials include metallic nanoparticles, mesoporous silica, nanoclays, and carbon-based nanostructures [160]. Given their favorable biocompatibility, gold and silver nanoparticles are commonly viewed as suitable candidates for antimicrobial agent design.

5.6.2.2 Polymeric Nanoparticles for AMPs

Polymeric nanoparticles, typically ranging from 50 to 100 nm, are among the most widely employed carriers for AMP delivery due to their outstanding characteristics. Frequently used polymeric materials encompass fibers, hydrogels, multilayer films, and functional linkers. The physicochemical properties of these polymers can be fine-tuned by modifying the monomer composition, adjusting chain lengths, or altering surfactants [161]. Moreover, polymeric nanoparticles can be metabolized into small, excretable molecules, facilitating clearance from the body. Consequently, integrating AMPs with polymeric nanoparticles provides a viable approach for enhancing their antimicrobial efficacy and structural stability.

References

1 Knight, G.M., Glover, R.E., Mcquaid, C.F. et al. (2021). Antimicrobial resistance and COVID-19: intersections and implications. *Elife* 10: e64139.

2 He, Q., Yang, Z., Zou, Z. et al. (2023). Combating Escherichia coli O157:H7 with functionalized chickpea-derived antimicrobial peptides. *Advance Science* 10 (6): e2205301.

3 Hadj Saadoun, J., Sogari, G., Bernini, V. et al. (2022). A critical review of intrinsic and extrinsic antimicrobial properties of insects. *Trends in Food Science & Technology* 122: 40–48.

4 Li, G., Deng, H., Xu, W. et al. (2025). Combating antibiotic-resistant bacterial infection using Coassembled dimeric antimicrobial peptide-based nanofibers. *ACS Nano* 19 (3): 3155–3171.

5 Baindara, P. and Mandal, S.M. (2022). Plant-derived antimicrobial peptides: novel preservatives for the food industry. *Foods* 11 (16): 2415.

6 Zou, F., Tan, C., Shinali, T.S. et al. (2023). Plant antimicrobial peptides: a comprehensive review of their classification, production, mode of action, functions, applications, and challenges. *Food & Function* 14 (12): 5492–5515.

7 Taveira, G.B., Carvalho, A.O., Rodrigues, R. et al. (2016). Thionin-like peptide from Capsicum annuum fruits: mechanism of action and synergism with fluconazole against Candida species. *BMC Microbiology* 16: 12.

8 Taveira, G.B., Mello, É.O., Carvalho, A.O. et al. (2017). Antimicrobial activity and mechanism of action of a thionin-like peptide from Capsicum annuum fruits and combinatorial treatment with fluconazole against fusarium solani. *Biopolymers* 108 (3): e23008.

9 Höng, K., Austerlitz, T., Bohlmann, T., and Bohlmann, H. (2021). The thionin family of antimicrobial peptides. *PLoS One* 16 (7): e0254549.

10 Luo, J.S., Yang, Y., Gu, T. et al. (2019). The Arabidopsis defensin gene AtPDF2.5 mediates cadmium tolerance and accumulation. *Plant, Cell & Environment* 42 (9): 2681–2695.

11 Omidvar, R., Vosseler, N., Abbas, A. et al. (2021). Analysis of a gene family for PDF-like peptides from Arabidopsis. *Scientific Reports* 11 (1): 18948.

12 Struyfs, C., Breukers, J., Spasic, D. et al. (2022). Multiplex analysis to unravel the mode of antifungal activity of the plant defensin HsAFP1 in single yeast cells. *International Journal of Molecular Sciences* 23 (3): 1515.

13 Silva, S., Rodrigues, C.F., Araújo, D. et al. (2017). *Candida* species biofilms' antifungal resistance. *Journal of Fungi* 3 (1): 8.

14 Osborn, R.W., De Samblanx, G.W., Thevissen, K. et al. (1995). Isolation and characterisation of plant defensins from seeds of Asteraceae, Fabaceae, Hippocastanaceae and Saxifragaceae. *FEBS Letters* 368 (2): 257–262.

15 Su, T., Han, M., Cao, D., and Xu, M. (2020). Molecular and biological properties of Snakins: the foremost cysteine-rich plant host defense peptides. *Journal of Fungi* 6 (4): 220.

16 Almasia, N.I., Nahirñak, V., Hopp, H.E., and Vazquez-Rovere, C. (2020). Potato Snakin-1: an antimicrobial player of the trade-off between host defense and development. *Plant Cell Reports* 39 (7): 839–849.

17 Teker, T., Albayrak, G., and Turan, K. (2023). Heterologous expression and initial in silico characterization of a novel Snakin-Z peptide. *International Journal of Peptide Research and Therapeutics* 29 (5): 84.

18 Oliveira-Lima, M., Benko-Iseppon, A.M., Neto, J. et al. (2017). Snakin: structure, roles and applications of a plant antimicrobial peptide. *Current Protein and Peptide Science* 18 (4): 368–374.

19 Wong, K.H., Tan, W.L., Kini, S.G. et al. (2017). Vaccatides: antifungal glutamine-rich Hevein-like peptides from *Vaccaria hispanica*. *Frontiers in Plant Science* 8: 1100.

20 Chávez, M.I., Vila-Perelló, M., Cañada, F.J. et al. (2010). Effect of a serine-to-aspartate replacement on the recognition of chitin oligosaccharides by truncated hevein. A 3D view by using NMR. *Carbohydrate Research* 345 (10): 1461–1468.

21 Odintsova, T.I., Vassilevski, A.A., Slavokhotova, A.A. et al. (2009). A novel antifungal hevein-type peptide from Triticum kiharae seeds with a unique 10-cysteine motif. *The FEBS Journal* 276 (15): 4266–4275.

22 Hwang, B., Hwang, J.S., Lee, J., and Lee, D.G. (2010). Antifungal properties and mode of action of psacotheasin, a novel knottin-type peptide derived from Psacothea hilaris. *Biochemical and Biophysical Research Communications* 400 (3): 352–357.

23 Kam, A., Loo, S., Dutta, B. et al. (2019). Plant-derived mitochondria-targeting cysteine-rich peptide modulates cellular bioenergetics. *The Journal of Biological Chemistry* 294 (11): 4000–4011.

24 Hsu, W.-H., Lai, Y.-J., and Wu, S.-C. (2017). Effects of the anti-microbial peptide pardaxin plus sodium erythorbate dissolved in different gels on the quality of Pacific white shrimp under refrigerated storage. *Food Control* 73: 712–719.

25 Rajanbabu, V. and Chen, J.Y. (2011). Antiviral function of tilapia hepcidin 1-5 and its modulation of immune-related gene expressions against infectious pancreatic necrosis virus (IPNV) in Chinook salmon embryo (CHSE)-214 cells. *Fish & Shellfish Immunology* 30 (1): 39–44.

26 Mcmillan, K. and a M, Coombs M R P. (2021). Investigating potential applications of the fish anti-microbial peptide Pleurocidin: A systematic review. *Pharmaceuticals (Basel)* 14 (7): 687.

27 Jin, Z., Shen, M., Wang, L. et al. (2023). Antibacterial and immunoregulatory activity of an antimicrobial peptide hepcidin in loach (Misgurnus anguillicaudatus). *International Journal of Biological Macromolecules* 242 (Pt 2): 124833.

28 Prior, B.S., Lange, M.D., Salger, S.A. et al. (2022). The effect of piscidin antimicrobial peptides on the formation of gram-negative bacterial biofilms. *Journal of Fish Diseases* 45 (1): 99–105.

29 Moore, K.S., Bevins, C.L., Brasseur, M.M. et al. (1991). Antimicrobial peptides in the stomach of Xenopus laevis. *The Journal of Biological Chemistry* 266 (29): 19851–19857.

30 Sehnal, F., Hansen, I., and Scheller, K. (2002). The cDNA-structure of the prothoracicotropic hormone (PTTH) of the silkmoth Hyalophora cecropia. *Insect Biochemistry and Molecular Biology* 32 (2): 233–237.

31 Barroso, C., Carvalho, P., Gonçalves, J.F.M. et al. (2021). Antimicrobial peptides: identification of two Beta-Defensins in a teleost fish, the European Sea bass (Dicentrarchus labrax). *Pharmaceuticals (Basel)* 14 (6): 566.

32 Subramanian, S., Ross, N.W., and Mackinnon, S.L. (2009). Myxinidin, a novel antimicrobial peptide from the epidermal mucus of hagfish, Myxine glutinosa L. *Marine Biotechnology (New York, N.Y.)* 11 (6): 748–757.

33 Park, C.B., Kim, M.S., and Kim, S.C. (1996). A novel antimicrobial peptide from Bufo bufo gargarizans. *Biochemical and Biophysical Research Communications* 218 (1): 408–413.

34 Nair, A., Sruthy, K.S., Chaithanya, E.R. et al. (2017). Molecular characterisation of a novel isoform of hepatic antimicrobial peptide, Hepcidin (Le-Hepc), from Leiognathus equulus and analysis of its functional properties in silico. *Probiotics and Antimicrobial Proteins* 9 (4): 473–482.

35 Liang, T., Ji, W., Zhang, G.R. et al. (2013). Molecular cloning and expression analysis of liver-expressed antimicrobial peptide 1 (LEAP-1) and LEAP-2 genes in the blunt snout bream (Megalobrama amblycephala). *Fish & Shellfish Immunology* 35 (2): 553–563.

36 Serna-Duque, J.A., Cuesta, A., and Esteban, M. (2022). Massive gene expansion of hepcidin, a host defense peptide, in gilthead seabream (*Sparus aurata*). *Fish & Shellfish Immunology* 124: 563–571.

37 Ortjohann, M. and Leippe, M. (2025). Characterization of NK-lysin A, a potent antimicrobial peptide from the zebrafish Danio rerio. *Developmental and Comparative Immunology* 162: 105266.

38 Travin, D.Y., Watson, Z.L., Metelev, M. et al. (2019). Structure of ribosome-bound azole-modified peptide phazolicin rationalizes its species-specific mode of bacterial translation inhibition. *Nature Communications* 10 (1): 4563.

39 Zukher, I., Pavlov, M., Tsibulskaya, D. et al. (2019). Reiterative synthesis by the ribosome and recognition of the N-terminal formyl group by biosynthetic machinery contribute to evolutionary conservation of the length of antibiotic microcin C peptide precursor. *mBio* 10 (2).

40 Bhosale, M., Pande, S., Kumar, A. et al. (2010). Characterization of two M17 family members in *Escherichia coli*, peptidase A and peptidase B. *Biochemical and Biophysical Research Communications* 395 (1): 76–81.

41 Thomas, X., Destoumieux-Garzón, D., Peduzzi, J. et al. (2004). Siderophore peptide, a new type of post-translationally modified antibacterial peptide with potent activity. *The Journal of Biological Chemistry* 279 (27): 28233–28242.

42 Saad, H., Majer, T., Bhattarai, K. et al. (2023). Bioinformatics-guided discovery of biaryl-linked lasso peptides. *Chemical Science* 14 (45): 13176–13183.

43 Jeanne Dit Fouque, K., Hegemann, J.D., Zirah, S. et al. (2019). Evidence of Cis/trans-isomerization at Pro7/Pro16 in the lasso peptide Microcin J25. *Journal of the American Society for Mass Spectrometry* 30 (6): 1038–1045.

44 Solbiati, J.O., Ciaccio, M., Farías, R.N. et al. (1999). Sequence analysis of the four plasmid genes required to produce the circular peptide antibiotic microcin J25. *Journal of Bacteriology* 181 (8): 2659–2662.

45 Hemu, X., Qiu, Y., Nguyen, G.K., and Tam, J.P. (2016). Total synthesis of circular Bacteriocins by Butelase 1. *Journal of the American Chemical Society* 138 (22): 6968–6971.

46 Komori, Y., Ozawa, N., Kuwahara, H. et al. (2025). Identification and characterization of a circular bacteriocin, garvicin SC, a novel garvicin ML variant, produced by Lactococcus garvieae ABG0038. *Journal of Bioscience and Bioengineering* 139 (2): 95–99.

47 Shi, C., Patel, V.A., Mitchell, D.A., and Zhao, H. (2024). Enterolyin S, a polythiazole-containing hemolytic peptide from *Enterococcus caccae. ChemBioChem* 25 (12): e202400212.

48 Collin, F. and Maxwell, A. (2019). The microbial toxin Microcin B17: prospects for the development of new antibacterial agents. *Journal of Molecular Biology* 431 (18): 3400–3426.

49 Reiners, J., Abts, A., Clemens, R. et al. (2017). Stoichiometry and structure of a lantibiotic maturation complex. *Scientific Reports* 7: 42163.

50 Acedo, J.Z., Chiorean, S., Vederas, J.C., and Van Belkum, M.J. (2018). The expanding structural variety among bacteriocins from gram-positive bacteria. *FEMS Microbiology Reviews* 42 (6): 805–828.

51 Tang, W. and Van Der Donk, W.A. (2013). The sequence of the enterococcal cytolysin imparts unusual lanthionine stereochemistry. *Nature Chemical Biology* 9 (3): 157–159.

52 Ishikawa, F., Nakamura, S., Nakanishi, I., and Tanabe, G. (2024). Recent progress in the reprogramming of nonribosomal peptide synthetases. *Journal of Peptide Science* 30 (3): e3545.

53 Agrawal, S., Acharya, D., Adholeya, A. et al. (2017). Nonribosomal peptides from marine microbes and their antimicrobial and anticancer potential. *Frontiers in Pharmacology* 8: 828.

54 Barsby, T., Kelly, M.T., Gagné, S.M., and Andersen, R.J. (2001). Bogorol A produced in culture by a marine Bacillus sp. reveals a novel template for cationic peptide antibiotics. *Organic Letters* 3 (3): 437–440.

55 Nagai, K., Kamigiri, K., Arao, N. et al. (2003). YM-266183 and YM-266184, novel thiopeptide antibiotics produced by Bacillus cereus isolated from a marine sponge. I. Taxonomy, fermentation, isolation, physico-chemical properties and biological properties. *Journal of Antibiotics* 56 (2): 123–128.

56 Mitova, M., Popov, S., and De Rosa, S. (2004). Cyclic peptides from a Ruegeria strain of bacteria associated with the sponge Suberites domuncula. *Journal of Natural Products* 67 (7): 1178–1181.

57 Desjardine, K., Pereira, A., Wright, H. et al. (2007). Tauramamide, a lipopeptide antibiotic produced in culture by Brevibacillus laterosporus isolated from a marine habitat: structure elucidation and synthesis. *Journal of Natural Products* 70 (12): 1850–1853.

58 Engelhardt, K., Degnes, K.F., Kemmler, M. et al. (2010). Production of a new thiopeptide antibiotic, TP-1161, by a marine Nocardiopsis species. *Applied and Environmental Microbiology* 76 (15): 4969–4976.

59 Sun, P., Maloney, K.N., Nam, S.J. et al. (2011). Fijimycins A-C, three antibacterial etamycin-class depsipeptides from a marine-derived Streptomyces sp. *Bioorganic & Medicinal Chemistry* 19 (22): 6557–6562.

60 Martín, J., Da, S.S.T., Crespo, G. et al. (2013). Kocurin, the true structure of PM181104, an anti-methicillin-resistant Staphylococcus aureus (MRSA) thiazolyl peptide from the marine-derived bacterium Kocuria palustris. *Marine Drugs* 11 (2): 387–398.

61 Amagata, T., Morinaka, B.I., Amagata, A. et al. (2006). A chemical study of cyclic depsipeptides produced by a sponge-derived fungus. *Journal of Natural Products* 69 (11): 1560–1565.

62 Oh, D.C., Kauffman, C.A., Jensen, P.R., and Fenical, W. (2007). Induced production of emericellamides A and B from the marine-derived fungus Emericella sp. in competing co-culture. *Journal of Natural Products* 70 (4): 515–520.

63 Pruksakorn, P., Arai, M., Kotoku, N. et al. (2010). Trichoderins, novel aminolipopeptides from a marine sponge-derived Trichoderma sp., are active against dormant mycobacteria. *Bioorganic & Medicinal Chemistry Letters* 20 (12): 3658–3663.

64 Kim, M.Y., Sohn, J.H., Ahn, J.S., and Oh, H. (2009). Alternaramide, a cyclic depsipeptide from the marine-derived fungus Alternaria sp. SF-5016. *Journal of Natural Products* 72 (11): 2065–2068.

65 Zheng, J., Xu, Z., Wang, Y. et al. (2010). Cyclic tripeptides from the halotolerant fungus aspergillus sclerotiorum PT06-1. *Journal of Natural Products* 73 (6): 1133–1137.

66 Macmillan, J.B. and Molinski, T.F. (2002). Lobocyclamide B from Lyngbya confervoides. Configuration and asymmetric synthesis of beta-hydroxy-alpha-amino acids by (−)-sparteine-mediated aldol addition. *Organic Letters* 4 (11): 1883–1886.

67 Müller, D., Krick, A., Kehraus, S. et al. (2006). Brunsvicamides A-C: sponge-related cyanobacterial peptides with mycobacterium tuberculosis protein tyrosine phosphatase inhibitory activity. *Journal of Medicinal Chemistry* 49 (16): 4871–4878.

68 Yu, L.P., Sun, B.G., Li, J., and Sun, L. (2013). Characterization of a c-type lysozyme of Scophthalmus maximus: expression, activity, and antibacterial effect. *Fish & Shellfish Immunology* 34 (1): 46–54.

69 Xu, Y., Niu, Y., Wu, C. et al. (2024). Synthesis, characterization, antifungal properties of quaternary ammonium salts derived from natural rosin. *Biomass Conversion and Biorefinery* 14 (16): 18803–18811.

70 Luo, X., Song, Y., Cao, Z. et al. (2022). Evaluation of the antimicrobial activities and mechanisms of synthetic antimicrobial peptide against food-borne pathogens. *Food Bioscience.* 49: 101903.

71 Benincasa, M., Skerlavaj, B., Gennaro, R. et al. (2003). *In vitro* and *in vivo* antimicrobial activity of two alpha-helical cathelicidin peptides and of their synthetic analogs. *Peptides* 24 (11): 1723–1731.

72 Carratalá, J.V., Serna, N., Villaverde, A. et al. (2020). Nanostructured antimicrobial peptides: the last push towards clinics. *Biotechnology Advances* 44: 107603.

73 Lee, H. and Yang, S. (2021). Dimerization of cell-penetrating buforin II enhances antimicrobial properties. *Journal of Analytical Science and Technology.* 12: 9.

74 Koehbach, J. and Craik, D.J. (2019). The vast structural diversity of antimicrobial peptides. *Trends in Pharmacological Sciences* 40 (7): 517–528.

75 Chen, N. and Jiang, C. (2023). Antimicrobial peptides: structure, mechanism, and modification. *European Journal of Medicinal Chemistry* 255: 115377.

76 Knappe, D., Piantavigna, S., Hansen, A. et al. (2010). Oncocin (VDKPPYLPRPRPPRRIYNR-NH2): a novel antibacterial peptide optimized against gram-negative human pathogens. *Journal of Medicinal Chemistry* 53 (14): 5240–5247.

77 Van Den Bergen, G., Stroet, M., Caron, B. et al. (2020). Curved or linear? Predicting the 3-dimensional structure of α-helical antimicrobial peptides in an amphipathic environment. *FEBS Letters* 594 (6): 1062–1080.

78 Järvå, M., Lay, F.T., Phan, T.K. et al. (2018). X-ray structure of a carpet-like antimicrobial defensin-phospholipid membrane disruption complex. *Nature Communications* 9 (1): 1962.

79 Mohanram, H. and Bhattacharjya, S. (2014). Resurrecting inactive antimicrobial peptides from the lipopolysaccharide trap. *Antimicrobial Agents and Chemotherapy* 58 (4): 1987–1996.

80 Liang, Y., Zhang, X., Yuan, Y. et al. (2020). Role and modulation of the secondary structure of antimicrobial peptides to improve selectivity. *Biomaterials Science* 8 (24): 6858–6866.

81 Della Pelle, G., Perà, G., Belardinelli, M.C. et al. (2020). Trematocine, a novel antimicrobial peptide from the Antarctic fish Trematomus bernacchii: identification and biological activity. *Antibiotics (Basel)* 9 (2): 66.

82 Song, C., De Groot, B.L., and Sansom, M.S.P. (2019). Lipid bilayer composition influences the activity of the antimicrobial peptide Dermcidin Channel. *Biophysical Journal* 116 (9): 1658–1666.

83 Barreteau, H., Kovac, A., Boniface, A. et al. (2008). Cytoplasmic steps of peptidoglycan biosynthesis. *FEMS Microbiology Reviews* 32 (2): 168–207.

84 Chen, J., Wang, W., Hu, X. et al. (2024). Medium-sized peptides from microbial sources with potential for antibacterial drug development. *Natural Product Reports* 41 (8): 1235–1263.

85 Chen, J., Wang, W., Hu, X. et al. (2024). Medium-sized peptides from microbial sources with potential for antibacterial drug development. *Natural Product Reports* 41 (8): 1235–1263.

86 Rackham, E.J., Grüschow, S., Ragab, A.E. et al. (2010). Pacidamycin biosynthesis: identification and heterologous expression of the first uridyl peptide antibiotic gene cluster. *ChemBioChem* 11 (12): 1700–1709.

87 Mcdonald, L.A., Barbieri, L.R., Carter, G.T. et al. (2002). Structures of the Muraymycins, novel peptidoglycan biosynthesis inhibitors. *Journal of the American Chemical Society* 124: 10260–10261.

88 Wiegmann, D., Koppermann, S., and Ducho, C. (2018). Aminoribosylated Analogues of Muraymycin Nucleoside Antibiotics. *Molecules.* 23 (12): 3085.

89 Chung, B.C., Zhao, J., Gillespie, R.A. et al. (2013). Crystal structure of MraY, an essential membrane enzyme for bacterial cell wall synthesis. *Science* 341 (6149): 1012–1016.

90 Chung, B.C., Mashalidis, E.H., Tanino, T. et al. (2016). Structural insights into inhibition of lipid I production in bacterial cell wall synthesis. *Nature* 533 (7604): 557–560.

91 Chu, J., Vila-Farres, X., Inoyama, D. et al. (2016). Discovery of MRSA active antibiotics using primary sequence from the human microbiome. *Nature Chemical Biology.* 12: 1004–1006.

92 Chu, J., Vila-Farres, X., Inoyama, D. et al. (2018). Human microbiome inspired antibiotics with improved β-lactam synergy against MDR Staphylococcus aureus. *ACS Infectious Diseases* 4 (1): 33–38.

93 Chu, J., Vila-Farres, X., Inoyama, D. et al. (2016). Discovery of MRSA active antibiotics using primary sequence from the human microbiome. *Nature Chemical Biology* 12 (12): 1004–1006.

94 Kohga, H., Lertpreedakorn, N., Miyazaki, R., and Tsukazaki, T. (2023). Critical residues of the antibiotic peptide LysM that inhibits lipid II flipping. *bioRxiv—microbiology* https://doi.org/10.1101/2023.05.22.541694.

95 Wang, X., Zhan, P., Zhang, Q. et al. (2024). Staphylococcus aureus acquires resistance to glycopeptide antibiotic vancomycin via CXCL10. *International Immunopharmacology* 132: 111780.

96 Gorrie, C., Higgs, C., Carter, G. et al. (2019). Genomics of vancomycin-resistant Enterococcus faecium. *Microbial Genomics* 5 (7): e000283.

97 Blaskovich, M.A.T., Hansford, K.A., Butler, M.S. et al. (2018). Developments in Glycopeptide antibiotics. *ACS Infectious Diseases* 4 (5): 715–735.

98 Wood, T.M., Zeronian, M.R., Buijs, N. et al. (2022). Mechanistic insights into the C(55)-P targeting lipopeptide antibiotics revealed by structure-activity studies and high-resolution crystal structures. *Chemical Science* 13 (10): 2985–2991.

99 Workman, S.D., Worrall, L.J., and Strynadka, N.C.J. (2018). Crystal structure of an intramembranal phosphatase central to bacterial cell-wall peptidoglycan biosynthesis and lipid recycling. *Nature Communications* 9 (1): 1159.

100 Sit, B., Srisuknimit, V., Bueno, E. et al. (2023). Undecaprenyl phosphate translocases confer conditional microbial fitness. *Nature* 613 (7945): 721–728.

101 Kleijn, L.H.J., Vlieg, H.C., Wood, T.M. et al. (2017). A high-resolution crystal structure that reveals molecular details of target recognition by the calcium-dependent Lipopeptide antibiotic Laspartomycin C. *Angewandte Chemie (International Ed. in English)* 56 (52): 16546–16549.

102 Singh, M.P., Petersen, P.J., Weiss, W.J. et al. (2003). Mannopeptimycins, new cyclic glycopeptide antibiotics produced by Streptomyces hygroscopicus LL-AC98: antibacterial and mechanistic activities. *Antimicrobial Agents and Chemotherapy* 47 (1): 62–69.

103 Wang, B., Liu, Y., Jiao, R. et al. (2016). Total synthesis of Mannopeptimycins α and β. *Journal of the American Chemical Society* 138 (11): 3926–3932.

104 Ruzin, A., Singh, G., Severin, A. et al. (2004). Mechanism of action of the mannopeptimycins, a novel class of glycopeptide antibiotics active against vancomycin-resistant gram-positive bacteria. *Antimicrobial Agents and Chemotherapy* 48 (3): 728–738.

105 Ling, L.L., Schneider, T., Peoples, A.J. et al. (2015). A new antibiotic kills pathogens without detectable resistance. *Nature* 517 (7535): 455–459.

106 Wirtz, D.A., Ludwig, K.C., Arts, M. et al. (2021). Biosynthesis and mechanism of action of the Cell Wall targeting antibiotic Hypeptin. *Angewandte Chemie (International Ed. in English)* 60 (24): 13579–13586.

107 Öster, C., Walkowiak, G.P., Hughes, D.E. et al. (2018). Structural studies suggest aggregation as one of the modes of action for teixobactin. *Chemical Science* 9 (47): 8850–8859.

108 Kong, F. and Carter, G.T. (2003). Structure determination of glycinocins A to D, further evidence for the cyclic structure of the amphomycin antibiotics. *Journal of Antibiotics (Tokyo)* 56 (6): 557–564.

109 Hover, B.M., Kim, S.H., Katz, M. et al. (2018). Culture-independent discovery of the malacidins as calcium-dependent antibiotics with activity against multidrug-resistant gram-positive pathogens. *Nature Microbiology* 3 (4): 415–422.

110 Wang, Z., Koirala, B., Hernandez, Y. et al. (2022). Bioinformatic prospecting and synthesis of a bifunctional lipopeptide antibiotic that evades resistance. *Science* 376 (6596): 991–996.

111 Rosenzweig, A.F., Wang, Z., Morales-Amador, A. et al. (2023). A family of antibiotics that evades resistance by binding Polyprenyl phosphates. *ACS Infectious Diseases* 9 (12): 2394–2400.

112 Wang, M., Buist, G., and Van Dijl, J.M. (2022). Staphylococcus aureus cell wall maintenance—the multifaceted roles of peptidoglycan hydrolases in bacterial growth, fitness, and virulence. *FEMS Microbiology Reviews* 46 (5): fuac025.

113 Culp, E.J., Waglechner, N., Wang, W. et al. (2020). Evolution-guided discovery of antibiotics that inhibit peptidoglycan remodelling. *Nature* 578 (7796): 582–587.

114 Kwon, Y.J., Kim, H.J., and Kim, W.G. (2015). Complestatin exerts antibacterial activity by the inhibition of fatty acid synthesis. *Biological & Pharmaceutical Bulletin* 38 (5): 715–721.

115 Leonard, A.C., Goncheva, M.I., Gilbert, S.E. et al. (2023). Autolysin-mediated peptidoglycan hydrolysis is required for the surface display of *Staphylococcus aureus* cell wall-anchored proteins. *Proceedings of the National Academy of Sciences of the United States of America* 120 (12): e2301414120.

116 Bush, N.G., Diez-Santos, I., Abbott, L.R., and Maxwell, A. (2020). Quinolones: mechanism, lethality and their contributions to antibiotic resistance. *Molecules* 25 (23).

117 Hooper, D.C. (2001). Mechanisms of action of antimicrobials: focus on fluoroquinolones. *Clinical Infectious Diseases* 32 (Suppl 1): S9–S15.

118 Imai, Y., Hauk, G., Quigley, J. et al. (2022). Evybactin is a DNA gyrase inhibitor that selectively kills mycobacterium tuberculosis. *Nature Chemical Biology* 18 (11): 1236–1244.

119 Arenz, S. and Wilson, D.N. (2016). Bacterial protein synthesis as a target for antibiotic inhibition. *Cold Spring Harbor Perspectives in Medicine* 6 (9): a025361.

120 Pantel, L., Florin, T., Dobosz-Bartoszek, M. et al. (2018). Odilorhabdins, antibacterial agents that cause miscoding by binding at a new ribosomal site. *Molecular Cell* 70 (1): 83–94.e7.

121 Metelev, M., Osterman, I.A., Ghilarov, D. et al. (2017). Klebsazolicin inhibits 70S ribosome by obstructing the peptide exit tunnel. *Nature Chemical Biology* 13 (10): 1129–1136.

122 Shimanaka, K., Takahashi, Y., Iinuma, H. et al. (1994). Novel antibiotics, amythiamicins. II. Structure elucidation of amythiamicin D. *Journal of Antibiotics (Tokyo)* 47 (10): 1145–1152.

123 Shimanaka, K., Kinoshita, N., Iinuma, H. et al. (1994). Novel antibiotics, amythiamicins. I. Taxonomy, fermentation, isolation, physico-chemical properties, and antimicrobial activity. *Journal of Antibiotics* 47 (6): 668–674.

124 Gross, S., Nguyen, F., Bierschenk, M. et al. (2013). Amythiamicin D and related thiopeptides as inhibitors of the bacterial elongation factor EF-Tu: modification of the amino acid at carbon atom C2 of ring C dramatically influences activity. *ChemMedChem* 8 (12): 1954–1962.

125 Parmeggiani, A., Krab, I.M., Okamura, S. et al. (2006). Structural basis of the action of pulvomycin and GE2270 A on elongation factor Tu. *Biochemistry* 45 (22): 6846–6857.

126 Macek, B., Forchhammer, K., Hardouin, J. et al. (2019). Protein post-translational modifications in bacteria. *Nature Reviews. Microbiology* 17 (11): 651–664.

127 Yoon, D., Moon, J.H., Cho, A. et al. (2023). Structure-based insight on the mechanism of N-glycosylation inhibition by Tunicamycin. *Molecules and Cells* 46 (6): 337–344.

128 Costa, S.R., Lira, B.O., Calixto, G.F.G. et al. (2025). Rational design of Chim3, a multifunctional peptide carrying a formyl peptide receptor 2 (FPR2) agonist module released by bacterial signal peptidase I (Spase I). *FEBS Journal* 292: 3242–3263.

129 Smith, P.A., Koehler, M.F.T., Girgis, H.S. et al. (2018). Optimized arylomycins are a new class of gram-negative antibiotics. *Nature* 561 (7722): 189–194.

130 Therien, A.G., Huber, J.L., Wilson, K.E. et al. (2012). Broadening the spectrum of β-lactam antibiotics through inhibition of signal peptidase type I. *Antimicrobial Agents and Chemotherapy* 56 (9): 4662–4670.

131 Pérez-Bonilla, M., Oves-Costales, D., González, I. et al. (2020). Krisynomycins, imipenem Potentiators against methicillin-resistant Staphylococcus aureus, produced by Streptomyces canus. *Journal of Natural Products* 83 (9): 2597–2606.

132 Bakelar, J., Buchanan, S.K., and Noinaj, N. (2016). The structure of the β-barrel assembly machinery complex. *Science* 351 (6269): 180–186.

133 Imai, Y., Meyer, K.J., Iinishi, A. et al. (2019). A new antibiotic selectively kills gram-negative pathogens. *Nature* 576 (7787): 459–464.

134 Miller, R.D., Iinishi, A., Modaresi, S.M. et al. (2022). Computational identification of a systemic antibiotic for gram-negative bacteria. *Nature Microbiology* 7 (10): 1661–1672.

135 Dong, S., Chen, H., Zhou, Q., and Liao, N. (2021). Protein degradation control and regulation of bacterial survival and pathogenicity: the role of protein degradation systems in bacteria. *Molecular Biology Reports* 48 (11): 7575–7585.

136 Li, L., Macintyre, L.W., Ali, T. et al. (2021). Biosynthetic interrogation of soil metagenomes reveals Metamarin, an uncommon Cyclomarin congener with activity against mycobacterium tuberculosis. *Journal of Natural Products* 84 (4): 1056–1066.

137 Gavrish, E., Sit, C.S., Cao, S. et al. (2014). Lassomycin, a ribosomally synthesized cyclic peptide, kills mycobacterium tuberculosis by targeting the ATP-dependent protease ClpC1P1P2. *Chemistry & Biology* 21 (4): 509–518.

138 Sohlenkamp, C. and Geiger, O. (2016). Bacterial membrane lipids: diversity in structures and pathways. *FEMS Microbiology Reviews* 40 (1): 133–159.

139 Zuttion, F., Colom, A., Matile, S. et al. (2020). High-speed atomic force microscopy highlights new molecular mechanism of daptomycin action. *Nature Communications* 11 (1): 6312.

140 Li, Y.X., Zhong, Z., Zhang, W.P., and Qian, P.Y. (2018). Discovery of cationic nonribosomal peptides as gram-negative antibiotics through global genome mining. *Nature Communications* 9 (1): 3273.

141 Lázár, V., Martins, A., Spohn, R. et al. (2018). Antibiotic-resistant bacteria show widespread collateral sensitivity to antimicrobial peptides. *Nature Microbiology* 3 (6): 718–731.

142 Wang, Z., Koirala, B., Hernandez, Y. et al. (2022). A naturally inspired antibiotic to target multidrug-resistant pathogens. *Nature* 601 (7894): 606–611.

143 Li, L., Koirala, B., Hernandez, Y. et al. (2022). Identification of structurally diverse menaquinone-binding antibiotics with in vivo activity against multidrug-resistant pathogens. *Nature Microbiology* 7 (1): 120–131.

144 Ahmed, T. and a E, Hammami R. (2019). Recent insights into structure-function relationships of antimicrobial peptides. *Journal of Food Biochemistry* 43 (1): e12546.

145 Gao, X., Chen, Y., Chen, Z. et al. (2019). Identification and antimicrobial activity evaluation of three peptides from laba garlic and the related mechanism. *Food & Function* 10 (8): 4486–4496.

146 Park, K.M., Yoon, S.G., Choi, T.H. et al. (2020). The bactericidal effect of a combination of food-grade compounds and their application as alternative antibacterial agents for food contact surfaces. *Food* 9 (1): 59.

147 Mahlapuu, M., Björn, C., and Ekblom, J. (2020). Antimicrobial peptides as therapeutic agents: opportunities and challenges. *Critical Reviews in Biotechnology* 40 (7): 978–992.

148 Marshall, E., Costa, L.M., and Gutierrez-Marcos, J. (2011). Cysteine-rich peptides (CRPs) mediate diverse aspects of cell-cell communication in plant reproduction and development. *Journal of Experimental Botany* 62 (5): 1677–1686.

149 Wang, S., Zeng, X., Yang, Q., and Qiao, S. (2016). Antimicrobial peptides as potential alternatives to antibiotics in food animal industry. *International Journal of Molecular Sciences* 17 (5): 603.

150 Tang, Z., Yin, Y., Zhang, Y. et al. (2009). Effects of dietary supplementation with an expressed fusion peptide bovine lactoferricin-lactoferrampin on performance, immune function and intestinal mucosal morphology in piglets weaned at age 21 d. *The British Journal of Nutrition* 101 (7): 998–1005.

151 Abushaheen, M.A., Muzaheed, F.A.J., Alosaimi, M. et al. (2020). Antimicrobial resistance, mechanisms and its clinical significance. *Disease-a-Month* 66 (6): 100971.

152 Gerlach, S.L., Rathinakumar, R., Chakravarty, G. et al. (2010). Anticancer and chemosensitizing abilities of cycloviolacin 02 from Viola odorata and psyle cyclotides from Psychotria leptothyrsa. *Biopolymers* 94 (5): 617–625.

153 Hancock, R.E., Haney, E.F., and Gill, E.E. (2016). The immunology of host defence peptides: beyond antimicrobial activity. *Nature Reviews Immunology* 16 (5): 321–334.

154 Chen, C.H. and Lu, T.K. (2020). Development and challenges of antimicrobial peptides for therapeutic applications. *Antibiotics (Basel)* 9 (1): 24.

155 Li, Z., Yu, Z., Wu, Y. et al. (2020). Self-sterilizing diblock polycation-enhanced polyamidoxime shape-stable blow-spun nanofibers for high-performance uranium capture from seawater. *Chemical Engineering Journal* 390: 124648.

156 Zhang, W., Li, J.-X., Tang, R.-C., and Zhai, A.-D. (2020). Hydrophilic and antibacterial surface functionalization of polyamide fabric by coating with polylysine biomolecule. *Progress in Organic Coatings* 142: 105571.

157 Gan, B.H., Gaynord, J., Rowe, S.M. et al. (2021). The multifaceted nature of antimicrobial peptides: current synthetic chemistry approaches and future directions. *Chemical Society Reviews* 50 (13): 7820–7880.

158 Yang, B., Good, D., Mosaiab, T. et al. (2020). Significance of LL-37 on immunomodulation and disease outcome. *BioMed Research International* 2020: 8349712.

159 Scott, R.W., Degrado, W.F., and Tew, G.N. (2008). De novo designed synthetic mimics of antimicrobial peptides. *Current Opinion in Biotechnology* 19 (6): 620–627.

160 Kumar, P., Kizhakkedathu, J.N., and Straus, S.K. (2018). Antimicrobial peptides: diversity, mechanism of action and strategies to improve the activity and biocompatibility in vivo. *Biomolecules* 8 (1).

161 Nordström, R. and Malmsten, M. (2017). Delivery systems for antimicrobial peptides. *Advances in Colloid and Interface Science* 242: 17–34.

6

Metal-Chelating Peptides

6.1 Introduction

Peptides have recently become a focal point of interest in the realm of nutrition and health. This surge in attention is primarily due to their potential as bioactive molecules that possess a wide range of physiological functions. One of the most significant applications of peptides lies in their unique ability to form stable complexes with metal ions. These complexes are known as metal-chelating peptides (MCPs) [1].

MCPs represent a special category of peptides. They can engage in chelation reactions with metal ions. Through the formation of coordination bonds, they can combine with metal ions to create metal–organic chelates that have cyclic structures. These MCPs are not just structurally interesting; they also possess a variety of biologically active functions [2].

Examining the origins of peptide–metal chelates, we can identify three main sources. Firstly, chelates can be synthesized through the deliberate coordination of artificially prepared MCPs with a variety of metal ions. This method allows for precise control over the composition and structure of the resulting chelates. Secondly, chelates can also be prepared by coordinating polypeptides, which have inherent chelating abilities, with metal ions. These polypeptides are obtained from the degradation of proteins derived from both animal and plant sources. This approach leverages the natural chelating potential of these polypeptides. Thirdly, natural peptide–metal ion chelates can be directly extracted from animal and plant tissues [3]. This method takes advantage of the existing chelates that occur naturally in biological systems.

Due to the diverse ligands of MCPs and the different metal ions that constitute the chelates, the resulting complexes exhibit a variety of compositions and structures. This diversity, in turn, leads to different biologically active functions. MCPs can enhance the bioavailability of metal ions by utilizing the absorption mechanisms of peptides within living organisms [4]. Moreover, they possess physiological and biochemical characteristics that are not found in inorganic metal ions.

In recent years, researchers have employed various methods to prepare and isolate MCPs. After chelating with different metal ions, these peptides have found applications in industrial production. They can be used as antibacterial agents, antioxidants, food additives, cosmetic

Food-Derived Bioactive Peptides: Green Preparation, Activity Mechanisms, and Health Applications, First Edition. Shao-Yun Wang, Xu Chen, and Xixi Cai.

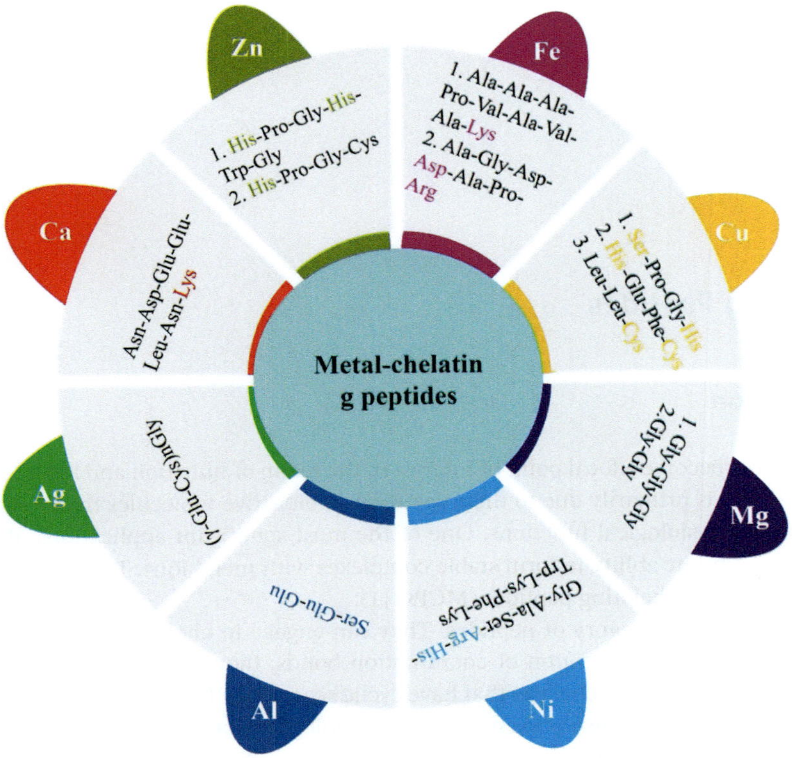

Figure 6.1 Common metal ions coordinated by chelating peptides and their corresponding peptide segments. His-Pro-Gly-His-Trp-Gly, His-Pro-Gly-Cys, Ser-Pro-Gly-His, His-Glu-Phe-Cys, Leu-Leu-Cys and Ser-Glu-Glu from Ref.[6]. Ala-Ala-Ala-Pro-Val-Ala-Val-Ala-Lys and Ala-Gly-Asp-Asp-Ala-Pro-Arg from Ref. [7]. Gly-Gly-Gly from Ref. [8]. Gly-Gly from Ref. [9]. Gly-Ala-Ser-Arg-His-Trp-Lys-Phe-Lys from Ref. [10]. (γ-Glu-Cys)nGly from Ref.[11]. Asn-Asp-Glu-Glu-Leu-Asn-Lys from Ref. [12].

additives, animal feed, organic fertilizers, and essential metal ion supplements for the human body. Compared with some inorganic antioxidants, natural antibacterial agents, or other types of essential metal ion supplements, MCPs offer several advantages [5]. They have almost no toxic side effects, superior biological valence, lower preparation costs, and simpler processes.

Given these benefits and their wide-ranging potential applications, the commercial value potential and research prospects of MCPs have increasingly become a research hotspot both domestically and internationally. Scientists and industries are continually exploring new ways to harness the power of these peptides to improve health, nutrition, and various industrial processes (Figure 6.1).

6.2 Preparation

Currently, the research on the preparation of peptide–metal ion chelates is primarily focused on two key aspects: the preparation of bioactive peptides and the chelation process between these peptides and metal elements. With the advancement of research on the comprehensive utilization of food resources, there is a growing trend among researchers to explore diverse raw materials for peptide extraction. These materials span a wide range of origins, including plant-based sources (such as soybeans, wheat, and legumes), animal-based sources (such as collagen and meat by-products), milk-derived sources (such as casein and whey proteins), and aquatic sources (such as fish, shrimp, and algae). Notably, there is a particular emphasis on utilizing by-products from food processing, such as fish scales, bones, and whey, as peptide sources [13]. This approach not only promotes the efficient utilization of waste resources, reducing environmental burdens, but also significantly enhances the added value of raw materials, contributing to a circular economy.

The chelation rate and biological activity of MCPs vary considerably depending on the preparation process conditions. Factors such as the source of raw materials, hydrolysis methods, and chelation parameters can lead to substantial differences in the final product's quality and functionality. Consequently, the study of the preparation process for MCPs holds profound significance. Optimizing these processes can lead to the development of highly effective MCPs with enhanced bioavailability, stability, and functional properties, which are crucial for applications in nutraceuticals, functional foods, and pharmaceuticals.

The preparation methods of MCPs mainly include enzymatic hydrolysis, solvent extraction, and chemical synthesis. Among these, solvent extraction and chemical synthesis methods, while effective in producing peptides and integrating them with metal ions, are associated with several drawbacks: these methods often involve high costs, complex procedures, and potential safety concerns, such as the presence of residual chemical substances that may pose health risks. In contrast, enzymatic hydrolysis has emerged as a preferred method due to its numerous advantages. It operates under mild conditions, ensuring the preservation of peptide integrity and bioactivity. Additionally, it is highly safe, cost-effective, and allows for precise control over the hydrolysis process, making it suitable for large-scale production.

Given these benefits, the trend of using enzymatic hydrolysis to prepare MCPs and subsequently produce peptide–metal ion chelates has become inevitable. Current research on enzymatic hydrolysis is primarily focused on three key areas: the selection of suitable enzyme species, the optimization of hydrolysis conditions, and the improvement of hydrolysis processes to achieve MCPs with a high integration activity. Different proteases, such as trypsin, pepsin, and alkaline protease, exhibit varying properties and specificity, leading to distinct hydrolysis outcomes. The efficiency of protein hydrolysis is influenced by multiple factors, including the liquid-to-solid ratio, hydrolysis temperature, pH, enzyme concentration, and hydrolysis duration [14]. Moreover, the interactions between these factors further complicate the optimization process.

The choice of enzyme and the optimal action conditions are highly dependent on the type of raw material used. For instance, aquatic raw materials, such as fish and algae, may require different enzymatic treatments compared with plant or dairy sources. As a result, determining the optimal conditions for the enzymatic hydrolysis of aquatic raw materials remains an area of extensive research. Studies have shown that the use of specific proteases, combined with tailored hydrolysis parameters, can significantly enhance the yield and functionality of MCPs. For example, alkaline protease has been found to be particularly effective in hydrolyzing certain aquatic proteins, resulting in peptides with high metal-chelating capacities.

In conclusion, the preparation of MCPs is a multifaceted process that requires careful consideration of raw material selection, enzymatic hydrolysis conditions, and chelation parameters. The shift toward enzymatic hydrolysis reflects its advantages in terms of safety, cost, and controllability [15]. However, further research is needed to fully optimize these processes, particularly for aquatic raw materials, to maximize the yield, functionality, and application potential of MCPs in various industries.

6.2.1 Enzymatic Hydrolysis

Enzymatic hydrolysis has gained widespread application in the preparation of metal ion-chelated peptides due to its mild reaction conditions, high specificity, safety, and environmental friendliness. Current research in this field primarily focuses on three key areas: the selection of appropriate enzyme types, the optimization of hydrolysis conditions, and the refinement of hydrolysis processes to achieve metal ion-chelated peptides with superior chelation activity. Proteases, the enzymes used in this process, exhibit diverse properties, which directly influence the hydrolysis outcomes. Factors such as the substrate-to-solvent ratio, hydrolysis temperature, pH, enzyme dosage, and hydrolysis duration, as well as the complex interactions among these variables, play a critical role in determining the efficiency and effectiveness of the hydrolysis process [16]. Consequently, identifying the optimal conditions for enzymatic hydrolysis of specific substrates remains an area requiring extensive investigation.

For instance, Lin Haiyan and colleagues successfully prepared a chelate of Antarctic krill peptides and ferrous ions, achieving a remarkable chelation rate of 77.25%. This study highlights the potential of enzymatic hydrolysis in enhancing the high-value utilization of underutilized marine resources. Similarly, Zhu conducted a comparative analysis of three proteases—alkaline protease, neutral protease, and trypsin—on defatted wheat germ protein hydrolysates. Their findings revealed that alkaline protease, under optimized conditions, yielded the highest degree of hydrolysis (15.61% ± 0.09%) and demonstrated a zinc chelation capacity of 69.62% ± 0.96%. These results underscore the importance of enzyme selection in maximizing both hydrolysis efficiency and functional properties of the resulting peptides [17].

Further advancements in enzymatic hydrolysis techniques have been demonstrated by Zhang Chunzhi and colleagues, who investigated the effects of five commercial proteases on the calcium chelation capacity of peanut hydrolysates. Their research indicated that alkaline protease-derived hydrolysates exhibited superior calcium chelation capacity compared with other enzymes. Moreover, the study introduced a novel approach of stepwise combined hydrolysis using ProteAXH, which significantly enhanced hydrolysis efficiency. This method resulted in peanut peptides with a calcium chelation capacity of 84.09 mg/g, demonstrating the synergistic effects of combining multiple proteases. Such findings suggest that the strategic use of enzyme combinations can not only improve the degree of hydrolysis but also enhance the functional properties of peptides [16], such as their metal-chelating capacity, within a shorter processing timeframe.

Additionally, variations in hydrolysis parameters, particularly duration and temperature, have been shown to significantly influence the calcium chelation capacity of peptides. For example, prolonged hydrolysis durations or suboptimal temperatures may lead to peptide degradation or reduced functionality. Therefore, meticulous optimization of these parameters is essential to achieve the desired peptide characteristics. In practical

Table 6.1 Some metal-chelating peptides and their chelating sites.

Metal chelating peptide	Protein family	Metal source	Chelation sites	Ref.
Zinc ion-chelating peptide	Oyster	$ZnSO_4{\cdot}7H_2O$	Zinc ion with carbonyl metal-acceptor mode, zinc ion with amino metal-donor mode, zinc ion with carboxyl charge–charge mode	[19]
	Tilapia skin	$ZnSO_4$	Amide bond, nitrogen atom of amino group, and nitrogen atom of carboxyl group	[20]
	Antarctic krill	$ZnSO_4$	Oxygen atom of peptide carboxyl group and amino acid atom	[21]
	Sea cucumber	$ZnSO_4{\cdot}7H_2O$	Hydroxyl group of -COOH at the C-terminal of the peptide chain	[22]
Iron ion-chelating peptide	Abalone vscera	$FeSO_4$	Amino group of peptide backbone, carboxylate terminal group, and peptide bond	[23]
	Fish scales	$FeCl_2{\cdot}4H_2O$	Amino group, carboxyl group, and peptide bond in the peptide	[24]
	Alaska salmon bone	$FeSO_4{\cdot}7H_2O$	Carboxylate group and peptide bond of the peptide	[25]
Calcium ion-chelating peptide	Marine fish bone collagen peptide	$CaCl_2$	Amino group and carboxyl group in the peptide	[26]
	Pacific cod bone	$CaCl_2$	Hydrogen atom of carboxyl group of Asp-3 and Lys-10, and nitrogen atom of side-chain amino group of Lys-10	[27]
	Anchovy juice	Oyster shell	$-NH_2$, COO−, N—H, C=O, C—H, and −OH groups	[28]

operations, selecting the appropriate protease and fine-tuning hydrolysis conditions are critical steps in obtaining peptides with the strongest chelation capacity.

In conclusion, enzymatic hydrolysis represents a highly effective and sustainable method for producing metal ion-chelated peptides. By leveraging the unique properties of different proteases and optimizing hydrolysis conditions, researchers can enhance both the yield and functional quality of these peptides. Future studies should continue to explore innovative enzyme combinations and process refinements to further advance the field and expand the applications of metal ion-chelated peptides in nutrition, health, and industry (Table 6.1) [18].

6.2.2 Microbial Fermentation Method

The microbial fermentation method is suitable for producing metal ion-chelated peptides from protein-rich raw materials. For example, Yang Liu and others used *Bacillus subtilis* to ferment defatted Antarctic krill powder, resulting in a high yield of peptides with good antioxidant activity. Wang and others successfully obtained peptides with excellent calcium chelation capacity by fermenting cucumber seeds using *B. subtilis* HYT. Similar to enzymatic hydrolysis, multiple microbial combinations can also be used in fermentation; however, careful selection of strains and control of fermentation time are necessary as many enzyme-producing strains may have harmful effects on the organism and the efficiency of protein utilization during production may be low, making it difficult to ensure the stability of biological activity [29]. Although the technology for preparing metal ion-chelated peptides through microbial fermentation is still in its developmental stage, it is gaining increasing attention due to its safety, economic benefits, and environmental characteristics. In industrial production, considering the time costs of the fermentation process, it is often combined with enzymatic hydrolysis to improve production efficiency.

6.2.3 Chemical Synthesis Method

Moreover, with the rapid development of bioinformatics, genetic engineering, and protein chemical modification technologies, chemical synthesis has become an effective means of preparing calcium-integrated peptides, aiming to maximize their application value. Currently, various metal ion-chelated peptides have been successfully synthesized, including calcium-chelated peptides such as osteopontin peptide DDVDDTDDSHOSDE, a seven-peptide derived from sea cucumber eggs NDEELNK, as well as desalting duck egg white peptides VHSS and VSEE. Zinc-chelated peptides include oyster protein peptide HLRQEEKEEVTVGSLK, which maintains a zinc-binding capacity of 85.98% of its original level, and a novel soybean peptide KYKRQRW with a zinc chelation capacity of 83.21%. Generally, the chemical synthesis of peptides is considered a promising strategy for exploring the relationship between the structure and function of metal ion-chelated peptides. However, the prerequisite for using this method is that the structure of the metal ion-chelated peptide has been fully identified. Chen and others also identified a peptide with the sequence HLRQEEKEEVTVGSLK from oyster protein hydrolysates, which maintained a zinc-binding capacity of 85.98% of its original level after simulated digestion. Zhu and others isolated a novel soybean peptide KYKRQRW with high zinc-binding capacity and zinc solubility from soybean isolate protein hydrolysates, achieving a zinc chelation capacity of 83.21%. Ke and others studied the zinc-binding activity of collagen protein hydrolysates using three enzymes to prepare the hydrolysate from tilapia skin collagen with the highest zinc-binding capacity, which was then purified using immobilized metal affinity chromatography (IMAC-Zn^{2+}) and reverse-phase high-performance liquid chromatography (HPLC), identifying GFLGSP as having the highest chelation activity.

6.3 Evaluation Method of Activity

The evaluation of MCPs' activity is a critical step in determining their efficacy in enhancing mineral bioavailability. A comprehensive assessment of MCPs involves analyzing their ability to chelate metal ions, maintain stability under physiological conditions, and facilitate efficient absorption in the gastrointestinal (GI) tract. Several key methods for evaluating MCPs' activity are outlined later, with a focus on three primary aspects: chelating capacity, physiological stability, and absorption efficiency [30]. These methods provide valuable insights into the functional performance of MCPs and their potential applications in improving mineral nutrition and health. What follows is the translated version of the chelating peptide evaluation methods in an English table format (Table 6.2).

Table 6.2 Evaluation methods for chelating peptides.

Category	Specific method	Technique/experimental approach	Primary purpose	Ref.
Structural characterization	Mass spectrometry	Matrix assisted laser desorption ionization time of flight mass spectrometry (MALDI-TOF), electrospray ionization mass spectrometry (ESI-MS)	Verify peptide sequence and determine metal binding ratio (e.g., 1:1 or 2:1)	[31, 32]
	Nuclear magnetic resonance (NMR)	1D/2D NMR	Analyze secondary/tertiary structures and identify metal-binding residues	[33]
	Circular dichroism	Wavelength scanning (190–250 nm)	Assess conformational changes (α-helix, β-sheet) upon metal binding	[33]
	Fourier transform infrared spectroscopy	Functional group vibration modes	Detect involvement of carboxyl, thiol, or amino groups in metal coordination	[34]
Metal-binding capacity	UV–Vis spectroscopy	Characteristic absorption peaks (e.g., Fe^{3+} at 280 nm)	Qualitative/quantitative analysis of metal binding	[35]
	Fluorescence spectroscopy	Fluorescence quenching/enhancement (Trp/Tyr)	Indirect evaluation of binding via fluorescence signal changes	[36]
	Isothermal titration calorimetry	Thermodynamic parameter measurement	Determine binding constant (Ka), stoichiometry (n), enthalpy (ΔH), and entropy (ΔS)	[37]
	Inductively coupled plasma mass spectrometry	Thermodynamic parameter measurement	Calculate chelation efficiency (e.g., Chelation% = [Initial-Free]/Initia×100%)	[38]
	Competitive chelation assay	EDTA (Ethylenediaminetetraacetic acid) or other chelators	Compare relative binding strength against known agents	[39]
Stability assessment	pH stability	Simulated gastric (pH 2.0)/intestinal (pH 7.4) fluids	Evaluate stability under physiological pH conditions	[40]
	Thermal stability	Heating (60 °C, 90 °C)	Detect dissociation of metal–peptide complexes after heat treatment	[41]
	Enzymatic stability	Pepsin/trypsin digestion	Test resistance to proteolytic degradation in biological systems	[42]

(*Continued*)

Table 6.2 (Continued)

Category	Specific method	Technique/experimental approach	Primary purpose	Ref.
Functional evaluation	Antioxidant capacity	DPPH/$ABTS^+$/FRAP assays	Measure free radical scavenging or oxidative stress inhibition	[43]
	Metal ion reduction ability	$Fe^{3+} \rightarrow Fe^{2+}$ reduction assay	Assess reductive capacity for high-valence metals	[44]
	Bioavailability	Caco-2 cell model	Evaluate metal ion absorption or excretion efficiency	[45]
	Toxicity assessment	MTT (Thiazolyl Blue) assay (cytotoxicity), animal studies	Validate biocompatibility of chelating peptides/complexes	[46]
Application simulation	In vitro digestion simulation	Simulated gastrointestinal environment	Analyze metal release behavior during digestion	[47]
	Environmental remediation	Heavy metal adsorption assays (Pb^{2+}, Cd^{2+})	Test adsorption capacity and efficiency for pollutants	[48]
	Drug delivery	Drug loading rate, release kinetics	Evaluate performance as a metal–drug carrier (e.g., platinum-based anticancer agents)	[49]
Computational modeling	Molecular docking	Software: AutoDock, Schrödinger	Predict metal-binding sites and interaction modes	[50]
	Molecular dynamics simulation	Software: GROMACS, AMBER	Analyze dynamic stability of metal–peptide complexes in solution	[51]
Comparison and optimization	Comparison with chelators	EDTA, glutathione, etc.	Benchmark performance (efficacy, biocompatibility)	[52]
	Structure–activity relationship	Site-directed mutagenesis, sequence optimization	Optimize chelation efficiency and selectivity	[53]
Final output	Comprehensive report	Data analysis and scenario adaptation	Summarize key metrics: efficiency, stability, safety, and application potential	[54]

6.3.1 Chelating Capacity Assessment

The chelating capacity of MCPs is a fundamental property that determines their ability to bind and transport metal ions. This is typically evaluated using a combination of spectroscopic and calorimetric techniques. UV–visible spectroscopy is employed to monitor changes in absorbance spectra, which can indicate the formation of peptide–metal complexes. Fourier transform infrared (FTIR) spectroscopy is used to identify specific functional groups involved in metal binding, such as carboxyl, amino, and hydroxyl groups. Isothermal titration calorimetry (ITC) provides detailed thermodynamic data, including binding constants, stoichiometry, and enthalpy changes, which are crucial for understanding the strength and specificity of peptide–metal interactions. Additionally, atomic absorption spectroscopy (AAS) is utilized to quantify the concentration of free metal ions remaining in solution after chelation, offering a direct measure of chelation efficiency. These techniques collectively provide a comprehensive understanding of the chelating capacity of MCPs [55].

The initial evaluation of MCPs focuses on quantifying their ability to bind metal ions, a fundamental property that underpins their functionality. This assessment is critical for understanding the potential of MCPs to enhance mineral bioavailability and their applications in nutrition and health. Several advanced analytical techniques are employed to evaluate the chelation capacity of MCPs, each providing unique insights into the binding mechanisms and efficiency of these peptides. UV–visible spectroscopy is one of the most commonly used techniques for monitoring changes in peptide absorbance upon metal ion binding. When MCPs interact with metal ions, shifts in the absorbance spectra can be observed, indicating the formation of peptide–metal complexes. These spectral changes provide indirect evidence of metal binding and can be used to estimate the binding affinity and capacity of MCPs. FTIR spectroscopy complements UV–visible spectroscopy by identifying specific functional groups involved in metal ion coordination. Peptides contain various functional groups, such as carboxyl (–COOH), amino ($-NH_2$), and hydroxyl (–OH) groups, which can participate in metal binding [56]. FTIR spectroscopy detects changes in the vibrational frequencies of these groups upon metal ion interaction, offering detailed information about the binding sites and the nature of the peptide–metal bonds.

For a more in-depth understanding of the interaction dynamics, ITC is employed. ITC measures the heat changes associated with peptide–metal binding, providing thermodynamic parameters such as binding constants, stoichiometry, enthalpy (ΔH), and entropy (ΔS). These parameters are crucial for characterizing the strength, specificity, and spontaneity of peptide–metal interactions. ITC is particularly valuable for comparing the binding affinities of different MCPs and optimizing their design for specific applications.

To quantify the chelation efficiency of MCPs, AAS is utilized. AAS measures the concentration of free metal ions remaining in solution after the chelation process. By comparing the initial and final metal ion concentrations, the chelation efficiency of MCPs can be precisely calculated. This method provides a direct and quantitative measure of the peptides' ability to sequester metal ions, making it an essential tool for evaluating MCP performance.

In addition to these techniques, other methods such as inductively coupled plasma mass spectrometry (ICP-MS) and nuclear magnetic resonance (NMR) spectroscopy can also be employed to further characterize the metal-binding properties of MCPs [57]. ICP-MS offers high sensitivity for detecting trace metal ions, while NMR spectroscopy provides detailed structural information about the peptide–metal complexes.

Together, these techniques form a comprehensive toolkit for assessing the chelation capacity of MCPs. By combining spectroscopic, calorimetric, and quantitative analytical methods, researchers can gain a holistic understanding of the metal-binding properties of MCPs, paving the way for their optimization and application in improving mineral nutrition and health. Future research should focus on integrating these methods with computational modeling to predict and design MCPs with enhanced chelation capacities and targeted functionalities.

6.3.2 Physiological Stability Assessment

The stability of MCPs under physiological conditions is essential for their bioactivity and functionality. In vitro digestion models, which simulate the harsh conditions of the GI tract, are commonly used to evaluate the stability of MCPs. The stability of MCPs in the GI tract is a critical factor in determining their bioactivity and functional efficacy. To ensure that MCPs can withstand the harsh conditions of the GI environment and reach their target sites in an active form, researchers employ a combination of in vitro digestion models and advanced analytical techniques. These methods simulate the physiological conditions of the stomach and intestine, providing valuable insights into the stability, integrity, and functionality of MCPs during digestion. These models typically involve the use of digestive enzymes such as pepsin and trypsin, as well as variations in pH, to mimic the stomach and intestinal environments.

HPLC and mass spectrometry (MS) are employed to analyze the peptide profiles before and after digestion, providing insights into the integrity and stability of MCPs. Zeta potential analysis is also used to assess the colloidal stability of MCPs and their metal complexes, which can influence their behavior in the GI tract. These methods help ensure that MCPs retain their functional properties during digestion and absorption.

6.3.2.1 High-Performance Liquid Chromatography

HPLC is a powerful analytical tool used to analyze the peptide profiles before and after in vitro digestion. This technique separates and quantifies the components of a peptide mixture based on their chemical properties, such as size, charge, and hydrophobicity. By comparing the chromatograms of undigested and digested MCPs, researchers can identify changes in peptide composition, such as the breakdown of larger peptides into smaller fragments or the formation of new peptide–metal complexes. HPLC provides a detailed assessment of the stability of MCPs and their ability to retain their metal-chelating activity after digestion.

6.3.2.2 Mass Spectrometry

MS complements HPLC by providing precise molecular weight and structural information about the peptides. MS can detect and characterize even minor changes in peptide structure, such as the loss of specific amino acids or the formation of covalent modifications [58]. By analyzing the mass spectra of MCPs before and after digestion, researchers can gain insights into the specific sites of enzymatic cleavage and the mechanisms of peptide degradation. This information is crucial for understanding how MCPs maintain their stability and functionality under GI conditions.

6.3.2.3 Zeta Potential Analysis

Zeta potential analysis is used to evaluate the colloidal stability of MCPs and their metal complexes in the GI environment. The zeta potential, which measures the surface charge of particles in a solution, is an indicator of the stability of colloidal dispersions. A high zeta potential (either positive or negative) suggests strong repulsive forces between particles, preventing aggregation and ensuring stability. In the context of MCPs, zeta potential analysis can reveal how changes in pH and enzymatic activity affect the surface charge and aggregation behavior of peptide–metal complexes [59]. This information is critical for predicting the behavior of MCPs in the GI tract and optimizing their formulation for enhanced stability and bioavailability.

6.3.2.4 Integration of Techniques for Comprehensive Assessment

The combination of in vitro digestion models, HPLC, MS, and zeta potential analysis provides a comprehensive evaluation of the stability of MCPs in the GI tract. These techniques collectively offer insights into the structural integrity, enzymatic resistance, and colloidal stability of MCPs, ensuring that they can effectively deliver metal ions to target sites in the body. Future research should focus on refining these methods and integrating them with in vivo studies to further validate the stability and bioactivity of MCPs in real physiological conditions.

In conclusion, the stability of MCPs in the GI tract is a key determinant of their functional efficacy. By employing advanced analytical techniques and in vitro models, researchers can ensure that MCPs maintain their structural and functional integrity during digestion, paving the way for their successful application in improving mineral nutrition and health.

6.3.3 Absorption Efficiency Assessment

The absorption efficiency of MCPs is a critical factor in determining their bioavailability and physiological effects. Caco-2 cell monolayers, a well-established in vitro model of the human intestinal epithelium, are widely used to study the transepithelial transport of MCPs and their metal complexes [60]. The quantification of transported metal ions to the basolateral side is typically achieved using inductively coupled plasma atomic emission spectrometry (ICP-AES) or HPLC. The everted gut sac model offers an *ex vivo* approach to study the absorption kinetics of MCPs, providing a more physiologically relevant assessment of their bioavailability. These models help elucidate the mechanisms of MCP absorption and their potential to enhance mineral uptake in the body.

The absorption studies of MCPs across intestinal cell models are essential for several reasons. Firstly, they help in identifying the most effective MCPs and their metal complexes for enhancing the bioavailability of essential minerals such as calcium, iron, and zinc. Secondly, these studies provide critical data for optimizing the formulation and delivery of MCP-based products, ensuring that they can effectively cross the intestinal barrier and exert their intended physiological effects [61].

Moreover, the insights gained from these absorption studies can guide the development of novel MCPs with improved absorption properties. For example, modifications to the peptide sequence or the incorporation of specific functional groups can be explored to enhance their binding affinity and transport efficiency.

In conclusion, the use of both in vitro (Caco-2 cell monolayers) and *ex vivo* (everted gut sac) models provides a comprehensive approach to studying the absorption of MCPs and their metal complexes. These models, combined with advanced analytical techniques such as ICP-AES and HPLC, offer valuable insights into the bioavailability and potential applications of MCPs in various fields, from nutrition to pharmaceuticals. Future research in this area is likely to focus on further refining these models and exploring new strategies to enhance the absorption and efficacy of MCPs.

6.3.4 Biological Activity Assessment

The biological activity of MCPs is evaluated through cell-based assays and animal models to determine their impact on cellular metabolism and overall health. Osteoblast-like MC3T3-E1 cells are commonly used to assess the proliferative and differentiative effects of MCPs, highlighting their potential in promoting mineralization and bone formation. Animal models, such as zebrafish and rodents, provide in vivo insights into the physiological effects of MCPs on mineral absorption and skeletal health [62]. These studies validate the efficacy of MCPs in ameliorating mineral deficiencies and support their potential applications in nutraceuticals and functional foods.

6.3.4.1 Cell-Based Assays: Osteoblast-Like MC3T3-E1 Cells

One of the key approaches to evaluating the impact of MCPs on cellular metabolism is through cell-based assays. Osteoblast-like MC3T3-E1 cells, derived from mouse calvarial bones, serve as a widely used model for studying bone formation and mineralization [63]. These cells exhibit many characteristics of osteoblasts, including the ability to proliferate, differentiate, and form mineralized extracellular matrices, making them an ideal tool for assessing the effects of MCPs on bone health.

When investigating the effects of MCPs on MC3T3-E1 cells, researchers typically focus on two main aspects: cell proliferation and differentiation. Cell proliferation assays, such as those using the MTT or BrdU incorporation methods, help determine whether MCPs stimulate or inhibit the growth of osteoblast-like cells [64]. This is an important factor, as a healthy rate of cell proliferation is essential for maintaining bone tissue.

Differentiation assays, on the other hand, provide insights into the ability of MCPs to promote the maturation of osteoblasts and the subsequent formation of mineralized bone matrix. This is often assessed through the detection of specific markers associated with osteoblast differentiation, such as alkaline phosphatase activity, collagen type I expression, and the formation of mineralized nodules. By evaluating these parameters, researchers can determine whether MCPs enhance the differentiation process, thereby contributing to improved bone formation and mineralization [65].

The use of MC3T3-E1 cells in these assays highlights the potential of MCPs in promoting mineral utilization and bone health. Positive results in these cell-based studies suggest that MCPs may have therapeutic applications in conditions such as osteoporosis, where bone mineral density is compromised.

6.3.4.2 *In Vivo* Studies: Animal Models

While cell-based assays provide valuable information on the direct effects of MCPs on osteoblast-like cells, in vivo studies are essential for understanding their physiological impact in a whole-organism context. Animal models, such as zebrafish and rodents, offer unique opportunities to investigate the broader effects of MCPs on mineral absorption and skeletal health.

Zebrafish, with their transparent embryos and rapid development, are particularly useful for visualizing the effects of MCPs on mineral uptake and bone formation in real time. Researchers can introduce MCPs into zebrafish larvae and observe the subsequent changes in mineral deposition and skeletal development using imaging techniques such as X-ray micro-computed tomography or fluorescent labeling. This allows for a detailed assessment of how MCPs influence mineral absorption and utilization in a living organism.

Rodent models, such as mice and rats, provide another valuable platform for studying the physiological effects of MCPs. These models enable researchers to investigate the long-term impact of MCPs on bone health, including changes in bone mineral density, bone strength, and overall skeletal structure. By feeding rodents diets supplemented with MCPs and monitoring their bone health over time, researchers can gather data on the efficacy of MCPs in ameliorating mineral deficiencies and preventing bone-related disorders.

In vivo studies in animal models are crucial for validating the findings from cell-based assays and providing a more comprehensive understanding of the biological effects of MCPs. These studies help bridge the gap between in vitro observations and potential clinical applications, ensuring that MCPs are effective and safe for use in humans.

In conclusion, the evaluation of MCPs requires a comprehensive approach that encompasses chemical, biochemical, and biological assays. These methods not only ascertain the chelation potential of MCPs but also simulate their journey from ingestion to absorption, providing a holistic understanding of their activity. Future research should aim to refine these methodologies and explore new ones to further elucidate the mechanisms underlying the activity of MCPs in various biological contexts.

6.3.4.3 Comprehensive Evaluation of MCPs

In conclusion, the evaluation of MCPs requires a comprehensive approach that integrates chemical, biochemical, and biological assays. Each method contributes unique insights into the activity and potential applications of MCPs (Table 6.3):

Table 6.3 Evaluation of mineral-chelating peptides.

Assay type	Techniques/models	Purpose	Key findings/outcomes
Chemical assays	UV–Visible Spectroscopy, FTIR Spectroscopy, ICP-AES	Determine the chelation capacity and stability of MCPs	Confirm MCPs' ability to bind and transport essential minerals effectively [25, 66]
Biochemical assays	Cell-based assays using osteoblast-like MC3T3-E1 cells	Assess proliferative and differentiative effects of MCPs on bone-forming cells	Demonstrate potential of MCPs in promoting mineralization and bone formation [63]
Biological assays	In vivo animal models such as zebrafish and rodents	Explore physiological effects of MCPs on mineral absorption and skeletal health	Validate efficacy of MCPs in reducing mineral deficiencies and enhancing overall bone health [2]

By combining these approaches, researchers can simulate the journey of MCPs from ingestion to absorption and provide a holistic understanding of their activity [67]. This comprehensive evaluation is essential for developing MCPs as effective therapeutic agents and nutritional supplements.

In summary, the comprehensive evaluation of MCPs through chemical, biochemical, and biological assays provides a solid foundation for understanding their potential in promoting mineral utilization and bone health. Future research should continue to refine these methodologies and explore new approaches to fully elucidate the mechanisms underlying the activity of MCPs, paving the way for their effective application in addressing mineral deficiencies and related health issues [68].

6.4 Chelation Mechanism

6.4.1 Chelation Mechanism of Metal-Chelating Peptides

The mechanism by which peptides chelate metal ions is a complex and fascinating process that involves the coordination of metal ions with specific amino acid residues within the peptide sequence. This interaction is primarily driven by the functional groups present in the amino acids, which can form stable bonds with metal ions, thereby facilitating their binding and transport.

6.4.2 Key Functional Groups in Metal Binding

Among the various functional groups in peptides, carboxylic and amino groups play crucial roles in metal binding. These groups are particularly abundant in residues such as aspartic acid, glutamic acid, and histidine [69]. The carboxylic groups (−COOH) and amino groups ($-NH_2$) can donate electrons to metal ions, forming coordinate covalent bonds. Additionally, the oxygen and sulfur atoms in the side chains of certain amino acids can also provide lone pairs of electrons, further enhancing the binding affinity for metal ions.

6.4.3 Advanced Techniques for Understanding Binding Modes

To gain a deeper understanding of the binding modes at the atomic level, researchers have employed advanced techniques such as molecular docking and density functional theory (DFT) calculations. Molecular docking simulations allow for the prediction of how peptides and metal ions interact in three-dimensional space, providing insights into the spatial arrangement and orientation of the binding sites. DFT calculations, on the other hand, offer detailed information on the electronic structure and energetics of the peptide–metal complexes, helping elucidate the nature of the coordinate covalent bonds formed during chelation.

These studies have revealed that the chelating mechanism often involves the formation of coordinate covalent bonds between the peptide's functional groups and the metal ions.

The nature of these bonds can vary depending on the peptide's structure and the metal ion's properties, with the possibility of bidentate or unidentate coordination. Bidentate coordination occurs when a single peptide residue can bind to a metal ion through two different atoms (both oxygens on a carboxylate group), while unidentate coordination involves binding through a single atom [70].

6.4.4 Specific Chelation Modes of Peptides with Metal Ions

6.4.4.1 Coordination Sites and Bond Formation

Research has identified that the mechanism of metal ion chelation with peptides involves coordination at amino, carboxyl, and peptide bond positions. Metal ions can form covalent bonds with these functional groups, and the oxygen and sulfur atoms in the side chains can provide lone pairs of electrons to bind with the metal ions. This interaction is essential for the stability and functionality of the peptide–metal complexes.

6.4.4.2 Main Chelation Modes for Marine Peptides

For marine peptides, three main chelation modes with metal ions have been extensively studied:

1) Monodentate mode: In this mode, a metal ion such as Ca^{2+} chelates with a single oxygen atom on the carboxylate group (COO^-). This type of binding is relatively weak but can still contribute to the overall stability of the complex.
2) Bidentate mode: Here, both oxygen atoms on the carboxylate group (COO^-) can bind with the metal ion (e.g., Ca^{2+}). This mode provides a stronger and more stable interaction compared to the monodentate mode, as the metal ion is coordinated by two atoms from the same functional group.
3) α-Chelation mode: This mode involves the binding of a metal ion (e.g., Ca^{2+}) with one oxygen atom from a carboxylate group and one suitable organic atom (O, N, S, etc.) from another calcium-binding group within the peptide. For example, Zhang et al. purified a calcium-binding decapeptide from the hydrolysate of Pacific cod bone gelatin, where Ca^{2+} binds with the carboxyl O atom and the side-chain amino N atom of Lys-10, forming a calcium chelated peptide through the α-chelation mode [62].

6.4.5 Factors Influencing Chelation

6.4.5.1 Amino Acid Composition and Sequence

The amino acid composition and sequence of a peptide are fundamental in defining its chelating properties. Different amino acids contribute distinct functional groups that can interact with metal ions. For instance, amino acids like histidine, aspartic acid, and glutamic acid are known for their high affinity for metal ions due to the presence of imidazole rings and carboxylate groups, respectively. The specific arrangement of these amino acids within the peptide sequence can create optimal binding sites for metal ions, thereby enhancing the chelation efficiency [5].

Moreover, the sequence also affects the peptide's conformation, which in turn influences its biological activity. For example, certain sequences may promote the formation of stable secondary structures (e.g., α-helices or β-sheets) that facilitate metal binding. These structural elements can enhance the peptide's stability and its ability to interact with biological targets, such as enzymes or cell membranes.

6.4.5.2 Hydrophilicity/Hydrophobicity Balance

The hydrophilicity or hydrophobicity of a peptide is another critical factor that impacts its biological activity, particularly its antibacterial properties. Peptides with a balanced hydrophilic/hydrophobic nature are often more effective in interacting with bacterial cell membranes, which are composed of both hydrophilic and hydrophobic regions. Hydrophobic regions of the peptide can anchor into the lipid bilayer, while hydrophilic regions can interact with the aqueous environment, facilitating the disruption of the bacterial membrane and leading to antibacterial effects.

In the context of metal chelation, the hydrophilic/hydrophobic balance can also influence the solubility and bioavailability of the peptide–metal complex. Hydrophilic peptides are more soluble in aqueous environments, which is beneficial for applications in biological systems. However, an optimal balance is often required to ensure that the peptide can effectively interact with both metal ions and biological targets.

6.4.5.3 Functional Groups and Metal Ion Chelation

The amino and carboxyl groups in amino acids are particularly effective at chelating metal ions due to their ability to donate electrons. The nitrogen atom in the amino group and the oxygen atoms in the carboxyl group can act as ligands, forming stable complexes with metal ions. This chelation process is critical for various biological functions, including the regulation of enzyme activity, the transport of essential minerals, and the maintenance of cellular homeostasis.

6.4.5.4 Influence of R Groups

In addition to the amino and carboxyl groups, the R groups (side chains) of amino acids also play a significant role in metal ion chelation. These side chains can alter the chemical environment of the ligand groups, thereby influencing the formation and stability of metal ion complexes. For example, the presence of sulfur-containing groups (as in cysteine) or aromatic rings (as in histidine) can enhance the binding affinity for certain metal ions by providing additional coordination sites or stabilizing the complex through π-electron interactions [71].

6.4.5.5 Amino Acids with High Chelation Activity

Current research has identified several amino acids that exhibit high chelation activity, making them particularly important in the formation of metal ion complexes. These amino acids include (Table 6.4):

Table 6.4 MCPs of key amino acids.

Amino acid	Abbreviation	Key functional group	Metal binding mechanism	Metal ions most effectively bound	Ref.
Histidine	His	Imidazole ring (contains nitrogen)	Nitrogen atoms can donate electrons to metal ions, forming coordination bonds	Zn^{2+}, Fe^{2+}	[72, 73]
Cysteine	Cys	Thiol group (–SH)	Forms strong bonds with metal ions, especially those with high sulfur affinity	Heavy metals (e.g., Hg^{2+}, Ag^{+}), essential minerals	[74, 75]
Aspartic Acid	Asp	Carboxylate group ($-COO^{-}$)	Carboxylate can form stable complexes with metals	Ca^{2+}, Mg^{2+}	[76, 77]
Serine	Ser	Hydroxyl group (–OH)	Hydroxyl group can participate in metal coordination	Various metal ions	[78]
Glutamic Acid	Glu	Carboxylate group ($-COO^{-}$)	Similar to Asp, coordinates via carboxylate groups	Ca^{2+}, other divalent cations	[79]

6.4.6 Conclusion

The structure of a metal-chelated peptide complex is intricately linked to its biological activity. The amino acid composition, sequence, and hydrophilicity/hydrophobicity balance are key factors that influence its antioxidant and antibacterial properties. While significant progress has been made in understanding these relationships, further research is needed to fully establish the structure–activity correlations and optimize peptide design for specific applications [80]. By leveraging advanced techniques and interdisciplinary approaches, researchers can unlock the full potential of metal-chelated peptides as versatile tools in biomedicine, nutrition, and beyond. The chelation of metal ions by peptides is a complex and dynamic process influenced by the amino acid composition, sequence, and side chain interactions. Amino acids such as histidine, cysteine, aspartic acid, serine, and glutamic acid play crucial roles in this process, with acidic amino acids like aspartic acid and glutamic acid being particularly effective in enhancing calcium chelation. The calcium-chelating peptide segment identified from tilapia-scale protein exemplifies the importance of these amino acids in forming stable metal complexes. Future research should focus on optimizing peptide design, elucidating chelation mechanisms, and exploring the diverse applications of MCPs in biology, medicine, and environmental science.

6.5 Applications

6.5.1 The Multifaceted Applications of Metal-Chelating Peptides

6.5.1.1 Applications in the Food Industry

The use of MCPs in the food industry is a rapidly growing area of interest. One of the primary applications is as natural mineral fortifiers in food products. By incorporating MCPs into various food items, manufacturers can significantly enhance the mineral content, thereby improving the overall nutritional value of these products [81]. This is particularly important in addressing widespread mineral deficiencies in populations with inadequate dietary intake of essential minerals such as calcium, iron, and zinc.

MCPs offer a unique advantage over traditional mineral supplements. Unlike inorganic minerals, which can be poorly absorbed and may interact negatively with other dietary components, MCPs form stable complexes with metal ions [82]. This chelation process increases the solubility and bioavailability of these ions, ensuring that they are more readily absorbed and utilized by the body. Additionally, MCPs can facilitate the transport of metal ions across cell membranes, further enhancing their absorption and utilization.

6.5.1.2 Applications in the Pharmaceutical Industry

In the pharmaceutical industry, MCPs are being explored for their potential in treating mineral deficiency-related diseases. Conditions such as osteoporosis, characterized by low bone mineral density, and anemia, often caused by iron deficiency, are significant public health concerns [83]. MCPs offer a promising therapeutic approach by improving the bioavailability and absorption of essential minerals, thereby addressing the underlying deficiencies.

Research has shown that MCPs can protect metal ions from degradation by digestive enzymes, a common issue with traditional mineral supplements. This protection ensures that the metal ions remain intact and available for absorption throughout the GI tract [84]. Furthermore, MCPs can promote the transport of these ions across intestinal cells, enhancing their uptake and utilization by the body.

6.5.1.3 Conclusion and Future Directions

In conclusion, the application of MCPs in enhancing mineral bioavailability and absorption represents a promising area of research with significant implications for public health and nutrition. MCPs offer a natural and effective means of improving mineral absorption, addressing widespread deficiencies, and treating related health issues [85]. As our understanding of the structure–function relationships of MCPs advances, so does the potential for developing novel therapeutic and nutritional strategies.

Future research should focus on several key areas:

1) Optimizing production: Developing efficient and scalable methods for producing MCPs from various sources, including food processing by-products, to ensure their availability and affordability.
2) Elucidating mechanisms: Further investigating the molecular mechanisms by which MCPs enhance mineral absorption, including their interactions with digestive enzymes, transporters, and receptors [86].

3) Exploring applications: Expanding the use of MCPs in various food systems and pharmaceutical formulations to maximize their benefits. This includes exploring their potential in functional foods, dietary supplements, and therapeutic agents.
4) Clinical trials: Conducting long-term clinical trials to assess the safety and efficacy of MCPs in human populations, particularly in diverse demographic groups with varying nutritional needs.

In summary, the development and application of MCPs represent a significant advancement in the field of food science and nutrition [87]. These peptides offer a natural and effective means of improving mineral bioavailability, with potential benefits for human health and nutrition. Further research is warranted to fully realize their potential and develop innovative solutions for addressing mineral deficiencies and related health issues.

References

1 Balint, R. and Boajă, I.P. (2024). Assisted phytoextraction as a nature-based solution for the sustainable remediation of metal(loid)-contaminated soils. *Integrated Environmental Assessment and Management* 20 (6): 2003–2022.

2 Bjørlie, M., Irankunda, R., Yesiltas, B. et al. (2024). Metal-chelating antioxidant peptides: biosensor screening methods as alternatives to the ferrozine assay. *Journal of the American Oil Chemists' Society* 101 (9): 827–837.

3 Bu, G., Ti, G., Zhao, X., and Duan, X. (2024). Isolation, identification, and chelation mechanism of ferrous-chelating peptide from peanut protein hydrolysate. *Journal of the Science of Food and Agriculture* 104 (15): 9368–9378.

4 Bu, G., Zhao, X., Wang, M. et al. (2024). Identification of calcium chelating peptides from peanut protein hydrolysate and absorption activity of peptide–calcium complex. *Journal of the Science of Food and Agriculture* 104 (11): 6676–6686.

5 Chai, K.F., Voo, A.Y.H., and Chen, W.N. (2020). Bioactive peptides from food fermentation: a comprehensive review of their sources, bioactivities, applications, and future development. *Comprehensive Reviews in Food Science and Food Safety* 19 (6): 3825–3885.

6 Barzinmehr, H., Ramezanpour, S., Shiri, P.M. et al. (2024). A review of fluorescent peptide-based chemosensors with selectivity for metal ions. *Coordination Chemistry Reviews* 518: 216055.

7 Rivero Pino, F., Pérez Gálvez, R., Espejo Carpio, F.J., and Guadix, E.M. (2020). Evaluation of Tenebrio molitor protein as a source of peptides for modulating physiological processes. *Food & Function* 11 (5): 4376–4386.

8 Case, D.R., Zubieta, J., Gonzalez, R., and Doyle, R.P. (2021). Synthesis and chemical and biological evaluation of a glycine tripeptide chelate of magnesium. *Molecules* 26 (9): 2419.

9 Schuette, S.A., Lashner, B.A., and Janghorbani, M. (1994). Bioavailability of magnesium diglycinate vs magnesium oxide in patients with ileal resection. *Journal of Parenteral and Enteral Nutrition* 18 (5): 430–435.

10 Lv, X.-L. and Luo, S.-Z. (2012). A fluorescence chemosensor based on peptidase for detecting nickel(II) with high selectivity and high sensitivity. *Analytical and Bioanalytical Chemistry* 402 (9): 2999–3002.

11 Mehra, R., Tran, K., Scott, G. et al. (1996). Ag(I)-binding to phytochelatins. *Journal of Inorganic Biochemistry* 61 (2): 125–142.

12 Cui, P., Lin, S., Han, W. et al. (2019). The formation mechanism of a sea cucumber ovum derived heptapeptide-calcium nanocomposite and its digestion/absorption behavior. *Food & Function* 10 (12): 8240–8249.

13 Chen, Q., Guo, L., Du, F. et al. (2017). The chelating peptide (GPAGPHGPPG) derived from Alaska Pollock skin enhances calcium, zinc and iron transport in Caco-2 cells. *International Journal of Food Science & Technology* 52 (5): 1283–1290.

14 Du, Q., Wang, R., Deng, Z. et al. (2024). Structural characterization and calcium absorption-promoting effect of sucrose-calcium chelate in Caco-2 monolayer cells and mice. *Journal of Food Science* 89 (3): 1773–1790.

15 Du, Y.N., Hong, J.N., Xu, S.Q. et al. (2022). Iron-chelating activity of large yellow croaker (pseudosciaena crocea) roe hydrolysates. *Journal of Food Processing and Preservation* 46 (11): e17080.

16 Durand, E., Beaubier, S., Fine, F. et al. (2021). High metal chelating properties from rapeseed meal proteins to counteract lipid oxidation in foods: controlled proteolysis and characterization. *European Journal of Lipid Science and Technology* 123 (6): 2000380.

17 Fadimu, G.J., Farahnaky, A., Gill, H., and Truong, T. (2022). Influence of ultrasonic pretreatment on structural properties and biological activities of lupin protein hydrolysate. *International Journal of Food Science & Technology* 57 (3): 1729–1738.

18 Fan, W., Wang, Z., Mu, Z. et al. (2020). Characterizations of a food decapeptide chelating with Zn(II). *eFood* 1 (4): 326–331.

19 Wang, S., Yang, X., Cheng, S., and Du, M. (2024). The structural characteristics of Crassostrea gigas peptide-zinc complexes and the bioactive mechanism of their enhanced zinc absorption. *Food Bioscience* 58: 103722.

20 Chen, L., Shen, X., and Xia, G. (2020). Effect of molecular weight of tilapia (Oreochromis Niloticus) skin collagen peptide fractions on zinc-chelating capacity and bioaccessibility of the zinc-peptide fractions complexes in vitro digestion. *Applied Sciences* 10 (6): 2041–2052.

21 Sun, R., Liu, X., Yu, Y. et al. (2021). Preparation process optimization, structural characterization and in vitro digestion stability analysis of Antarctic krill (Euphausia superba) peptides-zinc chelate. *Food Chemistry* 340: 128056.

22 Liu, X., Wang, Z., Yin, F. et al. (2019). Zinc-chelating mechanism of sea cucumber (Stichopus japonicus)-derived synthetic peptides. *Marine Drugs* 17 (8): 438–452.

23 Wu, W., Jia, J., Wen, C. et al. (2021). Optimization of ultrasound assisted extraction of abalone viscera protein and its effect on the iron-chelating activity. *Ultrasonics Sonochemistry* 77: 105670.

24 Zhao, Q., Liang, W., Xiong, Z. et al. (2024). Digestion and absorption characteristics of iron-chelating silver carp scale collagen peptide and insights into their chelation mechanism. *Food Research International* 190: 114612.

25 Ma, X., Liu, C., Song, W. et al. (2019). Evaluating the efficacy of a ferrous-ion-chelating peptide from Alaska pollock frame for the improvement of iron nutritional status in rats. *Food & Function* 10 (8): 4888–4896.

26 He, J., Guo, H., Zhang, M. et al. (2022). Purification and characterization of a novel calcium-binding heptapeptide from the hydrolysate of tilapia bone with its osteogenic activity. *Foods* 11 (3): 468–487.

27 Peng, Z., Hou, H., Zhang, K., and Li, B. (2017). Effect of calcium-binding peptide from pacific cod (Gadus macrocephalus) bone on calcium bioavailability in rats. *Food Chemistry* 221: 373–378.

28 Wang, Z., Cheng, S., Wu, D. et al. (2021). Hydrophobic peptides from oyster protein hydrolysates show better zinc-chelating ability. *Food Bioscience* 41: 100985.

29 Fletcher, N.L., Kempe, K., and Thurecht, K.J. (2020). Next-generation polymeric nanomedicines for oncology: perspectives and future directions. *Macromolecular Rapid Communications* 41 (18): e2000319.

30 Ge, H., Li, T., Yang, Q. et al. (2023). Egg white peptides administration in enhancing pathological immune response and regulating intestinal bacteria abundance: A new strategy for relieving young mice colitis. *Food Frontiers* 4 (2): 782–794.

31 Serebryakova, L., Palkeeva, M., Studneva, I. et al. (2019). Galanin and its N-terminal fragments reduce acute myocardial infarction in rats. *Peptides* 111: 127–131.

32 Abou-Zeid, L., Pell, A.G., Cortes, M. et al. (2023). Determination of the affinity of biomimetic peptides for uranium through the simultaneous coupling of HILIC to ESI-MS and ICP-MS. *Analytica Chimica Acta* 1242: 340773.

33 Adpressa, D.A., Reibarkh, M., Jiang, Y. et al. (2022). Interrogation of solution conformation of complex macrocyclic peptides utilizing a combined SEC-HDX-MS, circular dichroism, and NMR workflow. *The Analyst* 147 (2): 325–332.

34 Rezaeerod, K., Heinzmann, H., Torrence, A.V. et al. (2024). Qualitative monitoring of proto-peptide condensation by differential FTIR spectroscopy. *ACS Earth and Space Chemistry* 8 (5): 937–944.

35 Wojciechowski, J.P., Benge, T., Chen, K. et al. (2024). A facile method to quantify synthetic peptide concentrations on biomaterials. *ACS Applied Materials & Interfaces* 16 (37): 49880–49888.

36 Verdoliva, V., Digilio, G., Iaccarino, E., and Luca, S.D. (2025). Solvent-free procedure of an A9 peptide dimer exhibiting specific HER2 receptor binding: Fluorescence spectroscopy evaluation of the enhanced binding affinity. *Journal of Medicinal Chemistry* 68 (15): 16299–16305.

37 Hong, J.K.Y. and Schwendeman, S.P. (2020). Characterization of octreotide–PLGA binding by isothermal titration calorimetry. *Biomacromolecules* 21 (10): 4087–4093.

38 Jin, X., Yang, L., Yan, X., and Wang, Q. (2021). Screening platform based on inductively coupled plasma mass spectrometry for β-site amyloid protein cleaving enzyme 1 (BACE1) inhibitors. *ACS Chemical Neuroscience* 12 (7): 1093–1099.

39 Maisetta, G., Grassi, L., Esin, S. et al. (2017). The semi-synthetic peptide lin-SB056-1 in combination with EDTA exerts strong antimicrobial and antibiofilm activity against pseudomonas aeruginosa in conditions mimicking cystic fibrosis sputum. *International Journal of Molecular Sciences* 18 (9): 1994–2012.

40 Vishweshwaraiah, Y.L., Acharya, A., Hegde, V., and Prakash, B. (2021). Rational design of hyperstable antibacterial peptides for food preservation. *npj Science of Food* 5 (1): 26–42.

41 Sun, X.-X., Fan, J., Hou, Y.-N. et al. (2017). Fluorescence characterization of the thermal stability of collagen mimic peptides. *Chinese Chemical Letters* 28 (5): 963–967.

42 Gray, V.P. and Letteri, R.A. (2024). Designing coiled coils for heterochiral complexation to enhance binding and enzymatic stability. *Biomacromolecules* 25 (8): 5273–5280.

43 Coelho, M.S., Aquino, S.d.A., Latorres, J.M., and Salas-Mellado, M., M.d.l. (2019). In vitro and in vivo antioxidant capacity of chia protein hydrolysates and peptides. *Food Hydrocolloids* 91: 19–25.

44 Lower, B., Hochella, M.J., and Lower, S. (2005). Putative mineral-specific proteins synthesized by a metal reducing bacterium. *American Journal of Science* 305 (6–8): 687–710.

45 Yuanqing, H., Min, C., Lingling, S. et al. (2020). Ultrasound pretreatment increases the bioavailability of dietary proteins by dissociating protein structure and composition. *Food Biophysics* 15 (4): 409–415.

46 Kang, S.-M., Kim, T.-H., and Choi, J.-W. (2012). Cell chip to detect effects of graphene oxide nanopellet on human neural stem cell. *Journal of Nanoscience and Nanotechnology* 12 (7): 5185–5190.

47 de Matos, F.M., Rasera, G.B., and de Castro, R.J.S. (2024). Multifunctional properties of peptides derived from black cricket (Gryllus assimilis) and effects of in vitro digestion simulation on their bioactivities. *Food Research International* 196: 115134.

48 Huang, Y., Li, H., Xing, R. et al. (2025). The application of marine bioactive composite materials in environmental remediation. *Desalination* 602: 118629.

49 Yang, J., An, H.-W., and Wang, H. (2020). Self-assembled peptide drug delivery systems. *ACS Applied Bio Materials* 4 (1): 24–46.

50 Wei, Y., Wang, L., He, Y., and Ma, X. (2024). Identification and molecular docking of novel antioxidant peptides from Candida utilis. *Food Chemistry* 455: 139860.

51 Damjanovic, J., Miao, J., Huang, H., and Lin, Y.-S. (2021). Elucidating solution structures of cyclic Peptides using molecular dynamics simulations. *Chemical Reviews* 121 (4): 2292–2324.

52 Polyakov, N., Leshina, T., Fedenok, L. et al. (2018). Redox-active quinone chelators: properties, mechanisms of action, cell delivery, and cell toxicity. *Antioxidants & Redox Signaling* 28 (15): 1394–1403.

53 Xu, B., Dong, Q., Yu, C. et al. (2024). Advances in research on the activity evaluation, mechanism and structure-activity relationships of natural antioxidant peptides. *Antioxidants* 13 (4): 479–507.

54 Chen, L.C., Nersisyan, S., Wu, C.J. et al. (2022). On the peptide binding affinity changes in population-specific HLA repertoires to the SARS-CoV-2 variants Delta and Omicron. *Journal of Autoimmunity* 133: 102952.

55 Gu, H., Liang, L.,Wei, Y.,Wang, J. H.,Ma, W.,Fu, Y.,Fan, D.,Gao, W.,Yang, J., Zheng, X.,Chen,T. and Chen,Y. (2025) Novel insights into peptide-calcium chelates from lentinula edodes: preparation and its structure, stability, and calcium transport analysis. *Food Science & Nutrition*, 13(1): e4731.

56 Hao, L., Wang, X., Cao, Y., Xu, J. and Xue, C. (2022)A comprehensive review of oyster peptides: preparation, characterisation and bioactivities. *Reviews in Aquaculture*,14(1): 120–138.

57 Hardin, N.Z., Kocman, V., DiMauro, G.M. et al. (2019). Metal-chelated polymer nanodiscs for NMR studies. *Angewandte Chemie International Edition* 58 (48): 17246–17250.

58 Hu, G., Li, X., Su, R. et al. (2024). Effects of sheep bone peptide-chelated calcium on calcium absorption and bone deposition in rats fed a low-calcium diet. *Journal of Food Biochemistry* 2024 (1): 8434888.

59 Hu, Q., Jayasinghe-Arachchige, V.M., Sharma, G. et al. (2020). Mechanisms of peptide and phosphoester hydrolysis catalyzed by two promiscuous metalloenzymes (insulin degrading enzyme and glycerophosphodiesterase) and their synthetic analogues. *Wiley Interdisciplinary Reviews: Computational Molecular Science* 10 (4): e1466.

60 Hu, S., Wang, B., Zeng, R. et al. (2024). Recent advances, challenges and functional applications of protein glycosylation modification in food industry. *International Journal of Food Science & Technology* 59 (7): 4403–4419.

61 Huang, C.F., Liu, Y.H., and Tai, H.C. (2015). Synthesis of peptides containing 2-oxohistidine residues and their characterization by liquid chromatography-tandem mass spectrometry.Journal of. *Peptide Science* 21 (2): 114–119.

62 Hutschenreiter, S., Neumann, L., Rädler, U. et al. (2003). Metal-chelating amino acids as building blocks for synthetic receptors sensing metal ions and histidine-tagged proteins. *ChemBioChem* 4 (12): 1340–1344.

63 Kadam, D., Kadam, A., Koksel, F., and Aluko, R.E. (2024). Plant-derived bioactive peptides: a comprehensive review. *Sustainable Food Proteins* 2 (4): 183–214.

64 Kęska, P. and Stadnik, J. (2021). Effect of in vitro gastro-pancreatic digestion on antioxidant activity of low-molecular-weight (<3.5 kDa) peptides from dry-cured pork loins with probiotic strains of LAB. *International Journal of Food Science & Technology* 56 (12): 6268–6278.

65 Lafarga, T., Álvarez, C., and Hayes, M. (2017). Bioactive peptides derived from bovine and porcine co-products: a review. *Journal of Food Biochemistry* 41 (6): e12418.

66 Leone, L., DeFenza, M., Esposito, A. et al. (2024). Peptides and metal ions: a successful marriage for developing artificial metalloproteins.Journal of. *Peptide Science* 30 (10): e3606.

67 Li, G., Zhan, J., Hu, L. et al. (2022). Identification of novel antioxidant peptide from porcine plasma hydrolysate and its effect in in vitro digestion/HepG2cells model. *Journal of Food Biochemistry* 46 (4): e13853.

68 Liu, Y., Li, S., Huang, Z. et al. (2024). Dietary collagen peptide-chelated trace elements supplementation for breeder hens improves the intestinal health of chick offspring. *Journal of the Science of Food and Agriculture* 104 (1): 174–183.

69 Liu, Z., Zhang, M., Hao, Y. et al. (2024). Application of surface-modified functional packaging in food storage: a comprehensive review. *Comprehensive Reviews in Food Science and Food Safety* 23 (3): e13343.

70 Mondaca-Navarro, B.A., Ávila-Villa, L.A., González-Córdova, A.F. et al. (2017). Antioxidant and chelating capacity of Maillard reaction products in amino acid-sugar model systems: applications for food processing. *Journal of the Science of Food and Agriculture* 97 (11): 3522–3529.

71 Saupe, J., Roske, Y., Schillinger, C. et al. (2011). Discovery, structure–activity relationship studies, and crystal structure of nonpeptide inhibitors bound to the shank3 PDZ domain. *ChemMedChem* 6 (8): 1411–1422.

72 Boniecki, M.T., Freibert, S.A., Mühlenhoff, U. et al. (2017). Structure and functional dynamics of the mitochondrial Fe/S cluster synthesis complex. *Nature Communications* 8 (1): 1287–1302.

73 Zuccarello, L., Berthomieu, C., Boussac, A. et al. (2020). Protonation of the cysteine axial ligand investigated in His/Cys c-type cytochrome by UV–Vis and mid- and far-IR spectroscopy. *The Journal of Physical Chemistry Letters* 11 (10): 4198–4205.

74 Cui, Y., Hao, Y., Zhang, Y. et al. (2016). A water-soluble and retrievable ruthenium-based probe for colorimetric recognition of Hg(II) and Cys. *Spectrochimica Acta Part a-Molecular and Biomolecular Spectroscopy* 165: 150–154.

75 Song, Z., Liu, X., Fan, Y. et al. (2024). Dual-mode detection of cysteamine using Ag nanoparticle-riboflavin composites. *ACS Applied Nano Materials* 7 (7): 6995–7007.

76 Zheng, T. and Yi, H. (2021). Formation and stabilization of vaterite aggregate grooves with aspartic acid (Asp) by bubbling CO2 into a Ca(OH)2 suspension. *Crystal Research and Technology* 56 (11): 2100136.

77 Liu, X., Zhang, W., Liu, J. et al. (2020). Mg2+ inhibits heat-induced aggregation of BSA: The mechanism and its binding site. *Food Hydrocolloids* 101: 105450.

78 Szunyogh, D., Szokolai, H., Thulstrup, P.W. et al. (2015). Specificity of the metalloregulator cueR for monovalent metal Ions: Possible functional role of a coordinated thiol? *Angewandte Chemie International Edition* 54 (52): 15756–15761.

79 Jafari, M., Sagresti, L., Hu, J., and Merz, K.M. (2025). Ion-induced dipole interactions matter in metadynamics simulation of transition metal ion transporters. *Journal of Chemical Theory and Computation* 21 (8): 4221–4235.

80 Wang, Y., Kong, X., Du, M. et al. (2024). Preparation, identification, and digestive stability of antioxidant peptides from chlorella vulgaris. *Journal of Food Biochemistry* 1: 3828251.

81 Wang, Y., Wang, R., Bai, H. et al. (2024). Casein phosphopeptide calcium chelation: preparation optimization, in vitro gastrointestinal simulated digestion, and peptide fragment exploration. *Journal of the Science of Food and Agriculture* 104 (2): 788–796.

82 Wang, Z., Zhai, X., Xiao, X. et al. (2024). Novel peptides with calcium-binding capacity from antler bone hydrolysate, its bioactivity on MC3T3-E1 cells, and the possible chelating mode. *Food Science & Nutrition* 12 (11): 9069–9084.

83 Watanabe, M., Bhawal, U.K., Takemoto, S. et al. (2021). Bio-functionalized titanium surfaces with modified silk fibroin carrying titanium binding motif to enhance the ossific differentiation of MC3T3-E1. *Biotechnology and Bioengineering* 118 (7): 2585–2596.

84 Wei, S., Huo, J., Lin, J. et al. (2024). Egg yolk peptide-iron chelate: preparation, conformational structure, chelating mechanism and stability in vitro. *International Journal of Food Science and Technology* 59 (7): 4646–4659.

85 Xing, L., Liu, R., Cao, S. et al. (2019). Meat protein based bioactive peptides and their potential functional activity: a review. *International Journal of Food Science & Technology* 54 (6): 1956–1966.

86 Xing, L., Liu, R., Tang, C. et al. (2018). The antioxidant activity and transcellular pathway of asp-Leu-Glu-Glu in a Caco-2 cell monolayer. *International Journal of Food Science & Technology* 53 (10): 2405–2414.

87 Xiong, Y., Li, J.R., Peng, P.Z. et al. (2024). Positive effect of peptide-calcium chelates from grifola frondosa on a mouse model of senile osteoporosis. *Journal of Food Science* 89 (6): 3816–3828.

7

Antiaging Peptides

7.1 Definition of Aging and Antiaging Peptides

7.1.1 Biological Basis of Aging

Aging is a complex physiological process that occurs over time, and its biological basis involves multiple levels of cellular and molecular changes [1]. The key markers of cell senescence include telomere shortening, genomic instability, epigenetic changes, etc. [2]. Telomere is a protective structure at the end of the chromosome. As the number of cell divisions increases, telomeres will gradually shorten. When shortened to a certain extent, cells will enter senescence. The integrity of the genome is threatened by a variety of internal and external factors. With the increase of age, DNA repair capacity decreases and genomic instability increases, thus accelerating cell senescence. Epigenetics refers to the change of gene expression that does not involve the change of DNA sequence, but through DNA methylation, histone modification, and other mechanisms. In the process of aging, epigenetic modification will change, affect the expression pattern of genes, and then affect the function of cells and the process of aging [3].

Aging-related pathways include mTOR, AMPK, sirtuins, and insulin/IGF-1 signaling pathways. mTOR is an evolutionarily conserved serine–threonine kinase that senses and integrates a variety of environmental and intracellular signals. Inhibition of mTOR activity can inhibit cell senescence. AMPK plays a fundamental role in the energy metabolism of cells and organisms. With the increase of age, the activation ability of the AMPK signal decreases, which will accelerate the aging process. Sirtuins are a class of NAD (+)-dependent deacetylases, which are very important in regulating a variety of cellular processes. Members of the sirtuins family also interact with AMPK, mTOR, and other pathways to jointly regulate the aging process [4]. Insulin/IGF-1 signaling pathway is the earliest defined pathway related to aging and lifespan in model organisms. Reducing insulin/IGF-1 signal transduction can prolong the lifespan of many organisms [5].

These key markers of cell aging and aging-related pathways together constitute the biological basis of aging. With the continuous advancement of science, our understanding of aging is also deepening, providing more theoretical basis and potential therapeutic targets for antiaging research.

Food-Derived Bioactive Peptides: Green Preparation, Activity Mechanisms, and Health Applications,
First Edition. Shao-Yun Wang, Xu Chen, and Xixi Cai.

7.1.2 The Rise of Antiaging Peptides

Peptide molecules play an important role in the field of modern biomedicine, and their advantages are mainly reflected in three aspects: high specificity, low toxicity, and easy modification. Peptide molecules can accurately identify and combine specific targets to exert their biological effects. This high specificity enables peptide drugs to effectively reduce interference with non-targets and reduce the occurrence of side effects during treatment. Its low toxicity is due to the fact that the metabolites of peptide drugs are usually non-toxic amino acids, which will not accumulate in the body to produce toxicity, which makes peptide drugs have high safety in long-term use. In addition, the structure of peptide molecules is relatively simple, which can be prepared in large quantities by chemical synthesis or biosynthesis, and it is easy to introduce various modified groups in the synthesis process, thus changing their physical and chemical properties, such as improving stability and improving pharmacokinetic characteristics, which provides great convenience for the development and application of peptide drugs [6].

Antiaging peptides are a class of bioactive peptides with significant antiaging effects, which are widely used in food, medicine, cosmetics, and other fields. They delay or reverse the aging process through a variety of mechanisms and become an important direction of modern antiaging research [7–9]. These peptides repair natural aging by promoting collagen synthesis, inhibiting collagen degradation, and increasing antioxidant enzyme activity [10, 11]. The sources of antiaging peptides are diverse, mainly including animals, plants, and microorganisms, among which animals and plants are the main ones. [12] In addition, marine organisms and fungi have shown great potential in the extraction of new antiaging peptides, and researchers are increasingly interested in exploring and utilizing antiaging peptides from these two resources [13, 14].

Antiaging peptides are usually composed of amino acids connected by peptide bonds. They have the characteristics of small size and simple amino acid sequence, which makes them have high absorption efficiency, can cross the biological barrier and maintain biological activity in vivo. [15–17]. These peptides are usually rich in hydrophobic N-terminus, which is conducive to providing hydrogen donor interaction with other amino acids, thereby enhancing their antiaging effects [18, 19]. Antiaging peptides contain polar amino acids and aromatic amino acids. Polar amino acids can prevent free radical oxidation through the chelation of side chains, and aromatic residues are easy to release protons to electron-deficient free radicals, thus exhibiting their strong antiaging activity [20, 21]. For example, pentapeptide-3 (GPRPA) in snake venom and hexapeptide 11 (FVAPFP) in yeast can reduce skin roughness and wrinkles and improve skin elasticity. [22] Collagen peptide (CP) can promote the proliferation of fibroblasts and the synthesis of collagen and elastin by stimulating the nuclear factor κB signaling pathway and transforming growth factor β/Smad signaling pathway [23]. In the field of cosmetics, antiaging peptides are widely used. In addition, the application of polypeptide components in skin care products is also increasing, such as Yulan oil, Estee Lauder, SK-II, and other well-known brands have launched polypeptide cosmetics.

The mechanism of action of antiaging peptides includes anti-oxidation, inhibition of telomere shortening, reduction of mitochondrial damage, anti-glycosylation, and carnosine. Alan R Hipkiss et al. showed that carnosine can undergo a non-enzymatic reaction with the carbonyl group of the protein. This process is called "carnosine". The obvious antiaging effect of carnosine is partly due to its ability to react with glycosylation/oxidation proteins and carbonyl groups on other molecules [24]. Alessandra Zonari et al. found that Pep 14 is a new type of antiaging peptide. Antiaging peptides can affect cell senescence and FoxO signaling pathway by regulating the expression of PP2A subunit, reducing the expression of cell senescence markers, and showing significant antiaging effects in vivo and in vitro models [25]. In summary, antiaging peptides can not only promote the proliferation and repair of skin cells, but also alleviate aging-related diseases related to nerves, skin, and metabolic abnormalities by regulating signaling pathways related to oxidative stress, inflammation, apoptosis, and collagen synthesis. With the advancement of science and technology, the research and application of antiaging peptides will continue to deepen and become the next generation of interventions specifically targeting the root causes of aging, providing more possibilities for the extension of healthy human life expectancy.

7.2 Natural Sources and Synthetic Antiaging Peptides

7.2.1 Natural Antiaging Peptides

With the aging of the global population, antiaging research has become an important topic in the field of life science and medicine. Among many antiaging strategies, natural antiaging peptides have attracted more and more attention due to their advantages of wide source, high safety, and strong biological activity. The sources of natural antiaging peptides are very rich, mainly including animals, plants, and microorganisms. For example, soy protein isolate and sea cucumber protein are typical representatives of plant and animal proteins. They are rich in amino acids and are high-quality mother proteins for bioactive peptides. Collagen peptide (CP) is prepared by hydrolyzing collagen. It has the effects of promoting collagen synthesis, inhibiting collagen degradation, and increasing antioxidant enzyme activity, thereby delaying the natural aging of skin. In addition, milk-derived β-casomorphin-7, pumpkin seed protein peptide, rice bran antioxidant peptide, sunflower seed meal peptide, and bighead carp peptide all showed good antiaging activity. The antiaging peptides extracted from different sources and their antiaging mechanisms are shown in Table 7.1.

7.2.1.1 Animal-Derived Antiaging Peptides

With the increasing attention to health and antiaging, animal- and plant-derived antiaging peptides have attracted much attention due to their unique biological activity and nutritional value. These peptides have a wide range of types and diverse functions and have been widely used in the fields of medicine, health care products, and cosmetics.

Marine organisms are one of the important sources of antiaging peptides. Common marine biological extraction peptides include fish collagen peptides, sea cucumber peptides, oyster peptides, Antarctic krill peptides, etc. These peptides can be extracted from different parts of the animal's muscles, skin, bones, blood, and internal organs. For example, fish collagen peptides are usually extracted from fish scales, fish skins, or fish bones, while oyster peptides are derived from oyster meat. These antiaging peptides extracted from marine organisms have high antioxidant activity, which can effectively remove free radicals in the body and slow down the oxidative aging process of cells.

Table 7.1 Functional components related to antiaging and their mechanisms of action.

Source	Peptides	Mechanism	References
Animal	Whey peptide	It affects lipid metabolism in skeletal muscle and promotes fatty acid oxidation	[26]
	Nutraceutical collagen peptide	Proton-coupled oligopeptide transporters (such as peptide transporter 2, PEPT2) are transported into fibroblasts and affect cell behavior	[27]
	Swim bladder	It showed DPPH•, HO•, and O^{2-}• free radical scavenging ability, as well as high Fe2 + chelating activity. These activities indicate that these peptides can directly scavenge free radicals and chelate metal ions to reduce oxidative stress	[28]
	Marine collagen peptide	It could increase the activity of superoxide dismutase (SOD) and glutathione peroxidase (GSH-Px) and decrease the content of malondialdehyde (MDA) in serum	[29]
Botany	RP1	RP1 can significantly increase the activity of SOD and GSH-Px, decrease the level of MDA, and improve the content of ATP, thus inhibiting oxidative stress and memory impairment in aging mice	[30]
	ADWGGPLPH	By regulating the OPG/RANKL/RANK/TRAF6 signaling pathway, senile osteoporosis can be significantly improved, and it has the potential to be developed as an anti-senile osteoporosis nutritional therapy	[31]
	KF-8	The regulation of Nrf2 and NF-κB signaling pathways by KF-8 reduced oxidative stress and cell damage	[32]
	WHP	WHP may play a role by reducing the degeneration of hippocampal neurons and reducing the apoptosis of nerve cells.	[33]
Microorganism	*Aspergillus* sp. SCSIOW2	Inhibited the production of NO induced by LPS and INF-γ	[34]
	Cyanobacterium (*Anabaena torulosa*)	1) Induced apoptosis Affected mitochondrial functioning by promoting AMPK phosphorylation and inhibiting mTOR	[35]
	Isaria japonica	2) Enhanced the proliferation of astrocytes and activate genes related to neuroprotection and neuronal growth 3) Promoted the growth of dendrites and axons of primary hippocampal neurons Regulated the phenotype of microglia	[36]

Terrestrial animals are also important sources of antiaging peptides. Such as bovine bone peptide, bone marrow peptide, donkey-hide gelatin peptide, etc. In addition, there are peptides from poultry, livestock, and insects. These terrestrial animal-derived peptides also have good biocompatibility and low toxicity, and can be well absorbed and utilized by the human body. In the field of medicine, these antiaging peptides can be used to prepare antiaging drugs to assist in the treatment of aging-related diseases; in the field of health care products, they can be added to nutritional supplements to enhance the body's antioxidant capacity and delay aging, in the field of cosmetics, antiaging peptides are widely used in antiwrinkle, moisturizing, whitening, and other products to improve skin condition and delay skin aging. Antiaging peptides derived from animals and plants have broad application prospects due to their unique biological activities and diverse functions. Future research will further explore and develop the potential of these antiaging peptides to make greater contributions to human health and beauty.

7.2.1.2 Plant-Derived Antiaging Peptides

Plant-derived antiaging peptides are mainly derived from proteins in grains, polypeptides in plant medicines, natural antioxidants (glutathione and lycopene), specific plant components, etc. Grains such as soybeans, corn, and rice are rich in proteins. These proteins can be hydrolyzed to produce a variety of peptides with antiaging functions. These peptides can not only provide essential amino acids, but also improve skin health and delay the aging process through specific biological activities [37]. Some plant medicines, such as amethyst, cordyceps sinensis, and ganoderma lucidum, also contain rich polypeptide components. [38] These botanical drugs have been widely used in traditional medicine. Modern studies have found that the polypeptides in them have significant antioxidant and antiaging effects. The polypeptides in Cordyceps sinensis can enhance immunity and delay cell aging. The polypeptides in Ganoderma lucidum have anti-inflammatory and antioxidant functions, which help to protect cells from free radical damage. Many plant extracts contain natural antioxidants such as glutathione and lycopene [39].

Glutathione is an important antioxidant, which can remove excess reactive oxygen free radicals produced in the process of cell metabolism and reduce the damage to cells caused by membrane lipid peroxidation. The antiaging study of glutathione yeast extract on mice showed that the aging indexes of mice using glutathione yeast extract were lower than those of the placebo group, and the skin elasticity was significantly increased [40]. Lycopene is one of the strongest antioxidants in nature. It can effectively scavenge free radicals in the human body, maintain normal cell metabolism, prevent aging, quench singlet oxygen and capture peroxide-free radicals, prevent lipid peroxidation, and protect biofilms from free radical damage, thereby delaying aging. Studies using Drosophila survival models and aged rat models have shown that lycopene has certain antiaging, anti-oxidation, anti-inflammation, and anticoagulation effects. In the antiaging study of umbilical cord mesenchymal stem cells combined with lycopene, experiments have shown that the combination of these two methods can significantly delay aging [41].

Antiaging peptides are extracted from specific plants, the most typical of which is Heterophyllin B in Pseudostellaria heterophylla extract, which has a significant effect on skin antiaging. Heterophyllin B can inhibit the adhesion and invasion of ECA-109 human esophageal cancer cells by inhibiting the PI3K/AKT/β-catenin signaling pathway. Another study found that Heterophyllin B can alleviate inflammation and oxidative stress in RAW264.7 macrophages caused by lipopolysaccharide, which further indicates its potential in antiaging [42].

7.2.1.3 Microbial-Derived Antiaging Peptides

Antiaging peptides derived from microorganisms, as an important part of bioactive peptides, show significant antiaging properties. These peptides are mainly derived from microorganisms such as bacteria, fungi, and plankton. They can survive in extreme environments and produce new proteins to adapt to harsh living conditions. In particular, marine microorganisms, due to their unique living environment, are considered to be an important source of antiaging peptides. The environment in which marine microorganisms are located is complex and diverse, including extreme conditions such as high pressure, low temperature, and high salt. These special environmental factors have prompted marine microorganisms to evolve unique metabolic pathways and biosynthesis capabilities, which can produce antiaging peptides with novel structures and diverse activities.

Peptides isolated from marine microorganisms have a variety of beneficial properties, such as antioxidant, antibacterial and promoting skin regeneration. These characteristics make them have great application prospects in skin care product development and health industry. For example, some peptides derived from marine microorganisms can effectively scavenge free radicals in the body and reduce the damage of oxidative stress to cells, thereby delaying the aging process. There are also some peptides that can stimulate the proliferation and repair of skin cells, improve the elasticity and luster of the skin, and reduce the formation of wrinkles. In addition, these peptides also have a wide range of potential applications in food, medicine, and other fields, such as functional food additives, lead compounds for drug development, and so on.

Therefore, researchers are increasingly interested in using marine resources to explore peptides with antiaging properties. This study not only promotes the development of the antiaging industry, but also provides new ideas and directions for biomedical research and development. With the deepening of research, microbial sources, especially antiaging peptides in marine microorganisms, are expected to provide more possibilities for delaying aging, improving skin health, and developing new antiaging therapies, which has become a new hotspot in antiaging research and application.

7.2.2 Preparation Method of Natural Antiaging Peptides

The extraction, preparation, separation, and purification of antiaging peptides are the key steps in the study of bioactive peptides. First, the extraction methods of bioactive peptides include enzymatic hydrolysis, direct extraction, microbial fermentation, microwave-assisted extraction, and so on. Among them, enzymatic hydrolysis is widely used because of its mild, efficient, and safe advantages, and corresponding proteases can be selected for hydrolysis of different protein raw materials. Arrutia et al. [43] used trypsin to hydrolyze bovine serum albumin (BSA) to produce a mixture of polypeptide products with angiotensin converting enzyme (ACE) inhibition, dipeptidyl peptidase-4 (DPP-IV) inhibition, and antioxidant effects. Chang [44] et al. extracted peptides from the oil palm kernel hydrolysate (OPKH) obtained by pepsin-pancreatin hydrolysis of oil palm kernel, and purified

OPKH by ultrafiltration, reverse-phase high-performance liquid chromatography (HPLC), and semi-preparative HPLC. Then, three peptides VVGGDGDV, VPVTST, and LTTLDSE were identified by liquid chromatography-electrospray ionization/multi-stage mass spectrometry (LC-ESI-Q-TOF-MS/MS) and de novo sequencing. In order to improve the extraction efficiency and purity, new extraction technologies such as supercritical fluid extraction and ultrasonic-assisted extraction can also be used.

In the preparation of antiaging peptides, genetic engineering combined with *E.coli* fermentation process has shown unparalleled advantages, providing an efficient and stable way for large-scale production [45]. By introducing the gene sequence of the antiaging peptide into the expression vector of *E.coli*, a recombinant strain capable of efficiently expressing the target peptide was constructed. Using the strong reproductive capacity of *E.coli* and the precise control of the fermentation process, a large number of antiaging peptides can be obtained in a short time. This preparation method not only has high yield, but also has relatively low cost, which can meet the market demand. Specifically, the fermentation process of genetically engineered *E.coli* has the following advantages: First, as a prokaryote, *E.coli* has a clear genetic background and is easy to operate and culture, which provides convenient conditions for the construction and fermentation of genetically engineered bacteria. Second, by optimizing the fermentation conditions, such as temperature, pH, dissolved oxygen, etc., the yield and quality of antiaging peptides can be significantly improved. In addition, the *E.coli* fermentation process is scalable and can achieve a smooth transition from laboratory scale to industrial scale to meet the needs of large-scale production.

Separation and purification are the key steps in the preparation of antiaging peptides, which are directly related to the purity and activity of the products. Chemical extraction, salting-out, ultrafiltration, and chromatography are commonly used methods for separation and purification. Among them, ultrafiltration and gel filtration chromatography are often used for preliminary separation and purification. They can distinguish antiaging peptides from other impurity components according to differences in molecular size and shape. High-performance liquid chromatography is used to separate the final target components with its high resolution and high sensitivity to ensure that the purity of the product meets the standards for medicinal or cosmetic use [46]. In practical applications, these methods often need to be reasonably combined and optimized according to the characteristics and production requirements of antiaging peptides. For example, chemical extraction can be used as a preliminary extraction step to separate antiaging peptides from complex biological systems, the salting-out method can adjust the ionic strength of the solution to precipitate the peptide segment and further increase its concentration, ultrafiltration and gel filtration chromatography can effectively remove macromolecular impurities and small molecular impurities, and achieve preliminary purification of antiaging peptides. Finally, high-purity target peptides were obtained by high-performance liquid chromatography.

In the study of antiaging peptides, the identification of their structure and activity is an indispensable part. Ultra-high performance liquid chromatography–tandem mass spectrometry (UPLC-MS/MS) and nuclear magnetic resonance (NMR) techniques play an important role in this field [47, 48]. UPLC-MS/MS can accurately determine the molecular weight, amino acid sequence and modification of peptides, which provides strong support for the structural identification of peptides. NMR can reveal the spatial conformation of the peptide and its interaction mechanism with the receptor from the perspective of three-dimensional structure, which provides an important basis for further study of the mechanism of action of antiaging peptides. UPLC-MS/MS technology has the characteristics of high resolution, high sensitivity and rapid analysis, which can accurately detect the molecular characteristics of antiaging peptides, and can effectively identify target peptides even in complex mixtures. NMR technology can provide three-dimensional structure information of peptides in solution, help researchers understand their active sites and interaction with biomolecules, thus providing an important theoretical basis for structural optimization and functional research of antiaging peptides.

The preparation and identification of antiaging peptides is a complex multidisciplinary process, involving genetic engineering, fermentation engineering, separation and purification technology, analytical chemistry, and other fields. With the continuous progress of science and technology, we believe that in the near future, we can develop more efficient and safe antiaging peptide products and make greater contributions to the cause of human antiaging.

7.2.3 Engineering Synthetic Peptides

Antiaging peptides can simulate the function of the human body's own protein, effectively regulate physiological function, and achieve the purpose of delaying aging. However, natural extraction of antiaging peptides has low yield and high cost, which is difficult to meet market demand. The emergence of engineering synthesis technology has solved this problem. It has achieved efficient and stable production of antiaging peptides by means of genetic engineering, solid-phase peptide synthesis, and biological fermentation. This not only reduces the cost, but also opens the way for the wide application of antiaging peptides. In-depth study and optimization of engineering synthesis of antiaging peptides is of great significance to promote the development of antiaging industry and help human healthy aging.

7.2.3.1 Optimization Strategy Based on Computer-Aided Design

In the wave of modern biotechnology, the optimal design of engineered synthetic peptides is playing an indispensable key role, and the deep integration of computer-aided design (CAD) technology is driving this field to a new stage of precision and efficiency with unprecedented strength. Through CAD, researchers can accurately simulate and predict the structure and function of synthetic peptides, so as to select the most suitable amino acid sequence. This method not only improves the stability, solubility, and biological activity of synthetic peptides but also greatly shortens the development cycle and reduces the cost.

In the field of drug development, CAD has been widely used to design synthetic peptides with specific biological activities to meet therapeutic needs. For example, in the design of antiviral synthetic peptides, researchers use CAD to construct a model of the interaction between viral proteins and synthetic peptides. By simulating the binding of different amino acid sequences to the key proteins of the virus, the synthetic peptide sequence that can effectively block the virus replication or infection process is quickly screened. In the field of cancer therapy, CAD helps to design synthetic peptides that can accurately target specific antigens on the surface of tumor cells. These synthetic peptides can be used as drug carriers to efficiently deliver anti-tumor drugs to tumor cells, enhance the therapeutic effect, and reduce the toxic and side effects on normal tissues.

The advantages of CAD are not only reflected in the field of drug research and development, but also widely infiltrated into many industries such as food additives, cosmetics, and materials science. In the field of food additives, CAD is used to design synthetic peptides with specific antioxidant or antibacterial activity to ensure food safety and quality. In the cosmetics industry, synthetic peptides that promote skin cell regeneration and anti-oxidation are developed to meet consumers' demand for efficient skin care products. In materials science, synthetic peptides with specific self-assembly properties are designed for the construction of nanomaterials or biosensors.

With the vigorous development of cutting-edge technologies such as artificial intelligence and machine learning, the application prospect of CAD in the field of synthetic peptide optimization design is becoming more and more broad. Researchers are actively exploring the combination of deep learning algorithms and CAD. Through the learning and analysis of massive known synthetic peptide structure and function data, deep learning models can accurately predict the performance corresponding to new amino acid sequence combinations, so as to provide more intelligent and efficient solutions for the optimal design of synthetic peptides and promote this field to a new height.

7.2.3.2 Modification Techniques: Cyclization, D-Amino Acid Substitution, PEGylation

In order to further improve the performance of antiaging peptides, a variety of modification techniques have been widely used. Among them, cyclization technology, as an important modification method, has gradually emerged in the research of antiaging peptides in recent years.

Cyclization technology is the process of converting linear peptides into cyclic structures by chemical or enzymatic means. The chemical synthesis method has high flexibility and can carry out diversified chemical modification of peptide chains, which is suitable for the synthesis of cyclic peptides with special structural or functional requirements. The enzymatic synthesis method has the advantages of good biocompatibility, mild reaction conditions and high ligation efficiency, and is suitable for large-scale production and preparation of cyclic peptides with high requirements for biological activity.

The cyclized peptides have higher stability and biological activity, which makes them have unique advantages in the fields of drug delivery and biosensing. In terms of drug delivery, cyclic antiaging peptides can more effectively protect drugs from degradation by enzymes in vivo, improve the stability and bioavailability of drugs, and thus enhance the therapeutic effect of drugs. In the field of biosensors, the high stability and biological activity of cyclic peptides enable them to more accurately identify and bind to target molecules and improve the sensitivity and specificity of biosensors.

However, the application of cyclization technology in antiaging peptides still faces some challenges. For example, chemical synthesis may introduce unnecessary chemical groups, affecting the biological activity and safety of cyclic peptides; enzymatic synthesis requires high enzyme selection and reaction conditions, which need to be further optimized and explored. In addition, the pharmacokinetic characteristics, such as absorption, distribution, metabolism, and excretion of the cyclized antiaging peptide in vivo also need to be further studied to ensure its effectiveness and safety in practical applications. Cyclization technology provides a new way and strategy for improving the performance of antiaging peptides. With the deepening of research and the continuous advancement of technology, cyclic antiaging peptides are expected to play a more important role in the field of antiaging and make greater contributions to human health.

Second, with the continuous advancement of biotechnology and drug research and development, peptide drugs have gradually become a research hotspot due to their unique advantages. However, peptide drugs face many challenges in vivo, among which anti-enzymatic hydrolysis ability and biological stability are issues. As an important modification technology, D-amino acid substitution provides an effective way to solve this problem.

D-amino acid replacement, that is, the L-amino acid in the peptide is partially or completely replaced with D-amino acid. This substitution can significantly improve the anti-enzymatic hydrolysis ability and biological stability of peptides. The ability of natural proteases to recognize and degrade D-amino acids is weak, so that peptides replaced by D-amino acids can remain active in the body for a longer time. In drug research and development, this feature is of great significance for the application of antiaging peptides.

Antiaging peptides with D-amino acid substitution showed longer in vivo half-life and better therapeutic effect. In some experimental studies, the half-life of D-amino acid modified antiaging peptides in mice was extended from hours to days, which effectively improved their antiaging activity [49]. This significant improvement makes D-amino acid substitution one of the key strategies in the development of antiaging peptide drugs.

D-amino acid substitution not only improves the stability of antiaging peptides, but also provides a guarantee for their effective delivery and function in vivo. By extending the half-life, antiaging peptides can play a role in the body for a longer period of time, thereby better exerting their antiaging effects. This has an important role in promoting the development of antiaging peptide drugs with clinical application value.

Third, PEGylation technology also plays an important role in the development of antiaging peptide drugs and the application of biomaterials. PEGylation refers to the attachment of polyethylene glycol (PEG) molecules to peptides or other bioactive molecules through chemical reactions. This process not only significantly improves the water solubility of peptides, making them easier to disperse and transport in the physiological environment, thus making them more suitable for injection administration, but also greatly enhances the stability of peptides and reduces the risk of inactivation caused by enzymatic hydrolysis or physical and chemical factors in vivo.

In the development of antiaging peptide drugs, the application of PEGylation technology has solved many key problems. Unmodified peptide drugs often have poor water solubility, which not only limits their route of administration, but also may lead to uneven distribution of drugs in the body and affect the efficacy. PEGylation effectively improves the solubility of peptides by introducing hydrophilic PEG chains on the surface of peptide molecules, enabling them to form stable solutions or colloids in aqueous media, thus providing a more convenient and reliable administration method for clinical applications. In addition, the rapid clearance of peptide drugs in the body is also a major challenge, which makes it difficult for drugs to maintain effective therapeutic concentrations in the body and limits their therapeutic effects. PEGylation can significantly prolong the half-life of the peptide in the body, reduce the rate of its filtration and metabolism by the kidney, and prolong the retention time of the drug in the body by several times, thereby enhancing the therapeutic effect of the drug [50]. This prolonged in vivo retention time means that the drug can interact more fully with the target and exert its antiaging biological effects, thereby significantly improving the efficacy of the drug. At the same time, due to the prolonged action time of the drug in the body, patients do not need frequent administration when receiving treatment, which not only improves the patient's medication compliance, but also reduces the potential risks and discomfort caused by frequent administration.

In the field of biomaterials, PEGylation technology also shows great potential. By connecting PEG molecules to the surface of biomaterials or embedding them in their structures, materials can be endowed with better biocompatibility and stability. After implantation in vivo, this modified biomaterial can effectively reduce the immune response of tissues, prolong the service life of materials, and better interact with cells and tissues in vivo, providing a more ideal material choice for applications in the fields of tissue engineering and regenerative medicine. With the continuous research and innovation of PEGylation technology, its application in the field of medicine will be more extensive and in-depth and make greater contributions to the development of human health.

7.3 Structural Characteristics of Antiaging Peptides

Antiaging peptides show great potential in the field of antiaging due to their unique structural characteristics.

The N-terminus of antiaging peptides usually contains hydrophobic amino acids, such as leucine, isoleucine, and phenylalanine. These hydrophobic amino acids contribute to the interaction of the peptide with the cell membrane and enhance the cell permeability and stability of the peptide. The cell membrane is mainly composed of a phospholipid bilayer and has a certain degree of hydrophobicity. The presence of hydrophobic amino acids makes it easier for antiaging peptides to pass through the cell membrane, thereby better exerting their biological functions. In addition, the hydrophobic N-terminus can also promote the self-assembly of peptides to form nanostructures with specific functions and improve their biological activity. This self-assembled nanostructure can not only increase the stability and solubility of peptides in vivo, but also serve as a drug carrier to achieve targeted delivery of aging-related cells or tissues, thereby improving the antiaging effect.

First, the proportion of polar amino acids in antiaging peptides is high. These amino acids are hydrophilic and can form hydrogen bonds with water molecules through their side chains to improve the solubility and bioavailability of peptides. In the complex biological environment, good solubility makes these peptides more efficiently absorbed and utilized by the body, so as to give full play to their physiological functions. For example, in oral antiaging peptide products, highly soluble peptides can better pass through the intestinal mucosa, enter the blood circulation, and transmit antiaging benefits to various cell tissues throughout the body.

Second, polar amino acids such as serine, threonine, and tyrosine also have antioxidant activity, which can scavenge free radicals and reduce the damage of oxidative stress to cells, thereby delaying the aging process. Cells are the basic unit of life activities and the main place where aging occurs. Excessive free radicals produced by oxidative stress can attack unsaturated fatty acids in cell membranes, cause lipid peroxidation, destroy the integrity and fluidity of cell membranes, and then affect the normal function of cells. In addition, free radicals can also react with biological macromolecules such as proteins and DNA in cells, resulting in protein denaturation and inactivation, DNA damage mutation, cell senescence, and even apoptosis. Polar amino acids, with their antioxidant properties, can neutralize these harmful free radicals, protect cells from damage, and maintain the young state and normal physiological functions of cells.

Third, aromatic amino acids such as phenylalanine, tyrosine, and tryptophan are important components of antiaging peptides. These aromatic amino acids have a unique ring structure and can bind to free radicals through π-π interaction, thereby exerting antioxidant effects. This antioxidant mechanism can not only effectively remove excess free radicals in the body, reduce the damage of oxidative stress to skin cells, but also enhance the natural defense ability of the skin, making it more resistant to external environmental factors. Aromatic amino acids also play a role in regulating signaling pathways in antiaging. They can activate specific cell signaling pathways and promote collagen synthesis. Collagen is the main structural protein of the skin. With the increase of age, the content of collagen will gradually decrease, resulting in the loss of elasticity and tightness of the skin. By increasing the synthesis of collagen, aromatic amino acids help to repair and reconstruct the fiber network of the skin, so that the skin is restored to be compact and smooth, and the appearance of wrinkles is reduced, thereby achieving antiaging effects.

Fourth, antiaging peptides are usually composed of less than 20 amino acids, which makes them have a small molecular weight, easy to penetrate the cell membrane, and exert biological activity. Small molecular peptides are easy to be absorbed and utilized by the body, and can quickly enter the blood circulation and play an antiaging role. At the same time, the synthesis and purification of small molecular peptides are relatively simple, low cost, and have high application value. Antiaging peptides can simulate key molecules in cell signaling pathways, activate antiaging mechanisms in cells, and promote cell repair and regeneration. They can scavenge excessive free radicals in the body, reduce oxidative stress damage to cells, and thus delay the aging process of cells. In addition, antiaging peptides can also enhance the body's immunity by regulating the immune system, helping the body to better resist disease and aging. Antiaging peptides have broad application prospects in many fields, such as cosmetics, health products, and pharmaceuticals. In cosmetics, antiaging peptides can be added to products such as creams and essences to help reduce wrinkles, brighten skin color, and improve skin elasticity. In terms of health care products, antiaging peptides can be made into oral liquids, capsules, and other forms to provide additional antiaging support for the human body. In the field of pharmaceuticals, antiaging peptides are expected to become a new generation of drugs for the treatment of aging-related diseases. The molecular weight of antiaging peptides is usually less than 8 kDa, which makes them have high bioavailability and stability. Small molecular weight peptides are easy to be absorbed through the intestinal tract, enter the blood circulation, and play a role quickly. In addition, the metabolism and excretion of small molecular weight peptides in the body are relatively rapid, reducing potential side effects and improving their safety.

In summary, as shown in Table 7.2, the structural characteristics of antiaging peptides make them have significant biological activities in anti-oxidation, anti-inflammation, and promotion of collagen synthesis. By optimizing the amino acid composition and sequence,

Table 7.2 Structure of Antiaging Peptides.

Natural food-based antiaging Peptides	
High proportion of polar amino acids	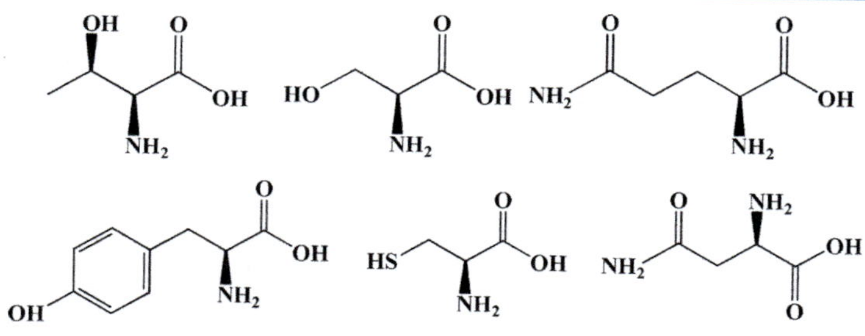
Aromatic amino acids	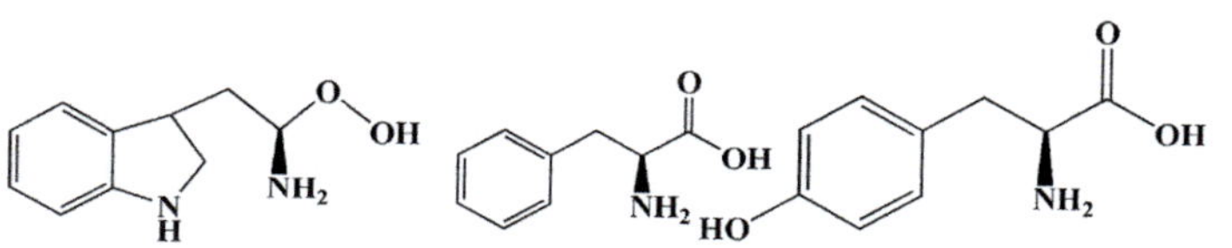
N-terminal is hydrophobic	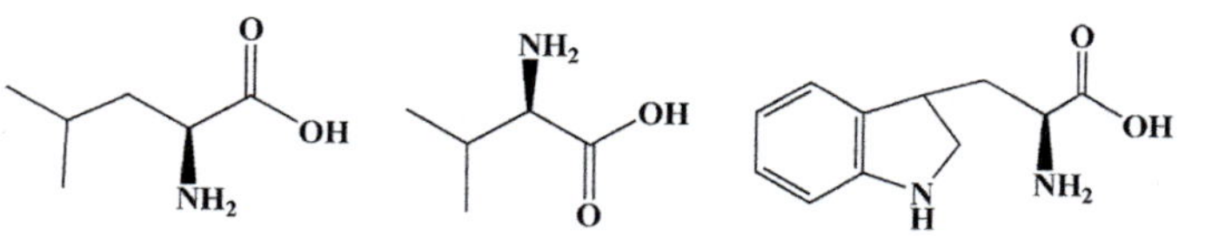

the functional properties of antiaging peptides can be further improved, providing a scientific basis for the development of efficient and safe antiaging products. With the deepening of research, the application prospect of antiaging peptides in the fields of medical treatment, beauty, and health care will be broader.

7.4 Mechanism of Action of Antiaging Peptides

Antiaging peptides are a class of bioactive peptides that delay or reverse the aging process through a variety of mechanisms. The following is a summary of the main mechanism of action of antiaging peptides as shown in Table 7.3:

1) **Antioxidant effect:** Antiaging peptides protect cells from oxidative damage by scavenging free radicals in the body and reducing oxidative stress. Free radicals are one of the main factors leading to cell damage. They can attack cell membranes, proteins, and DNA, triggering a series of oxidative reactions that accumulate to cause cell dysfunction and tissue aging. Antiaging peptides such as acetyl dipeptide-1 cetacean ester (AD-1) have shown significant antioxidant capacity in vitro and can reduce UVA-induced cell damage. UVA is the main component of solar ultraviolet rays. It penetrates deeper in the skin than UVB. It can cause changes in dermal connective tissue, produce oxidative stress, cause the skin to lose elasticity and toughness, and ultimately cause photoaging of

Table 7.3 The main mechanism of action of antiaging peptides.

Evaluation method	Content	Applicability
In vitro evaluation method		
Anti-oxidant activity	The antioxidant activity was determined by DPPH, ABTS free radical scavenging ability, reducing power, and Fe^{2+} chelating ability	It is suitable for rapid screening of antiaging peptides and provides preliminary activity data
Anti-collagenase activity	Evaluate the ability of peptides to inhibit collagenase activity and prevent collagen degradation	It is suitable for evaluating the potential of peptides in preventing skin aging
Anti-elastase activity	The inhibitory ability of peptides to elastase activity was evaluated to prevent elastin degradation	It is suitable for evaluating the role of peptides in maintaining skin elasticity
Anti-hyaluronidase activity	The inhibitory ability of peptides on hyaluronidase activity was evaluated to protect skin moisture	It is suitable for evaluating the role of peptides in moisturizing and anti-wrinkle
Anti-tyrosinase activity	Evaluate the inhibitory ability of peptides on tyrosinase activity and reduce pigmentation	It is suitable for evaluating the role of peptides in whitening and anti-spotting
Anti-inflammatory activity	The anti-inflammatory effect of the peptide was evaluated by measuring the expression levels of inflammatory factors (such as TNF-α, IL-6)	It is suitable for evaluating the role of peptides in reducing inflammatory response
Anti-glycation activity	The inhibitory ability of peptides on the formation of advanced glycation end products (AGEs) was evaluated.	It is suitable for evaluating the role of peptides in preventing protein glycosylation
Cell proliferation and antiaging effects	The effects of peptides on cell proliferation and antiaging were evaluated by MTT, CCK-8, and other cell proliferation experiments	It is suitable for evaluating the role of peptides in promoting cell growth and delaying cell senescence.
In vivo evaluation methods		
Animal model	Animal models such as mice and nematodes were used to evaluate the effects of peptides on life extension, antioxidant, and anti-inflammatory effects	Provide in vivo activity data to verify in vitro experimental results
Skin model	The effects of peptides on skin repair and anti-wrinkle were evaluated by animal skin model	It is suitable for evaluating the effect of peptides in practical applications
Clinical trial	A clinical trial was conducted in volunteers to evaluate the actual effect of peptides in skin care products	Provide the final safety and effectiveness data

the skin. As a natural antioxidant, antiaging peptide has relatively simple structure, easy absorption, good stability, and no immunoreactivity. It not only has strong antioxidant activity, but also has other health effects such as lowering blood pressure and anti-cancer. It has attracted more and more attention in the field of food and medical health products.

2) **Anti-glycosylation effect:** Antiaging peptides protect cells and tissues from glycosylation damage by inhibiting glycosylation reaction and reducing the formation of advanced glycation end products (AGEs). Glycosylation reaction refers to the non-enzymatic reaction of sugar molecules with proteins, lipids, or nucleic acids to form AGEs. This process is mainly carried out in organelles such as endoplasmic reticulum in cells and catalyzed by glycosyltransferase, which is an important part of post-translational modification of proteins. The accumulation of AGEs can lead to cell dysfunction and tissue damage, which is closely related to the occurrence and development of a variety of aging-related diseases, such as cardiovascular diseases, metabolic diseases, and nervous system diseases. The formation and accumulation of AGEs are closely related to the aging process and the acceleration of diabetic macroangiopathy. Persistent hyperglycemia can promote the non-enzymatic glycosylation reaction between protein and glucose and ultimately form AGEs.

   The formation of AGEs mainly has the following three biochemical mechanisms: the saccharification pathway, the polyol pathway, and the sugar oxidation pathway. AGEs can cause inflammation and procoagulant reactions, leading to tissue damage, and these reactions usually target specific organs. In summary, antiaging peptides play an important role in delaying aging by inhibiting glycosylation reaction, reducing the formation of AGEs, and protecting cells and tissues from glycosylation damage through various mechanisms. These mechanisms include direct inhibition of glycosylation, activation of autophagy to remove damaged proteins, and protection of collagen and elastin through anti-glycosylation to maintain skin and tissue health.

3) **Activation of autophagy:** Autophagy is a process of cleaning and recycling within cells, which is essential for maintaining normal cell function and homeostasis. During autophagy, cells form autophagosomes with a double-membrane structure. These autophagosomes can encapsulate damaged proteins and organelles, and then transport them to lysosomes for degradation and recycling.

   Antiaging peptides promote the removal of intracellular waste and maintain cell homeostasis by activating the autophagy pathway, thereby delaying aging. For example, acetyl dipeptide-1 cetyl ester (AD-1) can protect cells from photoaging damage by activating the autophagy pathway. The autophagy process can remove damaged proteins and organelles in cells and maintain the normal function of cells, which is essential for cell survival and health.

   The initiation of autophagy requires the ULK1 kinase complex, which is tightly regulated by AMPK and mTOR as activators and inhibitors, respectively. AMPK activates ULK1 by phosphorylation, and the activity of mTOR is affected by factors such as changes in nutrient content, changes in energy substance or growth factor levels, hypoxia, and stress.

   With age, the abundance of autophagy-related proteins gradually decreases, and the amount of goods transported to lysosomes decreases, which means that impaired autophagy is a major feature of body aging. The decrease of autophagy activity will lead to the accumulation of damaged proteins and organelles in cells, thus affecting the normal function of cells and accelerating the aging process of cells.

Therefore, by activating the autophagy pathway, antiaging peptides not only promote the removal of intracellular waste, but also help maintain cell homeostasis and function, which is of great significance for delaying the aging process. By promoting autophagy, antiaging peptides help to maintain cell cleanliness and health, reduce cell damage and dysfunction, and thus help to prolong the healthy life of cells and individuals.

4) **Promoting collagen synthesis:** Antiaging peptides play an important role in promoting the synthesis of collagen and other matrix proteins, thereby enhancing the elasticity and tightness of the skin. With age, the synthesis of collagen gradually decreases, leading to skin relaxation and wrinkle formation. Antiaging peptides, especially signaling peptides such as palmitoylpentapeptide-3, can effectively stimulate collagen production and help restore the structure and function of the skin.

   Palmitoyl pentapeptide-3 is a small molecule active peptide that can penetrate the dermis and directly act on the deep layer of the skin. By promoting the synthesis of collagen, elastin, and hyaluronic acid, it enhances the moisture content and water-holding capacity of the skin, thereby increasing the thickness and compactness of the skin. Studies have shown that palmitoyl pentapeptide-3 can not only improve the smoothness of the skin, but also effectively reduce the depth of fine lines and wrinkles and has a significant antiaging effect.

   Collagen is the main supporting component of the skin, accounting for 75% of the dermis. Its unique triple helix structure allows the skin to remain elastic and compact. With the increase of age, the loss of collagen will lead to the gradual weakening of the supporting structure of the skin, which will lead to relaxation and wrinkles. Therefore, supplementation of collagen synthesis is essential to maintain the young state of the skin.

   Antiaging peptides can fundamentally improve the structure of the skin and enhance its elasticity and tightness by stimulating the synthesis of collagen. In addition to palmitoyl pentapeptide-3, other signal peptides are also widely used in skin care products to help skin resist the effects of aging. With the development of skin care technology, more and more antiaging peptides have been developed and applied to the market to meet the needs of consumers for antiaging products.

   Antiaging peptides significantly enhance the elasticity and tightness of the skin by promoting the synthesis of collagen and other matrix proteins, thereby effectively delaying the process of skin aging. The application of these peptide components not only improves the appearance of the skin, but also improves the overall skin health.

5) **Neurotransmitter inhibition:** Antiaging peptides play an important role in reducing dynamic wrinkles, especially by inhibiting the release of neurotransmitters. Dynamic wrinkles are formed due to the frequent contraction of facial muscles, which are common in expression changes, such as head wrinkles and crow's feet. With the increase of age, muscle movement will lead to the deepening of wrinkles, and antiaging peptides can effectively slow down the formation of wrinkles by regulating this process.

   Taking acetyl hexapeptide-8 as an example, this peptide is called "smear botulinum toxin needle", and its structure mimics the N-terminal of SNAP-25 protein, which is composed of six amino acids. Acetylhexapeptide-8 inhibits the release of neurotransmitter acetylcholine (ACh) by competing with SNAP-25 to occupy the position in the SNARE complex. ACh is a key neurotransmitter that promotes muscle contraction. When its release is inhibited, the muscle contraction ability will be weakened, thereby reducing the generation of wrinkles.

Studies have shown that acetyl hexapeptide-8 can effectively relax facial muscles and reduce the occurrence of dynamic striae and the formation of fine lines. Compared with traditional botulinum toxin injection, acetylhexapeptide-8 provides a safer and more convenient alternative because it does not require injection and has fewer side effects. Through local application, acetyl hexapeptide-8 can achieve the effect of smoothing wrinkles without damaging the muscle.

In addition, acetyl hexapeptide-8 also has the ability to reorganize collagen elasticity, increase the activity of elastin, soften facial lines, and smooth wrinkles. Its anti-wrinkle effect has been verified in a number of clinical studies, showing its potential for wide application in antiaging skin care products.

Antiaging peptides can effectively smooth dynamic wrinkles and delay skin aging by inhibiting the release of neurotransmitters and reducing muscle contraction. This mechanism not only provides new ideas for the development of antiaging skin care products, but also provides consumers with safer and more effective skin care choices.

6) **Immunoregulation:** Antiaging peptides play an important role in delaying aging by regulating the immune system, reducing inflammatory response, and protecting cells from inflammatory damage. Inflammation is a natural response of the body to injury or infection, but chronic inflammation can lead to cell damage and dysfunction, which is related to a variety of aging-related diseases.

Carnosine, a dipeptide composed of β-alanine and L-histidine, is a naturally occurring antioxidant in the human body, especially in excitatory tissues such as muscle and brain. Carnosine has a variety of biological functions, including anti-oxidation, inhibition of telomere shortening, reduction of mitochondrial damage, anti-glycosylation, and anti-tumor. In skin antiaging, carnosine reduces skin aging by removing excess free radicals in the skin and increases skin elasticity, maintains skin moisture, and reduces pigmentation.

The anti-inflammatory effect of carnosine is to improve the symptoms of skin redness, swelling, and pain by inhibiting the inflammatory response. It can reduce the secretion of pro-inflammatory cytokines and restore the level of anti-inflammatory cytokines, thereby reducing inflammation. In addition, carnosine was also found to prevent Aβ-induced oxidative stress and inflammation in microglia by reducing intracellular NO and O^{2-} levels and iNOS and Nox enzyme expression to reduce oxidative stress, reduce the secretion of pro-inflammatory cytokines, and restore the level of anti-inflammatory cytokines.

In the experimental model of osteoarthritis, carnosine conjugates and hyaluronic acid (FidHycarn) showed protective effects on inflammatory response and cartilage degradation and showed potential therapeutic effects on osteoarthritis by reducing the levels of oxidative damage markers and pro-inflammatory cytokines.

In addition, carnosine has also been found to activate AKT2 signaling pathway through CD36 and RAGE, stimulate macrophage-mediated clearance of senescent skin cells, and enhance macrophage phagocytosis, thereby enhancing the ability of macrophages to eliminate senescent skin cells.

Antiaging peptides, especially carnosine, protect cells from inflammatory damage through their anti-inflammatory and immunomodulatory effects, providing an effective strategy for delaying aging.

7) **Protecting telomeres:** Telomere protection is an important part of the process of delaying aging. Telomeres are DNA fragments located at the end of chromosomes, which play a protective role in cell division and prevent chromosomes from fusing with each other. With the increase of age, each cell division will lead to the gradual shortening of telomeres. When telomeres are shortened to a certain extent, the cells will stop dividing and enter the aging state, eventually leading to cell death. Therefore, telomere length is regarded as a biomarker of cellular aging.

   Antiaging peptides help to prolong cell life by inhibiting telomere shortening. For example, carnosine is a dipeptide composed of β-alanine and L-histidine. It is naturally present in the muscles and brains of the human body and has a variety of biological functions, including anti-oxidation, anti-glycosylation, and anti-tumor. Studies have shown that carnosine can inhibit telomere shortening, thereby delaying cell senescence. In one study, fibroblast growth in a culture medium containing 20 mmol/L carnosine slowed telomere shortening and extended cell lifespan. When cells are in a state of proliferation inhibition, the presence of carnosine can reduce the accumulation of telomere fragments. Therefore, reducing telomere shortening and telomere DNA damage is one of the mechanisms by which carnosine can prolong cell life.

   In addition, telomerase is an enzyme that can prolong telomeres, and it is highly active in certain cell types, such as germ cells and stem cells. However, in most somatic cells, telomerase activity is low, resulting in telomere shortening as cells divide. By activating telomerase or protecting telomeres through other mechanisms, the life cycle of cells can be potentially prolonged, thereby delaying the aging process.

   It is worth noting that the relationship between telomere length and aging is not a simple linear relationship. Although telomere shortening is a sign of aging, long telomeres may also increase the risk of cancer. Therefore, antiaging strategies need to take into account a variety of factors, including telomere length, cell type, and individual genetic background.

   Antiaging peptides play an important role in delaying aging by protecting telomeres and reducing telomere shortening, which helps to prolong cell life. With the deepening of research on antiaging peptides, more effective antiaging interventions may be developed in the future.

8) **Maintaining protein homeostasis:** Antiaging peptides maintain intracellular protein homeostasis by regulating protein synthesis, folding, and degradation, preventing protein aggregation and loss of function. Protein homeostasis is an important guarantee for normal cell function. Any imbalance in protein synthesis, folding and degradation can lead to cell dysfunction and senescence.

   With the increase of age, the ability of protein homeostasis in cells gradually decreases, which is closely related to the aging process. Studies have shown that aging can lead to a decrease in the folding efficiency of newly synthesized proteins, and the ability of cells to degrade misfolded proteins is also weakened. This imbalance not only affects the normal function of cells, but also may lead to the accumulation of toxic proteins in cells, which in turn leads to a variety of age-related diseases, such as Alzheimer's disease and Parkinson's disease.

Antiaging peptides play a key role in maintaining protein homeostasis. They help newly synthesized proteins fold correctly and prevent misfolding and aggregation by promoting the activity of molecular chaperones. Molecular chaperone is an important protein quality control mechanism in cells, which can bind to incompletely folded or misfolded proteins to help them fold correctly and restore function. In addition, antiaging peptides can also promote the degradation of misfolded proteins by activating the ubiquitin-proteasome pathway and autophagy mechanisms, thereby removing harmful substances in cells and maintaining cell health.

For example, carnosine, as an antiaging peptide, has been shown to effectively inhibit protein misfolding and aggregation. It further promotes cell repair and regeneration by regulating intracellular pH and enhancing the antioxidant capacity of cells.

These characteristics of carnosine make it show a good prospect in delaying aging and improving cell function.

Antiaging peptides can effectively delay the aging process by regulating protein synthesis, folding, and degradation, maintaining intracellular protein homeostasis, preventing protein aggregation and loss of function. This mechanism not only provides a theoretical basis for the development of antiaging products, but also provides a new direction for clinical application.

9) **Protection of mitochondrial function:** Antiaging peptides play an important role in delaying aging by protecting mitochondrial function, reducing mitochondrial damage, and maintaining cell energy metabolism. Mitochondria, as the energy factory of cells, are responsible for producing ATP and maintaining the normal function of cells. Mitochondrial dysfunction can lead to energy metabolism disorders and decreased cell function, thereby accelerating the aging process.

Mitochondrial dysfunction is associated with a variety of aging-related diseases, including neurodegenerative diseases, cardiovascular diseases, and metabolic diseases. Antiaging peptides such as carnosine delay cell senescence by reducing mitochondrial damage and protecting cells. The antioxidant activity of carnosine can not only scavenge excessive free radicals and maintain the balance of the antioxidant system in vivo, but also protect mitochondria from oxidative damage by regulating mitochondrial function and inhibiting the excessive production of free radicals.

Mitochondrial dynamics, including mitochondrial fusion, fission, and autophagy, are essential for maintaining cell physiological function and homeostasis. The changes of mitochondrial fusion and fission are of great significance to the synthesis and supply of cell respiration and energy, and the changes of cell energy metabolism also affect mitochondrial dynamics. For example, mitochondrial fusion can increase cellular oxidative phosphorylation levels and ATP synthesis, while mitochondrial fission contributes to the proliferation of mitochondria and the transfer of ATP.

Mitophagy is a process of removing damaged mitochondria and is essential for maintaining mitochondrial quality and cell health. The activation of mitophagy can remove dysfunctional mitochondria and reduce abnormal ATP synthesis in the process of oxidative phosphorylation, thus protecting cells from mitochondrial damage. Studies have shown that the activation of mitophagy is associated with enhanced oxidative phosphorylation, and inhibition of mitophagy may lead to oxidative phosphorylation disorders.

In addition, mitochondria also regulate cell fate and function by interacting with other organelles, such as endoplasmic reticulum and lysosomes. Mitochondrial-endoplasmic reticulum interactions, such as mitochondrial-associated endoplasmic reticulum membranes (MAMs), are essential for lipid and calcium exchange, mtDNA replication and segregation, and mitochondrial fission. The interaction between mitochondria and lysosomes involves the process of mitophagy, removing damaged mitochondria and maintaining mitochondrial quality.

In summary, antiaging peptides play an important role in delaying aging by protecting mitochondrial function, reducing mitochondrial damage, and maintaining cell energy metabolism. These peptides protect mitochondria from damage and maintain cell energy metabolism by regulating mitochondrial dynamics and autophagy processes, thereby delaying cell senescence. With the deepening of research on antiaging peptides, more effective antiaging interventions may be developed in the future.

In summary, antiaging peptides delay or reverse the aging process through a variety of mechanisms, including anti-oxidation, anti-glycosylation, activation of autophagy, promotion of collagen synthesis, neurotransmitter inhibition, cell repair and regeneration, immune regulation, protection of telomeres, maintenance of protein homeostasis, and mitochondrial function protection.

These mechanisms work together to protect cells and tissues, delay aging, and improve healthy life. With the deepening of research, the application prospect of antiaging peptides in the fields of medical treatment, beauty, and health care will be broader.

With the aging of the global population, antiaging research has become a hot topic in the scientific and industrial circles. As a bioactive peptide with significant antiaging properties, antiaging peptides delay or reverse the aging process through a variety of mechanisms, including anti-oxidation, anti-glycosylation, activation of autophagy, promotion of collagen synthesis, neurotransmitter inhibition, cell repair and regeneration, immune regulation, protection of telomeres, maintenance of protein homeostasis, and mitochondrial function protection. The research and application of these peptides provide new possibilities for delaying aging and improving the quality of life.

7.5 Research Methods and Evaluation System of Antiaging Peptides

7.5.1 The Core Method of Antiaging Peptide Research

7.5.1.1 Molecular Interaction Analysis

To study the molecular interaction of antiaging peptides, surface plasmon resonance (SPR) [51] and isothermal titration calorimetry (ITC) can be used to detect peptide-target binding, and cryo-electron microscopy (Cryo-EM) can be used to analyze the structure of peptide-receptor complex. As a real-time, label-free biomolecular interaction analysis method, SPR technology can accurately measure the affinity, kinetic parameters, and conformational changes of intermolecular binding, and has unique advantages in studying the interaction between antiaging peptides and target proteins. ITC provides thermodynamic data of intermolecular binding by directly measuring the change of heat during the reaction, which is helpful to understand the driving force and stability of the binding process. The combination of these two methods can comprehensively analyze the interaction characteristics between antiaging peptides and targets from different perspectives.

With the continuous development of Cryo-EM technology, it plays an increasingly important role in the analysis of the structure of biological macromolecular complexes. The high-resolution three-dimensional structure of the peptide-receptor complex can be obtained by Cryo-EM, and the detailed information, such as the binding mode, key interaction sites, and conformational changes between the two, can be visually displayed, which provides strong support for understanding the mechanism of action of antiaging peptides from the structural level.

7.5.1.2 High-Throughput Screening Technology

Traditional screening methods are inefficient and costly, and it is difficult to meet the needs of large-scale screening. The emergence of high-throughput screening technology provides the possibility for the rapid discovery of antiaging peptides.

Screening autophagy activating peptides based on fluorescent reporter gene cell model is an important application of high-throughput screening technology in the study of antiaging peptides. Autophagy is a process of degradation and circulation in cells, which plays a key role in maintaining cell homeostasis and delaying senescence. Studies have found that many autophagy activators can regulate autophagy and interfere with aging-related diseases. By constructing a cell model with a fluorescent reporter gene, the level of autophagy can be monitored in real time, and peptides that can effectively activate autophagy can be quickly screened out.

The microfluidic chip platform accelerates the discovery of antiaging peptides, which is another important application of high-throughput screening technology. Antiaging peptides can scavenge free radicals in the body, reduce oxidative stress, and delay aging. Microfluidic chip has the advantages of high throughput, high sensitivity, and low sample consumption, which can quickly screen a large number of peptides and improve the efficiency of antiaging peptide discovery.

High-throughput screening technology has great potential in the study of antiaging peptides, which can quickly find peptides with specific biological activities and provide new ideas for the development of antiaging drugs and functional foods. In the future, with the continuous progress of technology, its application in the study of antiaging peptides will be more extensive and in-depth.

7.5.2 Evaluation Method of Antiaging Peptides

Antioxidant activity test is one of the important methods to evaluate the activity of antiaging peptides in vitro. The antioxidant capacity of antiaging peptides can be evaluated by measuring the scavenging ability of free radicals, such as DPPH, ABTS, hydroxyl, and superoxide anion radicals. For example, the scavenging rates of DPPH, superoxide anion, hydroxyl, and ABTS free radicals of milk-derived sleep-promoting peptides in vitro were 44.27, 18.37, 37.22, and 24.83%, respectively. [52] In addition, the anti-collagenase and anti-elastase activity tests can evaluate the protective effect of antiaging peptides on skin structural proteins, the anti-hyaluronidase activity test can evaluate the maintenance of skin moisturizing ability, the anti-tyrosinase activity test can evaluate its whitening effect, and the anti-inflammatory activity test can evaluate its alleviation effect on skin inflammation. The anti-glycosylation activity test can evaluate its antiaging effect on skin aging, and the cell proliferation and antiaging effect test can evaluate its improvement effect on cell viability and aging.

The comprehensive application of these activity evaluation methods provides a comprehensive scientific basis for the research and development of antiaging peptides, which is helpful to screen out peptides with potential antiaging effects and promote the development of antiaging drugs and cosmetics. At the same time, the rise of 3D organ technology has opened up a new way for antiaging research.

Organoids are three-dimensional cell cultures derived from pluripotent stem cells or adult stem cells, which can simulate the structural and functional characteristics of tissues or organs in vivo. For example, skin organoids can highly restore the physiological environment of the skin, including multi-layer structures, such as epidermis and dermis, and complex interactions between cells. By applying antiaging peptides to skin organoids, the effects of antiaging peptides on the regeneration process of skin cells, such as proliferation, differentiation, and migration can be monitored in real time, and the regeneration effect of antiaging peptides can be evaluated intuitively. Previous studies have found that some antiaging peptides can significantly promote the self-renewal and differentiation of skin stem cells, accelerate the repair of damaged skin tissues, and provide strong support for the development of antiaging products.

At the same time, organ chip technology also provides a powerful tool for studying the central delivery efficiency of antiaging peptides. The blood–brain barrier is an important protective barrier of the central nervous system, which strictly limits the entry of exogenous substances into the brain. Traditional in vitro models and animal experiments have many deficiencies in simulating the physiological characteristics of the blood–brain barrier, while the blood–brain barrier chip can accurately simulate its cell composition, structure, and function. In the study of antiaging peptides, the blood–brain barrier chip can observe the cross-barrier transport process of peptide molecules in real time and accurately determine their delivery efficiency. This helps to screen out antiaging peptides that can effectively cross the blood–brain barrier and play a role in the central nervous system, making it possible to develop therapeutic drugs for neurodegenerative diseases. 3D organoid and organ chip technology has great potential in the study of antiaging peptides. They can not only provide a model closer to the real physiological environment in vivo, but also explore the mechanism of action of antiaging peptides from different angles, providing new ideas and methods for the further development of antiaging research.

In the study of antiaging peptides, the development and application of in vitro and in vivo activity evaluation methods are crucial. These methods not only help to understand the mechanism of action of antiaging peptides, but also provide a scientific basis for drug screening and product development. In vivo evaluation methods include animal model studies and clinical trials. Animal models such as mice and nematodes are widely used to evaluate the effects of antiaging peptides on life extension, antioxidant, and anti-inflammatory effects. Clinical trials are conducted directly on the human body to evaluate the safety and efficacy of antiaging peptides. In addition, there are studies using computer screening methods to screen potential antiaging peptides from plant peptide libraries and conduct in vitro and in vivo tests to verify their antiaging activity.

Animal model research is one of the important means of in vivo evaluation. Caenorhabditis elegans is widely used as a model organism for evaluating antioxidant and antiaging activities due to its short life cycle, small size, easy culture, simple genetic material, and high similarity to human biochemical and genetic pathways. The antioxidant activity of active peptides can be reflected by measuring the changes of reactive oxygen species (ROS), antioxidant enzyme activities (SOD, CAT), and malondialdehyde (MDA) content in nematodes incubated with active peptides. The lifespan and mortality of experimental nematodes are the most intuitive indicators of the antiaging activity of the preparation.

Mice are also commonly used animal models. For example, the subacute aging model caused by D-galactose decreased the antioxidant capacity of mice and increased MDA. It was observed that PT extract could increase the level of glutathione (GSH) in aging mice, reduce the level of MDA and increase the total antioxidant capacity at high dose, indicating that PT extract could resist the damage of free radicals and had certain antiaging effect [53].

Clinical trials are important steps in assessing the safety and efficacy of antiaging peptides in humans. For example, through a double-blind, placebo-controlled clinical study, it was found that oral low-molecular-weight collagen peptides can improve the moisture content, elasticity, and wrinkle degree of human skin, showing their potential in anti-skin aging.

In addition, computer screening methods have also been applied to the study of antiaging peptides. By establishing a plant peptide library and using computer-aided screening technology, peptides with potential antiaging activity can be quickly screened out. These selected peptides were then subjected to in vitro and in vivo activity tests to verify their antiaging effects. This method not only improves the screening efficiency, but also reduces the experimental cost and time.

The comprehensive application of in vitro and in vivo activity evaluation methods provides a comprehensive scientific basis for the research and development of antiaging peptides, which is helpful to screen out peptides with potential antiaging effects and promote the development of antiaging drugs and cosmetics [1]. With the advancement of biotechnology and the development of new methods, the research of antiaging peptides will continue to deepen and contribute more possibilities to healthy human aging.

7.6 Application of Antiaging Peptides

As an important bioactive substance, the application value of antiaging peptides in the field of antiaging is increasingly prominent. First of all, antiaging peptides can promote cell proliferation, enhance cell viability, and delay the process of cell senescence. In the medical field, antiaging peptides can be used to treat senile diseases, such as cardiovascular diseases, Alzheimer's disease, etc., and improve the quality of life of patients. Second, antiaging peptides have antioxidant effects, which can effectively remove free radicals in the body, reduce oxidative stress, and protect cells from damage. In the field of health food, antiaging peptides help to delay human aging, enhance immunity, and improve disease resistance. In addition, antiaging peptides can also improve the skin state, improve skin elasticity, reduce wrinkles, and make the skin glow with youthful luster. In the field of beauty and skin care, antiaging peptides have become a key component of high-end skin care products and are favored by consumers.

7.6.1 Application in Food System

With the development of economy, the improvement of living standards and the aggravation of global population aging, people pay more and more attention to how to maintain a healthy quality of life and delay the aging process. Antiaging peptides attract food manufacturers to constantly seek innovative products to meet the needs of consumers. Adding antiaging peptides can be used as a way of food value-added, which makes antiaging peptides increasingly widely used in the food industry. However, there may be some defects and potential toxicity problems in the application of synthetic antiaging peptides. For example, many bioactive peptide amino acids are less, which may lead to low production efficiency, and may also encounter challenges such as low purity, poor stability, and poor permeability [54, 55]. During the synthesis and storage of synthetic peptide drugs, it is easy to form related structural impurities, such as amino acid loss, amino acid insertion, protective group residue, oxidation/reduction, etc. [56, 57].

Therefore, antiaging peptides with natural sources, stability, and high biosafety have brought new development opportunities to the food industry because of their unique biological activity. For example, bird's nest peptides have been reported to have the ability to resist skin aging and damage caused by light, improve skin quality, and down-regulate aging-related genes [58, 59]. Matsukawa et al. [60] simulated the menopausal human body by removing the ovarian rat model, and found that the thickness of the dermis caused by aging was improved after eating the bird's nest peptide prepared by trypsin F (Pancreatin F) enzymatic hydrolysis. Kim et al. [61] investigated the antiaging ability of bird's nest peptide prepared by Protamex® enzymatic hydrolysis in cells. It was found that bird's nest peptide could inhibit H_2O_2-induced cytotoxicity and scavenge intracellular ROS, down-regulate H_2O_2-induced MMP-1 mRNA expression, and reduce the activation of extracellular signal-regulated kinase (ERK) and c-Jun N-terminal kinase (JNK), thus playing an antiaging effect.

In recent years, antiaging peptides have several main application fields in the food industry, such as functional food, health food, food additives, and personalized nutritional food. Song et al. [62] found that sea cucumber peptides promoted cell proliferation by accelerating the transition of NIH 3T3 cells from G1 phase to S phase, and up-regulated the expression of type I collagen and tissue inhibitor of metalloproteinases-1 (TIMP-1) and inhibited the expression of matrix metalloproteinase-1 (MMP-1). (tissue inhibitor of metalloproteinases-1, TIMP-1) and inhibit the expression of matrix metalloproteinase-1 (MMP-1). Chen et al. [63] found that the antiaging peptide obtained by enzymatic hydrolysis of deep red sea bream scales can effectively prolong the average life span, maximum life span, and half-death time of Drosophila melanogaster. Ding et al. [64] extracted the antioxidant tetrapeptide Trp-Asp-Val-Lys from velvet antler by enzymatic hydrolysis and purification, which can effectively inhibit the formation of ROS in zebrafish and delay aging. Wenjing Qiu et al. [65] developed a dietary supplement called ERJ-CP, which exerted antiaging effect by enhancing antioxidant capacity. Experiments showed that it significantly prolonged the average life span and reduced oxidative damage in the Drosophila model.

In the development and application of antiaging peptides, the improvement of oral bioavailability is a key research direction. Due to the problems of easy decomposition and low absorption efficiency of peptide molecules in the digestive tract, the bioavailability after oral administration is limited. Therefore, researchers continue to explore new strategies to improve their oral bioavailability, such as the use of nanotechnology, microencapsulation technology, cyclization, and other methods to enhance the stability and absorption efficiency of peptides, so that they can better play the antiaging effect.

At the same time, regulation and safety assessment are also an indispensable part of antiaging peptides as nutritional and health supplements. Strict regulatory mechanisms and comprehensive safety assessments can ensure that products meet relevant standards and protect consumers' health rights. At present, the regulatory authorities of various countries have clear specifications on the quality, labeling, and efficacy of peptide products, which require enterprises to strictly abide by in the process of R&D and production, and to comprehensively evaluate the safety of products to ensure their safety and reliability in long-term use.

The application of antiaging peptides in the food industry has the great advantages of natural safety, multiple physiologically active functions, easy absorption by the human body, and high bioavailability, which makes the application prospect of antiaging peptides in the food industry very broad. In the future, with the development of science and technology, the problems of complex extraction process and high cost of antiaging peptides will be gradually solved, and the application of antiaging peptides in the food industry will be more extensive, bringing more healthy and delicious food to consumers.

7.6.2 Application in Pharmaceutical Industry

Aging is an unavoidable journey for human beings. Oxidative stress is one of the main causes of cell aging and death. For a long time, scientists have tried to change the process of aging, through the intervention of different biological systems, and signaling pathways and have changed the process of aging in a variety of animal models. Delay the weakening of aging-related body functions and delay the occurrence of various diseases such as cancer, cardiovascular disease, and neurodegenerative diseases. In the pharmaceutical industry, antiaging peptides have become a hot topic in the field of antiaging research due to their unique physiological functions and excellent biocompatibility. These peptides can delay the aging process by regulating cell signaling pathways and inhibiting inflammatory responses.

Yan et al. [66] studied the biological activity of GHK and its copper chelate GHK-Cu in vitro and in vivo. The study revealed that GHK, as a natural peptide with antiaging potential, promotes skin remodeling, wound healing, and regeneration, as well as antioxidant and anti-inflammatory effects, and partially reverses cognitive impairment in aged mice. Claude Dal Farra et al. [67] proposed an antiaging peptide, which was tested at the cellular level and could regulate the expression of survivin. It was used to prepare an antiaging drug composition with novel biomedical application potential.

At the same time, with the research and development of antiaging drugs and the aggravation of the problem of population aging, the market demand for antiaging drugs is growing. At present, researchers have found a variety of peptides with antiaging effects, which have significant antioxidant effects, can scavenge free radicals in the body, reduce oxidative stress levels, and thus protect cells from damage, such as collagen peptides, elastin peptides, and glutathione. Yiling Zhu et al. [68] studied the in vitro antioxidant activity and in vivo antiaging activity of different components (CP 30, CP 80) of Cordyceps cicadae polysaccharide (CP) and found that CP 70 had the strongest effect, which could prolong the lifespan of Drosophila melanogaster and improve the activity of antioxidant enzymes. Wenjing Qiu et al. [69] developed a dietary supplement called ERJ-CP, which exerted antiaging effect by enhancing antioxidant capacity. Experiments showed that it significantly prolonged the average life span and reduced oxidative damage in the Drosophila model.

Although the biomedical industry has invested a lot of time and resources in the treatment of aging-related chronic diseases, the treatment of many diseases (such as neurodegenerative diseases) is still very limited. However, with the in-depth study of the mechanism and physiological function of antiaging peptides, antiaging peptides have high biological activity and strong pertinence. Through gene detection and biomarker analysis of patients, individualized treatment of antiaging peptides can be achieved, and the curative effect can be improved.

Nady Braidy et al. [17] found that γ-glutamylcysteine (GGC), a precursor of glutathione (GSH), can improve oxidative damage and neuroinflammation in human astrocytes induced by Aβ40 oligomers, providing a new therapeutic strategy for Alzheimer's disease.

The application of antiaging peptides in the pharmaceutical industry has broad prospects. More and more antiaging peptides have been discovered and identified. The research and application of antiaging peptides will continue to deepen. These peptides provide rich resources for the development of new drugs and are expected to bring new breakthroughs in antiaging treatment.

7.6.3 Application in Skin Care Industry

With the improvement of living standards, people's pursuit of beauty and health is constantly improving, and antiaging has become an important research direction in the skin care industry. Antiaging peptides are a class of small molecule compounds formed by amino acids connected by peptide bonds. They have a variety of biologically active functions. Their main mechanisms of action include promoting collagen synthesis, anti-oxidation, anti-inflammatory, moisturizing, and repairing skin barriers. As an emerging bioactive component, it is gradually becoming a star component in the skin care industry due to its excellent antiaging effect and wide application prospects. American scientist Dr. Cohen discovered a new small molecule active polypeptide-epidermal growth factor (EGF), which can promote the growth of epidermal cells. In 1986, he won the Nobel Prize in Physiology and Medicine [70, 71]. E.A. Baptista Veiga et al. [72] focused on the application of antiaging peptides in advanced skin care products, especially the use of nano-delivery systems to improve their permeability and stability to enhance antiaging effects.

The application of antiaging peptides in skin care products has many technical advantages. First, the antiaging peptide has a small molecular weight and is easy to be absorbed by the skin. It can quickly penetrate into the deep layer of the skin and exert its biological activity. Second, antiaging peptides have high stability, are not easily destroyed by environmental factors, and can maintain long-term activity in skin care products. In addition, the preparation technology of antiaging peptides is becoming more and more mature. Through biological extraction, genetic engineering, and chemical synthesis, high-purity and high-activity antiaging peptides can be efficiently obtained. Israeli scientists Aaron Ciechanover, Avram Hershko and American scientist Irwin Rose were awarded the 2004 Nobel Prize in Chemistry for their discovery of ubiquitin-regulated protein degradation, and the resulting extract of Thalassiosira pteronyssinus is also a cosmetic peptide that restores skin to youthful beauty. This series of skin antiaging beauty cosmetics developed based on the theory of "ubiquitin-regulated protein degradation" has been tracked and pursued by researchers in the same field [73].

With the intensification of the global population aging trend and the increasing demand of consumers for antiaging products, antiaging peptides, as an efficient and safe bioactive component, play an antiaging role through a variety of mechanisms and have the technical advantages of easy absorption and high stability. It has shown great application potential in the skin care industry and has broad market prospects. In the future, the research and application technology of antiaging peptides will continue to progress, and new peptide components and preparation processes will continue to emerge, bringing more innovation and development opportunities to the skin care industry.

7.7 Challenges and Controversies in Antiaging Peptides

7.7.1 Technical Bottleneck

Antiaging peptides include natural peptides and synthetic peptides. Natural antiaging peptides are mainly derived from animal and plant tissues, microbial fermentation, etc. Synthetic antiaging peptides are prepared by chemical and biosynthetic methods. The content of natural antiaging peptides in organisms is usually low, which makes the extraction process face many challenges. For example, when extracting peptides from plants, a large number of plant materials need to be processed to obtain relatively few peptides, which not only increases the extraction cost, but also may lead to waste of resources and environmental pollution. In addition, natural antiaging peptides are easily degraded by enzymatic hydrolysis, oxidation, and other factors during the extraction process, resulting in reduced extraction efficiency and loss of active ingredients. For example, some heat-sensitive peptides may lose their activity during high-temperature extraction, thus affecting their antiaging effects.

Antiaging peptides can be prepared by chemical synthesis. Although the sequence and structure of peptides can be accurately controlled, there are still some areas for improvement. In recent years, researchers have significantly improved the yields of these two strategies and effectively controlled the occurrence of side reactions by optimizing reaction conditions and improving catalysts. For example, in the synthesis of peptides containing sensitive amino acids, the use of specific additives can reduce side reactions such as racemization and improve the purity of the target product. In the solid-phase synthesis process, the coupling reaction efficiency and purity of amino acids are the key issues. Side reactions such as racemization and dehydration may lead to abnormal stereostructure or increased impurities of peptides, thus affecting their biological activity and safety. Continuous flow chemistry technology provides new ideas and methods for the synthesis of long peptides. Compared with traditional batch synthesis, continuous flow chemistry has the advantages of rapid mass and heat transfer, good reaction selectivity, high space–time efficiency, and easy automatic control. In the continuous flow solid-phase peptide synthesis, the reactants flow continuously in the microreactor and react, which greatly shortens the reaction time and improves the production efficiency.

Biological synthesis of antiaging peptides is also a favorable means. Recombinant *E.coli* and yeast are commonly used bioengineering production systems, and codon optimization can improve the expression of target peptides. The codon usage preferences of different organisms are different. The codons in the expression vector are optimized to make them more in line with the preferences of the host organism, which can effectively improve the translation efficiency. For example, in the *E.coli* expression system, replacing the codons in the peptide gene with *E.coli*-preferred codons can significantly increase the yield of the target peptide.

Cell-free synthetic biology platform is an emerging peptide synthesis technology, which uses in vitro transcription and translation systems to synthesize peptides in a cell-free environment. The platform has the advantages of controllable reaction conditions and high product purity, but the cost is relatively high. At present, researchers are reducing the cost of cell-free synthesis platforms by optimizing the reaction system and improving the catalytic efficiency of enzymes to improve their competitiveness in peptide synthesis.

In addition, the purification of synthetic peptides is also a challenge. Although traditional purification methods such as HPLC can effectively separate and purify peptides, they are costly, complicated, and inefficient for large-scale production. For example, the synthesis and purification of some long-chain peptides require multiple HPLC operations, which not only increases production costs, but also may lead to loss and denaturation of peptides.

Multi-step chromatography is a key step in peptide purification. Ion exchange chromatography uses the difference in surface charge of polypeptide molecules for separation, and hydrophobic interaction chromatography is based on the different hydrophobicity of polypeptide molecules for purification. By reasonably combining these two chromatography techniques, impurities can be effectively removed and the purity of peptides can be improved. For example, in the purification process of some peptides, ion exchange chromatography was used for preliminary separation, and then further purified by hydrophobic interaction chromatography to obtain high-purity target peptides.

Mass spectrometry and NMR are important means to verify the purity of peptides. Liquid chromatography-mass spectrometry (LC–MS) can accurately determine the molecular weight and sequence of polypeptides, and NMR can provide three-dimensional structure information of polypeptides. Through the combination of these two technologies, the purity and structural integrity of peptides can be comprehensively evaluated. For example, in the process of synthesis and purification of enkephalin, a new method of LC–MS/MS separation and analysis was used to verify the purity of the crude peptide product to ensure the quality of the target product.

7.7.2 Stability and Half-Life of Antiaging Peptides

Peptides face the degradation of a variety of enzymes in the body, which greatly limits their stability and half-life. For example, after entering the human body, peptide drugs will be attacked by proteases, peptidases, etc.in the plasma, resulting in rapid decomposition. Taking the linear anti-angiogenic peptide P15 as an example, it will decompose within two hours in human serum. This rapid enzymatic degradation makes it difficult for peptide drugs to maintain activity in the body for a long time, thus affecting their therapeutic effects.

In order to improve the stability of peptides and prolong their half-life, researchers have adopted a series of chemical modification strategies. For example, the modification of antiaging peptides by PEG can shield the positive charge on the peptide and prevent it from binding to plasma proteins, thereby improving stability. Cyclic modification is also an effective method. Cyclic peptides are more difficult to be recognized and degraded by proteases than linear peptides due to their structural stability. For example, the integration of the P15 sequence into the main-chain cyclized trypsin inhibitor peptide not only significantly improved the stability of P15 to eight hours, but also improved its biological activity.

Long-acting sustained-release preparations can prolong the action time of peptides in vivo and improve their efficacy. poly [lactic-co-glycolic acid] (PLGA) microsphere is a commonly used sustained-release carrier, which has good biocompatibility and degradability. Encapsulating antiaging peptides in PLGA microspheres can achieve slow release and prolong the action time of peptides. For example, in some medical products, PLGA microspheres are used to encapsulate polypeptides, so that they are slowly released in the skin to achieve long-term antiaging effects.

7.7.3 Blood–Brain Barrier Penetration Challenge

The blood–brain barrier (BBB) is a major obstacle to drug delivery in the central nervous system. It is mainly composed of brain microvascular endothelial cells, which are interconnected by tight junctions to form a physical barrier, limiting the passage of macromolecular substances.

The blood–brain barrier is highly selective for the permeability of substances. Endothelial cells express a variety of transporters, such as P-glycoprotein, which can actively pump out substances entering the brain and further hinder the penetration of peptides. The special structural characteristics of the blood–brain barrier make it difficult for almost all macromolecular substances and 98% of small molecules to penetrate the blood–brain barrier into the brain parenchyma under such strict restrictions. For example, antibody-drug conjugates (ADCs) with a molecular weight of more than 150 kDa have limited penetration ability at the blood–brain barrier. In addition, the charge distribution of peptides also affects their ability to penetrate the blood–brain barrier. Positively charged peptides may interact with negatively charged components in the blood–brain barrier, thereby affecting their penetration efficiency.

In order to overcome this challenge, researchers have explored a variety of blood–brain barrier penetration strategies. At present, there are some prediction models of blood–brain barrier penetrating peptides based on machine learning and other methods, but the performance of these models still has some limitations. For example, the prediction accuracy of the model may be affected by the size and quality of the training data set. To evaluate the blood–brain barrier permeability of antiaging peptides, a variety of experimental methods are needed, such as in vitro cell experiments, animal experiments, etc. These methods have their own advantages and disadvantages, and the experimental results may be different, which increases the difficulty of accurate evaluation. At present, researchers have used the blood–brain barrier shuttle peptide as a carrier to deliver drugs to the brain. For example, SynB peptide is a cationic cell-penetrating peptide, which can enhance the transport of drugs through the blood–brain barrier through adsorption-mediated endocytosis. Studies have shown that after coupling the drug with SynB1 or SynB3, the uptake of the drug in the brain is significantly increased. On the other hand, nanotechnology has also been used to improve the blood–brain barrier penetration of peptide drugs. Liposome nanoparticles can encapsulate peptide drugs inside, and improve their penetration efficiency at the blood–brain barrier through surface modification and targeting of nanoparticles.

7.7.4 Security

The long-term use of antiaging peptides may bring some potential risks. The safety issues include cytotoxicity, immunogenicity, long-term use risk, insufficient clinical trial data, and evaluation complexity.

Good biocompatibility is the key to the application of antiaging peptides. Most natural peptides have good biocompatibility, but synthetic peptides may affect their biocompatibility due to their complex structure or the introduction of special chemical groups, causing cell stress response. Most antiaging peptides have no obvious toxicity to cells at a reasonable dose, but some peptides may produce cytotoxicity at high concentrations. Some peptides are metabolized slowly in the body, and long-term accumulation may damage important organs such as the liver and kidney. For example, some peptides containing specific amino acid sequences may interfere with the normal metabolic process of cells and affect cell survival and function. The long-term use of some metal-chelating peptides may affect the balance of metal ions, thereby interfering with enzyme activity and cell signal transduction. Some peptides that interact with hormone receptors may affect the normal secretion and function of hormones, leading to endocrine disorders and a series of health problems.

Antiaging peptides may have complex effects on the immune system. On the one hand, some peptides can regulate immune function and enhance the body's resistance; on the other hand, they may interfere with the immune system's ability to recognize and remove abnormal cells, increasing the risk of infection or cancer. Immunogenicity is also an important issue. Antiaging peptide drugs may be recognized as foreign antigens by the immune system in vivo, thus triggering an immune response. For example, although peptides containing D-amino acids can reduce immunogenicity, they may still be attacked by the immune system, leading to adverse reactions such as inflammation and allergies, affecting the safety of drugs and the health of patients.

Current safety assessment methods, such as cell experiments and animal experiments, can provide some references, but they are different from the real physiological environment of the human body, and it is difficult to fully and accurately predict the safety of antiaging peptides in the human body. The role of antiaging peptides in the body is complex, involving the interaction of various cells, tissues, and organs, which increases the difficulty of safety assessment. The clinical trial data of many antiaging peptides are limited, especially the lack of long-term large-scale clinical trial data, and it is difficult to fully evaluate their long-term safety and efficacy.

Different individuals have different responses to antiaging peptides, and some peptides may cause adverse reactions in some populations. The use of antiaging peptides needs to fully consider individual differences and develop personalized safety assessment programs.

In summary, antiaging peptides face many challenges and controversies in terms of technology, stability, and safety. Future research needs to comprehensively explore technological innovation, safety assessment, and ethical considerations, and conduct more in-depth research and strict clinical trials to ensure the safe application of antiaging peptides and promote the rational application and development of antiaging peptides.

7.8 Future Direction

7.8.1 Multi-Target Synergistic Strategy

With the aging of the global population, the incidence of aging-related diseases is increasing, which makes antiaging research one of the core issues in the field of contemporary life science. Exploring the mystery of aging at the cellular level, scientists have found that aging is not caused by a single factor, but a complex process involving many factors such as genes, metabolism, free radical damage, and protein homeostasis. This complexity determines that antiaging interventions with a single target often have limited effects, because they cannot fully address the multiple mechanisms of aging. Therefore, the multi-target synergistic antiaging strategy has gradually emerged and become a hot research direction. It emphasizes the simultaneous regulation of multiple aging-related pathways in order to achieve more significant and lasting antiaging effects.

Among many antiaging substances, antiaging peptides have attracted much attention due to their unique advantages. These bioactive peptides are usually derived from the hydrolysates or biosynthesis of natural proteins, which have the characteristics of small molecular weight, good biocompatibility, and diverse mechanisms of action. Antiaging peptides can exert antiaging effects through a variety of ways. For example, they can activate the antioxidant enzyme system in cells, reduce the production of free radicals, and enhance the resistance of cells to oxidative stress. It can also regulate cell signaling pathways, affect gene expression, and promote cell self-repair and regeneration. In addition, some antiaging peptides also show a regulatory effect on cellular immune function, which helps to maintain the homeostasis of the body.

As a key molecule in cell energy metabolism and DNA repair, NAD + precursor is an important part of multi-target antiaging strategy. NAD + (nicotinamide adenine dinucleotide) exists in the form of oxidized (NAD +) and reduced (NADH) in cells and plays a central role in various metabolic processes. In terms of cell energy metabolism, NAD + is involved in processes such as glycolysis and tricarboxylic acid cycle. As an electron carrier, it converts energy in food into ATP available to cells, providing power for normal physiological activities of cells. In the process of DNA repair, NAD+ is the substrate of repair enzymes such as PARP (poly ADP-ribose polymerase). When DNA is damaged, PARP will consume NAD+ to initiate the repair mechanism to ensure the integrity and stability of genetic information. However, with age, NAD+ levels gradually decrease in the body, which not only weakens the energy production capacity of cells, but also leads to a decrease in the efficiency of DNA damage repair, thereby accelerating the aging process.

The combination of antiaging peptides with NAD+ precursors is expected to achieve synergistic antiaging effects. On the one hand, antiaging peptides can enhance the antioxidant capacity and repair capacity of cells, creating a more conducive microenvironment for the role of NAD+ precursors in cells. On the other hand, NAD+ precursor can supplement the level of intracellular NAD +, promote the normal progress of cell energy metabolism, and the timely repair of DNA damage, which in turn may enhance the biological activity and stability of antiaging peptides and prolong their action time in vivo.

The theoretical basis of the combination of antiaging peptides and NAD+ precursors is their unique biological functions and mutual complementarity. Antiaging peptides can exert antiaging effects through a variety of ways, such as inhibiting proteolytic enzyme activity, regulating cell signaling pathways, and anti-oxidation. The NAD+ precursor can increase the level of intracellular NAD +, activate the Sirtuins family proteins, and then affect the process of cell metabolism, aging, and survival [74]. The combination of the two can interfere with the aging process from different levels to achieve a more comprehensive antiaging effect.

There are many ways in the specific combination strategy. A common way is to mix antiaging peptides with NAD+ precursors in a certain proportion to make a composite preparation. For example, some studies have combined Epitalon peptide with NR (nicotinamide ribose) and found that its effect in improving cell senescence model is better than using any component alone. In addition, antiaging peptides and NAD+ precursors can be co-encapsulated through delivery systems such as nanocarriers to improve their stability and bioavailability. Microneedle delivery technology has also been applied to the combination of the two to achieve more effective skin penetration. In addition, studies have explored the synergistic effect of antiaging peptides and NAD+ precursors at the level of gene regulation. For example, Tripeptide-32 can target PER1 gene and regulate circadian rhythm, while NAD+ precursor affects gene expression by activating sirtuins family proteins. The combination of the two shows synergistic advantages in regulating cell rhythm and metabolism [75].

The combination of antiaging peptides and NAD+ precursors provides new ideas and methods for antiaging research. Through multi-target synergy, the aging process can be intervened more comprehensively, and the antiaging effect can be improved. In the future, with the deepening of research and the development of technology, this combination strategy is expected to play a greater role in the development and clinical application of antiaging products.

7.8.2 Artificial Intelligence-Driven Design

Traditional peptide design and optimization methods are often time-consuming and laborious, and it is difficult to meet the rapid development of scientific research needs. In recent years, the rapid development of artificial intelligence technology has brought new breakthroughs to the research of antiaging peptides. In particular, deep learning has made significant progress in predicting the structure and function of peptides, providing a powerful tool for the efficient design and development of antiaging peptides.

The structure and function of antiaging peptides are closely related to their amino acid sequences, but the flexibility of the peptide chain and the complex biological environment make its structural prediction extremely challenging. Although traditional experimental methods such as NMR and X-ray crystallography can accurately determine the structure of peptides, there are problems such as high cost and long cycle. Deep learning technology can learn the relationship between peptide sequence and structure and function from a large amount of data by constructing a complex neural network model, so as to achieve rapid and accurate prediction. For example, the PepFlow model developed by the University of Toronto, which combines machine learning and physics principles, can capture the precise conformation of a peptide within minutes, providing a new solution for peptide structure prediction. In addition, methods such as PEP-FOLD3 and APPTEST also improve the accuracy of peptide structure prediction within a specific length range, further enriching the application of deep learning in this field [76].

Deep learning also plays an important role in peptide function prediction. Taking the Deep-ProBind model as an example, it can accurately predict protein-binding peptides and provide important support for drug development. The model uses advanced neural network architecture to extract features from peptide sequences, and classifies and predicts them through deep learning, which effectively improves the accuracy and efficiency of prediction. In addition, through the combination of machine learning and density functional theory (DFT) calculations, researchers have successfully designed peptide sequences with antioxidant activity. These peptides perform well in antioxidant capacity and anti-hemolytic activity, which provides a new idea for the functional expansion of antiaging peptides.

Artificial intelligence-driven antiaging peptide design not only achieves results in structural and functional prediction, but also shows great potential in actual peptide design and optimization. Combining artificial intelligence and molecular simulation techniques, researchers can discover candidate components of active peptides from the structure of molecular targets. Artificial intelligence technology, especially deep learning, plays an increasingly important role in the design and optimization of antiaging peptides.

By predicting the structure and function of peptides and combining molecular simulation techniques, researchers can design peptides with specific antiaging functions more efficiently. The development of these technologies not only accelerates the development of antiaging peptides, but also provides new possibilities for the development of antiaging products. In the future, with the continuous progress of artificial intelligence technology, its application in the field of antiaging peptides will be more extensive and in-depth, and it is expected to make greater contributions to human health.

7.8.3 Precision Antiaging Medicine

While pursuing longer life expectancy, people pay more attention to how to improve the quality of life and maintain youthful vitality and health. In this context, antiaging peptide precision antiaging medicine came into being, bringing new hope and dawn to the cause of human antiaging.

Antiaging peptide precision antiaging medicine is an individualized treatment plan based on biomarkers. It integrates cutting-edge scientific and technological achievements such as modern biotechnology, medical testing, multi-omics analysis, and antiaging peptides. It aims to customize the most effective antiaging treatment strategy for different individuals through accurate biomarker detection and analysis. This emerging field is gradually changing the traditional antiaging medicine model, leading us to a more personalized and accurate antiaging new era.

Biomarkers play a vital role in precision antiaging medicine of antiaging peptides. Biomarkers are indicators that can objectively measure and evaluate normal biological processes, pathological processes, or pharmacological responses to therapeutic interventions. By detecting biomarkers in samples such as blood, tissues, and cells, such as proteins, metabolites, and gene expression products, doctors and researchers can gain an in-depth understanding of the degree of individual aging, potential disease risks, and the functional status of various systems in the body. Based on the detection results of these biomarkers, the medical team can develop a precise antiaging treatment plan for the individual, including selecting the appropriate type of antiaging peptide, determining the dosage and course of treatment.

As an important active ingredient in precision antiaging medicine, antiaging peptides have unique advantages and mechanisms of action. They are short-chain proteins composed of multiple amino acids, which have the characteristics of small molecular weight, high biological activity, and easy absorption by the human body. Antiaging peptides can play a role in a variety of ways, such as promoting collagen production, reducing free radical damage, regulating immune function, and improving skin elasticity, thereby delaying the aging process and improving body function and appearance. Studies have shown that collagen peptides can repair natural aging skin by promoting collagen synthesis, inhibiting collagen degradation, and increasing antioxidant enzyme activity [77]. In addition, the bioactive peptide EQ-9 has also been found to have antioxidant, anti-glycosylation, and antiaging functions [78].

The core of precision antiaging medicine is to develop personalized treatment plans based on individual differences. Each person's genetic background, lifestyle, environmental exposure, and other factors are different, leading to the mechanism and performance of aging are also different. This multi-dimensional comprehensive evaluation makes antiaging treatment more accurate and effective. Antiaging peptide precision antiaging medicine is an individualized treatment plan based on biomarkers, which brings new hope and opportunity to the field of antiaging. Through in-depth research and application of antiaging peptides, combined with accurate biomarker detection, it is expected to achieve more personalized and efficient antiaging treatment, delay the aging process, and improve the quality of life.

Therefore, a unified antiaging treatment program is difficult to meet the needs of all people. Through biomarker detection and multi-omics analysis, precision antiaging medicine can reveal the internal mechanism and characteristics of individual aging, so as to design the most suitable antiaging strategy for everyone. For example, for individuals at high risk of some aging-related diseases, preventive intervention can be carried out in advance; for individuals with aging symptoms, targeted treatment can be carried out according to their specific conditions. This personalized treatment not only improves the effect of antiaging treatment, but also reduces unnecessary drug side effects and waste of medical resources.

Antiaging peptide precision antiaging medicine has broad development prospects, but it also faces some challenges. First of all, the screening and validation of biomarkers is a complex and lengthy process, which requires a lot of clinical research and data support. Second, there are many kinds of antiaging peptides, and their mechanism and effect need to be further studied and explored. In addition, how to closely integrate multi-omics technology with clinical practice to achieve standardization and normalization of precise antiaging treatment is also a problem that needs to be solved at present.

References

1 López-Otín, C., Blasco, M.A., Partridge, L. et al. (2013). The hallmarks of aging. *Cell* 153 (6): 1194–1217.

2 Shay, J.W. (2016). Role of telomeres and telomerase in aging and Cancer. *Cancer Discovery* 6 (6): 584–593.

3 Rando, T.A. and Chang, H.Y. (2012). Aging, rejuvenation, and epigenetic reprogramming: resetting the aging clock. *Cell* 148 (1–2): 46–57.

4 González, A., Hall, M.N., Lin, S.C., and Hardie, D.G. (2020). AMPK and TOR: the Yin and Yang of cellular nutrient sensing and growth control. *Cell Metabolism* 31 (3): 472–492.

5 Katic, M. and Kahn, C.R. (2005). The role of insulin and IGF-1 signaling in longevity. *Cellular and Molecular Life Sciences: CMLS* 62 (3): 320–343.

6 Zhang, S., Yang, J., Liu, M. et al. (2016). Synthesis of peptide dendrimers and their application in the drug delivery system. *Acta Chimica Sinica* 74 (5): 401–409.

7 Alexey, M., Zulfiya, G., Ines, D.A.L. et al. (2022). Targeting aging mechanisms: pharmacological perspectives. *Trends in Endocrinology and Metabolism* 33 (4): 266–280.

8 David, F.H., Eugenia, L.C., Diego, A.L. et al. (2021). Food-derived bioactive compounds with anti-aging potential for nutricosmetic and cosmeceutical products. *Critical Reviews in Food Science and Nutrition* 61 (22): 1–16.

9 Zihong, L., Jie, C., Ling, H. et al. (2021). Aging diagnostic probe for research on aging and evaluation of anti-aging drug efficacy. *Analytical Chemistry* 93 (41): 13800–13806.

10 Xinlu, G., Junjie, L., Jingyi, Q. et al. (2022). The role and mechanism of polysaccharides in anti-aging. *Nutrients* 14 (24): 5330.

11 Tao, L., QingMei, X., Yingshuang, L. et al. (2023). ErZhiFormula prevents UV-induced skin Photoaging by Nrf2/HO-1/NQO1 signaling: an in vitro and in vivo studies. *Journal of Ethnopharmacology* 309: 115935.

12 David, F.H., Eugenia, L.C., Diego, A.L. et al. (2021). Food-derived bioactive compounds with anti-aging potential for nutricosmetic and cosmeceutical products. *Critical Reviews in Food Science and Nutrition* 61 (22): 1–16.

13 Xiaomei, W., Zhongshan, Z., Shaoyong, Z. et al. (2021). Antiaging compounds from marine organisms. *Food Research International* 143: 110313.

14 Jacqueline L B, Merry G, Sabrina S et al. (2021). Microbiome and human aging: probiotic and prebiotic potentials in longevity, skin health and cellular senescence. *Nutrients* 13 (12): 4550.

15 Jiaheng, L., Kun, Z., Jiankang, L. et al. (2022). Injectable Protocatechuic acid based composite hydrogel with hemostatic and antioxidant properties for skin regeneration. *Materials & Design* 222: 111109.

16 Li, Z., Yuewei, X., Yumeng, X. et al. (2021). Injectable self-healing antibacterial bioactive polypeptide-based hybrid nanosystems for efficiently treating multidrug resistant infection, skin-tumor therapy, and enhancing wound healing. *Advanced Functional Materials* 31 (43): 1806883.

17 Qian, Z., Ka-Wing, C., Jianbo, X. et al. (2020). The multifunctional roles of flavonoids against the formation of advanced glycation end products (ages) and AGEs-induced harmful effects. *Trends in Food Science and Technology* 103: 333–347.

18 Gruber, C.W., Koehbach, J., and Muttenthaler, M. (2012). Exploring bioactive peptides from natural sources for oxytocin and vasopressin drug discovery. *Future Medicinal Chemistry* 4 (14): 1791–1798.

19 Mary, K.M., Haopei, W., Kengo, H. et al. (2020). Copper-mediated peptide arylation selective for the N-terminus. *Chemical Science* 11 (38): 10501–10505.

20 Yoon Sin, O. and Jun, H.-S. (2018). Effects of glucagon-like Peptide-1 on oxidative stress and Nrf2 signaling. *International Journal of Molecular Sciences* 19 (1).

21 Anthony, T., Bor-Sen, W., Yu-Wei, C. et al. (2021). Food-derived bioactive peptides with antioxidative capacity, xanthine oxidase and tyrosinase inhibitory activity. *Processes* 9 (5): 747.

22 Weidow, B.L., Vidosh, J., and Biggerstaff, J.P. (2006). The role of soluble fibrin and fibrin inhibitory peptides in Cancer metastasis. *Blood*, 2006 108 (11): 5200.

23 Zehua, L., Yuqi, L., Hongdong, S. et al. (2019). Collagen peptides promote photoaging skin cell repair by activating the TGF-β/Smad pathway and depressing collagen degradation. *Food & Function.* 10 (9): 6121–6134.

24 Hipkiss, A.R., Brownson, C., and Carrier, M.J. (2001). Carnosine, the anti-ageing, anti-oxidant dipeptide, may react with protein carbonyl groups. *Mechanisms of Ageing and Development.* 122: 1431–1445.

25 Zonari, A., Brace, L.E., Al-Katib, K. et al. (2023). Senotherapeutic peptide treatment reduces biological age and senescence burden in human skin models. *Npj Aging* 9: 10.

26 Ichinoseki-Sekine, N., Kakigi, R., Miura, S., and Naito, H. (2015). Whey peptide ingestion suppresses body fat accumulation in senescence-accelerated mouse prone 6 (SAMP6). *European Journal of Nutrition* 54: 551–556.

27 Mistry, K., van der Steen, B., Clifford, T. et al. (2021). Potentiating cutaneous wound healing in young and aged skin with nutraceutical collagen peptides. *Clinical and Experimental Dermatology* 46: 109–117.

28 Li, N., Lv, S., Ma, Y. et al. (2020). In vitro antioxidant and anti-aging properties of swim bladder peptides from Atlantic cod (Gadus morhua). *International Journal of Food Properties* 23: 1416–1429.

29 Pei, X., Yang, R., Zhang, Z. et al. (2010). Marine collagen peptide isolated from chum Salmon (Oncorhynchus keta) skin facilitates learning and memory in aged C57BL/6J mice. *Food Chemistry* 118: 333–340.

30 Qingzhi, D., Ricardo, A.W., Litao, Y. et al. (2019). Antioxidation and memory protection effects of solid-state-fermented rapeseed meal peptides on D-galactose-induced memory impairment in aging-mice. *Journal of Food Process Engineering* 42 (5): e13145.

31 Fang, W., Gaohong, Y., Yu, L. et al. (2022). A peptide from wheat germ abolishes the senile osteoporosis by regulating OPG/RANKL/RANK/TRAF6 signaling pathway. *Phytomedicine*, 2022 104: 154304.

32 Yuqian, W., Xiaoji, C., Qinlu, L. et al. (2020). Active peptide KF-8 from rice bran attenuates oxidative stress in a mouse model of aging induced by D-galactose. *Journal of Agricultural and Food Chemistry* 68 (44): 12271–12283.

33 Xiao-Chen, Y., Zhen, L., Xin-Ran, L. et al. (2021). The antioxidant effects of whey protein peptide on learning and memory improvement in aging mice models. *Nutrients* 13 (6): 2100.

34 Wang, L., Li, M., Lin, Y. et al. (2020). Inhibition of cellular inflammatory mediator production and amelioration of learning deficit in flies by deep sea aspergillus-derived cyclopenin. *The Journal of Antibiotics* 73: 622–629.

35 Alvariño, R., Alonso, E., Bornancin, L. et al. (2020). Biological activities of cyclic and acyclic B-type Laxaphycins in SH-SY5Y human neuroblastoma cells. *Marine Drugs* 18: 364.

36 Ishiguro, S., Shinada, T., Wu, Z. et al. (2021). A novel cyclic peptide (Naturido) modulates glia-neuron interactions in vitro and reverses ageing-related deficits in senescence-accelerated mice. *PLoS One* 16 (1): e0245235.

37 Sebastian, Z., Karina, V.D.L., Urs, L. et al. (2018). An Apoplastic peptide activates salicylic acid Signalling in maize. *Nature Plants* 4 (3): 172–180.

38 Qiang, W., Yong, L., Kuan, P. et al. (2019). Isolation and characterization of three antihypertension peptides from the mycelia of *Ganoderma lucidum* (Agaricomycetes). *Journal of Agricultural and Food Chemistry* 67 (29): 8149–8159.

39 Wenzhu, Z., Lijun, S., Zhipeng, Y. et al. (2022). Improved stability and controlled release of lycopene via self-assembled nanomicelles encapsulation. *LWT-Food Science and Technology* 155.

40 You, J.-G. and Tseng, W.-L. (2019). Peptide-induced aggregation of glutathione-capped gold nanoclusters: a new strategy for designing aggregation-induced enhanced emission probes. *Analytica Chimica Acta* 1078: 101–111.

41 Xi, Q., Yang, Y., Avani, V. et al. (2013). Effects of lycopene on protein expression in human primary prostatic epithelial cells. *Cancer Prevention Research* 6 (5): 419–427.

42 Mengqi, L., Liu, D., Hanting, D. et al. (2021). Potential mechanisms underlying the protective effects of Tricholoma matsutake singer peptides against LPS-induced inflammation in RAW264.7 macrophages. *Food Chemistry* 353: 129452.

43 Arrutia, F., Puente, Á., Riera, F.A. et al. (2016). Influence of heat pre-treatment on BSA tryptic hydrolysis and peptide release. *Food Chemistry* 202: 40–48.

44 Chang, S.K., Ismail, A., Yanagita, T. et al. (2015). Antioxidant peptides purified and identified from the oil palm (Elaeis guineensis Jacq.) kernel protein hydrolysate. *Journal of Functional Foods* 14: 63–75.

45 Wancen, X., Alberto, T., Baicang, L. et al. (2020). First exploration on poly (vinyl chloride) ultrafiltration membrane prepared by using the sustainable green solvent PolarClean. *ACS Sustainable Chemistry & Engineering* 8 (1): 91–101.

46 Bang, J.S., Jin, Y.J., and Choung, S.-Y. (2020). Low molecular polypeptide from oyster hydrolysate recovers Photoaging in SKH-1 hairless mice. *Toxicology and Applied Pharmacology* 386: 114844.

47 Freja, G., Stefan, S., Bente, G. et al. (2019). Development of a UPLC-IDA-ICP-MS/MS method for peptide quantitation in plasma by se-labelling, and comparison to S-detection of the native peptide. *Journal of Analytical Atomic Spectrometry* 34 (2): 375–383.

48 Gruber, C.W., Koehbach, J., and Muttenthaler, M. (2012). Exploring bioactive peptides from natural sources for oxytocin and vasopressin drug discovery. *Future Medicinal Chemistry* 4 (14): 1791–1798.

49 Jia-Yu, H., Ju, L., Mao-Song, X. et al. (2020). Effective strategies for improving the stability of peptides in vivo. *Acta Pharmaceutica Sinica* 55 (1): 25–32.

50 Liangyou, W. and Keliang, L. (2003). PEGylation of peptides and proteins. *Chinese Journal of Organic Chemistry* 23 (11): 1320–1323.

51 Yuan, Q., Ren, Q., Li, L. et al. (2022). A klotho-derived peptide protects against kidney fibrosis by targeting TGF-β signaling. *Nature Communications* 13 (1): 438.

52 Xingyu, L., Sufang, C., Qiuyi, Z. et al. (2022). Evaluation of antioxidant activity of milk-derived sleep-promoting peptide in vitro and antioxidant effect on *Caenorhabditis elegans* in vivo. *Food Science* 43 (5): 151–157.

53 Li, X., Yu, P., Zhao, D., and Yao, J. (2002). The influence of PT in traditional Chinese medicine on GSH, immunity function and element in aged little mouse. *Chinese Journal of Modern Applied Pharmacy* 5: 363–364.

54 Rahagir Salekeen, M.S., Lustgarten, U., Khan et al. (2023). Model organism life extending therapeutics modulate diverse nodes in the drug-gene-microbe tripartite human longevity interactome. *Journal of Biomolecular Structure and Dynamics* 1–19.

55 Mi, W., Zhenyuan, W., Tianqi, L. et al. (2023). Enhanced skin permeation of anti-aging peptides using bioactive ionic liquids at low concentrations. *Advanced Functional Materials* 33 (42).

56 Satoshi, O., Matthew, R.N., Chad, E.T. et al. (2019). Conformation and permeability: cyclic hexapeptide diastereomers. *Journal of Chemical Information and Modeling* 59 (6): 2952–2963.

57 D'Hondt, M., Bracke, N., Taevernier, L. et al. (2014). Related impurities in peptide medicines. *Journal of Pharma ceutical and Biomedical Analysis* 101: 2–30.

58 De Spiegeleer, B., Vergote, V., and Pezeshki, A. (2008). Impurity profiling quality control testing of synthetic peptides using liquid chromatography-photodiode Array-fluorescence and liquid Chroma tography-electrospray ionization-mass spectrometry: the Obestatin case. *Analytical Biochemistry* 376: 229–234.

59 N H, M L, P D, et al. (1975) Purification and chemical study of a Collocalia glycoprotein. *Biochimie*, 57(5): 603–608.

60 Noriko, M., Megumi, M., Wakoto, B. et al. (2011). Improvement of bone strength and dermal thickness due to dietary edible bird's Nest extract in Ovariectomized rats. *Bioscience, Biotechnology, and Biochemistry* 75: 590–592.

61 Ki, C.K., Kyoung, A.K., Chae, M.L. et al. (2012). Water extract of edible bird's Nest attenuated the oxidative stress-induced matrix Metalloproteinase-1 by regulating the mitogen-activated protein kinase and activator Protein-1 pathway in human keratinocytes. Journal of the Korean Society for Applied. *Biological Chemistry* 55 (3): 347–354.

62 Song, S.L., Wang, H., Liang, H. et al. (2017). Isolation and purification of sea cucumber peptide and its impacts on the secretion of collagen by NIH/3T3cells. *Modern Food Science & Technology* 33 (3): 7.

63 Chen, S.Y., Yang, Q., Chen, X. et al. (2020). Bioactive peptides derived from crimson snapper and in vivo anti-aging effects on fat diet-induced high fat *Drosophila melanogaster*. *Food & Function* 11 (1): 524–533.

64 Ding, Y.L., Ko, S.C., Moon, S.H. et al. (2019). Protective effects of novel antioxidant peptide purified from alcalase hydrolysate of velvet antler against oxidative stress in chang liver cells in vitro and in azebrafish model in vivo. *International Journal of Molecular Sciences* 20 (20): 5187.

65 Qiu, W., Chen, X., Tian, Y. et al. (2020). Protection against oxidative stress and anti-aging effect in Drosophila of royal jelly-collagen peptide. *Food and Chemical Toxicology* 135: 110881.

66 Yan, D., Amanda, L., Lida, Z. et al. (2020). The potential of GHK as an anti-aging peptide. *Aging Pathobiology and Therapeutics* 2 (1): 58–61.

67 Yu-Hsuan, C., Hsiao-Ling, C., Hueng-Chuen, F. et al. (2020). Anti-inflammatory, antioxidant and antifibrotic effects of kefir peptides on salt-induced renal vascular damage and dysfunction in aged stroke-prone spontaneously hypertensive rats. *Antioxidants* 9 (9): 790.

68 Yiling, Z., Xiaofeng, Y., Qi, G. et al. (2020). Antioxidant and anti-aging activities of polysaccharides from Cordyceps cicadae. *International Journal of Biological Macromolecules* 157: 394–400.

69 Wenjing, Q., Xu, C., Yongqi, T. et al. (2020). Protection against oxidative stress and anti-aging effect in Drosophila of Royal Jelly-Collagen Peptide. *Food and Chemical Toxicology*, 2020 135: 110881.

70 Nady, B., Martin, Z., Bat-Erdene, J. et al. (2019). The precursor to glutathione (GSH), Γ-glutamylcysteine (GGC), can ameliorate oxidative damage and neuroinflammation induced by Aβ40 oligomers in human astrocytes. *Frontiers in Aging Neuroscience* 11: 177.

71 Levi-Montalcini, R. and Cohen, S. (1956). In vitro effects of mouse sarcoma 180 on the spinal and sympathetic ganglia of the chick embryo. *Cancer Research* 16 (4): 351–355.

72 Veiga, E., Ferreira, L., Correia, M. et al. (2023). Anti-aging peptides for advanced skincare: focus on Nanodelivery systems. *Journal of Drug Delivery Science and Technology* 89: 105087.

73 Koyuncu, S., Loureiro, R., Lee, H.J. et al. (2021). Rewiring of the ubiquitinated proteome determines ageing in C. Elegans. *Nature* 596: 285–290.

74 Shixiong, Z., Jing, L., Shentao, G., and Xingfen, Y. (2023). Progress in research on mechanism of nicotinamide adenine dinucleotide and its precursors against aging-related diseases. *Chinese Journal of Public Health* 39 (6): 793–799.

75 Ferreira, M.S., Magalhães, M.C., and Sousa-Lobo, J.M. (2020). Trending anti-aging peptides. *Cosmetics* 7: 91.

76 Abramson, J., Adler, J., Dunger, J. et al. (2024). Accurate structure prediction of biomolecular interactions with AlphaFold 3. *Nature* 630 (8016): 493–500.

77 Liu, L., Yin, C.Y., Hu, H.Z. et al. (2023). The mechanism and clinical research progress of collagen peptide in anti-skin natural aging. *Chinese Journal of Dermatology* 56 (05): 455–458.

78 Wang Feifei, S., Wenrou, K.S. et al. (2023). A novel functional peptide, named EQ-9 (ESETRILLQ), identified by virtual screening from regenerative cell secretome and its potential anti-aging and restoration effects in topical applications. *Peptides* 169: 171078.

8

Antifatigue Peptides

8.1 Introduction of Antifatigue Peptides

Fatigue is a prevalent condition characterized by a state of exhaustion and diminished energy levels, significantly impacting the body's capacity to carry out daily tasks. It is generally classified as central or peripheral. Central fatigue refers to an adverse mental state caused by prolonged stress on the central nervous system (CNS) or blocked peripheral motor neuron signaling [1]. Peripheral fatigue is a physiological condition resulting from decreased muscle function due to overtraining [2]. Typically, rest can help alleviate normal fatigue symptoms. However, inadequate relief of fatigue can compromise the immune system and contribute to the onset of various diseases negatively impacting human health [3]. As a result, the scientific investigation of antifatigue has emerged as a critical subject in recent research.

The multifaceted causes of fatigue include inadequate energy supply and metabolite accumulation [4]. Strenuous exercise can disrupt the body's balance of oxidative and antioxidant systems, leading to peroxide production and fatigue. Antifatigue active substances have been investigated with three main mechanisms: (i) participating in metabolism to lower blood lactic acid (LD) and blood urea nitrogen (BUN); (ii) enhancing glucose homeostasis to increase glycogenolysis and energy supply; and (iii) mitigating the harm caused by an excessive amount of reactive oxygen species (ROS) through the restoration of oxidative stress-induced damage. Polysaccharides, proteins, peptides, as well as other bioactive compounds such as polyphenols, serve as the main reservoirs of antifatigue active constituents. Among them, antifatigue peptides have gained the attention of researchers in recent years due to their wide sources and low toxic effects [5].

8.2 Extraction and Purification of Antifatigue Peptides

8.2.1 Sources of Antifatigue Peptides

With the in-depth research and enormous application value of antifatigue peptides in recent years, scholars have been constantly trying to prepare and study biological peptides with antifatigue activity using different raw materials, making the research on antifatigue

Food-Derived Bioactive Peptides: Green Preparation, Activity Mechanisms, and Health Applications, First Edition. Shao-Yun Wang, Xu Chen, and Xixi Cai.

Table 8.1 The source of antifatigue peptides.

Category	Source	Reference
Animals	Sea cucumber	[6]
	Sea horse	[7]
	Tilapia	[8]
	Mackerel	[9]
	Hairtail	[10]
	Loach	[11]
	Croceine croaker	[12]
	Monkfish	[13]
	Cervus elaphus blood	[14]
	Egg white	[15]
	Milk proteins	[16]
Plants	Soybean	[17]
	Corn	[18]
	Pea	[19]
	Wheat	[20]
	Walnut	[21]
	Purple perilla seeds	[22]
	Hemp seeds	[23]
Microorganism	*Spirulina*	[24]

peptides very active. From the source, the peptides currently reported to have antifatigue activity are mainly prepared from animal and plant materials (Table 8.1).

In terms of animal sources, aquatic products are rich in protein content, and there have been many studies on using them as raw materials to prepare antifatigue peptides. Ye et al. [6] found that sea cucumber peptides (SCPs) prepared from sea cucumbers significantly improved the fatigue tolerance of rats. Guo et al. [7] found that hippocampal peptides have in vitro antioxidant activity and in vivo antifatigue effects. Ren et al. [8] found that collagen peptides derived from tilapia skin could significantly prolong the exhaustion swimming time of loaded mice. Wang et al. [9] found that mackerel peptides could alleviate fatigue by supplementing the concentration of blood glucose and enhancing the activity of antioxidant enzymes. Wang et al. [10] found that the hydrolysate of hairtail exhibited antifatigue activity in a mouse model of endurance swimming. You et al. [11] found that loach peptides (LPs) could increase the endurance of mice and promote fatigue recovery. Zhao et al. [12] found that swim bladder peptide components isolated from Croceine croaker protein hydrolysate have good antifatigue ability in mice by inhibiting oxidative reactions. Wang et al. [13] found that hydrolyzed low-molecular-weight peptides from monkfish have antifatigue and immune effects on mice. In addition to aquatic products, *Cervus elaphus* blood [14], egg whites [15], and milk proteins [16] are also sources of antifatigue peptides.

Among the peptides derived from plants, research on the antifatigue effect of soybean peptides was conducted earlier. Fang et al. [17] prepared soybean-active peptides by fermenting defatted soybean meal, and the study showed that soybean small-molecule peptides have a certain effect on enhancing the antifatigue ability of mice. Feng et al. [18] demonstrated the antifatigue effect of corn peptides through experiments. Feng et al. [19] conducted a study on the antiexercise fatigue of pea peptides, and the results showed that the peptide could significantly prolong the swimming time of mice, significantly reduce the levels of blood lactate (BLA) and serum urea nitrogen in mice after exercise, significantly increase the content of muscle glycogen and liver glycogen, and have antifatigue effects. Zheng et al. [20] enzymatically hydrolyzed wheat protein solution to obtain wheat peptides. Mouse experiments have shown that it has both antifatigue effects and can alleviate immune regulation disorders. Liu et al. [21] isolated small-molecule oligopeptides (WOPs) from walnuts, which can improve the exercise performance and antifatigue ability of mice. Liu et al. [22] and Ying et al. [23] found that plant seed peptides can also improve exercise fatigue in mice.

In addition to animal and plant sources, Chen et al. [24] found that the antioxidant activity of spirulina-derived peptides and their effects on the fat metabolism pathway in mice give them antifatigue activity.

8.2.2 Preparation of Antifatigue Peptides

8.2.2.1 Chemical Hydrolysis Method

The main principle of chemical hydrolysis method is to add an appropriate concentration of strong acid or strong base solution to the protein to be degraded under the condition of a water bath, destroy the peptide bond of the protein, and then control the degree of hydrolysis, terminate the reaction after a certain time, and separate and purify. Chemical hydrolysis method has the advantages of simple operation and low cost, but it has many disadvantages, such as alkali reagents easily destroy amino acids, the degree of hydrolysis is not easy to control, there are many by-products, and the alkali needs to be removed at the end of hydrolysis, which limits the application and popularization of this method.

8.2.2.2 Chemical Synthesis

The main principle of chemical synthesis is to synthesize bioactive peptides using appropriate chemical methods according to the original sequence of target bioactive peptides. According to the different synthetic mediums, chemical synthesis can be divided into solid-phase synthesis and liquid-phase synthesis. The application of liquid-phase synthesis is mainly to synthesize concentrated peptide fragments of small-molecule polypeptides. But this method requires separation, purification, or crystallization of the product after each peptide insertion to remove unreacted raw materials and by-products. Therefore, the steps are time-consuming and require high technical skills. The solid-phase carrier used in the solid-phase synthesis method is beneficial to the fixation, cyclization, deprotection, and purification of the peptide chain in the synthesis process. Therefore, the operation is simple, the reaction time is short, the yield is high, and it is easy to achieve automation, but it is mainly suitable for short-peptide synthesis. Currently, short antifatigue peptides with known amino acid sequences can be produced through solid-phase synthesis. For example,

Nam et al. [25] found that synthetic dipeptides Gly–Leu and Leu–Gly have significant antifatigue activity by modulating inflammation and oxidation responses.

8.2.2.3 Enzyme Synthesis Method

The principle of enzyme synthesis is to catalyze the synthesis of amino acids and their derivatives into bioactive peptides. Enzyme synthesis has many advantages, such as mild reaction conditions, which can control the direction of the catalytic position of the enzyme, but there are also many disadvantages, such as serious by-products produced during the reaction, low production efficiency, and lack of optimal enzymes, which adversely affect its development.

8.2.2.4 Microbial Fermentation Method

The principle of microbial fermentation is the process of using proteases produced during microbial fermentation to break down protein molecules into peptides or free amino acids. During fermentation, enzymes other than proteases are also produced to degrade complex carbohydrates and lipids. At the same time, microbial fermentation can reduce or eliminate bitterness. After separation and purification, the "debitter" step can be omitted, which can not only simplify the process, but also reduce the cost, and have less damage to the nutritional value, and have considerable potential in maintaining the nutritional value. According to the type of bacterial strain used, this method can be divided into single-strain fermentation and compound strain fermentation.

The preparation of bioactive peptides by single-strain fermentation has the advantages of simple process, easy control, and high product purity. Fang et al. [17] prepared fermented soybean protein peptides (FSPPs) by inoculating 0.15% *Lactobacillus acidophilus* into soybean protein hydrolysate and fermenting it (Table 8.2). FSPPs could reduce the levels of serum urea nitrogen and BLA in mice, increased the content of liver glycogen, and had the effect of relieving fatigue. Yu et al. [26] used *Bacillus subtilis* SHZ

Table 8.2 Amino acid composition (dry basis) of fermented soybean protein peptides.

Amino acid	Content (g 100/g)	Amino acids	Content (g 100/g)
Aspartic acid[a]	8.26 ± 0.04	Leucine	5.34 ± 0.07
Glutamine acid[b]	13.81 ± 0.15	Tyrosine	2.29 ± 0.08
Threonine	2.64 ± 0.05	Phenylalanine	3.61 ± 0.07
Serine	3.44 ± 0.11	Lysine	4.35 ± 0.13
Glycine	2.79 ± 0.08	Histidine	1.60 ± 0.05
Alanine	2.91 ± 0.06	Arginine	5.02 ± 0.04
Valine	3.38 ± 0.07	Proline	5.56 ± 0.03
Methionine	1.12 ± 0.03	Cysteine	0.90 ± 0.02
Isoleucine	3.34 ± 0.05	Tryptophan	0.23 ± 0.04

[a] Aspartic acid + asparagine.
[b] Glutamine acid + glutamine.
Source: [17]/John Wiley & Sons.

to ferment defatted soybean meal and produce soy peptides, purified them by ultrafiltration and gel chromatography, and studied the antifatigue effect of soybean peptides by swimming in mice.

Composite strain fermentation, through the synergistic effect of multiple strains, can improve substrate utilization, product types and biological activity, and enhance product flavor. However, the fermentation process is relatively complex, and product separation and purification are difficult. Cui et al. [27] inoculated *B. subtilis* GD1, *B. subtilis* N4, *Bacillus velezensis* GZ1, *Lactobacillus delbrueckii* subsp. *bulgaricus*, and *Hansen anomala* into soybeans for mixed culture and fermentation. The products showed strong antioxidant and antifatigue potential activities.

8.2.2.5 Enzymatic Hydrolysis Method

Enzymatic hydrolysis is the most commonly used method for preparing antifatigue peptides. Enzymatic hydrolysis is a method of producing peptide fragments by using proteolytic enzymes to hydrolyze specific peptide bonds in proteins. Compared with other methods, it not only has the advantages of low-cost, mild reaction conditions, easy control of reactions, and high product safety, but also can produce specific peptides through enzyme specificity to meet market demand. The rationality of protease selection is crucial in the enzymatic hydrolysis process, which not only requires the enzyme to digest at specific sites to obtain specific peptides, but also ensures their safety. The commonly used proteases in enzymatic hydrolysis include alkaline protease, trypsin, neutral protease, bromelain, pepsin, chymotrypsin, and papain. At present, the most commonly used methods are single-enzyme hydrolysis and complex enzyme hydrolysis.

Single-enzyme hydrolysis is a method of using a protease to hydrolyze proteins. Guo et al. [2] used hippocampal protein as raw material to prepare hippocampal antifatigue peptides through papain enzymatic hydrolysis and optimized the enzymatic hydrolysis conditions. The maximum activity was achieved after hydrolysis for 40 minutes at pH 6.0, 60 °C, enzyme dosage of 2000 U/g, and solid–liquid ratio of 1:15 in the reaction system. Huang et al. [28] prepared hairtail-hydrolyzed peptides by enzymatic hydrolysis of papain and studied the antifatigue activity of their ferrous chelates. The results showed that hairtail-hydrolyzed peptide ferrous chelates could significantly reduce the content of malondialdehyde (MDA) and BLA in mice and prolong the exhaustion swimming time of mice. You et al. [11] prepared LPs through papain digestion and measured in vitro antioxidant activity and in vivo antifatigue effect; it extended the time for mice to swim until exhaustion by 20%–28% compared with the control. Liu et al. [29] hydrolyzed whey protein hydrolysate (WPH) using trypsin (EC 3.4.21.1) and found that low-molecular-weight ($<$10 kDa) WPHs with higher free radical-scavenging and iron chelation activity exhibited higher antifatigue ability (Table 8.3).

Complex enzyme hydrolysis is a method of hydrolyzing proteins by mixing two or more enzymes. Pan et al. [16] used chymotrypsin and trypsin to hydrolyze milk protein to prepare peptide solutions, Orthogonal design experiments were used to optimize hydrolysis conditions (Table 8.4). The hydrolysis pH value has the largest R range among the four factors, indicating that the hydrolysis pH value has the greatest impact on peptide production, followed by milk powder concentration, hydrolysis time, and hydrolysis temperature. The final optimized hydrolysis conditions were temperature of 45 °C, pH value of 8.0, time of 2 hours,

Table 8.3 Antioxidant capacity of whey protein isolate (WPI), whey protein hydrolysate (WPH), and WPHs with different molecular weight (MW<5, 5–10, 10–30 and>30 kDa).

Group	Hydroxyl radical-scavenging ability (%)	DPPH (2,2-diphenyl-1-picrylhydrazyl)-scavenging ability (%)	Ferrous ion-chelating activity (%)	Copper ion-chelating activity (%)
WPI	81.37 ± 1.45y	65.15 ± 1.67w	59.18 ± 2.10y	70.13 ± 1.09
WPH	85.3 ± 4.98zy	82.54 ± 2.13y	69.24 ± 3.92z	74.17 ± 3.75
WPH < 5	92.47 ± 3.25z	93.18 ± 2.05z	68.13 ± 1.98z	71.87 ± 2.38
WPH 5–10	88.75 ± 2.81z	87.13 ± 1.97y	70.12 ± 3.41z	75.12 ± 1.18
WPH 10–30	80.64 ± 3.28y	77.63 ± 2.91x	61.23 ± 1.67y	73.25 ± 3.41
WPH > 30	63.21 ± 1.87x	64.18 ± 3.67w	61.73 ± 2.65y	70.15 ± 2.63

Source: [29]/John Wiley & Sons.
Data are mean ± SD (n = 20). Means in the same column with the same letter are not significantly different among groups ($P < 0.05$).

Table 8.4 Results of orthogonal design experiment.

Times	x_1	x_2	x_3	x_4	y
1	8	43	25	1	1.28 ± 0.01
2	8	45	20	1.5	1.43 ± 0.12
3	8	47	16.67	2	1.22 ± 0.01
4	8.5	43	16.67	2	1.24 ± 0.03
5	8.5	45	20	1	1.05 ± 0.21
6	8.5	47	25	1.5	1.20 ± 0.04
7	9	43	16.67	1.5	1.13 ± 0.01
8	9	45	25	2	1.33 ± 0.05
9	9	47	20	1	1.08 ± 0.02
K1	1.310	1.217	1.270	1.137	
K2	1.163	1.270	1.250	1.253	
K3	1.180	1.167	1.133	1.263	
R	0.147	0.103	0.137	0.126	

Hydrolysis pH (X_1), hydrolysis temperature (C) (X_2), milk powder concentration (%) (X_3), hydrolysis time (h) (X_4), and peptide concentration (mg/mL) (Y).
Source: [16]/John Wiley & Sons.

and milk powder concentration of 25%. And then weight swimming time, serum urea nitrogen, BLA, and liver glycogen were measured. It was found that it had good antifatigue activity. Feng et al. [18] used crude alkaline protease from *Bacillus licheniformis* (enzyme:substrate ratio = 1:100) to perform the first enzymatic hydrolysis of corn protein at 55 °C and pH 8.5. The obtained mixture was subjected to a second enzymatic hydrolysis using

Table 8.5 CP-based physicochemical composition.

Moisture (%)	Ash (%)	Total protein (%)	Acid-soluble protein (%)	Free amino acids (%)	Oligopeptides (%)
4.61	2.87	82.81	82.40	4.38	78.03

Source: [18]/John Wiley & Sons.

B. subtilis crude neutral protease (enzyme:substrate ratio = 1:100) at 45 °C and pH 7.0 to prepare corn antifatigue peptide (CP) and determine its physicochemical composition (Table 8.5).

8.2.3 Isolation and Purification of Antifatigue Peptides

The main purpose of separation and purification of bioactive peptides is to obtain samples with higher activity or purity, and increase the yield of the product as much as possible. In the study of bioactive peptides, the structural analysis and structure–activity relationship are more important parts, which need to use the separation and purification technology, so the separation and purification of bioactive peptides is very important. The main techniques of directional separation and purification are electrophoresis, chromatography and membrane separation.

8.2.3.1 Electrophoresis Technology

Electrophoresis is the phenomenon of charged particles moving in an electric field in the direction of an electrode opposite to their own charge. The main principle of protein electrophoresis separation and purification is to separate the protein samples by the electrophoretic mobility, electrostatic charge, and the molecular size and shape of each protein component. Among the separation techniques used in the laboratory, electrophoresis still plays an important role, including capillary electrophoresis, two-dimensional electrophoresis, pulse field electrophoresis, and so on.

8.2.3.2 Chromatography Technology

The most widely used chromatography technologies are macroporous resin adsorption chromatography, gel filtration chromatography, ion-exchange chromatography, affinity chromatography, reversed-phase high-performance liquid chromatography (RP-HPLC), and so on. RP-HPLC is a commonly used method for separating antifatigue peptides. Its principle is to utilize the hydrophobic or solvent-repellent properties of peptide molecules. Nonpolar or weakly polar peptide molecules will bind to the functional groups of the nonpolar stationary phase (octadecyl-bonded silica gel), and this association will decrease due to the change in polarity of the mobile phase, thereby causing solute molecules to be eluted and separated. RP-HPLC has high resolution, sensitivity, good separation effect, simple operation, and can automatically analyze target peptides in large quantities, but the instrument and equipment are expensive. Zhang et al. [30] purified peptide SH200 with a purity of over 99% from the hydrolysate of hippocampus using fast protein liquid chromatography and RP-HPLC.

Gel filtration chromatography is also a common method to separate antifatigue peptides. It is a separation technology based on the blocking effect of gel filling on the separated mixture due to different molecular sizes. The gel has the function of molecular sieve, so that different components can be separated due to different particle sizes, travel distances, and elution times. It can be used for the distribution and classification analysis of the relative molecular weight (MW) of polypeptides. Gel filtration chromatography has good separation effect and fast separation speed, and the product is not easy to denature, but its effect may not be good when used alone. When used in combination with other separation and purification methods, peptides with high purity or activity will often be obtained.

8.2.3.3 Membrane Separation Technology

Membrane separation technology refers to the selective separation of the mixture of molecules of different particle sizes at the molecular level through the semi-permeable membrane, also known as the separation membrane or filter membrane; the membrane wall is covered with small holes, and, according to the size of the aperture, can be divided into microfiltration, ultrafiltration, and nanofiltration. Among them, ultrafiltration technology plays an important role in the separation of antifatigue peptides. Ultrafiltration uses pressure difference as a driving force and the sieving principle of semi-permeable membranes to intercept substances with MWs between 1 and 500 kDa. The specific pore size of ultrafiltration membrane can effectively intercept and separate bioactive peptides based on their MW, achieving the classification of peptides with different MWs, thereby improving the purity of antifatigue peptides and preserving peptide activity. However, it is only suitable for initial separation and not suitable for separating specific peptides. Liu et al. [29] separated WPH fractions with different MWs through ultrafiltration and found that mice in the low MW (<10 kDa) WPH group had longer swimming time, higher concentrations of glucose, no esterified fatty acids, liver glycogen, superoxide dismutase (SOD), and glutathione peroxidase (GSH-Px), and lower concentrations of lactate (LA), indicating good antifatigue ability. Wang et al. [9] purified mackerel protein hydrolysate (MPH) through ultrafiltration membrane and studied its antifatigue effect and mechanism in mice. The results indicated that MPH could supplement the concentration of blood glucose and enhance the activity of antioxidant enzymes, thereby alleviating fatigue. Sun et al. [15] isolated four peptide components from egg white protein hydrolysate (EWPP) through ultrafiltration. Among them, the egg white peptide (EWP) component with an MW of 2–5 kDa (referred to as EWPs2) has the strongest antioxidant activity and exhibits antifatigue effects (Table 8.6).

8.2.3.4 Multiple Methods Combined

In practical applications, it is difficult to achieve the desired effect through only one method, so more is to use a combination of methods, not only to improve the separation effect, but also to achieve multistage separation of bioactive peptides. Zheng et al. [20] first collected fractions with MWs less than 1,000 Da through ultrafiltration membranes and freeze-dried them. Then, using C-18 chromatographic column, wheat peptides with MWs less than 1,000 Da were continuously separated by RP-HPLC. Pan et al. [16] separated six peptide components from milk protein hydrolysate by ultrafiltration and sephadex G-15 gel filtration

Table 8.6 In vitro antioxidant activity of crude EWPP, four EWPs fractions with different molecular weight (MW) ranges and glutathione (GSH).

Sample	Antioxidant activity (%)			
	DPPH radical scavenging	Hydroxyl radical scavenging	Superoxide anion radical scavenging	Lipid peroxidation inhibition
EWPP	91.18 ± 2.45^b	46.39 ± 1.47^b	46.57 ± 1.54^c	51.37 ± 1.25^c
EWPs1 (MW >5 kDa)	75.13 ± 1.67^c	41.13 ± 1.03^c	41.38 ± 1.27^d	45.26 ± 1.11^d
EWPs2 (MW 2–5 kDa)	98.93 ± 2.53^a	55.17 ± 2.19^a	64.78 ± 3.02^a	85.59 ± 2.41^a
EWPs3 (MW 1–2 kDa)	34.98 ± 0.95^e	17.21 ± 0.74^e	22.63 ± 1.03^f	27.75 ± 0.92^f
EWPs4 (MW <1 kDa)	65.54 ± 1.68^d	35.32 ± 0.98^d	38.55 ± 1.14^e	39.19 ± 1.04^e
GSH	9.97 ± 0.34^f	53.09 ± 2.01^a	52.77 ± 2.16^b	73.98 ± 1.35^b

Source: [15]/John Wiley & Sons.
Different letters in the same column indicate significant differences between samples ($P < 0.05$).

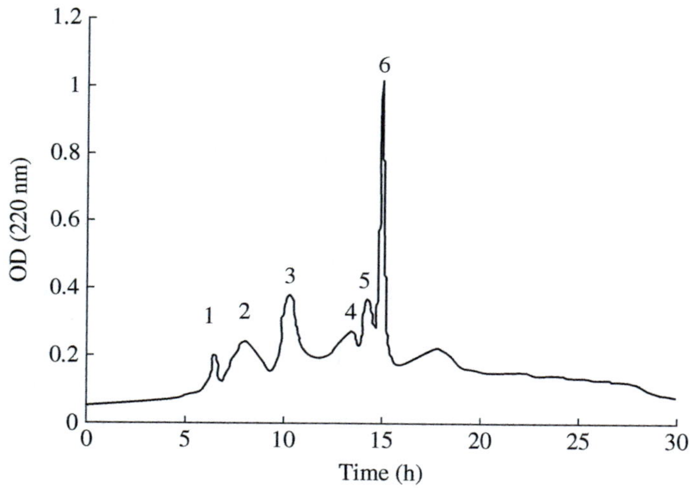

Figure 8.1 Sephadex G-15 gel filtration chromatography column profiles of peptide after ultra-filtered (<6 kDa). *Source:* [16]/John Wiley & Sons.

chromatography. Through swimming endurance testing to determine the antifatigue ability of each part, it was found that F3 peptide can significantly increase the load swimming time and liver glycogen of mice, and reduce SBUN and BLA after swimming (Figure 8.1).

8.2.4 Sequence Identification of Antifatigue Peptides

The sequence determination of antifatigue peptides is of great significance for the study of their functions and properties. Mass spectrometry is the most important method to determine the sequence of antifatigue peptides, and it is often used in conjunction with

RP-HPLC. In mass spectrometry, samples are separated after ionization based on the ion mass charge ratio (m/z), which is analyzed by measuring the intensity of different ion spectral peaks. In the early 1980s, matrix-assisted laser desorption ionization (MALDI) and electrospray ionization technology (ESI) received widespread attention. The upper limit of molecular mass determined by MALDI mass spectrometry is 400 ku, and the tolerance to salt is high. The upper limit of molecular mass determined by ESI mass spectrometry is 150 ku, which has high resolution and is often used in combination with liquid chromatography. In recent years, the development of series mass spectrometers has further strengthened the function of structural analysis of mass spectrometry, which can determine the amino acid sequence of polypeptides more quickly and sensitized, identify the modification site of the side chain, and determine the location of disulfide bonds. Zheng et al. [20] used matrix assisted laser desorption ionization time-of-flight mass spectrometry (MALDI-TOF-MS) with the highest active component F4 for peptide identification. Four peptide sequences were identified, Leu–Ala–Arg (LAR), Gln–Asp (QD), Gln–Pro (QP), and His–Gln–Gly–Ile (HQGI). Zhang et al. [30] found that the MW of the antifatigue peptide SH200 isolated from the hippocampus was 1909 Da by coupling the Q-TOF mass spectrometry with the ESI source. The complete amino acid sequence has been determined as LEDPFDKDDNWKS.

Wang et al. [31] measured the peptide sequences of sea cucumber-derived peptides SCP-1 (hydrolysis degree 5.09%) and SCP-2 (hydrolysis degree 13.31%) with different degrees of hydrolysis using liquid chromatography-mass spectrometry (LC–MS/MS), and studied specific amino acids [branched chain amino acids (BCAA), including leucine, isoleucine, and valine] closely related to antifatigue. In all sequences, 95.64% of the peptides in SCP-2 at <3,000 Da were composed of these three amino acids in all recognized sequences, while only 80.60% were present in SCP-1. Based on the amino acid composition and peptide profile, SCP-2 might have better antifatigue ability than SCP-1 due to its lower MW, shorter peptide length, and higher proportion of recognized sequences associated with BCAA.

Lv et al. [14] analyzed and identified three peptide components (Fr1, Fr2, and Fr4) of deer blood enzyme hydrolysate using HPLC-MS/MS. Thirty-one compounds were identified as glutamine or branched chain amino acids containing short peptides, among which Val–Ala–Asn, Val–Val–Ser–Ala, Leu (Ile)–Leu (Ile)–Val–Thr, Pro–His–Pro–Thr–Glu–Val–Ala Phe, and Val–Leu (Ile)–Asp–Ala–Phe are new peptides.

8.2.5 Correlation Between Structural Features and Antifatigue Function

8.2.5.1 Amino Acid Composition

Some amino acids also play important roles in regulating human metabolism related to muscle activity. Several amino acids, especially BCAA, can improve exercise ability and significantly delay the breakdown and metabolism of proteins in muscles during exercise. Miao et al. [32] found that lower MW fractions (< 6 kDa) of oyster hydrolysates (OH) have higher levels of BCAA, glutamic acid, taurine, aspartic acid, and arginine compared with higher MW fractions (> 6 kDa), which helps enhance antioxidant and antifatigue activity. You et al. [22] prepared purple perilla seed peptide PPSP and found that it is rich in BCAA

Table 8.7 The amino acid composition of PPSP.

Items	Names of amino acid	Content (g/100 g)
1	Asp (aspartic acid)	9.88 ± 0.06
2	Thr (threonine)	3.03 ± 0.04
3	Ser (serine)	4.00 ± 0.31
4	Glu (glutamate)	15.84 ± 0.89
5	Gly (glycine)	4.88 ± 0.11
6	Ala (alanine)	3.87 ± 0.11
7	Val (valine)	5.65 ± 0.22
8	Cys (cysteine)	2.70 ± 0.32
9	Met (methionine)	3.17 ± 0.06
10	Ile (isoleucine)	4.86 ± 0.15
11	Leu (leucine)	8.31 ± 0.21
12	Tyr (tyrosine)	3.09 ± 0.32
13	Phe (phenylalanine)	4.18 ± 0.04
14	His (histidine)	2.32 ± 0.18
15	Lys (lysine)	3.12 ± 0.10
16	Arg (arginine)	10.81 ± 0.29
17	Pro (proline)	3.30 ± 0.11

Source: [22]/John Wiley & Sons.

(18.82 g/100 g) and antifatigue amino acids (Table 8.7), including glutamic acid (Glu), aspartic acid (Asp), and arginine (Arg). Ye et al. [6] prepared SCPs with antifatigue activity. Amino acid composition analysis showed that SCP is rich in glycine, glutamic acid, and proline. The component EWPs2 isolated by Sun et al. [15] from the protein antifatigue peptide EWPP has strong antifatigue properties, and research has found that it has a high concentration of glutamic acid. Glutamate (GLU) is believed to have a positive effect on the nervous system and also helps alleviate mental and physical fatigue. EWPs2 contains a high proportion (10.51%) of aspartic acid, which has been found to be used for oxidative deamination and reducing blood ammonia concentration, thereby delaying the onset of fatigue. Feng et al. [18] found that the BCAA levels of corn antifatigue peptide CPs were very high (182.5 g/kg), including leucine (123.9 g/kg), valine (32.5 g/kg), and isoleucine (28.6 g/kg).

8.2.5.2 Sequence Features

Specific amino acid sequences in peptide chains may affect their binding ability to receptors, thereby affecting their antifatigue effects. Lv et al. [14] found that the antifatigue activity of deer blood enzyme-hydrolyzed peptides is attributed to the synergistic effect of different types of peptides. Components containing glutamine or valine short peptides, Ala–Gln,

Val–Gln, Val–Val–Ser–Ala, and Val–Leu (Ile) Ser improve exercise endurance by increasing hepatic glycogen (HG) storage. The peptide group containing Leu (Ile)–Leu (Ile), Asp–Gln, Phe–Leu (Ile), and Val–Val–Tyr–Pro helps reduce muscle lactate accumulation and increase HG. Wang et al. [31] prepared SCPs with different degrees of hydrolysis (SCP-1 and SCP-2), and SCP-2 exhibited better antifatigue performance than SCP-1. It has been identified that the proportion of sequences containing SCP-2 antifatigue amino acids (such as Leu, Ile, and Val) is higher than SCP-1.

8.2.5.3 Molecular Weight Size

Generally speaking, peptides with smaller MWs are more easily absorbed and utilized, thereby exerting antifatigue effects more quickly. Liu et al. [21] found that walnut peptide WOP containing 86.5% WOPs (180 < relative MW in WOP < 1,000) has good antifatigue activity. You et al. [11] reported a similar finding that compared with 58.27% of the peptides, and the hydrolysate of loach protein containing 65.41% of the 1,000 < MW peptide showed significant antifatigue potential. The MW range of the antifatigue peptide PPSP prepared by Liu et al. [22] is mainly below 1,000 Da (accounting for more than 78%), among which 150–1,000 Da peptides account for 66.21 ± 4.25% and < 150 Da account for 11.92 ± 1.04% (Table 8.8). Ye et al. [6] prepared the antifatigue peptide SCP, and its MW curve showed that the main peak was located in the low MW region. 72.6% of peptides are distributed in MWs less than 2 kDa, and 26.8% of peptides are distributed in the range of 2–5 kDa, indicating that SCP has a high content of low MW peptides (Figure 8.2).

8.2.5.4 Peptide Chain Length

Shorter peptide chains usually have better absorption, utilization, and biological activity, and can more effectively exert antifatigue effects. Wang et al. [31] prepared SCPs (SCP-1 and SCP-2) with different degrees of hydrolysis. Although there was no significant difference in amino acid composition between SCP-1 and SCP-2, the MW and peptide chain length of SCP-2 were smaller than those of SCP-1, indicating that SCP-2 exhibited better antifatigue performance than SCP-1.

Table 8.8 The molecular weight distribution of PPSPE.

Molecular weight range (Da)	Weight-average molecular weight (Da)	Peak area percentage
>10,000	11,231 ± 234	0.11 ± 0.02
5,000–10,000	6,577 ± 159	2.56 ± 0.13
3,000–5,000	3,829 ± 87	3.52 ± 0.15
2,000–3,000	2,437 ± 63	4.46 ± 0.21
1,000–2,000	1,404 ± 29	11.22 ± 1.96
150–1,000	421 ± 11	66.21 ± 4.25
<150	70 ± 5	11.92 ± 1.04

Source: [22]/John Wiley & Sons.

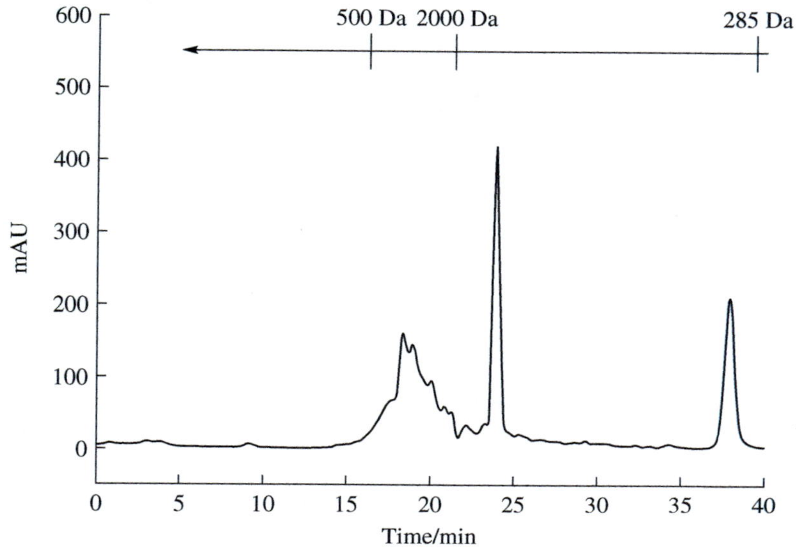

Figure 8.2 Molecular weight distribution of SCP. *Source:* [6]/John Wiley & Sons.

8.3 Detection and Evaluation of the Activity of Antifatigue Peptides

Exercise fatigue is a complex systemic syndrome that involves physiological changes in multiple systems, including changes in muscle energy metabolism, regulation of the nervous system, and response of the endocrine system [33]. It is challenging to comprehensively characterize the changes that occur in the body under fatigue by relying solely on in vitro indicators as these indicators frequently reflect only a local aspect of fatigue, thus providing an incomplete representation of the entire fatigue phenomenon [34]. For instance, the LA level in blood serves as an indicator of the anaerobic metabolic state of muscles; however, it does not directly provide a quantitative measurement of fatigue in the nervous system [35]. Consequently, the employment of animal models as a research platform has become a prevalent practice, owing to their capacity to more systematically and comprehensively assess the effects of antifatigue components [36]. These models can replicate the comprehensive physiological responses of humans in a fatigued state, thereby providing a more systematic and comprehensive experimental environment for the study of antifatigue components.

8.3.1 Mice Pole-Climbing Test

The duration of time that a mouse spends climbing a pole is a significant indicator of its level of fatigue. This is due to the fact that pole-climbing behavior necessitates the maintenance of elevated muscle tension and optimal neural coordination [37]. The experimental protocol entails the administration of a gavage to mice, followed by a period of observation during which the mice are subjected to a pole-climbing test. The experimental apparatus consists of a smooth Plexiglas rod placed at a specific height, with its upper end fixed by a designated device and the lower end suspended. Mice are then placed on the glass rod. During this process, the mice's muscles are under tension, and when the mice reach a state of fatigue, they will slip off the pole. The rationale behind employing the mice's time climbing the pole as a fatigue evaluation index stems from the fact that the duration of this activity provides a direct reflection of the mice's performance in terms of physical strength and endurance. For instance, in a study by Ding et al. [38], the antifatigue activity of a hydrolyzed peptide derived from jellyfish collagen was investigated. The findings revealed that the pole-climbing time of the peptide-treated mouse group exhibited a significant increase compared with the control group, thereby suggesting that jellyfish collagen-hydrolyzed peptide has the potential to enhance the physical strength and endurance of mice. Concurrently, the levels of lactic acid and urea nitrogen in the blood were reduced relative to the control group, while the glycogen content was elevated. These alterations in biochemical indicators further substantiated the antiexercise fatigue activity of the jellyfish collagen-hydrolyzed peptide. The accumulation of lactic acid is known to induce muscle soreness and fatigue, while the increase in urea nitrogen is indicative of enhanced protein metabolism, which is frequently associated with fatigue [39]. Conversely, the increase in glycogen content suggests that the mice had greater energy reserves to manage fatigue.

8.3.2 Mice Swimming Test

The time to exhaustion of weighted swimming in mice is also an important indicator for assessing the degree of fatigue in mice. The standard protocol for mouse swimming experiments involves the administration of a weight-bearing swimming test following a period of regular gavage in the mice [40]. In the experiment, an appropriately weighted lead block is affixed to the right hip of the mice, and then the mice are permitted to swim in a tank of appropriate size and water temperature. The time from the start of the mice's swimming to the time of force exhaustion is recorded. The criterion for the mice to exhaust themselves in swimming was the inability of the mice to rise to the surface of the water within 7 seconds after they were completely submerged. This criterion was established based on the physiological limits of mice that, under normal conditions, are capable of staying afloat and swimming in water; however, upon reaching a state of exhaustion, they are unable to maintain this ability. For instance, Wu et al. [41] investigated the effect of vitamin B on the antifatigue properties of mice by regularly gavaging them with ginsenosides. The results indicated that ginsenoside enhanced the swimming endurance of mice, suggesting an improvement in their physical strength and endurance. The study also noted a reduction in lactic acid and urea nitrogen levels in the blood, suggesting that ginsenosides may possess antifatigue properties. However, the antifatigue effect of ginsenosides was significantly reduced by B vitamins, thereby impairing their bioavailability. This phenomenon may be attributed to the interaction of B vitamins with ginsenosides during metabolic processes, thereby influencing the absorption and utilization of ginsenosides. Furthermore, Wang et al. [42] isolated a decapeptide (cms001) from the spleen of piglets and investigated its effect on weight-bearing swimming time in mice.

It was also found that regular gavage of the substance prolonged weight-bearing swimming time in mice, suggesting that CMS001 has a significant antifatigue effect. The study also demonstrated a reduction in BUN levels and lactic acid accumulation, further substantiating the antifatigue effect of CMS001 [42]. Furthermore, Nan et al. [25] investigated the antifatigue activity of glycine–leucine and leucine–glycine dipeptides in fermented pig placenta using a mouse exhaustion swimming model. Their findings indicated that these two dipeptides possess significant antifatigue activity, providing a substantial scientific foundation for the development of novel antifatigue agents. In the antifatigue agent market, these kinds of active peptides of natural origin have a promising application prospect because they usually have high safety and low side effects.

8.3.3 Mice Tail-Hanging Test

The mouse tail suspension test is a behavioral experimental method that is frequently utilized to evaluate an animal's resistance to fatigue [43]. Typically, this test is conducted on mice; however, to avoid potential damage to the tails of heavier rats, such as rats weighing over 450g, the use of rats of a greater weight is precluded. The experiment is based on the creation of an unavoidably oppressive environment in which the experimental animal undergoes a process from struggle to exhaustion to cessation of activity. The resistance to fatigue of each experimental animal is assessed by measuring the difference in the amount of time the animal attempts to sustain its evasive behavior [44]. This is due to the innate tendency of animals to attempt evasion through suspended movement, with the duration of this struggle reflecting their physical strength and endurance levels. In the experiment, the experimental animals were initially placed in a soundproof behavioral box and suspended by their tails via suspension bars or racks. Subsequent to this, video recording equipment is utilized to document the experimental animals' behavior during the experiment, which typically spans a duration of 6 minutes. The duration of this struggle is subsequently documented in the animal behavior analysis file. A prolonged resting period is indicative of reduced resistance to fatigue [45]. For instance, researches demonstrated that the fatigue level of mice under varying treatment conditions could be effectively assessed by the tail suspension experiment [46].

8.3.4 Mice Rotameter Test

The rotating rod test was originally developed for the purpose of evaluating the neurological effects of pharmaceutical agents. However, it has since gained widespread application in the measurement of fatigue levels and physical performance in animals. This is attributable to the capacity of the rotating rod test to evaluate both muscle coordination and physical endurance in animals [47]. Prior to the experiment, all mice are acclimatized for several days to ensure that they are familiar with the experimental environment and equipment. On the day of the test, the mice are placed on the rotating bar of the rotating bar apparatus to initiate the procedure. Statistical analysis is then conducted to assess quantitative indicators, such as fall time, which are used to determine the animals' fatigue status. The time spent by the mice on the rotating bar is indicative of their muscle coordination and physical endurance. For instance, Zhang et al. [30] demonstrated that the rotating rod experiment is a highly effective fatigue assessment method capable of detecting subtle variations in the fatigue state of mice under diverse treatment conditions with high sensitivity. The merits of this experimental method include its simplicity, intuitiveness, and capacity for simultaneous assessment of multiple mice, rendering it well-suited for large-scale experimental studies.

8.3.5 Common Antifatigue Biochemical Indexes

In the context of antifatigue research, the assessment of biochemical indicators constitutes a pivotal approach for evaluating the efficacy of antifatigue interventions. The most commonly used biochemical indicators include serum creatine kinase (CK), BLA, BUN, blood testosterone, and blood glucose (Glu) [39, 48, 49]. Changes in the levels of these indicators during periods of exercise fatigue can be used to assess the degree of fatigue and recovery. For instance, elevated serum CK typically reflects muscle damage, while the accumulation of BLA is indicative of anaerobic metabolism in muscles and serves as a critical marker of fatigue. Conversely, an increase in BUN is indicative of enhanced protein metabolism and is often concomitant with a state of fatigue. Changes in blood testosterone and blood glucose levels, in contrast, can be indicative of the endocrine regulation and energy supply status of the body [6, 50]. The antioxidant capacity of the organism is also an important indicator in the evaluation of the activity and mechanism study of antifatigue products because it can reflect the resistance of the organism to oxidative stress, thus indirectly reflecting the level of antifatigue [51]. The study of the antioxidant enzymes GSH-Px, SOD, and catalase (CAT) is pivotal in this regard [52]. The activities of these enzymes in the body serve as indirect indicators of the antifatigue effect of the test substance. Furthermore, the degree of tissue damage caused by free radicals can be determined by measuring the total antioxidant capacity of serum and muscle tissue.

Oxidative stress has been demonstrated to result in impaired cell membrane integrity, thereby triggering cellular damage. Consequently, the assessment of fatigue severity can be facilitated by the detection of markers indicative of oxidative damage to cells subsequent to the occurrence of fatigue [53]. MDA, a by-product of cell membrane lipid peroxidation, has been identified as a reliable indicator of fatigue status. Serum levels of enzymes such as alkaline phosphatase, CK, and lactate dehydrogenase (LDH) have been identified as markers of muscle damage, and a decline in these indicators suggests that the pharmaceutical agent under investigation exerts a protective effect on myocytes [54]. For instance, a study by Olivier et al. [55] demonstrated that certain antifatigue medications can mitigate muscle damage by reducing the levels of these enzymes. A comparable conclusion was reached in a study by Simon et al. [56]. In the event of heart muscle and liver cell damage from hypoxia, aspartate aminotransferase (AST) and alanine aminotransferase (ALT) leak into the serum. Consequently, fatigue can be analyzed by monitoring serum levels of these two enzymes.

In states of cellular fatigue, metabolic processes within the body undergo significant changes, leading to altered levels of a range of biomarkers. Two notable enzymes, AST and ALT, play a pivotal role in this process. Under typical conditions, these enzymes are predominantly located within the cells, particularly in liver and muscle cells. However, when cells are damaged by fatigue, these enzymes leak out of the cells and into the serum.

Consequently, the monitoring of serum levels of AST and ALT can serve as a surrogate marker for the assessment of cellular damage, thereby facilitating a comprehensive analysis of fatigue [57]. A seminal study demonstrated a marked elevation in serum levels of AST and ALT in cases of fatigue, with a strong correlation to the extent of cellular damage [55].

In addition, nuclear factor E2-related factor 2 (Nrf2) is a pivotal molecule in the regulation of cellular antioxidant responses and plays a central role in cellular response to oxidative stress. Nrf2 regulates the expression of a variety of downstream antioxidant enzymes and signaling pathways, thereby enhancing cellular antioxidant capacity. Specifically, Nrf2 modulates the signaling pathways of heme oxygenase-1 (HO-1), thioredoxin (Trx), and B-cell lymphoma-2 (Bcl2) [58]. The expression levels of these molecules in liver and muscle tissues are closely related to the body's resistance to fatigue. Enhanced expression of these molecules signifies an augmentation in the body's antioxidant capacity, thereby enabling enhanced resistance to fatigue [59].

8.4 Mechanisms of Antifatigue Peptides in Relieving Exercise Fatigue

Antifatigue active peptides have been demonstrated to play a significant role in alleviating exercise-induced fatigue. The underlying mechanism of these peptides primarily involves the regulation of energy metabolism, antioxidant activity, anti-inflammatory responses, and neuromodulation [38]. These peptides have been shown to promote glycogenolysis and fat oxidation, rapidly supplying muscle cells with energy and maintaining a balanced energy metabolism, thereby delaying the onset of fatigue [60]. Concurrently, these peptides have been observed to enhance the activity of antioxidant enzymes, such as SOD and GSH-Px, thereby mitigating the deleterious effects of oxidative stress. Furthermore, antifatigue peptides have been shown to reduce exercise-induced muscle inflammation and damage, promoting recovery by inhibiting the production of inflammatory cytokines [61]. Furthermore, antifatigue peptides have been shown to regulate neurotransmitter levels, improve mood, and enhance cognitive function, thereby promoting enhanced exercise endurance.

The antifatigue active peptides function synergistically through multiple mechanisms to effectively relieve exercise fatigue and enhance exercise endurance. These peptides have been shown to regulate energy metabolism, protect cells through antioxidant and anti-inflammatory mechanisms, and modulate neurotransmitter levels to enhance mood and cognitive function. The synergistic effects of these peptides make them effective in relieving exercise fatigue and enhancing exercise endurance, and they have a promising application, especially in the field of sports nutrition supplements and sports medicine.

8.4.1 Regulation of Energy Metabolism

Antifatigue active peptides have been demonstrated to possess remarkably intricate and multifaceted mechanisms of action in regulating energy metabolism. These peptides play an indispensable role in both exercise and physical labor. In circumstances where the human body is subjected to high-intensity or protracted physical exertion, there is a precipitous escalation in the energy demands of muscle cells [62]. In this regard, antifatigue active peptides have been shown to play a pivotal role in maintaining energy metabolism equilibrium by facilitating the breakdown and utilization of energy substrates. Specifically, these peptides promote the breakdown of glycogen in muscle and liver, rapidly supplying glucose for energy. This process involves the activation of glycogen phosphorylase, which is responsible for the breakdown of glycogen into glucose-1-phosphate [63]. Glucose-6-phosphate, the product of this process, subsequently enters the glycolytic pathway. Furthermore, antifatigue active peptides enhance the oxidation of fatty acids, thereby providing an additional source of energy for the cells [64]. This is achieved by promoting the activity of the hormone-sensitive lipase enzyme, which is responsible for breaking down triglycerides in fat cells and releasing free fatty acids (FFAs) into the circulation for oxidation and utilization by muscles and other tissues. Consequently, the antifatigue active peptide fulfills two primary functions: first, it addresses the elevated energy demands of muscle cells during periods of exercise; and second, it enhances the efficiency of energy metabolism by reducing the accumulation of metabolic by-products generated during the process. This, in turn, leads to a delay in the onset of fatigue [65].

During periods of high-intensity or prolonged physical exertion, there is a significant increase in the energy demands of muscle cells. Antifatigue active peptides have been shown to assist in maintaining a balanced energy metabolism by promoting the breakdown and utilization of energy substrates. Specifically, antifatigue active peptides promote the breakdown of glycogen in the muscle and liver to rapidly provide glucose for energy [66]. This process involves the activation of glycogen phosphorylase, which is responsible for the breakdown of glycogen into glucose-1-phosphate. Glucose-6-phosphate, the product of this process, subsequently enters the glycolytic pathway. Furthermore, antifatigue active peptides enhance the oxidation of fatty acids, thereby providing an additional source of energy for the cells. This is achieved by promoting the activity of the hormone-sensitive lipase enzyme, which is responsible for breaking down triglycerides in fat cells and releasing FFAs into the circulation for oxidation and utilization by muscles and other tissues. Consequently, the antifatigue active peptide fulfills two primary functions: first, it addresses the elevated energy demands of muscle cells during periods of exercise; and second, it enhances the efficiency of energy metabolism by reducing the accumulation of metabolic by-products generated during the process. This, in turn, leads to a delay in the onset of fatigue [14].

Antifatigue active peptides have been demonstrated to play an important role in regulating mitochondrial function and enhancing the efficiency of cellular energy production. Mitochondria, the energy factories within cells, directly determine cellular energy metabolism in terms of their number and functional status [67]. Antifatigue active peptides have been shown to significantly enhance mitochondrial function by promoting mitochondrial biosynthesis and reducing mitochondrial damage, thereby increasing the efficiency of ATP production. This enhancement of mitochondrial function not only augments the cell's energy supply capacity but also mitigates the production of deleterious metabolites during energy metabolism, thereby further delaying the onset of fatigue [68].

In regard to the energy-sensing signaling pathway, the antifatigue active peptide has been shown to activate key energy-sensing molecules, such as AMPK [69]. AMPK plays a central role in the regulation of cellular energy status, and its activation has been demonstrated to promote glycogen synthesis and fatty acid oxidation, while inhibiting unnecessary energy-consuming processes, such as fat synthesis and cholesterol synthesis. This intricate energy regulation mechanism contributes to the maintenance of energy balance, ensuring that cells can utilize energy efficiently and reduce energy expenditure during high-intensity or prolonged physical activities, thereby delaying the onset of fatigue.

Furthermore, the role of antifatigue active peptides in the resynthesis of energy substances after exercise is significant [70]. These peptides have been shown to promote glycogen and protein synthesis, aid in muscle cell repair and reconstruction, and thereby expedite the body's recovery from fatigue. This function is of particular relevance for athletes and individuals engaged in high-intensity physical labor as it not only enhances athletic performance but also substantially mitigates postexercise fatigue [71]. Consequently, the administration of antifatigue active peptides provides two distinct benefits: first, it offers immediate energy support during exercise, and second, it promotes rapid recovery of the body after exercise. This, in turn, enhances the body's antifatigue ability and exercise endurance.

8.4.2 Regulation of Inflammation

Antifatigue peptide, as a bioactive molecule, shows remarkable potential in modulating inflammatory response and promoting recovery of the body. After intense exercise or physical labor, the body may experience a transient inflammatory response as part of the body's repair and rebuilding process [72]. However, excessive inflammation may lead to muscle damage and increased fatigue. Antifatigue peptides regulate this process through a variety of mechanisms, thereby reducing the negative effects of inflammation [61].

The ingestion of antifatigue peptides has been demonstrated to directly inhibit the production of inflammatory cytokines. These peptides have been shown to reduce the expression of pro-inflammatory cytokines, such as tumor necrosis factor-alpha (TNF-α) and interleukin 6 (IL-6), by binding to specific receptors on immune cells. This inhibitory effect contributes to a reduction in inflammatory damage to muscles and tissues, thereby facilitating a more rapid recovery process [73]. Furthermore, antifatigue peptides have been shown to play a regulatory role in oxidative stress, a pivotal factor in the inflammatory response. They enhance the body's antioxidant capacity and protect cells from oxidative damage by increasing the activity of antioxidant enzymes such as SOD and GSH-Px and reducing the accumulation of free radicals [52].

Further studies have demonstrated that antifatigue peptides have the capacity to modulate the inflammatory response by affecting cell signaling pathways, such as the nuclear factor-κB (NF-κB) and mitogen-activated protein kinase (MAPK) signaling pathways [74]. Antifatigue peptides also act as anti-inflammatory agents, enhancing muscle repair by promoting muscle cell growth and differentiation. The potential mechanisms by which these peptides promote muscle growth and repair involve the activation of specific signaling molecules within muscle cells, such as the mammalian target of rapamycin (mTOR) signaling pathway. These signaling pathways play a central role in the expression of inflammatory factors, and the intervention of antifatigue peptides helps inhibit the activation of these pathways, thereby reducing inflammation.

The function of antifatigue peptides in the regulation of inflammatory responses is not only limited to the direct inhibition of pro-inflammatory cytokine production; it also attenuates oxidative stress by enhancing the body's antioxidant capacity. Oxidative stress, a pivotal component of the inflammatory response, can lead to excessive accumulation of free radicals, which in turn can trigger cellular damage, thereby exacerbating the inflammatory response. The antifatigue peptide effectively scavenges free radicals and reduces cell damage caused by oxidative stress by increasing the activity of antioxidant enzymes such as SOD and GSH-Px [52]. This antioxidant effect not only protects the cells from oxidative damage, but also indirectly reduces the inflammatory response, thereby accelerating the body's recovery process.

Furthermore, antifatigue peptides have been shown to exert their anti-inflammatory effects by modulating cell signaling pathways. The NF-κB and MAPK signaling pathways represent pivotal signaling pathways in the inflammatory response, occupying central roles in the regulation of inflammatory factor expression. Antifatigue peptides have been shown to modulate the activity of these signaling pathways, thereby inhibiting the expression of inflammatory factors. For instance, antifatigue peptides may curtail the transcription and translation of pro-inflammatory cytokines by impeding NF-κB activation, thereby diminishing the inflammatory response. Furthermore, antifatigue peptides may also affect the activation and migration of inflammatory cells by regulating the MAPK signaling pathway, further attenuating inflammatory injury [74].

The anti-inflammatory effect of the antifatigue peptide is also reflected in its ability to promote the repair and regeneration of muscle cells. In the aftermath of strenuous physical exertion or physical labor, muscle tissues may sustain damage. The antifatigue peptide promotes muscle cell growth and differentiation by activating specific signaling molecules in muscle cells, such as the mTOR signaling pathway, thereby enhancing muscle repair capacity. The mTOR signaling pathway plays a pivotal role in protein synthesis and cell growth. The antifatigue peptide promotes muscle protein synthesis and cellular repair through the activation of this pathway, thereby accelerating the body's recovery from fatigue.

8.4.3 Antioxidant Effect

Antifatigue active peptides fulfill a pivotal antioxidant role within the body, thereby significantly amplifying the activity of antioxidant enzymes, including SOD and GSH-Px [51]. These enzymes play a crucial role in scavenging free radicals and reducing oxidative stress. Oxidative stress, a state of the cell that arises in the face of various physiological and environmental stresses, is characterized by the production of ROS exceeding the antioxidant capacity of the cell, leading to cellular damage. The antifatigue active peptide effectively neutralizes these harmful free radicals by enhancing the activity of antioxidant enzymes, thereby reducing cell damage caused by oxidative stress [75].

The antifatigue active peptide has been demonstrated to enhance the activity of antioxidant enzymes. In addition, it has been shown to reduce the level of MDA, the end-product of lipid peroxidation. Lipid peroxidation is the hallmark of cell membrane lipid peroxidation, and the reduction of MDA indicates that the integrity of the cell membrane is better protected and the oxidative damage of the cell is effectively reduced. Furthermore, antifatigue active peptides modulate the activity of enzymes associated with energy metabolism, such as LDH. By optimizing the activity of these enzymes, antifatigue peptide enhances the efficiency of energy metabolism, thereby enabling cells to utilize energy more efficiently and, consequently, reducing fatigue [76].

Wheat peptides, for instance, have been demonstrated to elicit a substantial augmentation in SOD and GSH-Px activities, accompanied by a concomitant decline in MDA levels within the liver of murine subjects [77]. These changes have been demonstrated to reduce both cellular oxidative damage and LA and urea nitrogen levels in the blood following exercise, while concurrently increasing myo-glycogen content. The collective outcomes of these observations are the manifestation of substantial antioxidant and antifatigue properties in wheat peptides. Furthermore, antifatigue active peptides have been shown to protect cells from oxidative stress damage, promote cell repair and regeneration, and enhance the body's antifatigue ability. This, in turn, improves exercise endurance and reduces exercise-induced muscle fatigue. Wheat peptide has been demonstrated to further enhance mitochondrial fatty acid oxidation by activating the Nrf2 signaling pathway, a key transcription factor that regulates the expression of a variety of antioxidant enzymes, thereby enhancing cellular antioxidant capacity. Notably, the activation of the Nrf2 signaling pathway by wheat peptide not only reduces the oxidation of glucose but also promotes the synthesis of glycogen, which in turn leads to a significant enhancement in exercise capacity [78].

During the process of cellular respiration, particularly during periods of strenuous physical exertion, cells generate a variety of ROS, including hydroxyl radicals, superoxide anions, hydrogen peroxide, and nitric oxide. A substantial augmentation in the concentration of these ROS has the potential to impede the capacity of muscles to sustain contraction, consequently resulting in fatigue [79]. Antioxidant enzymes, such as SOD, CAT, and GSH-Px, play a pivotal role in safeguarding against oxidative stress. The ingestion of antifatigue peptides has been shown to elicit a substantial increase in the levels of these antioxidant enzymes in experimental animals, thereby augmenting the antioxidant capacity of cells. In addition, antifatigue-active peptides demonstrated strong free radical scavenging ability in vitro, suggesting that their antioxidant effects may be directly related to their antifatigue effects. The antifatigue active peptide's ability to scavenge free radicals has been shown to reduce cellular damage caused by oxidative stress and optimize cellular energy metabolism, thereby enhancing the organism's overall fatigue resistance and exercise endurance [80].

8.4.4 Regulation of Neurotransmitters

Antifatigue active peptides have been demonstrated to possess remarkable potential in combating fatigue by finely modulating neurotransmitter levels in the brain. These active peptides have been observed to induce a substantial increase in the concentration of excitatory neurotransmitters, such as dopamine (DA) and GLU, while concomitantly reducing the levels of inhibitory neurotransmitters, including gamma-aminobutyric acid and 5-hydroxytryptophan (5-HT) [81]. These neurotransmitters have been demonstrated to play a pivotal role in enhancing alertness and cognitive function [82]. DA, an essential neurotransmitter, is intimately associated with attention, motivation, and reward mechanisms [83]. Its increased levels have been shown to enhance an individual's focus and motivation. GLU, in contrast, is the predominant excitatory neurotransmitter in the CNS and plays a role in a variety of cognitive functions, including learning and memory [84]. By modulating the levels of these neurotransmitters, antifatigue active peptides have the potential to enhance the efficiency of neurotransmission in the brain, thereby strengthening the body's resistance to fatigue [60].

Furthermore, antifatigue active peptides may modulate neural signaling and regulation through a variety of mechanisms. For instance, they have the potential to enhance DA signaling by promoting DA synthesis or inhibiting its reuptake, thereby further enhancing the excitability and function of the nervous system [85]. Concurrently, these peptides may also modify the sensitivity of neurotransmitter receptors and influence neurotransmitter reuptake and metabolic processes. Collectively, these intricate regulatory mechanisms optimize the dynamic equilibrium of neurotransmitters and enhance the overall functional efficiency of the nervous system, thereby fortifying the body's resistance to fatigue.

The anti-inflammatory effects and neuroprotective abilities of antifatigue active peptides also provide significant support for the stabilization of the neurotransmitter system. The overproduction of inflammatory factors has the potential to disrupt the normal synthesis and delivery of neurotransmitters, thereby affecting the function of the nervous system. Antifatigue active peptides have been shown to mitigate the effects of inflammation on the neurotransmitter system by reducing the production of inflammatory factors [86]. Furthermore, they promote neuronal health and function by fostering the production of neurotrophic factors. Neurotrophic factors represent a class of proteins that are indispensable for the growth, survival, and functionality of neurons, and they promote neuronal repair and regeneration to maintain the stability and functional integrity of the nervous system. Consequently, the protective effects of antifatigue active peptides extend beyond the mitigation of inflammatory damage to the neurotransmitter system, enhancing its capacity to promote cognitive function and alleviate fatigue [60].

8.4.5 Improvement of Exercise Endurance

Antifatigue active peptides have been shown to exert significant antioxidant effects in vivo through a variety of molecular targeting mechanisms, thereby enhancing the body's ability to resist fatigue. These mechanisms involve several key biomolecules and signaling pathways, with a prominent role for carnitine acyltransferase (CrAT), a mitochondrial enzyme responsible for regulating the interconversion of acetyl coenzyme A and acetyl carnitine, a process that is critical for muscle energy utilization. The CrAT-mediated buffering of acetyl groups within the mitochondria has been suggested to be a key factor in optimizing athletic performance as it effectively regulates metabolic processes within the mitochondria, thereby enhancing the efficiency of energy utilization [87]. As it effectively regulates metabolic processes within the mitochondria, thereby increasing the efficiency of energy utilization [87]. L-carnitine, a notable nutritional supplement, has been demonstrated to augment CrAT mRNA expression [88]. This finding provides a new direction for improving exercise endurance. Studies have demonstrated that mice receiving L-carnitine supplementation prior to exercise exhibited a notable prolongation in time to exhaustion on a running platform, suggesting that L-carnitine enhances mitochondrial energy utilization efficiency by increasing CrAT activity, thereby improving exercise endurance. However, in the case of CrAT knockout mice, L-carnitine supplementation did not demonstrate a significant impact on exercise endurance [89]. This outcome serves to further substantiate the pivotal function of CrAT in the context of L-carnitine-enhanced exercise endurance.

Peroxisome proliferator-activated receptor (PPARδ) has been identified as an important molecular target in recent studies, which have demonstrated its pivotal role in combating exercise fatigue [90]. The findings suggest that this receptor and its coactivator, PGC-1α, play a crucial role in regulating various physiological processes, including mitochondrial biosynthesis, myofibrillar oxidative capacity, fatty acid oxidation, and exercise endurance [91].

A seminal study by Narkar et al. [90] further revealed the great potential of the PPARδ in improving exercise endurance. The researchers found that the PPARδ agonist GW1516, when used in combination with exercise training, produced a synergistic effect, leading to a significant increase in the proportion of oxidized muscle fibers in adult mice. This combination also resulted in a substantial enhancement in the mice's exercise endurance, as measured by a running exercise test. Specifically, mice treated with GW1516 and subjected to exercise training exhibited a 70% increase in exercise endurance compared with mice that received only exercise training. This finding indicates that the combination of a PPARδ agonist and exercise training can significantly enhance the body's exercise endurance, providing a significant theoretical foundation for the development of novel antifatigue strategies. The mechanism of action of PPARδ in regulating mitochondrial biosynthesis and myofibrillar oxidative capacity is intricate. PPARδ promotes mitochondrial biosynthesis through the activation of a series of downstream signaling pathways, thereby increasing the number and function of mitochondria. As intracellular energy factories, the increase in the number and function of mitochondria directly enhances the energy metabolism of the cell, enabling muscle cells to more efficiently utilize energy substrates such as fatty acids, thereby improving exercise endurance. Moreover, the PPARδ has been shown to further optimize the energy metabolism process by regulating the expression of genes related to fatty acid oxidation and enhancing fatty acid oxidation capacity. PGC-1α also plays an indispensable role in this process as a coactivator of PPARδ. PGC-1α is able to synergize with PPARδ to enhance its ability to regulate downstream genes, thereby promoting mitochondrial biosynthesis and fatty acid oxidation more effectively. This synergistic effect not only improves mitochondrial function but also enhances the oxidative capacity of muscle fibers, allowing muscles to utilize oxygen more efficiently and reduce fatigue during prolonged exercise.

Narkar et al. also demonstrated that the combination of the PPARδ agonist GW1516 with exercise training significantly increased the proportion of oxidized muscle fibers. Oxidized muscle fibers, which include type I and type IIa muscle fibers, primarily rely on aerobic metabolism for energy supply. These fibers have higher mitochondrial content and fatty acid oxidation capacity, resulting in greater endurance during prolonged exercise. Consequently, the augmentation of oxidized muscle fibers, induced by the combination of the PPARδ agonist GW1516 and exercise training, has the dual benefits of enhancing muscle endurance and optimizing the energy metabolism pattern of muscles. This enhanced metabolic efficiency reduces fatigue during prolonged exercise.

8.5 Current Status of Research and Its Applications

When the human body engages in high-intensity physical labor, continuous exercise, or long-term heavy mental activities for a period of time, the body generates a large number of free radicals, causing oxidative damage to cell membranes and leading to temporary reduction in work capacity or cognitive impairment. This state is called fatigue, which is a normal physiological phenomenon in humans. For a long time, due to the lack of research on the mechanisms of fatigue, past antifatigue research mainly focused on nutritional supplementation and the use of stimulant drugs, such as caffeine, mepemoline, and modafinil [25]. However, chemical drugs are far from satisfactory due to their significant side effects and limited efficacy. Therefore, in recent years, nutritional intervention using natural products or their components with fewer side effects has gradually become a research hotspot in the field of antifatigue. Antifatigue peptides can not only provide essential amino acids (EAAs) for the human body but also serve as an energy source for human activities [6]. They can maintain the stability of the body's internal environment and enzyme activity, enhance the body's antioxidant capacity, and prevent neurotransmitter imbalances, etc. These functions are incomparable to other antifatigue factors.

The main mechanisms by which antifatigue peptides delay or eliminate fatigue involve either eliminating the accumulation of fatigue-inducing substances or serving as a rapidly replenishable energy source, both of which enhance exercise endurance and promote physical recovery [92]. Thanks to their high absorbability, antifatigue peptides hold significant relevance for individuals engaged in physically or mentally demanding work, who are prone to fatigue. These potential benefits position anti-fatigue peptides to have broad applications in sports nutrition, functional foods, and clinical nutrition, thereby supporting improvements in quality of life and work efficiency.

For decades, dietary proteins have been recognized as a source of bioactive peptides capable of broadly regulating physiological functions. With the deepening research and development of anti-fatigue peptides in recent years, their substantial application value has been increasingly uncovered. Numerous peptides with proven antifatigue activity are already known and utilized, while researchers continue to explore diverse raw materials for producing bioactive peptides with antifatigue properties, keeping this field of study highly dynamic [93].

8.5.1 Animal-Derived Peptides

Seafood is a high-quality source of protein, with low cost and abundant availability. Bioactive peptides derived from marine organisms have attracted significant attention due to their multiple biological functions, such as antioxidant and antifatigue activities.

Chronic fatigue syndrome is a disorder characterized by persistent and recurrent fatigue, with its etiology remaining unclear. However, oxidative stress is believed to play a significant role in its pathogenesis. Several theories exist regarding the mechanisms of exercise-induced fatigue, including the exhaustion theory, radical theory, clogging theory, mutation theory, homeostasis theory, and protective inhibition theory [10]. The exhaustion theory posits that energy sources such as muscle and liver glycogen are depleted during prolonged exercise, accompanied by a drop in blood glucose levels, leading to fatigue. The clogging theory, in particular, emphasizes that during intense exercise, metabolites like lactic acid and BUN may not be cleared promptly. Their excessive accumulation disrupts metabolism and reduces muscle function, thereby inducing physical fatigue. The radical theory describes how the accumulation of reactive free radicals in the body creates an imbalance between the oxidative and antioxidant systems. Vigorous exercise generates free radicals, and if not neutralized in a timely manner, these radicals can damage macromolecules and cellular organelles, causing lipid peroxidation and cellular injury. Based on these theories, researchers are exploring natural bioactive compounds to mitigate oxidative damage and combat fatigue. Antioxidant compounds targeting exercise-induced oxidative stress help enhance the body's antioxidant defenses, with exogenous dietary antioxidants interacting with endogenous antioxidants to form a collaborative network of cellular protection.

The forced swimming test, a classic behavioral experiment reflecting physical exercise capacity, can evaluate the antifatigue effect of peptides and has been widely used in functional assessments of antifatigue peptides [47]. The swimming time of mice is recorded: the longer the swimming time, the better the antifatigue effect. The most widely used test in antifatigue peptide research is the exhaustive swimming test, which uses mice or rats to swim continuously until they can no longer float on the water's surface, reflecting the animals' ability to tolerate exercise-induced fatigue. The general procedure for the exhaustive swimming test is as follows: (i) before the experiment, mice or rats receive gastric perfusion of antifatigue peptides for a certain period to ensure the drug takes full effect. (ii) During the experiment, animals are placed in a water tank with an appropriate amount of water, and lead plates of suitable weight are attached to their legs to make them swim. (iii) The time from when the animal enters the water until exhaustion is recorded (exhaustion is typically defined as the animal sinking into the water and being unable to surface for more than 10 seconds). The forced swimming test usually ends after 20–30 minutes. The effect of antifatigue peptides is evaluated by changes in biochemical indicators of mice or rats after exercising at similar intensity and duration.

8.5.1.1 Oyster Peptides

Oysters are the world's most widely farmed bivalve mollusks. Beyond their culinary use, they are recognized for their excellent health benefits and potential to treat chronic diseases due to their rich content of protein, taurine, and trace elements, making them suitable for developing functional foods.

Hao and colleagues hydrolyzed oyster meat using neutral protease and collected the permeate (MW cutoff <4 kDa) via an ultrafiltration system, yielding oyster peptides (OPs) [94]. The study evaluated the impact of these antifatigue peptides on animal endurance through exhaustive swimming tests. Glycogen, a critical energy reserve in the body, plays a role in delaying fatigue—higher glycogen levels correlate with extended exercise capacity. Excessive exercise depletes glycogen and accumulates metabolic by-products, both of which contribute to physical fatigue. Results showed that compared to the control group, OP, oyster meat, and oyster protein significantly prolonged mouse swimming time by 78%, 72%, and 44%, respectively. Glycogen content was measured in the liver and gastrocnemius muscle—the latter being the primary muscle group engaged during mouse swimming, with energy metabolism closely linked to exercise endurance. As the main energy storage molecule in muscles, glycogen is broken down into glucose during exercise to fuel muscle contraction. The gastrocnemius exhibits more active glycogen metabolism than other muscles (e.g., cardiac muscle or slow-twitch fibers), making its glycogen levels a robust indicator of energy consumption and reserve during exercise. OP increased glycogen content in the liver and gastrocnemius by 100% and 55.6%, respectively. Additionally, serum levels of lactic acid and BUN were analyzed. Lactic acid is a key marker of muscle fatigue: during intense exercise, hypoxia drives pyruvate reduction to lactic acid via glycolysis, elevating BLA levels. Accumulated lactic acid lowers muscle pH, disrupts acid–base balance, impairs metabolism, and induces muscle soreness, fatigue, and reduced exercise capacity. Elevated BUN reflects enhanced protein metabolism and fatigue severity. OP intake significantly reduced serum lactic acid and BUN levels by 24.8% and 11.2%, respectively.

Miao and colleagues investigated the antifatigue and antioxidant potential of OH prepared using complex proteases [32]. The OP solution was fractionated via a 6 kDa ultrafiltration membrane to yield two hydrolysates with distinct MW distributions: OH-I (<6 kDa) and OH-II (≥6 kDa). This study first compared the antifatigue effects of the OH, OH-I, and OH-II fractions in the OH, then further characterized the antioxidant activities of OH-I and OH-II. Antifatigue assay results demonstrated that the OP fraction OH-I exhibited the most potent antifatigue activity among all tested groups, evidenced by its significant capacity to scavenge free radicals and enhance antioxidant enzyme activities. Amino acid composition was hypothesized to play a critical role in mediating the antifatigue and antioxidant effects of OH. Compared with the control group, the OH-I treatment significantly prolonged swimming time (67.79% increase), elevated muscle glycogen (45.65%) and liver glycogen (49.01%) stores, and reduced BUN levels (18.44% decrease). Mice administered the full OH solution also exhibited significantly extended swimming endurance compared to controls, indicating that peptides within the hydrolysate effectively enhance exercise tolerance and exert robust antifatigue effects. Collectively, these findings highlight OPs as promising functional food ingredients for developing nutraceuticals with targeted antifatigue and antioxidant properties.

8.5.1.2 Jellyfish Peptides

The pole-climbing test plays a pivotal role in antifatigue research. This test measures the duration mice maintain balance on a rotating rod; as exercise time prolongs, muscular and neurological fatigue gradually develops, eventually causing the animal to fall due to lost equilibrium. The pole-climbing test thus directly reflects exercise endurance and antifatigue capacity, simulates fatigue states, provides objective quantitative metrics, and can be integrated with other biochemical indicators (e.g., BLA, BUN, liver glycogen, muscle glycogen) to comprehensively elucidate the antifatigue effects and mechanisms of test substances. Jellyfish, rich in collagen, possess high dietary and medicinal value, including fatigue-relieving properties. Collagen constitutes approximately 50% of jellyfish dry weight, making it a low-calorie, high-protein natural health food. Ding and colleagues prepared jellyfish collagen hydrolysate (JCH) via Protamex enzyme hydrolysis at 50 °C for 8 hours, yielding a mixture primarily composed of peptides and amino acids [39]. This study evaluated JCH's in vivo antifatigue activity via pole-climbing tests and biochemical assays after 6 weeks of gavage, and its in vivo antioxidant activity using a D-galactose-induced aging model and biochemical indicators. Four experimental groups were established: control, low-dose (5 mL/kg/day), high-dose (10 mL/kg/day), and positive-control (spirulina 800 mg/kg/day).

Results showed that compared with the control group, the low-dose, high-dose, and positive-control groups exhibited significantly prolonged pole-climbing times (74.54%, 95.12%, and 114.86% increases, respectively), indicating JCH significantly enhances exercise endurance. All treatment groups had significantly lower BLA levels than the control group [44.91% (low), 39.20% (high), and 41.82% (positive) reductions]. As a key fatigue marker, decreased BLA suggests JCH effectively delays lactate accumulation and postpones fatigue onset. BUN levels were significantly lower in all treatment groups. As a product of protein and amino acid metabolism during exercise, reduced BUN reflects diminished protein catabolism and enhanced physical endurance. Liver and muscle glycogen contents were significantly higher in all treatment groups compared with the control group. As critical energy sources during exercise, elevated glycogen stores help maintain blood glucose levels and delay fatigue. Collectively, these results demonstrate JCH's pronounced antifatigue effects. In vitro antioxidant assays (DPPH, ABTS, ORAC) and HepG2 cell-based antioxidant activity tests further confirmed JCH's significant antioxidant capacity across multiple biological levels. These findings highlight JCH's potential as a natural bioactive substance for developing antioxidant functional foods or pharmaceuticals.

8.5.1.3 Sea Cucumber Peptides

Sea cucumbers, marine invertebrates commonly found in benthic and deep-sea regions, have traditionally been regarded as ideal nutritionally dense tonic foods due to their high protein content and low fat levels. Beyond serving as protein sources, sea cucumbers exhibit multiple physiological activities. As a renowned luxury seafood and nutritional supplement in Asia, the sea cucumber *Stichopus japonicus* has garnered increasing attention in recent years for its nutritional and bioactive properties. Belonging to the family Holothuroidea, *S. japonicus* is primarily distributed along the coastal regions of China and Japan, representing a species of high commercial value and remarkable nutritive significance. Comprising approximately 70% (w/w) protein, it serves as an abundant source for producing high-quality peptides.

Ye and colleagues prepared low-molecular-weight (<2 kDa) SCPs rich in glycine, glutamic acid, and proline via neutral protease hydrolysis of *S. japonicus* [6]. SCP alleviates oxidative damage by activating antioxidant defense pathways. Following gastric administration of the test substances, mice underwent daily swimming training for 150 minutes in 30 °C water. Subsequent exhaustive swimming tests revealed that SCP significantly prolonged average swimming time, with the highest dose group achieving a 116.91% increase. FFAs, products of lipid breakdown, are released from adipose tissues (e.g., subcutaneous fat) into the bloodstream and utilized as energy sources by tissues such as muscles and the liver. During prolonged exercise, FFAs are decomposed into glycerol and fatty acids, which generate ATP via β-oxidation. Excessive exercise may lead to FFA accumulation, inhibiting muscle glucose uptake and inducing fatigue. Moderate elevations in FFAs reflect increased fat consumption to meet energy demands. LDH, an enzyme widely distributed in human tissues (e.g., muscle, liver, heart), catalyzes the reversible conversion between LA and pyruvate, participating in glycolysis and cellular respiration. Thus, LDH serves as a marker of muscle damage and excessive glycolysis. After 4 weeks of gavage, rats exhibited sharp increases in BLA, BUN, LDH, and FFA concentrations, as well as CK activity. Conversely, glucose, liver glycogen, and muscle glycogen concentrations significantly decreased.

The model control group (MC) showed markedly elevated FFA levels, indicating lipid metabolic disorders. In the high-dose SCP group, FFA levels dropped to 1,102.4 μmol/L (a 17% reduction), suggesting that SCP optimized lipid metabolism and mitigated energy imbalance. SCP protected muscle cells against damage via antioxidant and anti-inflammatory effects, reducing LDH release. Additionally, SCP promoted glycogen storage and decreased reliance on lipolysis, thereby lowering FFA levels and avoiding metabolic derangements.

Wang and colleagues prepared two SCP fractions (SCP-1 and SCP-2) with different degrees of hydrolysis (DH) [31]. SCP-2 exhibited lower MW, shorter peptide chain length, and a higher proportion of BCAA-related sequences. Compared with the control group, both SCP-1 and SCP-2 significantly prolonged exhaustive swimming time and enhanced forelimb grip strength in mice, with SCP-2 demonstrating superior efficacy over SCP-1. In the SCP-treated groups, serum levels of LA, BUN, ammonia, and LDH were significantly reduced, muscle damage markers (e.g., CK) decreased, and muscle glycogen and energy metabolism indicators (e.g., ATP) increased. In this study, SCP-treated groups (particularly the SCP-2H group) significantly upregulated NRF2 protein expression and downregulated Keap1 protein expression compared with the control, indicating that SCPs activate the NRF2 signaling pathway. Activated NRF2 translocates to the nucleus to bind antioxidant response elements, promoting the expression of antioxidant enzymes (e.g., HO-1). SCP treatment significantly increased HO-1 protein levels in mouse skeletal muscle, enhancing antioxidant capacity, reducing oxidative damage, and facilitating fatigue alleviation. Additionally, SCP-treated groups (especially SCP-2H) significantly upregulated p-AMPK/AMPK expression and increased the expression of PGC-1α, NRF1, TFAM, and COX IV proteins, suggesting that SCPs activate the AMPK signaling pathway to improve mitochondrial function. PGC-1α serves as a key regulator of mitochondrial biogenesis, while NRF1 and TFAM are involved in mitochondrial DNA transcription and replication. Elevated COX IV expression—an essential protein in the mitochondrial inner membrane—indicates enhanced mitochondrial quantity and function. Strengthened mitochondrial function boosts ATP production to meet energy demands during exercise, thereby relieving fatigue. The synergistic action of NRF2 and AMPK signaling pathways enables SCPs to exert dual effects in inhibiting oxidative stress and enhancing energy metabolism, significantly improving exercise performance and antifatigue capacity in mice. Collectively, these findings demonstrate that SCPs effectively mitigate fatigue by activating NRF2 and AMPK signaling pathways, thereby enhancing antioxidant capacity and energy metabolism.

8.5.1.4 Freshwater Fish Peptides

Mackerel is one of China's most important economic fish species, characterized by abundant availability, low cost, and rich content of high-quality protein. Proteins derived from mackerel seafood resources exhibit a well-balanced amino acid composition. Hydrolysates of mackerel proteins are rich in peptides with excellent in vitro antioxidant activity, which correlates with antifatigue effects. Wang and colleagues prepared mackerel peptides (MPH, <2.5 kDa) by hydrolyzing mackerel proteins with Neutrase at 39.55 °C for 5.5 hours, followed by purification via ultrafiltration membrane [9]. This study utilized six mouse groups for comparison: (i) nonswimming control (normal), (ii) negative control dose group (NC), (iii) American ginseng-positive control dose group (PC), (iv) low-dose MPH treatment (MPH-L), (v) medium-dose MPH treatment (MPH-M), and (vi) high-dose MPH treatment

(MPH-H). Gavage administration was used to investigate in vivo antifatigue effects and mechanisms. MPH's antioxidant activity was evaluated via DPPH radical scavenging, hydroxyl radical scavenging, superoxide radical scavenging, and reducing power assays. ROS are associated with multiple diseases, including cancer, diabetes, arthritis, Alzheimer's disease, lung diseases, and autoimmune disorders. Results demonstrated that MPH exhibited not only antioxidant activity but also significant antifatigue effects in mice. Swimming time was significantly prolonged in MPH-treated groups, particularly in the high-dose group (MPH-H), which showed a 3.8-fold increase compared with the NC. Concurrently, liver glycogen levels increased by 33%–35%, indicating improved metabolic control during exercise and activated energy metabolism via enhanced glycogen storage. Levels of BUN, LA, and MDA were reduced by 16%–17%, 14%–19%, and 16%–31%, respectively. MPH improved the enzymatic antioxidant system by increasing the activities of SOD and GSH-Px, replenished blood glucose concentrations, and eliminated fatigue-related metabolites. These findings indicate that MPH alleviates fatigue by scavenging free radicals and enhancing antioxidant enzyme activity.

Peng and colleagues extracted WOPs from tilapia (*Oreochromis mossambicus*) using papain, yielding SMOT (180–1,000 Da) [95]. Their study revealed that SMOT intervention reduced BLA, BUN, and MDA levels in mice, significantly improved liver glycogen storage, enhanced the activities of SOD and GSH-Px, increased the activity of immune factor IL-2, stimulated secretory immunoglobulin A (sIgA) secretion, and regulated inflammatory cytokine levels. SMOT exhibited significant antifatigue effects, with mechanisms potentially involving enhanced energy reserves, elimination of metabolite accumulation, reduced oxidative stress, and improved immune function. Ren and colleagues demonstrated that tilapia skin collagen peptides derived from *Oreochromis nilotica* L. exert antifatigue effects through multiple mechanisms, including reducing serum LA, serum urea nitrogen, and CK activity, increasing liver and muscle glycogen content, and enhancing endogenous SOD activity. You and colleagues prepared LPs via papain hydrolysis. In vitro assays showed that LP inhibited lipid peroxidation and exhibited high DPPH radical scavenging, hydroxyl radical scavenging, and copper ion-chelating activities. In vivo studies revealed that LP elevated blood glucose and liver glycogen levels, reduced LA and BUN levels, enhanced antioxidant enzyme activity, and prolonged swimming time.

8.5.1.5 Marine Fish Peptides

The global fishing industry generates substantial by-products, with monkfish (*Lophius* spp.) being one of the largest. As a specialty of East China and the Yellow Sea, monkfish has become an important commercial fish species. By-products from monkfish processing can account for up to 60% of total fish weight, of which 10% is liver waste. Monkfish liver is rich in protein but highly perishable, necessitating conversion of its proteins into hydrolyzed peptides. Xu and colleagues prepared monkfish liver hydrolysate (MLH) via Protamex enzyme hydrolysis, yielding a product primarily composed of free amino acids and small-molecule peptides [96]. The antifatigue effects of MLH were evaluated using pole-climbing endurance tests, comparing mouse climbing duration and serum biochemical markers. MLH's DPPH radical-scavenging capacity increased with hydrolysis time, reaching a maximum clearance rate of26.78%. In aging model mice, MLH significantly enhanced SOD activity in liver and kidney tissues while reducing MDA content, indicating pronounced antioxidant effects. MLH also significantly prolonged mouse pole-climbing time: the low-dose, high-dose, and positive-control groups showed 24.76%, 43.39%, and 48.03% increases over the control group, respectively. The amino acid composition of MLH provides insight into its antioxidant mechanisms. MLH is rich in hydrophilic amino acids such as aspartic acid, glutamic acid, serine, arginine, and cystine (comprising over 83%), which are known to enhance exercise capacity and delay fatigue.

Oligopeptides with high Fischer ratios prolong swimming time in exercised mice, increase liver glycogen content, and reduce BUN and LA levels. The seafood processing industry generates substantial by-products, from which protein hydrolysates and peptides can be prepared. Zhao and colleagues hydrolyzed swim bladder proteins from *Pseudosciaena crocea* using different proteases, isolating and purifying the SBP-III-3 peptide fraction (5–10 kDa) [12]. SBP-III-3 is rich in antioxidant amino acids such as glycine (Gly), alanine (Ala), and proline (Pro), comprising 59.84% of total amino acids. Additionally, EAAs account for 15.7% of SBP-III-3, potentially contributing to its antifatigue capacity. Hydroxyl radicals (HO•) are highly destructive species in free radical pathology, attacking nearly all molecules in living cells. Scavenging HO• represents one of the most effective defense mechanisms against various diseases. In vitro assays showed that SBP-III-3 exhibited potent free radical-scavenging activity, significantly neutralizing HO•, DPPH•, O_2•, and ABTS+•. In vivo studies demonstrated that SBP-III-3 significantly enhanced antioxidant enzyme activity: SOD, CAT, and GSH-Px activities increased by 44.2%, 159.5%, and 240.5%, respectively. SBP-III-3 protected supercoiled pBR322 plasmid DNA from HO•-mediated fragmentation by scavenging hydroxyl radicals generated via iron-catalyzed H_2O_2 decomposition, demonstrating its ability to reduce DNA strand breaks through radical scavenging. Furthermore, mice treated with SBP-III-3 exhibited significantly elevated activities of endogenous antioxidant enzymes (SOD, GSH-Px, and CAT).

8.5.1.6 Livestock-Derived Peptides

Livestock products contain abundant proteins in various forms, such as eggs, milk, and bovine bones. Egg white is a rich source of high-quality protein. Sun and colleagues prepared EWPs via pepsin digestion, which contain 17 amino acids potentially contributing to their antioxidant and antifatigue activities [15]. Results showed that EWP prolonged exhaustive swimming time in mice, improved blood glucose regulation, activated energy metabolism, and reduced metabolic by-product accumulation. Milk proteins, recognized as Generally Regarded As Safe natural food ingredients, serve as abundant sources of bioactive peptides. As by-products of the dairy industry, they are widely available and cost-effective.

Pan and colleagues hydrolyzed milk proteins using trypsin and chymotrypsin [16]. Supplementation with the resulting peptides had no significant effect on BLA levels in mice before swimming, but significantly reduced BLA levels after a 20-minute post-swim rest, suggesting potential to accelerate post-exercise LA clearance. Peptides derived from milk proteins with antifatigue and antioxidant activities exhibit no adverse side effects, making them suitable for functional foods or beverages designed for populations affected by fatigue, ROS, and free radical-induced diseases. As a major by-product of yak processing, yak bones are often discarded or processed into low-value products, highlighting the substantial potential of yak bone collagen peptides (YBP). Feng and colleagues supplemented Balb/c mice with YBP at 1 g/kg body weight for 18 weeks and found that YBP significantly increased blood SOD activity, LDH content, GSH levels, and glucose concentration, while prolonging swimming time. These results indicate that YBP enhances energy supply, reduces metabolic by-product accumulation, mitigates exercise-induced oxidative stress, and thereby improves exercise endurance.

8.5.2 Plant-Derived Peptides

Xu and colleagues hydrolyzed highly denatured soybean meal (43% protein content) with Neutrase after heat or ultrasonic treatment, followed by ultrafiltration to fractionate the hydrolysate into three components: I (>10 kDa), II (5–10 kDa), and III (<5 kDa) [97]. Mice were gavaged with different MW hydrolysates for 30 days, followed by exhaustive swimming tests and biochemical marker analyses. Results showed that the hydrolysate exhibited high DPPH radical-scavenging activity and reducing power, with the <5 kDa fraction (III) demonstrating the highest antioxidant activity. All hydrolysates significantly increased hemoglobin content, reduced BUN and BLA levels, and elevated liver glycogen stores in mice. The <5 kDa fraction (III) showed the most pronounced antifatigue effect. These findings indicate that highly denatured soybean meal hydrolysates alleviate fatigue by scavenging free radicals, inhibiting lipid peroxidation, and regulating energy metabolism.

Wheat, one of the world's major cereal crops, generates by-products during starch processing that consist primarily of gliadin and glutenin, collectively known as wheat proteins. The poor water solubility of wheat proteins is attributed to their high content of hydrophobic and uncharged amino acids, which create extensive intramolecular hydrophobic interactions and limit their applications. Enzymatic modification improves wheat protein solubility, thereby enhancing their functional and processing properties while generating peptide materials with diverse MWs. Similar to soybean peptides, wheat peptides—derived from plant protein hydrolysates—exhibit antifatigue functions.

Zheng and colleagues prepared wheat peptides by hydrolyzing wheat proteins with alkaline protease and flavor protease, followed by ultrafiltration and RP-HPLC to isolate short peptides with MWs <1,000 Da, which are more readily absorbed and exhibit higher bioactivity [20]. Amino acid composition analysis revealed the highest levels of glutamic acid and glutamine, while Pro, His, and Tyr accounted for 10.5% of total amino acids, potentially contributing to the peptides' robust antioxidant capacity. Notably, the Leu–Ala–Arg (LAR) peptide showed high DPPH radical-scavenging activity. Using a rat model of incremental-load swimming exercise, the study demonstrated that wheat peptide supplementation significantly prolonged exhaustive swimming time, though it had no significant effect on skeletal muscle glycogen storage. Wheat peptides enhanced the activities of SOD and GSH-Px in rat skeletal muscle while reducing MDA content, indicating improved antioxidant capacity and mitigation of oxidative stress damage. Additionally, wheat peptides increased intestinal sIgA levels, strengthened intestinal mucosal immune barrier function, reduced intestinal epithelial cell apoptosis, and protected intestinal barrier integrity. The peptides also significantly decreased intestinal and serum levels of IL-6 and IL-8, alleviating inflammatory responses. However, wheat peptide supplementation did not significantly affect intestinal antioxidant capacity during exercise or prevent bacterial translocation.

Prolonged exercise-induced stress can lead to CNS dysfunction. Serotonin (5-HT), a neurotransmitter, regulates mood, sleep, energy, and appetite, with lower levels linked to anxiety, depression, and impulsive behavior. Acetylcholine, widely distributed in both the CNS and peripheral nervous system, is critical for autonomic and motor nerve transmission, being released upon neural stimulation. Acetylcholinesterase (AChE), a key enzyme in neurotransmission, is localized in synaptic clefts of cholinergic neurons and hydrolyzes acetylcholine to terminate synaptic excitation and ensure proper neural signaling. Wheat peptides increased brain 5-HT levels while reducing AChE activity in rats, suggesting they may alleviate exercise stress by regulating CNS neurotransmitter levels to improve mood stability and neural transmission.

Peptides with low MW also have antifatigue effects. Liu and colleagues prepared low-molecular-weight (<5 kDa) walnut oligopeptides (WOPs) from walnuts using complex proteases [21]. WOPs significantly enhanced exercise endurance in mice, thereby delaying fatigue onset. In WOPs-treated mice, BLALDH, CK, and BUN levels were significantly reduced, indicating that WOPs help mitigate exercise-induced muscle damage and metabolic by-product accumulation. Additionally, WOPs significantly increased glycogen content in the liver and skeletal muscle, improving energy supply during exercise, and enhanced the activities of SOD and GSH-Px while reducing MDA content. These results suggest that WOPs effectively alleviate oxidative stress, protect cells from free radical damage, and potentially enhance energy metabolism by improving mitochondrial function to combat fatigue.

Feng and colleagues prepared pea peptides from *Pisum sativum L.* using complex proteases [19]. Compared with the control group, mice supplemented with pea peptides exhibited significantly prolonged swimming time, reduced BUN and BLA levels, and increased muscle and liver glycogen content. The peptides also demonstrated potent antioxidant activity, significantly decreasing MDA levels while enhancing SOD and GSH-Px activities. Additionally, pea peptides significantly enhanced phagocyte activity and sIgA levels, while reducing pro-inflammatory cytokines IL-6 and TNF-α. These results indicate that pea peptides not only strengthen immune defense but also mitigate exercise-induced inflammation, thereby facilitating fatigue recovery.

8.5.3 Yeast-Derived Peptides

Cai and colleagues isolated and purified two peptides with high antioxidant activity, YPLP (Tyr–Pro–Leu–Pro) and AGPL (Ala–Gly–Pro–Leu), from the fermentation products of the yeast *Yamadazyma triangularis* XHY69 [98]. These peptides showed no cytotoxicity in C2C12 cells, significantly reduced the activity of LDH, and enhanced the activities of succinate dehydrogenase and malate dehydrogenase. Through immunofluorescence and protein expression analysis, the yeast-derived peptides increased the proportion of slow-twitch fibers while decreasing that of fast-twitch fibers. This suggests that the yeast-derived peptides may contribute to improving muscle endurance and antifatigue capacity. The peptides also elevated the mitochondrial membrane potential ($\Delta\Psi m$) in C2C12 myotubes, increased the expression of mitochondrial DNA-related genes (such as cytochrome b and cytochrome c oxidase II) in the cells, and enhanced the intracellular ATP content. These findings indicate that the yeast-derived peptides promote mitochondrial biogenesis and function. The YPLP peptide activated AMPK through phosphorylation, which in turn upregulated the expression of downstream molecules such as PGC-1α, Sirt1, and NRF1. The activation of this signaling pathway helps regulate energy metabolism and mitochondrial biogenesis, thereby enhancing muscle function. Molecular docking studies revealed that YPLP has a strong binding affinity for AMPK, suggesting that YPLP may directly interact with AMPK and regulate muscle function via the AMPK/PGC-1α signaling pathway. In an ICR mouse model, YPLP significantly prolonged the exercise time on the treadmill, indicating its potential to improve exercise endurance. YPLP also reduced the levels of LA and urea nitrogen in the blood of mice after exercise, suggesting that it helps reduce the accumulation of exercise-induced metabolic by-products and thus alleviates fatigue. Moreover, YPLP increased the glycogen content in the muscles and liver of mice, providing more energy sources for exercise and further enhancing exercise capacity.

8.5.4 Productions

Taking the Chinese market as an example, data from the State Food and Drug Administration (SFDA) show that China has approved 751 imported health products and 15,842 domestic health products. Among these, 2,160 products (13.63% of the total) are registered for alleviating physical fatigue (including the former "antifatigue" function), comprising 2,094 domestic and 66 imported products. With the acceleration of social life rhythms and the enhancement of health awareness among Chinese consumers, the market demand for antifatigue health products is expected to expand significantly.

Most of the 2,160 registered antifatigue health products in SFDA use expensive Chinese medicinal materials such as *Panax quinquefolius, Rhodiola rosea*, and *Cordyceps sinensis* as raw materials, resulting in high costs and prices. Notably, 15 products utilize animal peptides as primary ingredients, categorized into five types based on origin: five products each use marine fish skin collagen oligopeptide powder and whey protein peptides, while 1 product each uses salmon protein peptides, OPs, and pig spleen peptides. These offer new pathways for developing cost-effective antifatigue health products. Exploring natural, safe, and highly effective animal-derived bioactive peptide functional factors—combined with other Chinese medicinal materials and nutrients (taurine, vitamin C, zinc citrate, zinc gluconate, ribose, etc.)—enables the development of products with prominent health benefits, including not only physical fatigue alleviation but also antiaging, immune enhancement, and hypoxia tolerance.

Compared with existing animal peptide-based health products, these patents cover more diverse raw materials, with 66.67% derived from marine animals, providing valuable references for developing antifatigue peptide products. The patents address a wide range of health functions, most of which involve multiple benefits such as promoting metabolism, enhancing immunity, and antioxidant activity. Authorized patents include traditional formulations like oral liquids, granules, capsules, and powders, as well as food forms such as beverages and biscuits.

China's rich animal species and resources enable the production of animal-derived antifatigue peptides as functional food bases for developing health foods with fatigue-alleviating properties. Currently, a series of antifatigue products derived from animals—such as velvet antler oral liquid, velvet antler lozenges, silkworm moth wine, and oyster soft capsules—have been launched. As bioactive peptides are more readily absorbed by the human body and exhibit higher biological potency than animal proteins, developing functional foods with antifatigue peptides promises greater health benefits and market prospects [99].

References

1 Lim, E.J., Ahn, Y.C., Jang, E.S. et al. (2020). Systematic review and meta-analysis of the prevalence of chronic fatigue syndrome/myalgic encephalomyelitis (CFS/ME). *Journal of Translational Medicine* 18 (`1): 100.

2 Shen, W., Li, X., Deng, Y. et al. (2021). Polygonatum cyrtonema hua polysaccharide exhibits anti-fatigue activity via regulating osteocalcin signaling. *International Journal of Biological Macromolecules* 175: 235–241.

3 Caldwell, J.A., Caldwell, J.L., Thompson, L.A., and Lieberman, H.R. (2019). Fatigue and its management in the workplace. *Neuroscience & Biobehavioral Reviews* 96: 272–289.

4 Song, J., Wu, Y., Jiang, G. et al. (2019). Sulfated polysaccharides from rhodiola sachalinensis reduce d-gal-induced oxidative stress in NIH 3t3 cells. *International Journal of Biological Macromolecules* 140: 288–293.

5 Daliri, E.B., Lee, B.H., and Oh, D.H. (2018). Current trends and perspectives of bioactive peptides. *Critical Reviews in Food Science And Nutrition* 58 (13): 2273–2284.

6 Ye, J., Shen, C., Huang, Y. et al. (2017). Anti-fatigue activity of sea cucumber peptides prepared from*stichopus japonicus*in an endurance swimming rat model. *Journal of the Science of Food and Agriculture* 97 (13): 4548–4556.

7 Guo, Z., Lin, D., Guo, J. et al. (2017). In vitro antioxidant activity and in vivo anti-fatigue effect of sea horse (hippocampus) peptides. *Molecules* 22 (3): 482.

8 Ren, Y., Wu, H., Chi, Y. et al. (2020). Structural characterization, erythrocyte protection, and antifatigue effect of antioxidant collagen peptides from tilapia (oreochromis nilotica l.) skin. *Food & Function* 11 (11): 10149–10160.

9 Wang, X., Xing, R., Chen, Z. et al. (2014). Effect and mechanism of mackerel (pneumatophorus japonicus) peptides for anti-fatigue. *Food & Function* 5 (9): 2113–2119.

10 Wang, P., Zeng, H., Lin, S. et al. (2020). Anti-fatigue activities of hairtail (trichiurus lepturus) hydrolysate in an endurance swimming mice model. *Journal of Functional Foods* 74: 104207.

11 You, L., Zhao, M., Regenstein, J.M., and Ren, J. (2011). In vitro antioxidant activity and in vivo anti-fatigue effect of loach (misgurnus anguillicaudatus) peptides prepared by papain digestion. *Food Chemistry* 124 (1): 188–194.

12 Zhao, Y., Zeng, L., Yang, Z. et al. (2016). Anti-fatigue effect by peptide fraction from protein hydrolysate of croceine croaker (*Pseudosciaena crocea*) swim bladder through inhibiting the oxidative reactions including DNA damage. *Marine Drugs* 14 (12): 221.

13 Wang, X., Yu, H., Xing, R. et al. (2023). Structural properties, anti-fatigue and immunological effect of low molecular weight peptide from monkfish. *Journal of Functional Foods* 105: 105546.

14 Lv, J., Liu, Y., Zeng, X. et al. (2021). Anti-fatigue peptides from the enzymatic hydrolysates of cervus elaphus blood. *Molecules* 26 (24): 7614.

15 Sun, S., Niu, H., Yang, T. et al. (2014). Antioxidant and anti-fatigue activities of egg white peptides prepared by pepsin digestion. *Journal of the Science of Food and Agriculture* 94 (15): 3195–3200.

16 Pan, D., Guo, Y., and Jiang, X. (2011). Anti-fatigue and antioxidative activities of peptides isolated from milk proteins. *Journal of Food Biochemistry* 35 (4): 1130–1144.

17 Fang, L., Zhang, R.X., Wei, Y. et al. (2022). Anti-fatigue effects of fermented soybean protein peptides in mice. *Journal of the Science of Food and Agriculture* 102 (7): 2693–2703.

18 Feng, Z., Wei, Y., Xu, Y. et al. (2022). The anti-fatigue activity of corn peptides and their effect on gut bacteria. *Journal of the Science of Food and Agriculture* 102 (8): 3456–3466.

19 Feng, T., Huang, Y., Tang, Z. et al. (2021). Anti-fatigue effects of pea (pisum sativum l.) peptides prepared by compound protease. *Journal of Food Science and Technology* 58 (6): 2265–2272.

20 Zheng, Z., Yang, X., Liu, J. et al. (2017). Effects of wheat peptide supplementation on anti-fatigue and immunoregulation during incremental swimming exercise in rats. *RSC Advances* 7 (69): 43345–43355.

21 Liu, R., Wu, L., Du, Q. et al. (2019). Small molecule oligopeptides isolated from walnut (*juglans regia* L.) and their anti-fatigue effects in mice. *Molecules* 24 (1): 45.

22 Liu, Y., Li, D., Wei, Y. et al. (2020). Hydrolyzed peptides from purple perilla (*perilla frutescens* l. Britt.) seeds improve muscle synthesis and exercise performance in mice. *Journal of Food Biochemistry* 44 (11): e13461.

23 Ying, Z., Huang, Y.Y., Li, M.Y. et al. (2023). High fischer ratio peptide of hemp seed: preparation and anti-fatigue evaluation in vivo and in vitro. *Food Research International* 165: 112534.

24 Chen, Y., Wang, F., Zhou, J. et al. (2020). In vivo antifatigue activity of spirulina peptides achieved by their antioxidant activity and by acting on fat metabolism pathway in mice. *Natural Product Communications* 15 (8).

25 Nam, S., Kim, H., and Jeong, H. (2016). Anti-fatigue effect by active dipeptides of fermented porcine placenta through inhibiting the inflammatory and oxidative reactions. *Biomedicine & Pharmacotherapy* 84: 51–59.

26 Yu, B., Lu, Z., Bie, X. et al. (2008). Scavenging and anti-fatigue activity of fermented defatted soybean peptides. *European Food Research and Technology* 226 (3): 415–421.

27 Cui, J., Xia, P., Zhang, L. et al. (2020). A novel fermented soybean, inculated with selected bacillus, lactobacillus and hansenula strains, showed strong antioxidant and anti-fatigue potential activity. *Food Chemistry* 333: 127527.

28 Huang, S., Lin, H., and Deng, S. (2015). Study of anti-fatigue effect in rats of ferrous chelates including hairtail protein hydrolysates. *Nutrients* 7 (12): 9860–9871.

29 Liu, J., Wang, X., and Zhao, Z. (2014). Effect of whey protein hydrolysates with different molecular weight on fatigue induced by swimming exercise in mice. *Journal of the Science of Food and Agriculture* 94 (1): 126–130.

30 Zhang, Y., Ryu, B., Cui, Y. et al. (2019). A peptide isolated from hippocampus abdominalis improves exercise performance and exerts anti-fatigue effects via AMPK/PGC-1α pathway in mice. *Journal of Functional Foods* 61: 103489.

31 Wang, Q., Shi, J., Zhong, H. et al. (2021). High-degree hydrolysis sea cucumber peptides improve exercise performance and exert antifatigue effect via activating the NRF2 and AMPK signaling pathways in mice. *Journal of Functional Foods* 86: 104677.

32 Miao, J., Liao, W., Kang, M. et al. (2018). Anti-fatigue and anti-oxidant activities of oyster (ostrea rivularis) hydrolysate prepared by compound protease. *Food & Function* 9 (12): 6577–6585.

33 Davis, J.M. and Bailey, S.P. (1997). Possible mechanisms of central nervous system fatigue during exercise. *Medicine and Science in Sports and Exercise* 29 (1): 45–57.

34 Fitts, R.H. (1994). Cellular mechanisms of muscle fatigue. *Physiological Reviews* 74 (1): 49–94.

35 Brooks, G.A. (1985). Anaerobic threshold: review of the concept and directions for future research. *Medicine and Science in Sports and Exercise* 17 (1): 22–34.

36 Rezaee, Z. (2025). Guidelines for animal models of endurance and resistance exercise. *Journal of Neuroscience Methods* 416: 110362.

37 Stutz, P.V., Golani, L.K., and Witkin, J.M. (2019). Animal models of fatigue in major depressive disorder. *Physiology & Behavior* 199: 300–305.

38 Zhuang, Y., Sun, L., Zhao, X. et al. (2009). Antioxidant and melanogenesis-inhibitory activities of collagen peptide from jellyfish (*rhopilema esculentum*). *Journal of the Science of Food and Agriculture* 89: 1722–1727.

39 Ding, J., Li, Y., Xu, J. et al. (2011). Study on effect of jellyfish collagen hydrolysate on anti-fatigue and anti-oxidation. *Food Hydrocolloids* 25 (5): 1350–1353.

40 Can, A., Dao, D., Arad, M. et al. (2012). The mouse forced swim test. *Journal of Visualized Experiments* 29 (59): e3638.

41 Wu, Y.C., Lian, Y.Z., Zhao, H. et al. (2025). Ginsenosides, salidroside, and syringin complex exhibits anti-fatigue in exhaustive exercise rats. *International Journal of Medical Sciences* 22 (1): 17.

42 Wang, L., Zhang, H., Lu, R. et al. (2008). The decapeptide CMS001 enhances swimming endurance in mice. *Peptides* 29 (7): 1176–1182.

43 Steru, L., Chermat, R., Thierry, B., and Simon, P. (1985). The tail suspension test: a new method for screening antidepressants in mice. *Psychopharmacology* 85 (3): 367–370.

44 Moon, P., Kim, K., Rew, K. et al. (2014). Anti-fatigue effects of porcine placenta and its amino acids in a behavioral test on mice. *Canadian Journal of Physiology and Pharmacology* 92 (11): 937–944.

45 Zhang, X., Jing, S., Lin, H. et al. (2019). Anti-fatigue effect of anwulignan via the NRF2 and PGC-1α signaling pathway in mice. *Food & Function* 10 (12): 7755–7766.

46 Jun, G. (2024). Tail suspension test. In: *The ECPH Encyclopedia of Psychology*, 1484–1485. Singapore: Springer Nature Singapore.

47 Wang, P., Wang, D., Hu, J. et al. (2021). Natural bioactive peptides to beat exercise-induced fatigue: a review. *Food Bioscience* 43: 101298.

48 Daniel, H. and Zietek, T. (2015). Taste and move: glucose and peptide transporters in the gastrointestinal tract. *Experimental Physiology* 100 (12): 1441–1450.

49 Xu, C., Lv, J., You, S. et al. (2013). Supplementation with oat protein ameliorates exercise-induced fatigue in mice. *Food & Function* 4 (2): 303–309.

50 Ren, J., Zhao, M., Wang, H. et al. (2011). Effects of supplementation with grass carp protein versus peptide on swimming endurance in mice. *Nutrition* 27 (7): 789–795.

51 Matés, J.M. (2000). Effects of antioxidant enzymes in the molecular control of reactive oxygen species toxicology. *Toxicology* 153 (1): 83–104.

52 Kankofer, M., Kolm, G., Aurich, J., and Aurich, C. (2005). Activity of glutathione peroxidase, superoxide dismutase and catalase and lipid peroxidation intensity in stallion semen during storage at 5°C. *Theriogenology* 63 (5): 1354–1365.

53 Marrocco, I., Altieri, F., and Peluso, I. (2017). Measurement and clinical significance of biomarkers of oxidative stress in humans. *Oxidative Medicine and Cellular Longevity* 1–2: 1–32.

54 Lim, A.K. (2020). Abnormal liver function tests associated with severe rhabdomyolysis. *World Journal of Gastroenterology* 26 (10): 1020–1028.

55 Olivier, D., Wafa, D., Dimitri, T. et al. (2018). An evidence-based approach for choosing post-exercise recovery techniques to reduce markers of muscle damage, soreness, fatigue, and inflammation: a systematic review with meta-analysis. *Frontiers in Physiology* 9: 403.

56 Simon, M.C., Liu, L., Barnhart, B.C., and Young, R.M. (2008). Hypoxia-induced signaling in the cardiovascular system. *Annual Review of Physiology* 70 (1): 51–71.

57 Yang, X., Schnackenberg, L.K., Shi, Q., and Salminen, W.F. (2014). Chapter 13- hepatic toxicity biomarkers. In: *Biomarkers in Toxicology* (ed. R.C. Gupta), 241–259. Elsevier.

58 Ma, Q. (2013). Role of Nrf2 in oxidative stress and toxicity. *Annual Review of Pharmacology and Toxicology* 53: 401–426.

59 Muthusamy, V.R., Kannan, S., Sadhaasivam, K. et al. (2012). Acute exercise stress activates Nrf2/are signaling and promotes antioxidant mechanisms in the myocardium. *Free Radical Biology and Medicine* 52 (2): 366–376.

60 Cunha, S.A. and Pintado, M.E. (2022). Bioactive peptides derived from marine sources: biological and functional properties. *Trends in Food Science & Technology* 119: 348–370.

61 Peake, J.M., Neubauer, O., Della, G.P.A., and Nosaka, K. (2017). Muscle damage and inflammation during recovery from exercise. *Journal of Applied Physiology* 122 (3): 559–570.

62 König, D., Kohl, J., Jerger, S., and Centner, C. (2021). Potential relevance of bioactive peptides in sports nutrition. *Nutrients* 13 (11): 3997.

63 Richter, E.A. and Hargreaves, M. (2013). Exercise, GLUT4, and skeletal muscle glucose uptake. *Physiological Reviews* 93 (3): 993–1017.

64 Jeppesen, J. and Kiens, B. (2012). Regulation and limitations to fatty acid oxidation during exercise. *The Journal of Physiology* 590: 1059–1068.

65 Burke, L.M., Hawley, J.A., Jeukendrup, A. et al. (2018). Toward a common understanding of diet–exercise strategies to manipulate fuel availability for training and competition preparation in endurance sport. *International Journal of Sport Nutrition and Exercise Metabolism* 28 (5): 451–463.

66 Chen, W., Huang, W., Chiu, C. et al. (2014). Whey protein improves exercise performance and biochemical profiles in trained mice. *Medicine & Science in Sports & Exercise* 46 (8): 1517–1524.

67 Woodhead, J.S.T. and Merry, T.L. (2021). Mitochondrial-derived peptides and exercise. *Biochimica et Biophysica Acta (BBA)—General Subjects* 1865 (12): 130011.

68 Lu, X., Wang, M., Yue, H. et al. (2024). Novel peptides from sea cucumber intestines hydrolyzed by neutral protease alleviate exercise-induced fatigue via upregulating the glutaminemediated Ca^{2+}/calcineurin signaling pathway in mice. *Journal of Food Science* 89 (3): 1727–1738.

69 Yu, Y., Wu, G., Jiang, Y. et al. (2020). Sea cucumber peptides improved the mitochondrial capacity of mice: a potential mechanism to enhance gluconeogenesis and fat catabolism during exercise for improved antifatigue property. *Oxidative Medicine and Cellular Longevity* 2020: 1–17.

70 Ivy, J.L., Katz, A.L., Cutler, C.L. et al. (1988). Muscle glycogen synthesis after exercise: effect of time of carbohydrate ingestion. *Journal of Applied Physiology* 64 (4): 1480–1485.

71 Lemon, P.W.R. (1997). Dietary protein requirements in athletes. *The Journal of Nutritional Biochemistry* 8 (2): 52–60.

72 Jegani, K.T., Balde, A., and Nazeer, R.A. (2025). A review on anti-inflammatory and antioxidant peptides derived from marine organisms: mechanism of action and therapeutic applications. *Food Bioscience* 63: 105745.

73 Maes, M., Twisk, F.N.M., and Ringel, K. (2012). Inflammatory and cell-mediated immune biomarkers in myalgic encephalomyelitis/chronic fatigue syndrome and depression: inflammatory markers are higher in myalgic encephalomyelitis/chronic fatigue syndrome than in depression. *Psychotherapy and Psychosomatics* 81 (5): 286–295.

74 Majumder, K., Mine, Y., and Wu, J. (2016). The potential of food protein-derived anti-inflammatory peptides against various chronic inflammatory diseases. *Journal of the Science of Food and Agriculture* 96 (7): 2303–2311.

75 Li, H., Liu, E.Q., Wu, Y.H., and Chen, S.L. (2012). The anti-fatigue effect of black soybean peptide in mice. *Advanced Materials Research* 554-556: 1475–1482.

76 Vibhute, P., Radhakrishnan, A. andJeyachandran, S. (2023) Chapter 2- antioxidant properties of marine proteins and peptides. *Marine Antioxidants*,11–25, Elsevier.

77 Weng, Z., Chen, Y., Liang, T. et al. (2023). A review on processing methods and functions of wheat germ-derived bioactive peptides. *Critical Reviews in Food Science and Nutrition* 63 (22): 18.

78 Qi, Y., Wu, D., Fang, L. et al. (2023). Anti-inflammatory effect of walnut-derived peptide via the activation of Nrf2/keap1 pathway against oxidative stress. *Journal of Functional Foods* 110: 105839.

79 Powers, S.K., Deminice, R., Ozdemir, M. et al. (2020). Exercise-induced oxidative stress: friend or foe? *Journal of Sport and Health Science* 9 (5): 415–425.

80 Zou, T.B., He, T.P., Li, H.B. et al. (2016). The structure-activity relationship of the antioxidant peptides from natural proteins. *Molecules* 21.

81 Meeusen, R. and Roelands, B. (2010). Central fatigue and neurotransmitters, can thermoregulation be manipulated? *Scandinavian Journal of Medicine & Science in Sports* 20: 19–28.

82 Wu, S., Bekhit, A.E.A., Wu, Q. et al. (2021). Bioactive peptides and gut microbiota: candidates for a novel strategy for reduction and control of neurodegenerative diseases. *Trends in Food Science & Technology.* 108: 164–176.

83 Schultz, W. (2016). Dopamine reward prediction-error signalling: a two-component response. *Nature Reviews Neuroscience* 17 (3): 183–195.

84 Zhou, Y. and Danbolt, N.C. (2014). Glutamate as a neurotransmitter in the healthy brain. *Journal of Neural Transmission* 121 (8): 799–817.

85 Fernstrom, J.D. and Fernstrom, M.H. (2007). Tyrosine, phenylalanine, and catecholamine synthesis and function in the brain123. *The Journal of Nutrition* 137 (6): 1539S–1547S.

86 Chanajon, P., Tian, F., Noisa, P. et al. (2023). Corn gluten meal peptides inhibit prolyl oligopeptidase and modulate α-synuclein aggregation in KCl-treated SH-SY5y cells. *Journal of Functional Foods* 104: 105501.

87 Seiler, S.E., Koves, T.R., Gooding, J.R. et al. (2015). Carnitine acetyltransferase mitigates metabolic inertia and muscle fatigue during exercise. *Cell Metabolism* 22 (1): 65–76.

88 Muoio, D.M., Noland, R.C., Kovalik, J. et al. (2012). Muscle-specific deletion of carnitine acetyltransferase compromises glucose tolerance and metabolic flexibility. *Cell Metabolism* 15 (5): 764–777.

89 Wall, B.T., Stephens, F.B., Constantin-Teodosiu, D. et al. (2011). Chronic oral ingestion of l-carnitine and carbohydrate increases muscle carnitine content and alters muscle fuel metabolism during exercise in humans. *The Journal of Physiology* 589 (4): 963–973.

90 Narkar, V.A., Downes, M., Yu, R.T. et al. (2008). AMPK and PPARδ agonists are exercise mimetics. *Cell* 134 (3): 405–415.

91 Liu, J., Liang, X., and Gan, Z. (2015). Transcriptional regulatory circuits controlling muscle fiber type switching. *Science China Life Sciences* 58 (4): 321–327.

92 Karami, Z. and Akbari-Adergani, B. (2019). Bioactive food derived peptides a review on correlation between structure of bioactive peptides and their functional properties. *Journal of Food Science and Technology* 56 (2): 535–547.

93 Sarmadi, B.H. and Ismail, A. (2010). Antioxidative peptides from food proteins: a review. *Peptides*, 2010 31: 1949–1956.

94 Hao, G., Cao, W., Hao, J., and Zhang, C. (2013). In vitro antioxidant activity and in vivo anti-fatigue effects of oyster (ostrea plicatula gmelin) peptides prepared using neutral proteinase. *Food Science and Technology Research* 19: 623–631.

95 Peng, S., Ji, H., Song, W. et al. (2022). Anti-fatigue effect of small molecule oligopeptides from tilapia (*oreochromis mossambicus*) in mice. *Food Science and Technology* 42: 93021.

96 Xu, J., Li, Y., Regensteinc, J., and Su, X. (2017). In vitro and in vivo anti-oxidation and anti-fatigue effect of monkfish liver hydrolysate. *Food Bioscience* 18: 9–14.

97 Xu, J., Zhao, Q., Qu, Y., and Ye, F. (2015). Antioxidant activity and anti-exercise-fatigue effect of highlydenatured soybean meal hydrolysate prepared using neutrase. *Journal of Food Science and Technology* 4 (52): 1982–1992.

98 Cai, J., Tao, Y., Xing, L. et al. (2024). Effect of yamadazyma triangularis derived peptide XHY69AP on muscle fatigue via regulation of AMPK/PGC-1α and Nrf2/keap1 pathways. *Food Bioscience* 58: 103793.

99 Lu, Z., Sun, N., Dong, L. et al. (2022). Production of bioactive peptides from sea cucumber and itspotential health benefits a comprehensive review. *Journal of Agricultural and Food Chemistry* 70: 7607–7625.

9

Immune Regulatory Peptides

9.1 Introduction

9.1.1 Molecular Basis of Immunomodulation and Host Defense Mechanisms

In recent years, the escalating prevalence of health threats such as chronic diseases and microbial infections has propelled the regulatory mechanisms of the immune system into a prominent research focus. As the core defense against pathogenic invasion, the immune system operates through a dual defense system comprising innate and adaptive immunity: the former establishes rapid-response networks via physical barriers, antimicrobial substances, and innate immune cells, while the latter employs lymphoid organs (e.g., thymus and spleen) and T/B lymphocytes to develop antigen-specific recognition and long-term immunological memory [1]. These two immunomodulatory paradigms exhibit both functional specialization and intricate interconnection, collectively maintaining host defense homeostasis.

Driven by the imperative for precise immune regulation, immunomodulators have emerged as bioactive agents capable of directing immune response intensity and polarity. This category encompasses clinical immunosuppressants like cyclosporine and tacrolimus, as well as bidirectional immunomodulatory agents including glucocorticoids and levamisole [2, 3]. Their therapeutic scope has expanded beyond conventional immunosuppression to encompass immunoenhancement and precision immunoregulation.

Endogenous antimicrobial immunomodulatory molecules (EAIMs), serving dual roles as fundamental regulators of immune equilibrium and critical antimicrobial effectors, represent an essential class of immunomodulators [1]. While antimicrobial peptides (AMPs) were traditionally considered primary antimicrobial agents, their exclusive microbicidal activity fails to explain the sophisticated coexistence of antimicrobial and immunoregulatory functions in the host defense. Given the multivalent pathogenic strategies employed by invasive microbes, hosts require more versatile defense mechanisms. EAIMs address this limitation by integrating molecular diversity across size dimensions (small molecules to proteins) and structural classes (peptides, lipids, carbohydrates, nucleic acids, organic small molecules, and metal ions) [1, 4]. This molecular heterogeneity

Food-Derived Bioactive Peptides: Green Preparation, Activity Mechanisms, and Health Applications, First Edition. Shao-Yun Wang, Xu Chen, and Xixi Cai.

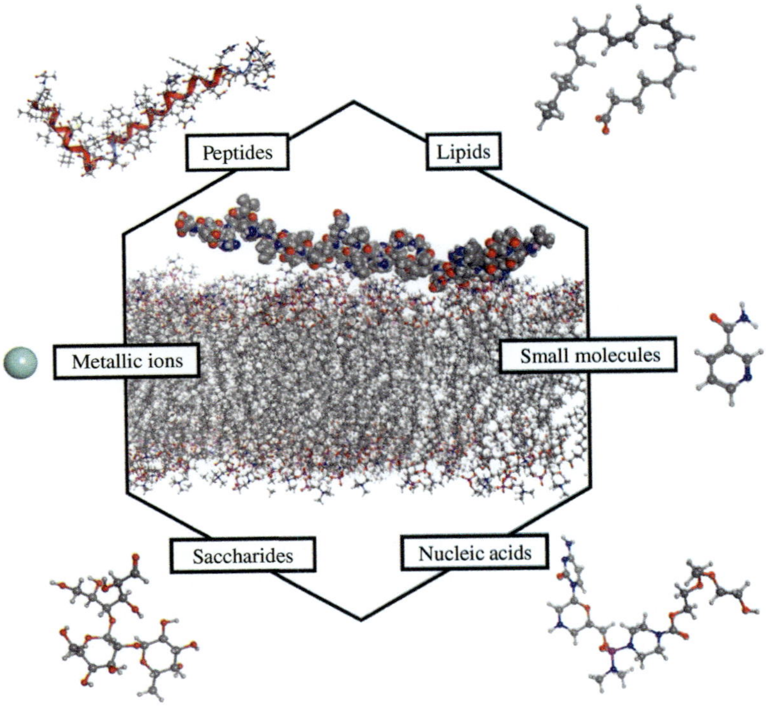

Figure 9.1 Endogenous antimicrobial-immunomodulatory molecules. *Source:* Reproduced with permission of [1]/John Wiley & Sons/CC BY 4.0.

enables simultaneous pathogen eradication and immune modulation while preserving physiological homeostasis (Figure 9.1).

Investigating the multifaceted defense mechanisms of EAIMs reveals key components such as immunomodulatory peptides. These peptidic molecules exemplify the evolutionary transition from simplistic pathogen elimination to sophisticated immune regulation—a critical breakthrough in the development of advanced host defense strategies [1, 5]. Their capacity to decode how biological defense systems evolved from "pathogen killing" to "immune response programming" offers crucial insights into host-pathogen coevolution dynamics.

9.2 Introduction to Immunomodulatory Peptides

The immune system, a cornerstone of human health, orchestrates a delicate balance between defense against pathogens and prevention of excessive inflammation. Central to this equilibrium are immunomodulatory peptides—short chains of amino acids that exhibit the capacity to regulate immune responses through diverse mechanisms. These peptides, derived from natural sources or engineered synthetically, have emerged as pivotal players in both innate and adaptive immunity, offering therapeutic potential across a spectrum of diseases, from inflammatory disorders to cancer [2, 6]. Their ability to modulate cytokine production, influence immune cell activity, and mitigate oxidative stress positions them as versatile agents in modern biomedical research and clinical applications.

9.3 Immunomodulatory Peptides

9.3.1 Mechanisms and Biological Roles

Immunomodulatory peptides function through multifaceted pathways, often targeting key signaling nodes in immune regulation. For instance, the synthetic peptide RDP58, derived from human MHC class I sequences, inhibits the formation of the TRAF6-MyD88-IRAK complex, thereby suppressing activation of proinflammatory pathways such as NF-κB and MAPK [6]. This dual capacity to downregulate cytokines like TNF-α, IL-6, and IL-12 while upregulating antioxidant enzymes such as heme oxygenase-1 (HO-1) underscores its therapeutic promise in conditions like contact dermatitis and inflammatory bowel disease [6]. Similarly, melittin, a 26-amino acid peptide constituting 40–60% of honeybee venom, disrupts cancer cell membranes via hydrophobic interactions, inducing apoptosis through mitochondrial pathways and inhibiting oncogenic signaling via PI3K/Akt/mTOR [7]. Its selective cytotoxicity toward malignant cells, coupled with anti-inflammatory properties in low doses, highlights the nuanced balance between therapeutic efficacy and safety challenges, such as hemolytic activity [7].

The immunomodulatory effects of these peptides extend beyond direct pathogen neutralization. AMPs like LL-37, the sole human cathelicidin, exemplify this duality. While LL-37 exhibits minimal direct antimicrobial activity under physiological conditions, it modulates immune cell functions by enhancing neutrophil recruitment, regulating dendritic cell (DC) differentiation, and suppressing LPS-induced inflammation [8, 9]. Such immunoregulatory roles are further exemplified by its ability to synergize with cytokines at mucosal surfaces, priming the immune response without provoking detrimental inflammation [9]. This shift from antimicrobial to immunomodulatory paradigms reflects a broader recognition of host defense peptides as critical coordinators of immune homeostasis [5].

9.4 Applications in Disease Management

9.4.1 Inflammatory and Autoimmune Disorders

Chronic inflammation, driven by dysregulated cytokine networks, is a hallmark of autoimmune diseases. Immunomodulatory peptides offer targeted strategies to restore balance. For example, the topical application of RDP58 in murine models of phorbol ester-induced dermatitis significantly reduces skin inflammation by dampening keratinocyte-derived cytokines and oxidative stress [6]. Similarly, vasoactive intestinal peptide (VIP) and pituitary adenylate cyclase-activating peptide (PACAP), two neuropeptides with potent anti-inflammatory properties, mitigate neuroinflammation in conditions like multiple sclerosis by suppressing microglial activation and promoting neuronal survival [10]. These examples illustrate how peptides can intervene at specific inflammatory checkpoints, offering precision unattainable with broad-spectrum immunosuppressants.

9.4.2 Cancer Immunotherapy

The advent of cancer immunotherapy has revolutionized oncology, with peptides emerging as key contributors. Peptide-based vaccines, such as those targeting tumor-associated antigens, activate cytotoxic T lymphocytes to recognize and eliminate malignant cells. Immune checkpoint inhibitors like pembrolizumab and nivolumab, though antibody-based, underscore the potential of peptide mimics designed to block PD-1/PD-L1 interactions with enhanced tissue penetration and lower production costs [2]. Melittin's ability to selectively lyse cancer cells while sparing normal tissues—mediated through membrane disruption and inhibition of survival pathways—positions it as a promising candidate for targeted therapies, particularly when delivered via nanoparticle systems to mitigate systemic toxicity [7]. Furthermore, self-assembling peptides and peptide-functionalized nanocarriers address challenges like enzymatic degradation and poor bioavailability, enhancing delivery to tumor microenvironments [2].

9.4.3 Infectious Diseases and Antimicrobial Resistance

The global rise in antibiotic-resistant pathogens has reignited interest in AMPs with dual antimicrobial and immunomodulatory functions. Fowlicidin-3, a chicken-derived cathelicidin, demonstrates the potent antibacterial activity against multidrug-resistant strains while neutralizing LPS-induced inflammation, making it a viable candidate for sepsis management [5]. Similarly, catestatin, a chromogranin A-derived peptide, not only exhibits microbicidal properties but also recruits and activates mast cells, bridging neuroendocrine and immune responses during cutaneous infections [11]. These peptides exemplify the potential to combat infections while modulating host immunity, reducing reliance on conventional antibiotics.

9.4.4 Challenges and Innovations in Peptide Therapeutics

Despite their promise, the clinical translation of immunomodulatory peptides faces hurdles. Stability issues, exemplified by the rapid degradation of LL-37 by serum proteases, necessitate innovations in delivery systems. Nanoparticle encapsulation, as explored for melittin, enhances tumor targeting while minimizing off-target effects [7]. Similarly, structural modifications—such as D-amino acid substitutions in RDP58—improve protease resistance without compromising bioactivity [6].

Another critical challenge lies in balancing efficacy with safety. Melittin's hemolytic activity underscores the need for precision engineering, while the immunosuppressive phase of sepsis complicates the timing of anticytokine therapies [7, 12]. Combinatorial approaches, such as coadministering peptides with chemotherapy or radiotherapy, may enhance therapeutic windows [7]. Additionally, the exploration of postbiotics—nonviable microbial metabolites—offers a safer alternative to probiotics for modulating gut immunity, circumventing risks associated with live bacteria [13].

9.4.5 Future Perspectives

The integration of omics technologies and computational modeling holds promise for de novo peptide design. For instance, activity-based screening and rational drug design, as employed in RDP58 development, could identify novel peptides with tailored immunomodulatory profiles [6]. Advances in nanomedicine, such as peptide self-assemblies and functionalized carriers, may further overcome delivery barriers, enabling systemic administration with localized effects [2].

Moreover, the gut-immune axis presents a frontier for peptide research. Bioactive peptides derived from dietary proteins, such as milk hydrolysates, modulate gut-associated lymphoid tissues (GALT), influencing systemic immunity and chronic disease outcomes [4, 14]. Harnessing these interactions could lead to nutraceuticals that prevent or ameliorate conditions like inflammatory bowel disease and obesity [13].

9.5 Conclusion

Immunomodulatory peptides represent a convergence of natural defense mechanisms and synthetic ingenuity. Their ability to precisely regulate immune responses—whether by silencing cytokine storms, redirecting antitumor immunity, or neutralizing pathogens—positions them at the forefront of therapeutic innovation. While challenges in stability, delivery, and safety persist, interdisciplinary approaches combining peptide engineering, nanotechnology, and systems biology are poised to unlock their full potential. As research continues to unravel the complexities of immune regulation, these peptides will undoubtedly play an expanding role in personalized medicine, offering hope for diseases once deemed intractable.

The present study will proceed to elaborate on immunomodulatory peptides through systematic examination of four key aspects: **preparation, evaluation method of activity, mechanism consideration, and applications**.

9.6 Preparation

9.6.1 Preparation Method of Immunomodulatory Peptide

Immunomodulatory peptides are prepared in a variety of ways and can be classified into the following categories based on source and process characteristics:

9.6.1.1 Organic Solvent Extraction

Organic solvent extraction is commonly used to extract immunomodulatory peptides from microbial fermentation products based on the differences in solubility of the target peptides in different solvents to achieve a preliminary separation. Immunomodulatory peptides and impurities, among other substances, have different solubilities in the organic solvent phase and the aqueous phase. When a mixed solution containing immunomodulatory peptides is mixed with an organic solvent, the immunomodulatory peptides will be distributed between the two phases according to their own chemical properties. If the solubility of the immunomodulatory peptide in a certain organic solvent is greater than that in water, it will transfer from the aqueous phase to the organic phase. After promoting mass transfer through shaking or stirring, the mixture is left to stand for stratification, separating the organic phase from the aqueous phase. Eventually, the immunomodulatory peptide is enriched in the organic phase. As an example, the preparation of silver cyclic peptides (Argyrins) was fermented using Mucor sp. Mx.DK1622, and the strain was cultured in 0.1%

L-α-aminobutyric acid medium to promote peptide chain elongation using precursor amino acids. Fermentation was completed to collect the bacterium. Next, ethyl acetate was added to the fermentation broth precipitate containing the bacteria and Amberlite XAD-16 resin for extraction. Ethyl acetate can effectively dissolve the silver cyclic peptide, and the extract was filtered and then dried by rotary evaporator to remove the ethyl acetate to obtain the crude extract, which was dissolved in methanol and then analyzed by High-Performance Liquid Chromatography-Mass Spectrometry (HPLC-MS). The crude extract was dissolved in methanol and analyzed by HPLC-MS [15]. Organic solvent extraction is easy to operate and suitable for the initial enrichment of fat-soluble peptides from microbial sources, but the purity of the extract is limited and may affect the peptide activity.

9.6.1.2 Tissue Homogenization Extraction

Tissue homogenization releases intracellular peptides by mechanical fragmentation and is suitable for the extraction of active ingredients from plant or animal tissues. In the extraction of cyclic peptides (Cliotides) from butterfly beans (*Clitoria ternatea*), fresh plants were chopped and homogenized with ethanol, which both disrupts the phospholipid bilayer structure of the cell membrane and inhibits oxidase activity, thus preventing peptide oxidation. Initial enrichment of the cyclic peptides was achieved by utilizing ethanol's ability to solubilize a variety of components in butterfly beans, including the target cyclic peptides [16]. Tissue homogenization is also commonly used in animal studies to extract immunomodulatory peptides from immune organs such as spleen and thymus. By homogenizing spleen tissues, possible immunomodulatory peptides can be obtained and used to study their mechanism of action on the immune system.

9.6.1.3 Enzymatic Hydrolysis

Enzymatic hydrolysis of protein substrates by proteases to release active peptides has become an important means of preparing immunomodulatory peptides of plant and animal origin because of the advantages of mild reaction conditions and high retention of product activity. Nowadays, the reported enzymatic methods mainly include multienzyme complex digestion and directed enzymatic digestion.

Multienzyme complex hydrolysis can significantly enhance the efficiency of substrate degradation by combining proteases with different cleavage sites. For example, by utilizing alkaline protease to cut aromatic amino acids, neutral protease to cut hydrophobic residues, and papain to cut lysine or arginine, the efficiency of protein hydrolysis can be improved and the yield of target peptides can be increased [17]. Similarly, the researchers used an enzymatic method combined with various analytical techniques to extract immunomodulatory peptides (Se-EWP) from selenium-enriched egg whites, in which the selenium-enriched egg white solution was first denatured sufficiently, followed by successive enzymatic hydrolysis using alkaline protease and neutral protease. In this process, the proteins in selenium-enriched egg white were hydrolyzed into peptide fragments by controlling the enzymatic conditions to obtain a mixture containing immunomodulatory peptides [18]. In addition to this, other studies have utilized enzymatic hydrolysis of pine nut proteins to break down complex protein macromolecules into smaller peptides. The hydrolyzed products were then preliminarily separated by ultrafiltration technology based on the molecular size to remove macromolecular impurities and incompletely hydrolyzed proteins. Finally, purification is carried out to obtain relatively pure QWFM tetrapeptide [19]. Immunomodulatory peptides from *Sipunculus nudus* L were extracted using an enzymatic method combined with various separation techniques, in which animal proteases and flavored proteases were used to enzymatically digest *S. nudus* proteins, followed by separation and purification using ultrafiltration, gel filtration chromatography (GFC), and reversed-phase HPLC, in order to obtain the target immunomodulatory peptides [20]. The step-by-step enzymatic digestion by pepsin and trypsin mimics the digestive environment of the human gastrointestinal tract, generating physiologically active peptides. In the hydrolysis of colostrum whey protein, pepsin was first added to hydrolyze the protein, and pepsin specifically cleaved hydrophobic residues such as phenylalanine and tyrosine to generate medium-length peptides. Subsequently, trypsin was added to continue the hydrolysis, and trypsin and chymotrypsin in trypsin cleaved arginine/lysine and aromatic amino acids, respectively, which were further degraded into small molecule peptides. Finally, the hydrolysis was terminated by heating [21]. The immunomodulatory peptides were obtained by simulating the in vivo digestive process, and the small molecular weight peptides generated by this method not only possessed immunomodulatory activity, but also had amino acid sequences closer to the intestinal absorbed form, and the bioavailability was significantly improved.

Targeted zymography is a technical means of hydrolyzing proteins under specific conditions using a specific protease to obtain peptides with specific properties or targets. Directed enzymatic hydrolysis of immunomodulatory peptides in conch meat was used. Conch meat was filtered by removing the shell, grinded, washed, and soaked in distilled water, then incubated with trypsin, heated to inactivate the enzyme, centrifuged, and the concentration was determined by the Bradford method, and lyophilized to obtain the hydrolyzed peptides of conch [22]. Trypsin has specific enzyme cleavage sites and can specifically act on the proteins in conch meat to hydrolyze them into peptides. This specificity makes the hydrolyzed product relatively pure with less impurity content, and the subsequent isolation and purification process is relatively simple, which helps to improve the purity and quality of immunomodulatory peptides.

9.6.1.4 Chemical Synthesis

The solid-phase peptide synthesis technique in the chemical synthesis method has unique advantages in the field of immunomodulatory peptide preparation, which can precisely control the peptide chain sequence and provides a powerful means for the production of structurally defined immunomodulatory peptides. First, the carboxyl end of the first amino acid is attached to an insoluble solid phase carrier (e.g., polystyrene resin) by covalent bonding, and the amino end of this amino acid is protected with a protecting group to prevent it from reacting unnecessarily in subsequent reactions. Then, through a series of chemical reactions, other amino acids are gradually attached sequentially to the amino acid already immobilized on the carrier in a predetermined sequence. Each time a new amino acid is attached, the protective group on the amino group of the previous amino acid needs to be removed, and then the carboxyl group of the new amino acid and the amino group of the already attached amino acid under the action of a condensing agent undergoes a condensation reaction to form a peptide bond. This process is repeated until a peptide chain of the desired length and sequence is synthesized. Finally, the synthesized peptide chain is cleaved from the solid phase carrier, and all protective groups are removed to obtain the target immunomodulatory peptide. The following is an example of the synthesis of ApoE (133–150)-related peptides and the construction of a transformable supramolecular self-assembling peptide, TEP-FFG-CRApY: The researchers used solid-phase 9-fluorenylmethoxycarbonyl (Fmoc) chemical synthesis to prepare ApoE(133–150)-related peptides, such as ApoE(133–150), COG-133, and Ac-ApoE(133–150)-NH_2. These peptides are immunomodulatory peptides derived from human ApoE, and their synthesis is useful for investigating their antimicrobial and immunomodulatory activities. In the synthesis process, amino acids were progressively linked to form peptide chains of specific sequences following the conventional steps of Fmoc chemical synthesis. First, the carboxyl group of the first amino acid was attached to a solid-phase carrier and its amino group was protected with a Fmoc protecting group, then the amino group was exposed by removing the Fmoc group, and then the condensation reaction was carried out with the next activated amino acid, and so on until the target peptide was synthesized. The synthesized peptides were purified using reversed-phase HPLC to achieve a purity of greater than 95%, and the quality of the peptides was confirmed by mass spectrometry. Through this solid-phase peptide synthesis technique, the amino acid sequences of the peptides were precisely controlled, which provided a material basis for the subsequent study of the antibacterial activity of these peptides against different bacteria, the cytotoxicity in different cell lines, and the modulating effect on immune cells. In the preparation of the translatable supramolecular self-assembling peptide TEP-FFG-CRApY for cancer immunotherapy, N3-FFG-CRApY and N3-CRApY were first synthesized using a conventional Fmoc-based solid phase peptide synthesis protocol. Specific sequences in these peptides, such as the GPX4 protein-targeting peptide sequences, enable the synthesized TEP-FFG-CRApY to possess targeting functions such as inducing iron death in tumor cells. The synthesized peptide was conjugated to TEP via a copper-catalyzed "click" reaction to obtain the final target peptide [23]. The structural accuracy of the peptide is ensured by this precise synthesis method, thus ensuring that it can perform its intended function in subsequent cancer immunotherapy studies, providing a powerful tool for cancer therapy research.

Recombinant technology combined with chemical cleavage is a method commonly used in the preparation of immunomodulatory peptides, which combines the advantages of large-scale protein expression by recombinant technology and the precise acquisition of target peptides by chemical cleavage. For example, in the preparation of the recombinant peptide rApoE$_{PM}$ (133–150), recombinant technology combined with chemical cleavage was used. First, recombinant technology was applied to synthesize the gene sequence encoding the target immunomodulatory peptide ApoE (133–150) and cloned into an expression vector, which was fused with the optimized vector for frog ribonuclease onconase (ONC), ONC-DCless-H6. Next, the constructed vector was transformed into *E. coli* BL21 (DE3) for expression. During the growth of *E. coli*, the fusion protein ONC-DCless-H6-(PM) ApoE was expressed in large quantities as the organism multiplied. Since the AMP may be toxic to the host cell, the fusion expression can avoid the effect of this toxicity on cell growth and improve the expression efficiency. After the end of expression, the bacteriophage containing the fusion protein was collected, and the inclusion bodies were obtained by centrifugation and other operations. Afterward, the fusion protein was initially isolated by purification using affinity chromatography, such as using Ni Sepharose TM 6

Fast Flow. Next, chemical cleavage was performed to separate the target peptide ApoE (133–150) from the carrier ONC-DCless-H6 by hydrolyzing the linker peptide (Asp-Pro-Met) in the fusion protein using acid (e.g., 0.5 M HCl) under acidic conditions (pH 2). The carrier ONC-DCless-H6 was precipitated by adjusting the pH to neutral or alkaline (pH 7–7.2), and the recombinant peptide $rApoE_{PM}$ (133–150) was finally obtained by repeated centrifugation and lyophilization [24].

9.6.1.5 Microbial Fermentation

Microorganisms are diverse and can synthesize and secrete a variety of bioactive substances during their growth and metabolism. Some microorganisms can convert nutrients in the culture medium into immunomodulatory peptides under specific culture conditions. Some microorganisms are able to synthesize peptides through their own metabolic pathways using amino acids, sugars, and other substances. Some lactic acid bacteria can produce short peptides with immunomodulatory functions during fermentation. This metabolic synthesis ability of microorganisms can be further optimized by genetic engineering techniques. By introducing genes encoding specific immunomodulatory peptides into microbial cells, they are made to synthesize the target peptides according to people's needs. The synthetic immunomodulatory peptide genes are introduced into *E. coli* or yeast, and the rapid reproduction of these microorganisms is utilized to produce the target immunomodulatory peptides in large quantities. In this process, microorganisms are like miniature factories, producing immunomodulatory peptides under the joint action of their own metabolism and the imported genes, using the culture medium as raw materials. Using microbial metabolism or secretion function to prepare immunomodulatory peptides has unique advantages. In the preparation of Spirulina immunomodulatory peptides [25], the researchers utilized a mixture of Lactobacillus plantarum B7 and Bacillus subtilis 168 to ferment dried Spirulina powder. During the fermentation process, the microorganisms utilized the nutrients in Spirulina to carry out metabolic activities, which led to the decomposition of the cellular structure of Spirulina, the hydrolysis of proteins into peptide segments, and the increased accumulation of low molecular metabolites, thus generating peptides with immunomodulatory activity. Experiments showed that fermented Spirulina and its peptide fractions performed better in promoting lymphocyte proliferation and regulating cytokine secretion, and had higher immunomodulatory activity than unfermented Spirulina. Some studies have introduced genes encoding specific immunomodulatory peptides into Saccharomyces cerevisiae and utilized the secretion system of Saccharomyces cerevisiae to produce immunomodulatory peptides. Saccharomyces cerevisiae has the advantages of being easy to transform, fast-growing, and easy to cultivate on a large scale, and is able to express and secrete exogenous proteins efficiently. The immunomodulatory peptides produced in this way, after isolation and purification, can be used in the study of immunomodulatory mechanisms as well as in the development of related drugs and nutraceuticals.

9.6.1.6 Genetic Engineering

Genetic recombination technology has significant advantages in constructing engineered strains to express target immunomodulatory peptides, which can effectively meet the demand for high yield immunomodulatory peptides in scientific research and production.

Genetic recombination technology recombines exogenous genes with vector DNA in vitro and introduces the recombinant DNA into recipient cells, ultimately realizing the stable expression of the exogenous genes. In constructing an engineered strain expressing an immunomodulatory peptide, it is first necessary to obtain the gene encoding the target immunomodulatory peptide. The corresponding DNA sequence can be designed and synthesized by artificial synthesis based on the known amino acid sequence of the immunomodulatory peptide; or the relevant gene can be cloned from the biological genome. The acquired target gene is ligated with a suitable vector. The vector is usually a plasmid, which has self-replicating ability and multiple restriction enzyme cleavage sites to facilitate the insertion of exogenous genes. The vector and the target gene are cut using restriction endonuclease, and then, they are joined together by DNA ligase to form a recombinant plasmid. The recombinant plasmid is introduced into a recipient cell, such as *E. coli*, yeast, etc. Recipient cells into which the recombinant plasmid is introduced are screened and characterized to obtain engineered strains capable of stably expressing the target immunomodulatory peptide. The screening process usually utilizes resistance genes carried on the vector, such as the ampicillin resistance gene, and the recipient cells are cultured on medium containing ampicillin, so that only the cells that have successfully introduced the recombinant plasmid can grow. Taking the optimization of the immunomodulatory peptide Argyrins as an example [26], researchers constructed a heterologous production system using recombinant technology in order to optimize the production of the immunomodulatory peptide Argyrins. They introduced the biosynthetic gene cluster encoding Argyrins into Myxococcus xanthus DK1622ΔmchA-tet host bacterium, and increased the total production of Argyrins to 350–400 mg/L by engineering the transcription and translation initiation. In this process, through the recombination technology, not only the Argyrins were realized to be expressed in a heterologous host in a heterologous host, but also significantly increased the yield by optimizing the expression regulation of related genes, which provided a sufficient material basis for the subsequent research and development of Ajmalicin. In the preparation of certain human immunomodulatory peptides, the researchers introduced the genes encoding these peptides into *E. coli* or yeast, and constructed engineering strains for expression. High-yield production of these immunomodulatory peptides was achieved through the optimization of fermentation conditions, such as the regulation of medium composition, temperature, and pH. These high-yield immunomodulatory peptides can be used for further research on their immunomodulatory mechanisms, development of novel immunotherapeutic drugs, and as functional food additives.

9.6.2 Isolation and Purification Method of Immunomodulatory Peptides

Separation and purification is a key step in obtaining high-purity immunomodulatory peptides, and the commonly used techniques are as follows:

9.6.2.1 Ultrafiltration

Separation of peptides using ultrafiltration membranes with different molecular weight cutoffs based on molecular weight differences is an important method in the separation and purification process of immunomodulatory peptides. In the extraction of immunomodulatory peptides (Se-SPep) from selenium-enriched soybeans, an ultrafiltration membrane with a molecular weight cutoff of 3 kDa was used. After enzymatic hydrolysis of selenium-enriched soy protein, the hydrolysate was first passed through this ultrafiltration membrane, and selenium-enriched soy peptides smaller than 3 kDa were collected through the membrane, whereas large impurities and proteins that were not completely hydrolyzed were retained, achieving the preliminary separation and purification of the target peptides [17].

9.6.2.2 Chromatography

Reversed-phase HPLC (RP-HPLC) is a key technique for precise separation based on peptide hydrophobicity difference, which is widely used in immunomodulatory peptide research. The stationary phase of RP-HPLC is usually silica gel with nonpolar alkyl chains (e.g., C18, C8, etc.) bonded to the surface, while the mobile phase is a mixed solution consisting of water and organic solvents (e.g., acetonitrile and methanol), and the aqueous phase often contains certain buffer salts. When the sample containing immunomodulatory peptides is injected into the system, the peptides with strong hydrophobicity will have strong interaction with the nonpolar alkyl chains of the stationary phase and will be retained in the column for a long time; whereas the peptides with weak hydrophobicity will have weak interaction with the stationary phase and will pass through the column faster under the driving force of the mobile phase. By gradually increasing the proportion of organic solvent in the mobile phase and changing the polarity of the mobile phase, the immunomodulatory peptides with different hydrophobicity were sequentially eluted from the stationary phase, thus realizing the separation. In the preparation of naked square starling peptide (SNLP), SNLP was eluted by gradient elution on a C18 column to obtain a single active fraction [20].

Gel filtration chromatography, also known as molecular sieve chromatography, plays an important role in the separation and purification of immunomodulatory peptides, and its core principle is to separate peptides based on molecular size differences. The stationary phase for GFC is gel particles with a porous mesh structure, and these gel particles have many pores of different sizes inside. When the sample solution containing immunomodulatory peptides of different sizes passes through the gel column, the peptide molecules which are larger than the pores of the gel cannot enter into the pores, and can only flow in the gaps between the gel particles, so they have a short flow path inside the column, and flow out of the chromatographic column first; and the smaller peptide molecules are able to enter into the pores of the gel, and have a long flow path inside the column, and flow out of the chromatographic column later. In this way, immunomodulatory peptides of different sizes can be separated in order of molecular size. Like objects of different sizes passing through a sieve with a specific pore size, larger objects pass through first and smaller objects pass through second. For example, in the isolation of immunomodulatory peptides from fermentation [25], Sephadex G-25 GFC was utilized, where larger proteins and impurities could not enter the gel pores and were eluted first, while smaller immunomodulatory peptides entered the gel pores and were subsequently eluted, resulting in the initial isolation of peptides of different sizes.

Ion exchange chromatography (IEC) is a commonly used separation technique that plays an important role in the separation and purification, analysis, and detection of immunomodulatory peptides, based on the difference in electrostatic interactions between different peptides and ion exchangers. The stationary phase of IEC is the ion exchanger, which consists of an inert support skeleton, ion exchange groups, and balancing ions. The ion exchange groups can undergo an exchange reaction with the ions in solution. Based on the nature of the ion exchange groups, ion exchangers are classified as cation exchangers and anion exchangers. Cation exchangers with acidic groups can exchange with cations in solution, and anion exchangers with basic groups can exchange with anions in solution. When the immunomodulatory peptide solution containing different charges passes through the ion exchange column, the positively charged peptide will bind to the cation exchanger, and the negatively charged peptide will bind to the anion exchanger. The ability of different peptides to bind to the ion exchanger depends on the amount of charge they carry, the distribution of the charge, and the nature of the ion exchanger. By changing the ionic strength, pH, and other conditions of the eluent, the peptides bound to the ion exchanger can be sequentially eluted to achieve separation. Pterostilbene cyclic peptides are passed through a strong cation exchange column to remove pigments and impurities. IEC allows separation of peptides based on their charge characteristics [16].

9.6.2.3 Solvent Extraction

Selective extraction of target peptides by polar solvents is a method of separation based on the difference in solubility of target peptides in different polar solvents. Selective extraction of target peptides was by polar solvents. To extract immunomodulatory peptides from Agaricus, Agaricus was degreased by petroleum ether and the water-soluble peptide fractions were graded by ultrafiltration [27].

9.6.2.4 Membrane Separation Continuous Coupling Technology

The immobilized enzyme membrane reactor (CIEH-MS) enables continuous production by synchronizing enzymatic hydrolysis with membrane separation. The CIEH-MS system mainly consists of a feed tank, a thermostatic hydrolysis reactor with immobilized neutral protease, and a membrane module connected to a product tank for collection of the final product. In a study on the preparation of hydrolyzed peptides of type II collagen from chicken sternal cartilage [28], immobilized neutral protease was prepared by immobilizing L-arginine on the surface of a resin with amino groups by EDC/NHS coupling. Soluble type II collagen (CII) was dissolved in phosphate buffer solution and prepared as 1.0 mg/mL solution in the feed tank. The CII solution was pumped into the hydrolysis reactor at a constant rate using a gear pump, and when the temperature reached 50 °C, the immobilized neutral protease was added, and its total activity was set at 40–160 U/mg of substrate according to the experimental demand, and the CII was prehydrolyzed for 10 minutes After that, the hydrolysis mixture was continuously pumped into a membrane separation unit with a molecular weight cutoff of 5 kDa. In the hydrolysis reactor, immobilized neutral protease catalyzed the enzymatic reaction of CII to generate immunomodulatory peptides. At the same time, the membrane module separates the reaction products, and the small molecule immunomodulatory peptide passes through the membrane into the permeate, which is collected in the product tank; the unreacted CII, enzyme, and large molecule impurities, etc. are retained on the other side of the membrane, and returned to the feed tank through the recirculation pump, and continued to participate in hydrolysis reaction. The schematic diagram of the immobilized enzyme membrane reactor is shown in Figure 9.2.

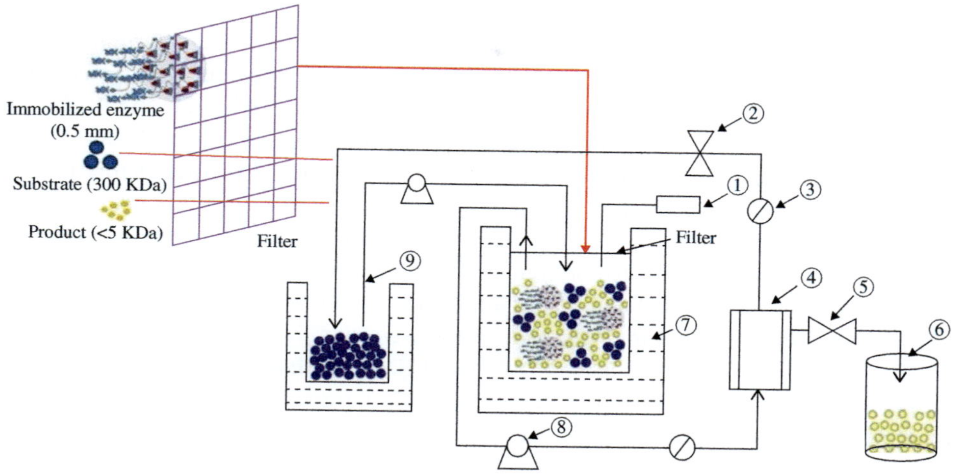

Figure 9.2 1.6 Schematic diagram of the continuous CIEH-MS system loaded with immobilized neutral protease (1, pH controller; 2, retentate valve; 3, pressure transducer; 4, UF membrane module; 5, Permeate valve; 6, permeate [product tank]; 7, immobilized enzyme reactor with thermostat-controlled water bath; 8, recycling pump; 9, feed tank). *Source:* [28]/John Wiley & Sons.

9.7 Evaluation Method of Activity

In the study of immunomodulatory peptides, the precise evaluation of their immunological activity is a key link to reveal their mechanism of action and potential application value. From the microscopic cellular level to the macroscopic animal level, various evaluation methods complement each other to provide a comprehensive perspective for the in-depth understanding of immunomodulatory peptides. In the following, the evaluation of immunomodulatory peptides from "in vitro cell model" and "in vivo animal model" will be described in detail.

9.7.1 *In Vitro* Cell Models

Cell models have an irreplaceable role in the evaluation of immunoregulatory peptide immunoreactivity. Cells are the basic unit of immune response, and different types of cells have specific functions in the immune process, and their responses to immunomodulatory peptides can visually reflect their immune activities. The use of cell modeling has the advantages of simple operation, short experimental period, and easy control of conditions, which can be used to explore the role of immunomodulatory peptides in a relatively stable environment, and provide important theoretical basis and data support for subsequent in vivo studies.

RAW264.7 cells were used to assess the immunoreactivity of immunomodulatory peptides [29]. Researchers used the neutral red method to determine the phagocytic ability of RAW264.7 cells, and the results showed that the phagocytic activity of cells in the pea protein hydrolysate-treated group was significantly higher than that of the blank control group (CG) and the positive CG, which indicates that pea protein hydrolysate can effectively enhance the phagocytic function of cells. The phagocytosis of macrophages is a key link in the body's immune defense, and the detection of phagocytic activity can directly reflect the effect of pea protein hydrolysate on the immune function of macrophages. Using the Giess method to measure NO production, the NO production of the pea protein hydrolysate-treated group was significantly increased, and as an important immunoregulatory molecule, the change of its production can deeply reflect the intrinsic mechanism of the immunoregulatory function of the pea protein hydrolysate on macrophages. In addition, the secretion and gene expression of inflammatory factors were detected by enzyme-linked immunosorbent assay (ELISA) and qPCR, and it was found that the secretion and gene expression of TNF-α and IL-6 in the treated groups were significantly increased in a dose-dependent manner, and the changes of TNF-α and IL-6, which are important proinflammatory cytokines, could directly reflect the effect of immune modulation. These experiments comprehensively and thoroughly evaluated the effects of pea protein hydrolysate on the immunomodulatory function of macrophages from multiple dimensions.

Human monocyte cell line THP-1 was applied to study the immunomodulatory activity of immunomodulatory peptides [16]. The content of cytokines IL-8 and TNF-α in the cells was detected by ELISA, and it was found that the secretion levels of IL-8 and TNF-α in the cells of the cyclic peptide-treated group increased in both the absence of LPS prestimulation and after LPS prestimulation. Meanwhile, the Bio-Plex cytokine assay was used to quantify 26 cytokines in the same cell culture supernatant samples, and the results showed that different cyclic peptide treatments led to changes in the secretion levels of different cytokines and chemokines. Cytokines and chemokines play a central role in the immune response, and the detection of the secretion levels of these factors can comprehensively assess the regulatory effect of cyclic peptides on the function of immune cells, and then accurately determine the strength of their immunostimulatory activity.

Human immortalized keratinocytes (HaCaT), undifferentiated monocyte (THP-1), and THP-1 cells differentiated by fobol ester (PMA) have played an important role in the study of the immunomodulatory properties of the novel cationic AMP ApoE (133–150) derived from human apolipoprotein E [24]. The researchers first pretreated the cells with *Pseudomonas aeruginosa* (*P. aeruginosa*) LPS, after which the cells were treated with 3 μM of rApoE$_{PM}$(133–150), sApoE(133–150), COG-133, Ac-ApoE(133–150)-NH_2 or buffer, respectively. As detected by qRT-PCR, in LPS-stimulated PMA-differentiated THP-1 cells, the rApoE$_{PM}$(133–150)-treated group of cells showed a decrease in the mRNA expression of IL-8 and Cox-2. rApoE$_{PM}$(133–150) was able to significantly downregulate, in LPS-pretreated HaCaT cells, the COX-2 and IL-8 and Cox-2 are important cytokines related to inflammatory response, and detecting their expression changes at the mRNA level can visualize the regulatory effect of ApoE(133–150) on cellular inflammatory response.

Human peripheral blood mononuclear cells (PBMC) have been widely used in assessing the immunoreactivity of immunomodulatory peptides, especially in the study of the effect of gluten's whey solubilizing protein (gliadin) on the immune response [30]. Researchers used pepsin-trypsin-digested whey soluble proteins or synthetic peptides to stimulate PBMC, and some of the samples were pretreated with anti-CXCR3 monoclonal antibody before stimulation. Detection of cytokines revealed differences in the levels of cytokines produced by PBMC from different individuals after PT-methanolysin stimulation, with

IL-8 production occurring only in some individuals and possibly at higher levels in patients with CD-GFD than in healthy controls. The production of IL-8 and some cytokines (e.g., IL-6) in PBMC from CD-GFD patients was inhibited by the addition of the anti-CXCR3 antibody. PBMC contain a variety of immune cells, which are important participants in the body's immune response, and the detection of the level of cytokine production after their stimulation can directly reflect the activation effect of maltolysin on the immune cells as well as the body's immune response, and at the same time, it can help to explore the mechanism of CXCR3 in this immune response.

The human mast cell line LAD2 and human peripheral blood-derived mast cells were used to study the effects of the neuroendocrine AMP Catestatin on human mast cells [11]. In β-hexosaminidase release assay experiments, researchers found that wild-type Catestatin and its variants were able to significantly induce β-hexosaminidase release at certain concentrations, whereas β-hexosaminidase release was inhibited when an inhibitor acting on the pathway associated with Catestatin-induced mast cell activation was added. β-hexosaminidase is coreleased with histamine and is a mast cell degranulation The release of β-hexosaminidase, together with histamine, is a marker for mast cell degranulation, and by detecting its release, it is possible to visualize the effect of Catestatin on the mast cell degranulation function, and thus to assess its modulatory effect on mast cell immunoreactivity. By enzyme immunoassay (EIA) and ELISA, it was found that wild-type Catestatin and its variants were able to promote the release of LTC4, PGD2, and PGE2 from mast cells in a dose-dependent manner, as well as selectively increase the release of cytokines and cytotoxicity factors such as GM-CSF, MCP-1/CCL2, MIP-1α/CCL3, and MIP-1β/CCL4. CCL4 and other cytokines and chemokines production. These inflammatory mediators and cytokines and chemokines play key roles in immune regulation and inflammatory responses, and detecting the level of their release provides a comprehensive understanding of the effects of Catestatin on mast cell immunoreactivity.

Mouse splenocytes are commonly used to assess the effects of immunomodulatory peptides on lymphocyte proliferation [28, 31, 32]. In studying the immunomodulatory activity of hazelnut hydrolyzed peptides (HHPs), researchers aseptically isolated mouse splenocytes, coincubated them with concomitant cutinoglobulin A (ConA), and found that the proliferative capacity of splenic lymphocytes was significantly increased by each of the medium- and low-molecular-weight HHPs dosage groups, compared with the negative CG, using the MTT assay [32]. In the study of immunomodulatory peptides prepared serially from type II collagen (CII) using a novel immobilized enzyme membrane reactor, splenocyte lymphocytes were added to a 96-well plate, and CII peptide at a concentration of 2 mg/mL or ConA at a concentration of 5 μg/mL was added as a stimulus, respectively, and it was detected by the MTT assay and found that lymphocyte proliferation rate in the CII peptide-treated group was higher than that in the unstimulated group [28]. Lymphocytes are an important component of the immune system, and their proliferation ability is one of the key indicators of immune activity. The MTT method reflects cell proliferation by detecting the activity of succinate dehydrogenase in cell mitochondria, which can visually assess the effect of the immunomodulatory peptide on the proliferation of lymphocytes, and thus determine its immunomodulatory activity.

Although the pancreatic β-cell line RIN5fm is mainly used to study islet amyloid polypeptide (IAPP)-related mechanisms, immunoreactivity-related information is also involved in assessing the cytoprotective effects of LL-37[33]. Researchers used the MTT reduction method to assess cell viability, if LL-37 can effectively inhibit the formation of IAPP cytotoxic assembly, then the cell viability of the experimental group with the addition of LL-37 should be higher compared with that of the group treated with IAPP alone. The MTT reduction method can intuitively reflect the survival state of the cells, and the inhibition of cellular damage caused by IAPP by LL-37 can be indirectly judged by this method. An effect which is closely related to the function of protecting cells from damage in immune regulation.

9.7.2 *In Vivo* Animal Models

Animal models have unique value in the evaluation of immunomodulatory peptides. The complex and complete physiological environment in animals can reflect the comprehensive immunomodulatory effects of immunomodulatory peptides in vivo, covering the interactions between immunomodulatory peptides and various tissues and organs as well as the overall immune response process. This provides an important reference for evaluating the potential clinical application of immunomodulatory peptides and helps to understand their mechanism of action under real physiological conditions. However, animal experiments are susceptible to a variety of factors, such as individual differences in animals, the breeding environment, etc., and the high cost and long period of the experiments, which puts forward higher requirements for the design and implementation of experiments.

In studying the effects of whey protein hydrolysate (WPH) on the growth and immunomodulation of captive mouse pups, mouse pups were chosen as an animal model [34]. The researchers measured the thymus weight of mouse pups and found that the thymus weight of mouse pups in the group containing WPH was higher than that of the group without WPH and close to that of the group fed by mothers. As an important immune organ for the development and maturation of T-lymphocytes, changes in the weight of the thymus reflect the developmental state of the immune organ and the strength of the immune function to a certain extent. By measuring the weight of the thymus, the effect of WPH on the development of immune organs can be visualized. A phenotypic analysis of thymocytes and splenocytes in mouse pups revealed that the number of certain immune cells (e.g., $CD3^+CD8^+$, $B220^+CD19^+$, IA/IE^+CD11c^+, and $CD11b^+$ cells) was higher in the WPH-added artificial feeding group than that in the unadded WPH group, which suggests that WPH may have the effect of promoting the proliferation or differentiation of immune cells, and thus enhancement of immune activity.

In evaluating the immunoreactivity of selenium-enriched egg white peptide (Se-EWP), cyclophosphamide-induced immunosuppressed mice were selected as an animal model [18]. The immune organ index was measured, and the results showed that the immune organ index of the Se-EWP-treated group was higher than that of the model group (MG), which indicated that Se-EWP could slow down the atrophy of immune organs due to immunosuppression, and help to maintain or enhance the function of immune organs. Immune organs, such as spleen and thymus, are important sites of immune response, and their weight changes can directly reflect the immune status of the body. Detection of serum cytokine levels revealed that serum IL-6, IL-2, and TNF-α levels were elevated in the

Se-EWP-treated group compared with the MG, and the levels of these cytokines in the medium-dose Se-EWP group (Se-MDG) may be higher than those in the high-dose Se-EWP group (Se-HDG) and medium-dose normal egg white peptide group (MDG). Cytokines play a key regulatory role in the immune system, and the detection of these cytokine levels can visualize the effect of Se-EWP on immune cell function and thus determine its immunomodulatory activity. Detection of liver selenium levels revealed that the liver selenium levels in the Se-EWP-treated group were higher than those in the normal egg white peptide (EWP)-treated group, and there was a dose-effect relationship. Selenium is an essential trace element to maintain the function of selenium-dependent enzymes, and it is closely related to immune function. Detecting liver selenium content can understand the effect of selenium supplementation of Se-EWP in vivo, which in turn indirectly reflects its effect on immune function. When blood routine tests were performed, the total number of leukocytes in the MG was significantly lower than that in the CG, while in the Se-EWP-treated group, the number of leukocytes (WBC), neutrophils (Neu), and lymphocytes (Lym) in the Se-MDG was higher than that in the MG and the EWP-treated group. Leukocytes are an important part of the body's defense system, and the detection of these cell counts can visually reflect the ability of the body's immune system to fight pathogens, and assess the effect of Se-EWP on the number of immune cells in immunosuppressed mice, which in turn determines its immunomodulatory activity.

In the study of the immunomodulatory function of Tc52 virulence factor in Trypanosoma cruzi, Balb/c mice and CBA mice were selected as animal models [31]. In this study, a mouse model infected with Trypanosoma cruzi and immunized with a synthetic peptide was constructed to investigate the mechanisms related to the immunomodulatory function of Trypanosoma cruzi Tc52 virulence factor. In the immunization and antibody production experiments, researchers immunized BALB/c mice with synthetic peptides (P1, P2, P3, or P5) coupled to OVA and combined with adjuvants, and found by the immunoblotting assay that sera of mice immunized with P1, P2, or P3-OVA showed strong signals in response to the natural protein of Tc52, whereas sera of mice immunized with P5-OVA showed no obvious Tc52 response. This suggests that P5-OVA is less immunogenic and may have an effect on immunoreactivity. Immunoblotting can visualize the production of antibodies against specific antigens in animals, and thus determine the strength of the immune response. In vivo immunization experiments and lymphocyte proliferation assays revealed differences in serum levels of antibodies against Tc52 and proliferation of draining lymph node cells in response to stimulation in mice immunized with different synthetic peptides. Sera from P1-, P2-, and P3-OVA-immunized mice should have strong signals in response to Tc52, and the draining lymph node cells should proliferate markedly when stimulated with the corresponding peptide or Tc52, while sera from P5-OVA-immunized mice should have a strong signal in response to Tc52. OVA-immunized mice had weak serum and Tc52 response signals, and the proliferation of draining lymph node cells was not significant in response to P5-OVA stimulation, suggesting that the immunogenicity of P5-OVA is different from that of the other peptides, which in turn affects the immune activity.

In studying the immunomodulatory effects of peptides obtained by protease fermentation of milk proteins, healthy male Kunming (KM) mice were used as animal models, and a lipopolysaccharide (LPS)-induced acute inflammation model of KM mice was constructed [35]. The mouse immune model was constructed and acute inflammation was induced by intraperitoneal injection of 0.02 mg/kg of LPS solution in all groups of mice except the blank group, which was injected with 0.9% saline. Subsequently, several tests were performed to determine the splenic index, and it was found that the splenic index of the mice in the LPS group decreased compared with that of the normal group, whereas the splenic index of the mice in the peptide group given MHP was higher than that of the LPS group, and that there might be an optimal dose that would result in the most pronounced effect of the elevation of the splenic index. As an important immune organ, the spleen index can reflect the development and functional status of the spleen, and its changes can visualize the effects of immunomodulatory substances on immune organs. When peripheral blood cytometry was performed, the LYM% of the LPS group was lower than that of the blank group, and the number of leukocytes was also reduced; in the peptide group given MHP, the LYM% and the number of leukocytes were higher than that of the LPS group, and the group of the medium agent 200 mg/kg might have the best effect. Lymphocytes are an important part of the immune system, and the detection of these indicators can visualize the effect of immunomodulatory peptides on the body's immune cells and assess their immunomodulatory activity. The levels of IgG, IL-6, and IL-10 in serum were measured by double-antibody sandwich ELISA, and it was found that compared with the LPS group and the blank group, the IgG content in the medium-dose 200 mg/kg peptide group was significantly increased; the content of IL-6 in the peptide group was lower than that of the LPS group, and the content of IL-10 was higher than that of the LPS group. IgG is an important immunoglobulin in humoral immunity, IL-6 is a proinflammatory cytokine, IL-10 is a proinflammatory cytokine, and IL-10 is an important immunomodulatory activity. Among the cytokines, IL-10 is an anti-inflammatory cytokine. Detecting the levels of these immunoglobulins and cytokines can provide insights into the effects of immunomodulatory peptides on the body's immunoregulatory network, and clarify their specific mechanisms of action in immunomodulation.

A mouse infection model was used for assessing the effects of AMPs and immunomodulatory peptides in the study of the protective effects of immunomodulatory peptides in animal infection models [36]. The animals were divided into experimental and CGs, and the animals in the experimental group were given immunomodulatory peptides (e.g., IDR-1), while those in the CG were given saline or other control substances, after which the animals were infected with pathogens (e.g., methicillin-resistant *Staphylococcus aureus*, vancomycin-resistant *Enterococcus faecium*, *Salmonella*, etc.). It was observed that animals in the experimental group treated with immunomodulatory peptides had increased survival rates, reduced symptoms of infection, and histopathological changes showing reduced inflammatory responses and significantly lower bacterial loads in the tissues, compared with the CG. The animal infection model can more realistically simulate human infections and comprehensively assess the immunomodulatory effects of immunomodulatory peptides in vivo, and these indexes can intuitively reflect the immunomodulatory peptide's ability to resist pathogen infections, as well as its impact on the overall immune function of the organism.

9.8 Mechanism Consideration

Immunomodulatory peptides play a pivotal role in regulating the functionality of the immune system, serving as crucial molecular mediators in both innate and adaptive immunity [37]. These bioactive peptides exert their biological effects through a variety of sophisticated molecular mechanisms, directly or indirectly influencing the functions of diverse immune cells, including T cells, B cells, macrophages, DCs, and natural killer (NK) cells [38]. The immunomodulatory effects of functional peptides are primarily achieved through the following mechanisms:

9.8.1 Direct Regulation of Immune Cell Function

Peptides can directly modulate immune cell activity by interacting with specific receptors on the cell surface, serving as key molecular switches that regulate immune responses. For instance, they may bind to receptors such as Toll-like receptors (TLRs), which are critical for recognizing pathogen-associated molecular patterns (PAMPs) and initiating innate immune responses [37]. Upon binding, these peptides can trigger downstream signaling pathways, such as the NF-κB or MAPK pathways, leading to the production of proinflammatory cytokines and the activation of immune cells like macrophages and DCs. Additionally, functional peptides can interact with cytokine receptors, such as those for interleukins (ILs) or interferons (IFNs), to amplify or suppress cytokine-mediated signaling. This can influence the differentiation of T cells into specific subsets, such as Th1, Th2, or regulatory T cells (Tregs), thereby tailoring the immune response to the specific threat. Furthermore, peptides that bind to G protein-coupled receptors (GPCRs) can modulate intracellular signaling cascades involving cAMP, calcium, or other secondary messengers, affecting immune cell migration [38].

9.8.2 Modulation of Cytokine Networks

Peptides play a pivotal role in influencing the production and secretion of cytokines, which are critical mediators of immune responses. By promoting or inhibiting the release of specific cytokines, such as ILs, IFNs, or tumor necrosis factor (TNF), these peptides can modulate the balance between proinflammatory and anti-inflammatory responses. For example, certain peptides may enhance the secretion of IL-10, an anti-inflammatory cytokine, to suppress excessive inflammation and prevent tissue damage, while others may stimulate the production of IL-6 or TNF-α to bolster the immune defense against infections [39]. This precise regulation of cytokine networks allows functional peptides to fine-tune immune responses, ensuring an appropriate reaction to various challenges, such as infections, autoimmune diseases, and other immune-related conditions. By modulating cytokine levels, these peptides can also influence the activation, differentiation, and function of immune cells, such as T cells, B cells, and macrophages, thereby shaping the overall immune response to maintain homeostasis and promote effective immunity.

9.8.3 Regulation of Signal Transduction Pathways

Functional peptides play a critical role in modulating key signaling pathways that govern immune cell activation and function. These include the nuclear factor-kappa B (NF-κB) pathway, which regulates the expression of proinflammatory cytokines and immune cell survival; the mitogen-activated protein kinase (MAPK) pathway, which controls cell proliferation, differentiation, and stress responses; and the Janus kinase-signal transducer and activator of transcription (JAK–STAT) pathway, which mediates cytokine signaling and immune cell development. By targeting these pathways, functional peptides can influence immune cell survival, proliferation, differentiation, and effector functions, such as cytokine production, phagocytosis, and cytotoxicity. This regulation allows peptides to fine-tune immune responses, ensuring an appropriate balance between immune activation and suppression, and ultimately shaping the overall immune response to infections, inflammation, and other immune-related conditions.

9.8.4 Regulation of Gut Immune Homeostasis

Functional peptides play a crucial role in maintaining gut immune homeostasis by modulating the intestinal barrier function and influencing the composition of the gut microbiota. The GALT, a critical component of the immune system, is highly responsive to these peptides, which can enhance the integrity of the intestinal epithelial barrier by promoting the expression of tight junction proteins, such as occludin and zonula occludens-1 (ZO-1). This strengthening of the barrier prevents the translocation of harmful pathogens and toxins into the bloodstream, reducing systemic inflammation. Additionally, functional peptides can exert anti-inflammatory effects by suppressing the production of proinflammatory cytokines, such as TNF-α and IL-6, while promoting the secretion of anti-inflammatory cytokines like IL-10. Furthermore, these peptides can influence the gut microbiota by selectively promoting the growth of beneficial bacteria. This regulation of gut immune homeostasis can have systemic effects, influencing overall immune function and protecting against inflammatory and autoimmune diseases.

9.8.5 Direct Regulation of Immune Cells by Functional Peptides

9.8.5.1 Regulation of T Cells

T cells are the core cells of adaptive immunity, and their functional abnormalities are closely related to various immune-related diseases. Functional peptides play a crucial role in immune regulation by modulating the activation, differentiation, and function of T cells. Immunomodulatory peptides can regulate immune responses by influencing the differentiation and activation of T cells. For instance, certain peptides can promote the differentiation of Th1 or Th2 cells by binding to receptors on the T cell surface, thereby biasing the immune response [40]. For example, IFN-γ-inducing peptides can enhance Th1 cell activation, boosting cellular immune responses. Immunomodulatory peptides can directly regulate T cell activation and function by binding to specific receptors (e.g., T cell receptor [TCR], costimulatory molecules) on the T cell surface. For instance, a peptide that inhibits systemic lupus erythematosus (SLE)-associated T cell responses upregulates the negative regulators Foxj1 and Foxo3a, suppressing excessive T cell activation and alleviating autoimmune reactions. (The negative regulators Foxj1 and Foxo3a are upregulated by a peptide that inhibits SLE-associated T cell responses) [41]. Additionally, the VIP peptide indirectly modulates T cell function by inhibiting the proinflammatory polarization of macrophages, thereby reducing inflammatory responses (VIP impairs acquisition of the macrophage proinflammatory polarization profile) [41].

Some immunomodulatory peptides play a pivotal role in regulating T cell responses through various mechanisms, making them promising candidates for cancer immunotherapy and treatment of infectious diseases. One key function is their ability to mimic antigens by binding directly to the TCR, thereby activating specific T cell clones. Immunomodulatory peptides can also regulate T cell activation by interacting with costimulatory molecules such as CD28 and CTLA-4 on the T cell surface [42]. For example, certain peptides are designed to inhibit CTLA-4, a negative regulator of T cell activation, thereby enhancing T cell responses. For instance, certain peptides can inhibit CTLA-4 activity, enhancing T cell activation and thereby improving antitumor immune responses.

9.8.5.2 Regulation of B Cells

Immunomodulatory peptides can influence the proliferation and antibody secretion of B cells. For instance, certain peptides can promote antibody class switching by binding to receptors on the B cell surface, thereby regulating humoral immune responses. Through indirect regulation of T cells, immunomodulatory peptides can activate helper [43] T cells (Th cells), particularly Th2 cells, which secrete cytokines (e.g., IL-4, IL-5, and IL-6) that act on B cells to promote their proliferation and antibody production. Regulation of antigen presentation: immunomodulatory peptides can indirectly enhance B cell recognition and response to antigens by influencing the function of antigen-presenting cells (e.g., DCs). Modulation of the cytokine microenvironment: immunomodulatory peptides can regulate the balance of cytokines in the immune system, such as promoting the secretion of anti-inflammatory factors like IL-10 and TGF-β, inhibiting excessive immune responses, and promoting moderate activation of B cells. Activation of the complement system: immunomodulatory peptides can enhance B cell recognition and response to antigens by activating the complement system and forming immune complexes.

9.8.5.3 Regulation of Macrophage Polarization

Macrophages are critical effector cells in the immune system, and their functional states can be classified into proinflammatory M1 and anti-inflammatory M2 types depending on the microenvironment. The polarization state of macrophages (M1/M2) plays a key role in inflammatory responses. Functional peptides exert significant immunomodulatory effects by regulating macrophage polarization. For example, mussel-derived bioclickable peptide mimics promote M2 macrophage polarization, reducing the release of proinflammatory cytokines and creating a favorable immune microenvironment for bone tissue repair (Mussel-Derived and Bioclickable Peptide Mimic). Similarly, lactoferrin and its derived peptides inhibit M1 macrophage activation by modulating the NF-κB signaling pathway, thereby reducing inflammatory responses (the immunomodulatory effects of lactoferrin and its derived peptides on NF-κB signaling pathway).

9.8.5.4 Regulation of Natural Killer Cells

Immunomodulatory peptides can enhance the phagocytic and cytotoxic functions of NK cells by activating them. For example, AMPs can directly bind to receptors on the surface of macrophages, promoting their ability to phagocytose pathogens. The NF-κB is an essential intracellular transcription factor that plays a critical role in regulating immune and inflammatory responses. Abnormal activation of NF-κB is closely associated with the development of various immune-related diseases. Studies have shown that functional peptides can exert immunomodulatory effects by regulating the NF-κB signaling pathway. For instance, hydrolysates of Rapana venosa (a type of sea snail peptide) can restore intestinal immune homeostasis in mice with DSS-induced colitis by inhibiting the NF-κB pathway [44]. Similarly, lactoferrin and its derived peptides have also been demonstrated to modulate the NF-κB signaling pathway and exert immunoregulatory effects.

9.8.6 Modulation of Cytokine Networks

Immunomodulatory peptides play a significant role in shaping immune responses by regulating the production of cytokines, which are essential signaling molecules in the immune system. Cytokines facilitate communication between immune cells and determine the nature, intensity, and direction of immune responses. For instance, certain peptides can promote the secretion of proinflammatory cytokines like IL-2 and IFN-γ, enhancing T cell activation and antitumor immunity.

9.8.6.1 Regulation of Proinflammatory Cytokines

Certain immunomodulatory peptides can mitigate inflammatory responses by inhibiting the production of proinflammatory cytokines such as TNF-α, IL-1β, and IL-6. For instance, specific functional peptides can reduce the release of these cytokines by inhibiting the NF-κB signaling pathway. Sea snail peptide hydrolysate significantly lowers the levels of TNF-α and IL-1β in a DSS-induced colitis mouse model through the suppression of the NF-κB pathway [39]. Similarly, lactoferrin and its derived peptides inhibit the production of TNF-α and IL-1β by modulating the NF-κB signaling pathway [40]. IL-6 and IL-8 are pivotal proinflammatory cytokines involved in the pathogenesis of various inflammatory diseases. VIP notably decreases the secretion of IL-6 and IL-8 by inhibiting the polarization of macrophages toward the proinflammatory M1 phenotype [42]. Additionally, shrimp peptide hydrolysate restores immune function in a cyclophosphamide-induced immunosuppression mouse model by regulating the levels of IL-6 and IL-8 [45].

9.8.6.2 Regulation of Anti-Inflammatory Cytokines

Immunomodulatory peptides can also promote the production of anti-inflammatory cytokines such as IL-10 and TGF-β, thereby suppressing excessive immune responses and maintaining immune homeostasis. IL-10 is a critical anti-inflammatory cytokine that can inhibit excessive inflammatory reactions. Functional peptides enhance the production of IL-10 through various mechanisms. For example, *Lunasin* peptide promotes the secretion of IL-10 by modulating the function of immune cells in the human gastrointestinal tract [46]. Moreover, oyster peptide exerts immunoprotective effects in a cyclophosphamide-induced immunosuppression rat model by enhancing IL-10 levels [47]. TGF-β plays a vital role in immune tolerance and homeostasis. Certain functional peptides can inhibit excessive immune responses by promoting the production of TGF-β. For instance, *Rapana venosa* peptide hydrolysate restores intestinal immune homeostasis by regulating the NF-κB pathway and promoting the TGF-β expression [22].

9.8.6.3 Modulation of Cytokine Network Balance by Functional Peptides

The balance of Th1/Th2 cytokines is crucial in immune responses. Functional peptides maintain immune homeostasis by modulating the production of Th1 and Th2 cytokines.

For example, VIP regulates the Th1/Th2 balance by inhibiting the production of Th1 cytokines (e.g., IFN-γ) and promoting the secretion of Th2 cytokines (e.g., IL-4) [42]. The balance of Treg/Th17 cytokines is significant in autoimmune diseases. Functional peptides suppress excessive immune responses by modulating the production of Treg and Th17 cytokines. For instance, certain peptides modulate the Treg/Th17 balance by upregulating Foxj1 and Foxo3a, thereby inhibiting T cell responses associated with SLE [48].

9.8.7 Regulation of Signaling Pathways by Functional Peptides

9.8.7.1 NF-κB Signaling Pathway

The NF-κB signaling pathway is a pivotal regulator of inflammatory responses, playing a critical role in immune homeostasis. Sea conch (*Rapana venosa*) hydrolysate has been shown to restore intestinal immune balance in DSS-induced colitis mouse models by effectively inhibiting the activation of the NF-κB pathway, thereby reducing inflammation and promoting tissue repair [22]. Similarly, lactoferrin and its derived peptides demonstrate significant anti-inflammatory and immunomodulatory properties by targeting the NF-κB pathway [49]. These peptides inhibit NF-κB activation, reducing the production of proinflammatory cytokines and alleviating inflammation. This mechanism highlights their potential in managing inflammatory conditions and enhancing immune regulation [50].

9.8.7.2 MAPK and JAK–STAT Pathways

Certain functional peptides play a crucial role in modulating immune cell proliferation and differentiation by regulating key signaling pathways such as MAPK and JAK–STAT. These pathways are essential for transmitting extracellular signals into the nucleus, thereby influencing gene expression and cellular responses. For instance, shrimp peptide hydrolysate has been shown to restore immune function in cyclophosphamide-induced immunosuppression mouse models by activating the MAPK and JAK–STAT pathway [45].

9.8.8 Role of Functional Peptides in Intestinal Immune Regulation

The intestinal immune system is a critical component of systemic immunity. Functional peptides affect intestinal immune homeostasis by modulating intestinal barrier function and the gut microbiota. For instance, *Lunasin* peptide exerts immunomodulatory effects in the human gastrointestinal tract by regulating the responses of intestinal immune cells [46]. Additionally, sea conch peptide hydrolysate alleviates inflammatory responses in a mouse colitis model by inhibiting the NF-κB pathway [22].

These immunomodulatory peptides achieve multidimensional regulation of the immune system through complex signaling networks, including the MAPK, NF-κB, and JAK–STAT pathways. They not only play a vital role in maintaining immune balance but also demonstrate broad application prospects in the immunotherapy of various diseases, such as autoimmune disorders, infectious diseases, and cancer. A deeper understanding of the mechanisms of action of immunomodulatory peptides will provide a critical theoretical foundation for the development of novel immunotherapies.

9.9 Applications

Immunomodulatory peptides have potential applications in the treatment of various diseases, primarily including the following aspects:

Autoimmune Diseases: Immunomodulatory peptides can alleviate the symptoms of autoimmune diseases by suppressing excessive immune responses. For instance, certain peptides can mitigate conditions such as rheumatoid arthritis and multiple sclerosis by inducing Treg cells or inhibiting the differentiation of Th17 cells. For example, some peptides can upregulate Foxj1 and Foxo3a, thereby inhibiting T cell responses associated with SLE [48].

Tumor Immunotherapy: Immunomodulatory peptides can enhance the body's antitumor immune responses, promoting tumor cell apoptosis or inhibiting their proliferation [38]. For example, certain peptides can serve as adjuvants in tumor vaccines, enhancing T cell recognition and attack against tumor antigens.

Allergic Diseases: Immunomodulatory peptides can alleviate allergic reactions by regulating the Th1/Th2 balance. For instance, some peptides can reduce the production of IgE, thereby mitigating conditions such as asthma and allergic rhinitis.

9.9.1 Application of Functional Peptides in Pulp Therapy

9.9.1.1 Synergistic Antibacterial Effects and Dentin Regeneration

Pulp therapy is an essential area of research in dentistry, aiming to preserve and regenerate dental pulp tissue in response to injury or infection. Functional peptides, known for their antibacterial and immunomodulatory properties, have shown great promise in this field. Studies indicate that AMPs, when combined with functionalized dentin matrix hydrogels, can enhance pulp regeneration through a synergistic approach that targets both infection control and tissue repair. The AMPs help inhibit the growth of pathogenic bacteria that often compromise pulp health, while the functionalized hydrogels provide a scaffold that supports dentin formation and cellular migration [51]. Beyond their antibacterial action, these peptides also play a key role in modulating local immune responses, promoting an environment conducive to tissue healing and regeneration. By balancing immune activation and tissue repair, this combined strategy not only accelerates pulp regeneration but also improves the long-term outcomes of pulp therapy. Overall, the incorporation of functional peptides into pulp therapy offers a promising approach to enhancing the regenerative potential of dental pulp and improving treatment efficacy.

9.9.1.2 Immunomodulatory Mechanisms

Functional peptides play a crucial role in modulating the immune response within the dental pulp. By regulating the activity of immune cells such as macrophages and T cells, these peptides help mitigate inflammatory responses, thus preventing excessive tissue damage. This creates a more favorable microenvironment for pulp repair [41]. Additionally, functional peptides promote the differentiation of dental pulp stem cells, enhancing their regenerative potential. They also stimulate the secretion of extracellular matrix components, which are essential for the reconstruction of pulp tissue. Overall, functional peptides contribute significantly to accelerating pulp tissue regeneration and improving healing outcomes in dental therapies.

9.9.2 Application of Functional Peptides in Tissue Engineering

9.9.2.1 Serum-Free Expansion of Artificial Mesenchymal Stem Cells

Tissue engineering is a vital branch of regenerative medicine that focuses on restoring or replacing damaged tissues using biological materials and advanced cell culture techniques. Functional peptides play a crucial role in this field by supporting cell growth, differentiation, and immune regulation [43]. A novel synthetic Xeno-free biomimetic surface material has recently been developed to expand human mesenchymal stem cells (hMSCs) without relying on serum, addressing safety and ethical concerns associated with animal-derived components [52]. This innovative material optimizes the cell culture environment by leveraging the immunomodulatory properties of functional peptides, which help to regulate inflammatory responses and enhance cell viability. By creating a more biocompatible and controlled microenvironment, this biomimetic surface promotes the efficient expansion of hMSCs while maintaining their regenerative potential. This advancement provides a safe and scalable cell source for tissue engineering applications, paving the way for improved therapies in regenerative medicine, including bone and cartilage repair, wound healing, and organ regeneration. The integration of functional peptides into biomaterials marks a significant step forward in developing safer and more effective strategies for tissue engineering and regenerative therapies.

9.9.2.2 Tissue Repair and Regeneration

Functional peptides can significantly accelerate tissue repair by modulating local immune responses and promoting cell proliferation. One of their key actions is reducing tissue damage and enhancing regeneration by inhibiting the expression of inflammatory factors. For instance, certain functional peptides target pathways like NF-κB, a central regulator of inflammation, thereby reducing the production of proinflammatory cytokines. By suppressing these inflammatory signals, functional peptides create a more balanced immune response, reducing excessive inflammation and promoting a conducive environment for tissue healing. This ability to regulate inflammation makes functional peptides valuable in accelerating the repair and regeneration of damaged tissues [52].

9.9.3 Application of Functional Peptides in Immune Enhancement

9.9.3.1 Peptide Nanofibers as Adjuvants

Biotin-functionalized self-assembling peptide nanofibers have emerged as a novel class of immune adjuvants with significant potential in enhancing immune responses. These peptide nanofibers possess unique properties that enable them to activate key immune cells, such as DCs and T cells. By interacting with these cells, the nanofibers enhance antigen presentation, thereby boosting both the innate and adaptive immune responses [41]. Furthermore, the nanostructure of the peptide fibers is highly effective at embedding antigen molecules, which not only stabilizes the antigens but also increases their immunogenicity. This increased stability ensures that the antigens remain effective for longer periods, allowing for stronger and more sustained immune activation. As a result, biotin-functionalized peptide nanofibers can significantly improve the efficacy of vaccines, making them a promising candidate for developing more effective vaccine formulations.

Their ability to enhance antigen presentation and immune response paves the way for advanced vaccine technologies and immunotherapies [41].

9.9.3.2 Effects of Peptide Hydrolysates on Immune Function

The peptide hydrolysate of Rapana venosa (sea snail) has shown promising effects in restoring intestinal immune homeostasis, particularly in inflammatory conditions. Research indicates that this peptide hydrolysate modulates the NF-κB signaling pathway, a key regulator of inflammation, by inhibiting the release of inflammatory factors. This action helps restore intestinal barrier integrity, thereby improving gut health and enhancing immune function in mice with DSS-induced colitis [22]. Additionally, shrimp peptide hydrolysate has demonstrated significant immune-boosting effects in cyclophosphamide-induced immunosuppressed mouse models. It achieves this by regulating the proportion and function of various immune cells, leading to improved immune responses. These findings suggest that both Rapana venosa and shrimp peptide hydrolysates are potential natural agents for supporting immune health and combating inflammation-related disorders [45].

9.9.4 Application of Functional Peptides in Immune Suppression

9.9.4.1 Peptides in Systemic Lupus Erythematosus

Foxj1 and Foxo3a, as negative regulatory factors, play crucial roles in controlling T cell responses, particularly in the context of SLE. These transcription factors are involved in regulating immune cell function, preventing excessive immune activation. Studies have shown that certain peptides can upregulate the expression of Foxj1 and Foxo3a, thereby enhancing their regulatory effects. By increasing the levels of these factors, the peptides help inhibit the abnormal activation of T cells that are commonly observed in SLE, which in turn reduces the hyperactive immune responses. This regulatory effect can alleviate the immunopathological manifestations of SLE, providing a potential therapeutic approach for managing the disease. These peptides offer a promising strategy for modulating the immune system and improving immune tolerance in autoimmune conditions like SLE [48].

9.9.4.2 Immunosuppressive Effects of VIP Peptide

Vasoactive intestinal peptide has demonstrated significant immunomodulatory effects, particularly in inhibiting the proinflammatory polarization of macrophages. Macrophages, as key players in the immune system, can undergo polarization into proinflammatory (M1) or anti-inflammatory (M2) phenotypes. VIP plays a crucial role in regulating this polarization process, shifting macrophages away from the M1 proinflammatory state toward the M2 anti-inflammatory phenotype [42]. By doing so, VIP suppresses the release of inflammatory cytokines and other immune mediators that contribute to tissue damage and chronic inflammation. Research has shown that VIP can significantly reduce inflammation in various experimental models, demonstrating its potential as a therapeutic agent for managing inflammatory diseases. Its ability to modulate macrophage function and mitigate excessive immune responses makes VIP a promising candidate for the treatment of conditions such as autoimmune disorders, inflammatory bowel disease, and other chronic inflammatory diseases, offering a novel approach to immune regulation and inflammation control.

References

1 Weaver, D.F. (2024). *Endogenous antimicrobial-immunomodulatory molecules: networking biomolecules of innate immunity. ChemBioChem* 25 (12): e202400089.

2 Song, Y. et al. (2025). *Immunomodulatory peptides for tumor treatment. Advanced Healthcare Materials* 14 (5).

3 Cicero, A.F.G., Fogacci, F., and Colletti, A. (2017). *Potential role of bioactive peptides in prevention and treatment of chronic diseases: a narrative review. British Journal of Pharmacology* 174 (11): 1378–1394.

4 Santiago-López, L., Hernández-Mendoza, A., Vallejo-Cordoba, B. et al. (2016). *Food-derived immunomodulatory peptides. Journal of the Science of Food and Agriculture* 96 (11): 3631–3641.

5 Bommineni, Y.R., Dai, H., Gong, Y.X. et al. (2006). *Fowlicidin-3 is an α-helical cationic host defense peptide with potent antibacterial and lipopolysaccharide-neutralizing activities. The FEBS Journal* 274 (2): 418–428.

6 De Vry, C.G., Valdez, M., Lazarov, M. et al. (2005). *Topical application of a novel immunomodulatory peptide, RDP58, reduces skin inflammation in the Phorbol Ester-induced dermatitis model. Journal of Investigative Dermatology* 125 (3): 473–481.

7 Jafari, Z., Sadeghi, S., Dehaghi, M.M. et al. (2024). *Immunomodulatory activities and biomedical applications of melittin and its recent advances. Archiv der Pharmazie* 357 (4): e2300569.

8 Alalwani, S.M., Sierigk, J., Herr, C. et al. (2010). *The antimicrobial peptide LL-37 modulates the inflammatory and host defense response of human neutrophils. European Journal of Immunology* 40 (4): 1118–1126.

9 Bowdish, D.M.E., Davidson, D.J., Lau, Y.E. et al. (2005). *Impact of LL-37 on anti-infective immunity. Journal of Leukocyte Biology* 77 (4): 451–459.

10 Waschek, J.A. (2013). *VIP and PACAP: neuropeptide modulators of CNS inflammation, injury, and repair. British Journal of Pharmacology* 169 (3): 512–523.

11 Aung, G., Niyonsaba, F., Ushio, H. et al. (2011). *Catestatin, a neuroendocrine antimicrobial peptide, induces human mast cell migration, degranulation and production of cytokines and chemokines. Immunology* 132 (4): 527–539.

12 Aziz, M., Jacob, A., Yang, W.L. et al. (2013). *Current trends in inflammatory and immunomodulatory mediators in sepsis. Journal of Leukocyte Biology* 93 (3): 329–342.

13 Asif, A., Afzaal, M., Shahid, H. et al. (2023). *Probing the functional and therapeutic properties of postbiotics in relation to their industrial application. Food Science & Nutrition* 11 (8): 4472–4484.

14 Reyes-Díaz, A., González-Córdova, A.F., Hernández-Mendoza, A. et al. (2018). *Immunomodulation by hydrolysates and peptides derived from milk proteins. International Journal of Dairy Technology* 71 (1): 1–9.

15 Pogorevc, D. and Müller, R. (2022). *Biotechnological production optimization of argyrins – a potent immunomodulatory natural product class. Microbial Biotechnology* 15 (1): 353–369.

16 Nguyen, K.N.T., Nguyen, G.K.T., Nguyen, P.Q.T. et al. (2016). *Immunostimulating and gram-negative-specific antibacterial cyclotides from the butterfly pea (Clitoria ternatea). The FEBS Journal* 283 (11): 2067–2090.

17 Zhang, J., Gao, S., Li, H. et al. (2021). *Immunomodulatory effects of selenium-enriched peptides from soybean in cyclophosphamide-induced immunosuppressed mice. Food Science & Nutrition* 9 (11): 6322–6334.

18 Chen, M., Zhang, F., Su, Y. et al. (2022). *Identification and immunomodulatory effect on immunosuppressed mice of selenium-enriched peptides of egg white. Journal of Agricultural and Food Chemistry* 70 (39): 12663–12671.

19 Sun, L., Zhang, S., Yuan, L. et al. (2020). *Glutamine and methionine targeted pulsed electric field treatment for enhanced immune activity in pine nut Gln-Trp-Phe-met peptides. International Journal of Food Science & Technology* 55 (8): 2954–2961.

20 Lin, H., Li, W., Sun, R. et al. (2023). *Purification and characterization of a novel immunoregulatory peptide from Sipunculus nudus L. protein. Food Science & Nutrition* 11 (12): 7779–7790.

21 Fajardo-Espinoza, F.S., Ordaz-Pichardo, C., Sankar, U. et al. (2021). In vitro *cytomodulatory and immunomodulatory effects of bovine colostrum whey protein hydrolysates. International Journal of Food Science & Technology* 56 (5): 2109–2121.

22 Ullah, H., Alioui, Y., Ali, M. et al. (2024). *Sea conch (Rapana venosa) peptide hydrolysate regulates NF-κB pathway and restores intestinal immune homeostasis in DSS-induced colitis mice. Food Science & Nutrition* 12 (12): 10070–10086.

23 Wang, H., Jiao, D., Feng, D. et al. (2024). *Transformable supramolecular self-assembled peptides for Cascade self-enhanced Ferroptosis primed Cancer immunotherapy. Advanced Materials* 36 (21).

24 Pane, K., Sgambati, V., Zanfardino, A. et al. (2016). *A new cryptic cationic antimicrobial peptide from human apolipoprotein E with antibacterial activity and immunomodulatory effects on human cells. The FEBS Journal* 283 (11): 2115–2131.

25 An, J., Feng, Y., Zheng, J. et al. (2020). *The immune-enhancing potential of peptide fractions from fermented* Spirulina platensis *by mixed probiotics. Journal of Food Biochemistry* 44 (7): e13245.

26 Kozutsumi, D., Tsunematsu, M., Yamaji, T. et al. (2006). *PS80 interferes with the antiallergic effect of cry-consensus peptide, a novel recombinant peptide for immunotherapy of Japanese cedar pollinosis, at very low concentration through modulation of Th1/Th2 balance. Immunology* 118 (3): 392–401.

27 Zhang, G., Guo, F., Zeng, M. et al. (2022). *The immune-enhancing effect and* in vitro *antioxidant ability of different fractions separated from Colla corii asini. Journal of Food Biochemistry* 46 (11): e14174.

28 Cao, H., Cao, J., Zhang, Y. et al. (2019). *Continuous preparation and characterization of immunomodulatory peptides from type II collagen by a novel immobilized enzyme membrane reactor with improved performance. Journal of Food Biochemistry* 43 (7): e12862.

29 Liu, H., Wang, J., Liu, Y. et al. (2024). *Characterisation of functional pea protein hydrolysates and their immunomodulatory activity. International Journal of Food Science & Technology* 59 (5): 3317–3330.

30 Lammers, K.M., Khandelwal, S., Chaudhry, F. et al. (2011). *Identification of a novel immunomodulatory gliadin peptide that causes interleukin-8 release in a chemokine receptor CXCR3-dependent manner only in patients with coeliac disease. Immunology* 132 (3): 432–440.

31 Borges, M., da Silva, A.C., Sereno, D., and Ouaissi, A. (2003). *Peptide-based analysis of the amino acid sequence important to the immunoregulatory function of Trypanosoma cruzi Tc52 virulence factor. Immunology* 109 (1): 147–155.

32 Ren, D., Wang, M., Shen, M. et al. (2016). In vivo *assessment of immunomodulatory activity of hydrolysed peptides from Corylus heterophylla Fisch. Journal of the Science of Food and Agriculture* 96 (10): 3508–3514.

33 Armiento, V., Hille, K., Naltsas, D. et al. (2020). *The human host-defense peptide cathelicidin LL-37 is a Nanomolar inhibitor of amyloid self-assembly of islet amyloid polypeptide (IAPP). Angewandte Chemie International Edition* 59 (31): 12837–12841.

34 Takeda, S., Harauma, A., Okamoto, M. et al. (2020). *Effects of whey protein hydrolysate on growth promotion and immunomodulation in mouse pups in artificial rearing system. Animal Science Journal* 91 (1): e13395.

35 Xie, Y., Li, M., and Zhang, S. (2022). *Identification of peptides from protease-fermented milk protein and immunomodulatory effect* in vivo *against lipopolysaccharide-induced inflammation. International Journal of Food Science & Technology* 57 (10): 6503–6511.

36 Haney, E.F. and Hancock, R.E.W. (2013). *Peptide design for antimicrobial and immunomodulatory applications. Peptide Science* 100 (6): 572–583.

37 Ye, X., Wang, J., Xu, P. et al. (2023). Peptide MSI-1 inhibited MCR-1 and regulated outer membrane vesicles to combat immune evasion of Escherichia coli. *Microbial Biotechnology* 16 (9): 1755–1773.

38 Wu, Y., Li, Y., Hu, Z. et al. (2024). Extracellular matrix-trapped bioinspired lipoprotein prolongs tumor retention to potentiate antitumor immunity. *Advanced materials (Weinheim)* 36 (15): e2310982.

39 Nairz, M., Schroll, A., Haschka, D. et al. (2015). Lipocalin-2 ensures host defense against Salmonella typhimurium by controlling macrophage iron homeostasis and immune response. *European Journal of Immunology* 45 (11): 3073–3086.

40 Prakash, P.K., Eligar, S.M., Prakruthi, M., and Jyothi Lakshmi, A.S. (2022). Comparative assessment of antioxidant and immunomodulatory properties of skimmed milk protein hydrolysates and their incorporation in beverage mix. *Journal of the Science of Food and Agriculture* 102 (14): 6414–6422.

41 Demircan, M.B., Tohumeken, S., Gunduz, N. et al. (2020). Biotin functionalized self-assembled peptide nanofiber as an adjuvant for immunomodulatory response. *Biotechnology Journal* 15 (12): e2000100.

42 Carrion, M., Perez-Garcia, S., Martinez, C. et al. (2016). VIP impairs acquisition of the macrophage proinflammatory polarization profile. *Journal of Leukocyte Biology* 100 (6): 1385–1393.

43 Zhou, W., Liu, Y., Dong, J. et al. (2024). Mussel-derived and bioclickable peptide mimic for enhanced interfacial Osseointegration via synergistic immunomodulation and vascularized bone regeneration. *Advance Science (Weinh)* 11 (32): e2401833.

44 Fontes, L.B., Dos, S.D.D., de Carvalho, L.S. et al. (2014). Immunomodulatory effects of licochalcone a on experimental autoimmune encephalomyelitis. *The Journal of Pharmacy and Pharmacology* 66 (6): 886–894.

45 Khan, A.I., Rehman, A.U., Farooqui, N.A. et al. (2022). Shrimp peptide hydrolysate modulates the immune response in cyclophosphamide immunosuppressed mice model. *Journal of Food Biochemistry* 46 (9): e14251.

46 Fernandez-Tome, S., Indiano-Romacho, P., Mora-Gutierrez, I. et al. (2021). Lunasin peptide is a modulator of the immune response in the human gastrointestinal tract. *Molecular Nutrition & Food Research* 65 (12): e2001034.

47 Dai, Y., Ren, Z., Li, P. et al. (2024). Selective adsorption of volatile compounds of oyster peptides by V-type starch for effective deodorization. *Food Hydrocolloids* 147: 109295.

48 Sela, U., Dayan, M., Hershkoviz, R. et al. (2006). The negative regulators Foxj1 and Foxo3a are up-regulated by a peptide that inhibits systemic lupus erythematosus-associated T cell responses. *European Journal of Immunology* 36 (11): 2971–2980.

49 Yami, H.A., Tahmoorespur, M., Javadmanesh, A. et al. (2023). The immunomodulatory effects of lactoferrin and its derived peptides on NF-κB signaling pathway: a systematic review and meta-analysis. *Immunity, Inflammation and Disease* 11 (8): e972.

50 Li, T., Chen, X., Huang, Z. et al. (2019). Pectin oligosaccharide from hawthorn fruit ameliorates hepatic inflammation via NF-κB inactivation in high-fat diet fed mice. *Journal of Functional Foods* 57: 345–350.

51 Xie, Z., Jiang, W., Liu, H. et al. (2024). Antimicrobial peptide- and dentin matrix-functionalized hydrogel for vital pulp therapy via synergistic bacteriostasis, immunomodulation, and dentinogenesis. *Advanced Healthcare Materials* 13 (18): e2303709.

52 Thamm, K., Mobus, K., Towers, R. et al. (2020). A novel synthetic, Xeno-free biomimetic surface for serum-free expansion of human mesenchymal stromal cells. *Advanced Biosystems* 4 (8): e2000008.

10

Flavor Peptides

10.1 Introduction

Salty and umami are two of the essential taste modalities in humans, predominantly elicited by sodium chloride (NaCl) and monosodium glutamate (MSG), respectively. NaCl plays a crucial role in maintaining physiological balance and has been incorporated into human diets for thousands of years. In the modern food industry, its widespread application is attributed to its affordability and versatility. It serves multiple functions, including as a flavor enhancer, texture modifier, and preservative, significantly elevating the commercial value of raw ingredients. According to Zhu et al. [1], varying NaCl concentrations influence the microenvironment of myofibrillar proteins, leading to modifications in the texture and quality of thermally processed meats. This study proposed a novel strategy combining salt concentration adjustment with temperature control to improve food quality. In another example, Li and Zhong et al. [2] found that regulating pH and adding NaCl during frozen storage of egg yolks successfully inhibited yolk gelation by dissociating granules, thereby preserving the functional properties of the yolk. Similarly, Chen and Yang et al. [3] further demonstrated that the antimicrobial performance of D-tryptophan against *Pseudomonas fluorescens* and *Shewanella baltica* was significantly modulated by NaCl levels, with D-tryptophan addition markedly prolonging the shelf life of refrigerated seafood.

MSG, the sodium salt derived from L-glutamic acid, was originally discovered in wheat gluten in 1866 and subsequently linked to the umami taste in 1908 [4]. Beyond contributing its own umami character, MSG can enhance overall flavor by interacting synergistically with NaCl to suppress undesirable taste attributes such as off-flavors [5]. A study by Panzarini, Menis-Henrique, and Conti-Silva [6] indicated that increasing MSG percentages could improve the salty flavor in snacks, especially when formulated at an optimal 1.2% NaCl to 0.6% MSG ratio, delivering a strong umami profile and enhanced sensory properties. These findings underscore the dependency of the food industry on NaCl, MSG, and a wide range of seasonings to develop products that meet consumer taste preferences.

Nevertheless, excessive reliance on sodium-based flavorings has raised public health concerns, as elevated sodium ion (Na^+) intake is strongly associated with hypertension, a major risk factor for cardiovascular diseases [7]. The World Health Organization currently

Food-Derived Bioactive Peptides: Green Preparation, Activity Mechanisms, and Health Applications, First Edition. Shao-Yun Wang, Xu Chen, and Xixi Cai.

recommends limiting daily NaCl intake to under 5 g; however, average consumption often far exceeds this benchmark, typically doubling the suggested value [8]. Despite efforts by individuals to reduce their sodium intake, it remains difficult to avoid high levels of sodium due to its pervasive use in various condiments [9]. As a result, finding effective strategies to lower Na^+ in foods—either through reducing sodium salt usage or by employing substitutes—has become critical. One approach involves physical methods for salt reduction, which aim to enhance the solubility and delivery of salt to the palate via techniques such as ultrasound, high-pressure processing, or spray drying. Pietrasik, Gaudette, and Johnston [10] demonstrated that replacing a portion of NaCl with potassium chloride (KCl), coupled with high-pressure processing, significantly extended the shelf life of restructured ham without sensory deterioration. Additionally, Chen et al. [11] developed a method using spray drying to create hollow salt particles made from *quillaja* saponin, which were less than 10 μm in size and served as a vehicle for new encapsulated flavor compounds to reduce salt use. To mitigate the bitterness caused by high KCl concentrations, Lu and Hu et al. [12] introduced a bitterness-masking strategy that employs kappa-carrageenan, a natural polysaccharide that binds to potassium ions (K^+), reducing the unpleasant bitter aftertaste. This represents an innovative salt-reduction technique that could be widely applied in the food industry.

An alternative approach involves substituting Na^+ with other ionic or non-ionic compounds, typically referred to as non-sodium salt substitutes. These often include metal chlorides such as KCl, calcium chloride ($CaCl_2$), and magnesium chloride ($MgCl_2$), which share physicochemical similarities with NaCl and can partially fulfill its sensory and preservative roles. Vidal et al. [13] assessed the substitution of 50% NaCl with various combinations of KCl and $CaCl_2$, demonstrating substantial sodium reduction without significantly compromising the perceived saltiness. However, calcium chloride was associated with bitterness and rancid aromas, highlighting a drawback of calcium salts, such as astringency [14]. Non-metallic salt substitutes include compounds like nucleotides and peptides. Compounds such as inosine monophosphate (IMP) and guanosine monophosphate (GMP) are known for enhancing umami intensity, particularly when used in conjunction with MSG or NaCl [15]. Flavor peptides, classified into categories such as salty, umami, bitter, and sour, along with kokumi peptides, have also emerged as potential solutions for reducing sodium intake from seasonings. Among these, salty, umami, and kokumi peptides show promise in decreasing the need for NaCl and MSG [16].

Salty peptides, in particular, deliver a salty taste without the presence of Na^+. In the 1970s, researchers synthesized the salty peptides, including H-L-Orn-β-Ala-OH-HCl and H-L-Orn-Tau-HCl, which were found to generate a salt-like taste that rivaled or exceeded that of NaCl [17]. Peptides that enhance salty taste, when combined with NaCl, can also greatly amplify the perception of saltiness. Umami peptides such as Glu-Asp, Glu-Thr, Glu-Ser, and Glu-Glu, identified in 1973, impart umami characteristics but exhibit lower intensity than MSG [18]. Kokumi peptides are noted for their ability to provide sensations of richness, depth, and fullness [19]. These peptides, when interacting with salts such as NaCl, KCl, or $CaCl_2$, can enhance both salty and umami perceptions [20]. As interest in natural foods grows, peptides with sodium-reducing effects have gained traction, with many being derived from natural protein sources [8], as summarized in Table 10.1. The sensory properties of these peptides are closely related to their amino acid composition,

Table 10.1 Overview of the food-derived sources and corresponding taste characteristics of bioactive flavor peptides.

Peptide sequence	Source	Type	Matrix	References
AF IF	Salt-free soy sauce	Salt peptides	—	Zhu et al. [21]
DDD FI SPE ED DD	Yeast extract	Salt peptides	Water/beef broth/ vegetable broth	Zheng et al. [22]
EDEGEQPRPF	Chinese commercial fermented soybean curds	Salt peptide/Salty taste-enhancing peptides	Water	Chen et al. [23]
RM/MR RV/VR RQ/QR RD/DR RK/KR RP/PR RA/AR RS/SR RG/GR RR	Lysozyme	Salty taste-enhancing peptides	A cheese taste matrix prepared by mixing all key taste compounds identified in a Gouda cheese	Harth et al. [24]
MR/RM RR RG/GR RP VR/RV RQ AR/RA RD RI/IR	Protein digests	Salty taste-enhancing peptides	Water/model broth	Schindler et al. [68]
ALDPLPT GR DTDEPN VP LDYY RP LPPP PV EEL SP PPP PS RG	Soy sauce	Salty taste-enhancing peptides	Partially reconstituted solution of soy sauce	Jünger et al. [25]
RG/GR TR/RT AR/RA RI ER/RE RP RS/SR	Fermented fish sauces	Salty taste-enhancing peptides	Water/model broth	Schindler et al. [26]
RGENESEEEGAIVT TESSSE EDG DQR NNP EGF	Peanut protein isolate hydrolysate	Umami peptides	Water	Zhang et al. [27]
SGCVNEL SEPSHF ESCAPQL PHEMQ EPLCNQ	*Stropharia rugoso-annulata* mushroom	Umami peptides	Water	Chen et al. [28]
HHYE DGF KCGQ	*Boletus edulis*	Umami peptides	Water	Song et al. [29]

(*Continued*)

Table 10.1 (Continued)

Peptide sequence		Source	Type	Matrix	References
HGEDKEGE SEASNNK		Hydrolysates of Sanhuang chicken	Umami peptides	Water	Chen et al. [30]
AGDDAPR IGPGLGR			Umami taste-enhancing peptides	Water containing 0.35% (w/v) MSG	
DGV RD	DR DK	Dry-cured Spanish mackerels (*Scomberomorus niphonius*)	Umami peptides/ Umami taste-enhancing peptides	Water/MSG solution	Wang et al. [31]
FN			Umami taste-enhancing peptides	MSG solution	
γ-Glu-Leu γ-Glu-Gln γ-Glu-Glu	γ-Glu-Met γ-Glu-Gly γ-Glu-His	Soy sauces	Kokumi peptides	Water	Jünger et al. [25]

sequence, and three-dimensional conformation. For example, glutamate-containing peptides with γ-spiral structure are especially prone to inducing kokumi sensations [32]. Understanding the mechanisms by which these peptides are perceived in the oral cavity is crucial for their development and application in the food industry.

The molecular basis of salty taste remains complex, with ongoing investigations into the receptors and signaling mechanisms involved. Epithelial sodium ion channels (ENaCs), primarily found in amiloride-sensitive taste bud cells, are the primary sodium taste receptors. Emerging evidence suggests that TMC4 may function as an anion channel linked to high-salt taste perception [33]. Continued research into salty taste receptors could pave the way for better use of salty peptides in food products.

Umami perception, on the other hand, is mediated by G-protein coupled receptors, including type 1 member 1 (T1R1)/type 1 member 3 (T1R3) and metabotropic glutamate receptors 1 and 4 (mGluR1/4) [34, 35]. Kokumi sensation is primarily related to calcium-sensing receptors (CaSRs), which are key in detecting kokumi peptides that enhance both salty and umami flavors. Advances in computational tools, such as molecular docking, have facilitated the identification of a wide array of peptides with sodium-reducing potential, many of which also exhibit bioactivity relevant to human health. Furthermore, since these peptides often interact with taste receptors that also regulate biological functions, many flavor peptides possess bioactive properties, allowing them to be applied in functional foods with promising future prospects.

The scopes of this chapter are to: (i) provide an in-depth overview of current techniques for producing sodium-reducing flavor peptides; (ii) elucidate the role of amino acid sequences and structures in determining flavor perception; and (iii) evaluate the practical applications and challenges of integrating such peptides into food products designed to reduce sodium content.

10.2 Preparation of Flavor Peptides

Flavor peptides are derived from diverse biological sources, which are generally classified according to origin into animal-based, plant-based, and microbial-based categories. Those derived from animals are mainly isolated from protein-rich tissues such as meats, poultry, fish, and their processing by-products. Plant-sourced flavor peptides frequently originate from legumes, whereas microbial-origin peptides are commonly obtained from edible fungi and yeast strains. The primary strategies for producing these peptides include hydrolysis, microbial fermentation, direct extraction, and chemical or biological synthesis, as summarized in Figure 10.1.

10.2.1 Methods for Hydrolysis and Microbial Fermentation

Hydrolysis breaks down the protein structure, reducing its molecular weight and increasing the concentration of free amino acids. Hydrolysis, primarily carried out through chemical or enzymatic means, remains the most frequently employed method for flavor peptide preparation.

Hydrolysis facilitates the disruption of protein structures, leading to reduced molecular weight and elevated levels of free amino acids. This process is predominantly achieved through chemical or enzymatic means and remains the most frequently employed method for flavor peptide preparation. Chemical hydrolysis, which typically involves the use of acids and alkalis, is seldom applied in the food industry due to its lack of control and potentially damaging effects. In contrast, enzymatic hydrolysis is favored for its mild reaction conditions, safety, and cost-effectiveness [36]. Gao et al. [37] explored the use of four different proteases to hydrolyze cheese proteins and noted a significant enhancement in flavor due to protease treatment. However, hydrolysis can sometimes result in undesirable odors that negatively affect the sensory quality of products. The application of ultra-high

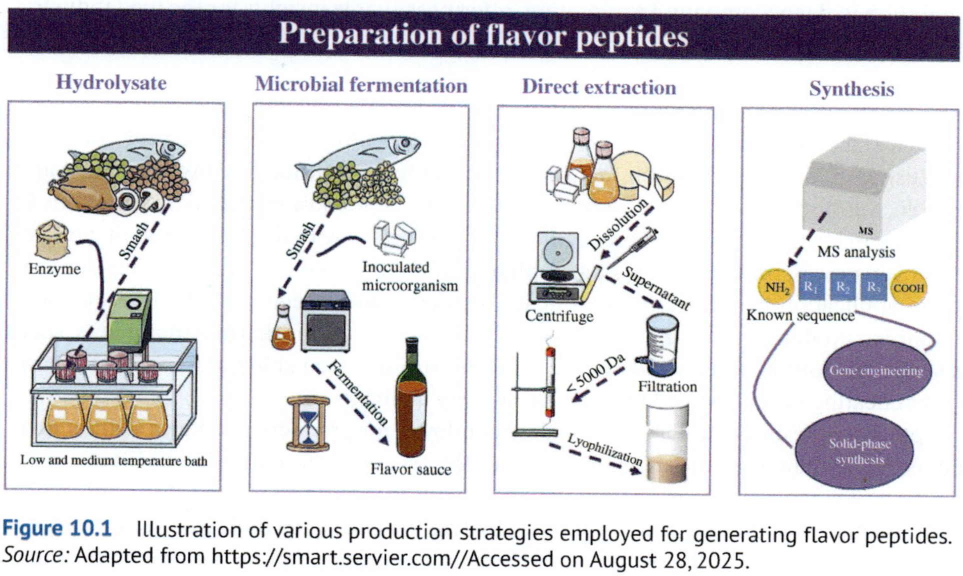

Figure 10.1 Illustration of various production strategies employed for generating flavor peptides. *Source:* Adapted from https://smart.servier.com//Accessed on August 28, 2025.

pressure (UHP) has been evidenced to enhance the sensory characteristics of hydrolysates via altering protein conformation and exposing more peptides, thereby boosting enzymatic efficiency. Zhu et al. [36] documented that UHP-assisted hydrolysis considerably increased the diversity of volatile flavor molecules. Additionally, combining microwave treatment with ultrasonic pretreatment has been demonstrated to improve hydrolysate quality and flavor [38]. Given the central role of proteases in defining hydrolysate composition and characteristics, it is crucial to carefully control the processing parameters to achieve the desired flavor and properties [39].

Microbial fermentation utilizes the metabolic capabilities of microorganisms that produce complex enzyme systems tailored to proteolysis and release distinctive flavor peptides. Commonly utilized microbes in food fermentation include lactic acid bacteria, yeast, and mold, all contributing to the generation of free amino acids and peptides in various food matrices [40]. For instance, Chen and Gao et al. [41] evaluated multiple microbial strains for shiitake mushroom fermentation and identified *Lactobacillus plantarum* as the most effective in imparting umami taste. Correspondingly, Bai et al. [42] illustrated that the same bacterium promoted myofibrillar protein degradation and umami peptide formation in dry-cured mackerel. Another study by Chen et al. [43] further established that lactic acid bacteria sourced from Harbin dry sausages degraded pork sarcoplasmic protein, yielding notable flavor compounds. Zhao and Gänzle [44] observed that *Lactobacillus reuteri* fermentation led to markedly elevated levels of γ-glutamyl dipeptides relative to chemically acidified counterparts. They also noted that employing glutamate- and kokumi peptide-enhancing *Lactobacillus* strains improved bread taste without negatively affecting other quality parameters. Fermented soybean derivatives have also gained recognition as valuable sources of flavor peptides with sodium-replacement potential. Through taste reconstitution experiments, Jünger et al. [25] revealed that the abundance of amino acids and peptides in fermented soy products contributes to umami, saltiness, and kokumi perceptions. These findings underscore the significant role of microbial fermentation in flavor compound exploration, offering valuable insights for the food industry.

10.2.2 Methods for Direct Extraction and Synthesis

Direct extraction refers to the isolation of native peptides directly from biological raw materials using various fractionation and separation methods. Although considered safe and reliable, this approach is often constrained by the inherently low peptide concentrations in most raw matrices, making it more applicable to substrates with degraded protein content. Moreover, processing techniques such as high-temperature cooking, microwaving, or ultrasound application can substantially impact the flavor and integrity of the final peptide-containing product. Luo et al. [45] assessed the impact of various thermal treatments, such as steaming, boiling, frying, and microwaving, on surimi-based gels and found that microwave reheating best preserved the original sensory quality.

Peptide synthesis technologies are widely employed for producing peptides with known sequences. Two main methods are widely adopted: (i) chemical synthesis and (ii) recombinant expression through molecular biology. Song et al. [29] successfully synthesized umami-active peptides (e.g., Asp-Gly-Phe, His-His-Tyr-Glu) and conducted sensory analyses that validated low umami threshold values predicted through molecular docking.

Separately, Zhao et al. [46] applied genetic engineering in *Escherichia coli* to express a heptapeptide (Leu-Gly-Ala-Gly-Gly-Ser-Leu-Ala), offering potential for scalable production of umami peptides. Despite their advantages, large-scale peptide synthesis faces challenges such as high costs, the presence of residual compounds, and potential side reactions. Moreover, reagents such as acylation and coupling agents may pose environmental hazards [47]. Given these limitations, synthetic peptide production is primarily used in laboratory research to verify taste characteristics. Despite these challenges, low doses of synthetic peptides have been shown to exhibit strong salty or umami taste recognition thresholds, as reported in several studies [23, 25]. As such, there is still considerable interest in developing environmentally friendly peptide synthesis strategies.

10.3 Taste Characteristics of Flavor Peptides

10.3.1 The Effect of Amino Acids

Flavor peptides are structurally diverse molecules whose sensory properties are primarily modulated by their amino acid composition, sequences, and spatial structures (Figure 10.2a). The flavor profile of a peptide is primarily governed by the specific amino acids that comprise it [48, 49]. Among the 20 standard amino acids, the majority tend to impart bitterness, while only a smaller number contribute to sour or umami flavors [25]. Although individual amino acids rarely exhibit a salty profile, particular combinations of amino acids can result in peptides capable of evoking salty sensations. Several amino acids, such as glutamate (Glu), glutamine (Gln), aspartate (Asp), serine (Ser), and methionine (Met), commonly contributed to both umami and salty peptides. For instance, Chen and Wang et al. [23] identified a decapeptide sequence, Glu-Asp-Glu-Gly-Glu-Gln-Pro-Phe, which demonstrated both salty and umami sensory properties. Similarly, a whey protein hydrolysate rich in umami amino acid residues (>20%) was also reported to exhibit a combination of salty and umami flavors [50]. Moreover, salty peptides often exhibit umami qualities, with solutions of umami peptides also producing a detectable salty taste [23]. This overlap may be due to the positive association between umami and salty perception [50]. In fact, it can be challenging to differentiate between these two tastes in solutions rich in umami amino acids [16], highlighting the importance of considering umami-related amino acids in the search for novel salty peptides.

Peptides containing arginine (Arg) or proline (Pro) appear to enhance salty flavors, particularly in their dipeptide forms. Dipeptides such as Arg-Ala, Arg-Val, and Ser-Val were found to significantly influence salty taste in both aqueous media and model broth systems [26]. Proline-containing peptides naturally present in soy sauce (e.g., Val-Pro, Ser-Pro, and Leu-Pro-Pro-Pro) also contributed to enhanced saltiness [25]. These arginine and proline peptides were identified as having high dose-over-threshold scores in soy sauce, emphasizing their strong impact on flavor profile and indicating that saltiness in soy sauce is not governed solely by NaCl concentration [25]. However, not all arginine or proline peptides enhance salty taste. Due to the presence of other flavor-masking components, 19 arginine dipeptides with inhibitory effects on salty taste have been identified in lysozyme and casein hydrolysates (Figure 10.2c), consistent with observations by

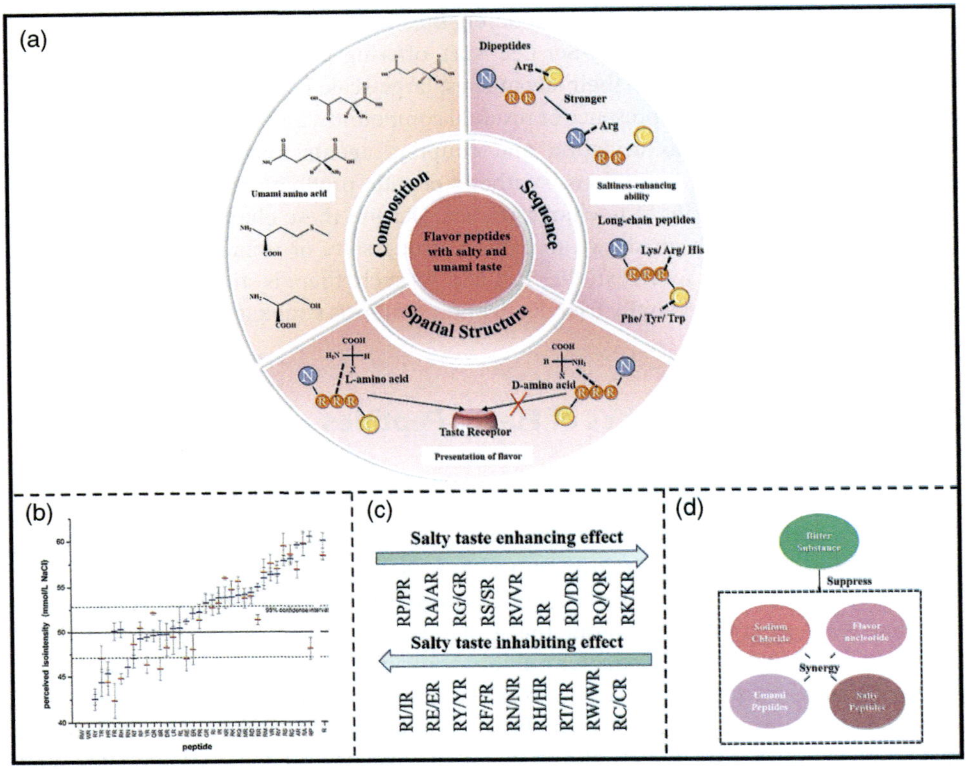

Figure 10.2 Physicochemical and sensory properties of flavor peptides. (a) Influence of specific amino acids on flavor peptides; (b) Enhancement of saltiness perception by arginyl-containing dipeptides, assessed in aqueous solution (red bars) and model broth systems (blue bars), both adjusted to 50 mmol/L NaCl. *Source:* [26] American Chemical Society; (c) Relative contribution of distinct arginyl dipeptides to salty taste intensity. *Source:* Adapted from [24]; (d) The interaction among flavors.

Schindler et al. [24, 26]. Many inhibitory Arg dipeptides carry bitter residues and are perceived as bitter, whereas saltiness-enhancing counterparts tend to include sweet-tasting amino acids. Chen and Wang et al. [23] further concluded that saltiness-enhancing peptides commonly feature hydrophilic residues with acidic side chains (e.g., Glu and Asp) or possess an overall excess of negative over positive charges. These findings provide practical guidelines for selecting candidate peptides containing three or more amino acids with such characteristics.

The primary sequence of amino acids plays another crucial role in flavor expression. Even with identical amino acid compositions, peptides differing in sequence can result in markedly distinct sensory flavors. A typical example is the peptide Arg-Gly, which enhances saltiness in NaCl-containing systems, whereas Gly-Arg does not produce the same effect (Figure 10.2b). Similar patterns are observed in dipeptides such as Arg-Ser vs.

Ser-Arg and Arg-Gln vs. Gln-Arg. Dipeptides with arginine at the C-terminal often present higher saltiness compared to when Arg is positioned at the N-terminal (Figure 10.2a). For longer chain peptides, a C-terminal hydrophobic residue (e.g., Phe, Tyr, Trp) or a positively charged residue positioned near the C-terminus (Lys, Arg, or His) is more likely to enhance salty taste [23].

In addition to the primary sequence, the three-dimensional conformation of peptides critically affects flavor perception. This is largely due to its influence on how peptides interact with gustatory receptors [51]. Consequently, structural changes, especially in long-chain peptides, can modulate or even eliminate taste functionality. Natural flavor peptides are predominantly composed of L-type amino acids. Incorporation with D-type amino acids alters the conformation and often leads to diminished taste. For instance, L-glutamic acid produces a strong umami taste, but its D-form lacks this functionality. Thus, during the formulation and application of flavor peptides in food systems, it is crucial to control factors such as heating temperatures or ultrasonic treatments to avoid undesirable conformational changes that could compromise the intended flavor [52].

10.3.2 The Effect of Flavor and Matrix Interactions

In actual food systems, flavor peptides rarely act in isolation. They frequently interact with other volatile and non-volatile compounds, resulting in synergistic enhancement or taste masking. One illustrative case is the decapeptide Glu-Asp-Glu-Gly-Glu-Gln-Pro-Arg-Pro-Phe (EDEGEQPRPF), which imparts a strong salty and umami profile. Remarkably, just 0.2 mg/mL of this peptide added to a 50 mmol/L NaCl solution significantly elevated the perceived saltiness. Co-application of flavor peptides with small nucleotides, such as guanylate and inosinate, has been shown to greatly improve salty and umami tastes in solutions [53]. Additionally, peptides that undergo Maillard-type Amadori rearrangements (e.g., Glu-Pro and Arg-Phe-Pro-His-Ala-Asp-Phe) exhibit improved umami intensity, especially when used in combination with MSG or GMP [54]. On the other hand, bitter compounds tend to mask salty and umami flavors [55].

Matrix effects on peptide taste partly reflect these intercomponent interactions [56]. However, some studies suggest that the surrounding medium may exert little direct influence. A study reported by Zheng et al. [22] showed that adding peptides like Asp-Asp, Glu-Asp, or Ser-Pro-Glu into various bases (ultrapure water, beef broth, and vegetable broth) yielded no significant difference in taste intensity, suggesting that the matrix had minimal effect. Similarly, Schindler et al. [26] evaluated multiple arginine dipeptides in aqueous and model-broth systems containing 50 mmol/L NaCl and found minimal substrate effects for most cases (Figure 10.2b). However, in some cases, the matrix did influence the flavor characteristics. For instance, Arg-Pro enhanced saltiness in broth yet suppressed it in aqueous solution, and Asp-Asp displayed a much weaker saltiness-enhancing effect in NaCl solution than in broth. These differences suggest that further investigation is needed to understand the underlying mechanisms.

10.4 Taste Mechanism of Flavor Peptides

10.4.1 Release and Migration of Flavor Peptides

The sensory perception of flavor peptides inherently relies on their release and subsequent migration. Typically, these peptides are generally either distributed along the surface of food items or embedded within the food matrix as part of an aqueous solution or an oil–water mixture [57]. When food is chewed, its structure is broken down, resulting in smaller particles that are released into the oral cavity and subsequently mixed with saliva. Saliva serves as a critical transmission medium that facilitates the transfer of peptides from food to the taste buds located on the surface of the tongue. Receptors situated on the cell surfaces of these taste bud cells recognize the chemical stimuli and initiate a cascade of biochemical events, which propagate neural signals via specific gustatory nerves to the central nervous system, ultimately culminating in the perception of taste [58]. This intricate signal transduction pathway involves diverse types of cells and receptor systems, the complexity of which still challenges the current depth of scientific elucidation.

10.4.2 Taste Buds and Taste Cells

Taste buds are predominantly distributed across three morphological types of papillae on the tongue, namely fungiform, foliate, and circumvallate, comprising approximately 250 taste cells each [59]. At the apex of a taste bud lies a taste pore, which allows flavor compounds to interact with microvilli extending from taste bud cells, thereby triggering flavor perception [60]. Based on their ultrastructural characteristics, taste bud cells can be categorized into four distinct categories. Type I cells constitute nearly 50% of the cellular population within a taste bud. These cells are capable of releasing enzymes or transport proteins involved in the clearance of extracellular neurotransmitters. Functionally analogous to glial cells in the nervous system, Type I cells provide structural and metabolic support within the taste bud environment [59, 61]. Notably, studies have shown that Type I cells do not exhibit excitability in response to gustatory stimuli, suggesting their irrelevance in the flavor perception process [61].

Type II taste cells, characterized by their larger size compared to Type I cells, represent approximately one-third of the cellular constituents in a typical taste bud. These cells are preferentially located at the periphery of the taste bud and function as chemosensory receptors, specifically detecting umami, sweet, and bitter compounds [59, 64]. Type III cells, comprising 2–20% of the taste bud cell population, are distinguished by their elongated morphology with prominent synapses, contributing to the perception of sour taste [59, 65]. Lastly, Type IV cells function as proliferative stem cells situated at the base of the taste bud. These undifferentiated cells act as precursors that give rise to the other specialized taste cell types through cellular differentiation processes [66].

10.4.3 Taste Receptors and Conductive Pathway

The perceptions of saltiness and umami are primarily associated with responses occurring within taste cells, where membrane receptors recognize chemical cues and convert them into electrical signals that are relayed along gustatory nerves. Although the salty taste is typically associated with NaCl, the molecular mechanisms that encode saltiness are not fully resolved. Current evidence in mammals supports the presence of two functionally distinct salt detection pathways: one responsive to elevated salt concentrations (high-salt pathway) and a sodium-selective pathway [67, 68]. The high-salt pathway engages aversive taste mechanisms, which appear to have evolved as a response to acid and bitter tastes, aimed at preventing potential health hazards [68]. Importantly, the sensation of high salt is not exclusive to sodium-based salts; other cationic salts can evoke comparable taste responses. Notably, amiloride-insensitive taste bud cells, which operate via mechanisms independent of epithelial sodium channels, are involved in the detection of elevated NaCl levels [69]. TMC4 has been identified as a mediator of high salt pathways when exposed to high concentrations of NaCl [70]. TMC4 is characterized as a novel chloride ion channel embedded in the cell membrane, composed of eight transmembrane domains, with both its N- and C-terminal regions residing intracellularly. It is strongly expressed in the circumvallate and foliate papillae and projects toward the glossopharyngeal nerve [70, 71]. When taste receptor cells encounter high concentrations of salt, cation influx through selective channels induces membrane depolarization. This depolarization triggers an action potential, initiating the gustatory signal transmission cascade. During this activation process, TMC4 facilitates the influx of extracellular chloride ions, which aids in repolarizing the membrane and restoring the resting potential of taste cells. Thus, TMC4 may accelerate the frequency of action potential generation in response to salty stimuli, ultimately enhancing the intensity and fidelity of salt taste perception [70].

Sodium-induced taste, typically activated at lower NaCl concentrations (10–150 mmol/L), demonstrates selectivity toward sodium salts and predominantly occurs in taste cells that exhibit sensitivity to amiloride [64, 67]. In rodents, roughly 65% of the taste cells located within the fungiform papillae and about 35% in the foliate papillae respond to amiloride [66]. In contrast, taste cells located in the circumvallate papillae demonstrate complete insensitivity to this compound [66]. This differential sensitivity has been widely interpreted as supporting the critical role of ENaCs, which are sensitive to amiloride, in mediating salt taste perception [59, 68, 72]. Experimental evidence from rodents has shown that amiloride attenuates both the strength of afferent nerve signals and the behavioral preference for NaCl solutions [72, 73]. Nevertheless, similar inhibitory effects have not been consistently observed in humans, suggesting insufficient evidence to confirm ENaCs as the principal salt taste receptors in human taste perception [66]. Despite this, the identification of ENaCs in taste perception research remains significant for advancing the study of salty peptides.

Compared to inorganic salts, the biochemical pathways by which flavor peptides evoke salty taste responses remain largely unexplored. It remains unclear which receptors are targeted by salty peptides and the specific mechanisms through which they stimulate taste cells. Although ENaCs are well-documented candidates as primary salt taste receptors, their potential interaction with bioactive flavor peptides requires further systematic investigation (Figure 10.3a). In one study, the dipeptide Ala-Arg (AR), identified as a salt taste-enhancing peptide [26], significantly increased NaCl-induced responses in cultured human fungiform papillae via ENaC-mediated mechanisms [66]. Unfortunately, the precise mechanism by which ENaCs enhance salty sensations following AR stimulation remains unknown.

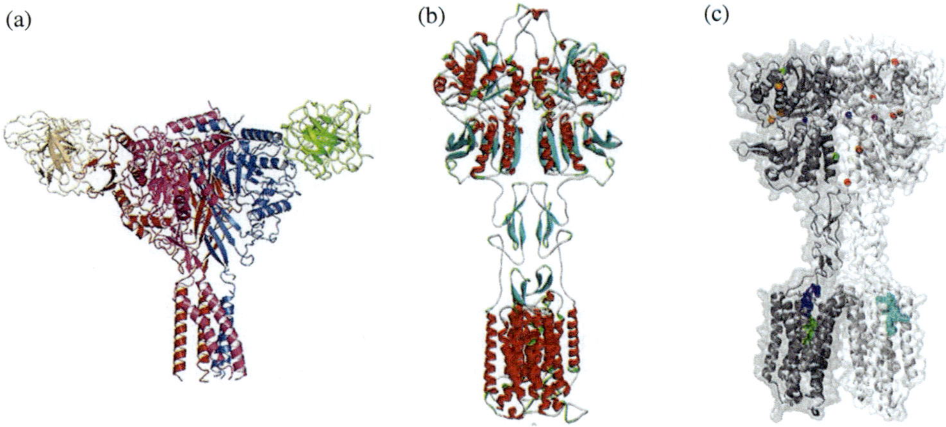

Figure 10.3 Architecture of (a) sodium taste receptor ENaC. Source [62]/eLife Sciences Publications Ltd/CC BY 4.0; (b) umami taste receptor T1R1/3. *Source:* Adapted from [51]; (c) kokumi taste receptor CaSR. *Source:* [63]/American Chemical Society.

A proposed hypothesis suggests that salty peptides and related taste enhancers may interact with the structural domains of ENaCs, thereby increasing sodium ion permeability across various ion channels, which in turn generates intracellular depolarization and action potentials, culminating in a cascade that ultimately leads to the perception of saltiness [64, 66]. However, this hypothesis requires additional validation. Shen et al. [71] employed computational modeling to design TMC4-based systems capable of screening for salty taste-enhancing peptides. The identified candidates were later validated through sensory assessments, suggesting a promising strategy for discovering novel salt taste modulators. In parallel, investigations into the role of kokumi peptides, particularly glutathione and γ-glutamyl-valyl-glycine, have shown that they can enhance salt perception, revealing that the enhancement was not mediated by ENaC [74]. These findings point to the involvement of multiple receptor systems in mediating salty taste enhancement by peptides.

Umami receptors are primarily composed of a heterodimer formed by two G protein-coupled receptors (GPCRs), T1R1 and T1R3. Umami peptides interact specifically with the venus flytrap (VFT) domains of these receptor subunits (Figure 10.3b) [75, 76]. Umami peptides attach to these receptors, sensing the umami flavor through the formation of hydrogen bonds and hydrophobic interactions [51]. Substantial efforts have been made to identify umami-binding peptides through molecular docking approaches targeting T1R1/3. Zhang and Cui et al. [51] characterized the interaction between T1R1-VFT and meaty peptides, revealing a 1:1 ligand-receptor binding stoichiometry with relatively weak affinity. These findings contribute to the understanding of ligand recognition mechanisms by umami receptors. Additional studies suggest that T1R3 may act as a primary target in docking-based analyses. Notably, the VFT domain of T1R1 closes upon ligand binding, while the corresponding domain in T1R3 remains open, indicating that T1R3 might offer a more favorable binding conformation for umami peptides [77, 78].

Kokumi peptides constitute a unique class of flavor-active compounds that do not produce a distinct taste on their own but are known to amplify the intensity and complexity of existing flavors [79]. Among these, γ-Glutamyl peptides (γ-GPs) are prominent examples, interacting allosterically with the extracellular domains of CaSRs (Figure 10.3c). This interaction enhances a comprehensive flavor profile that includes both salty and umami tastes [47]. Broadhead et al. [80] proposed that γ-GPs engage with the extracellular VFT regions of CaSRs, activating phospholipase C β2 (PLCβ2) in type II taste cells and initiating a cascade of intracellular signaling events. Consequently, constructing the structure of CaSR could aid in identifying functionally potent kokumi peptides through computer simulations. Dellafiora et al. [81] computationally screened a library of 400 γ-Glutamyl tripeptides and identified γ-Glu-Pro-Ala and γ-Glu-Pro-Ser as strong kokumi candidates. Such computational strategies offer powerful tools for the efficient selection and rational development of kokumi modulators.

The transduction pathways involved in salty and umami taste perception are illustrated in Figure 10.4. Salivary transport facilitates the delivery of flavor compounds to the taste buds, where they traverse the taste pore to reach receptor-rich taste cells. Each taste modality is

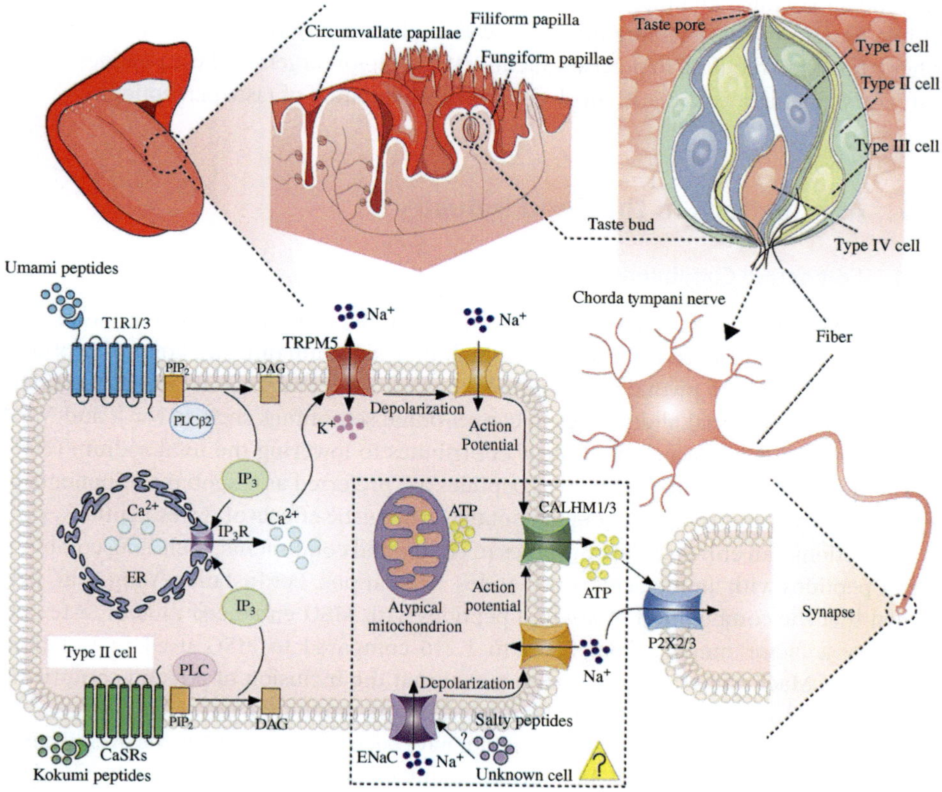

Figure 10.4 Morphology of taste buds located on lingual papillae and the intracellular signaling cascades initiated by flavor peptide–taste receptor interactions. PLC, phospholipase C; PLCβ2, phospholipase Cβ2; IP_3, inositol trisphosphate; IP_3R, inositol trisphosphate receptor; DAG, diacylglycerol; ER, endoplasmic reticulum; TRPM, transient receptor potential melastatin; CALHM, calcium homeostasis modulator; P2X2/3, ATP receptors.

detected by specific subtypes of taste cells. For sodium detection, ENaCs are predominantly implicated, although the precise cell subtypes remain under investigation. Sodium taste signaling begins with Na^+ influx through ENaCs, triggering membrane depolarization and action potential initiation. These electrical signals are subsequently transmitted to the central nervous system via afferent nerve fibers, particularly through the chorda tympani nerve [67]. Salty peptides appear to enhance the response by increasing Na^+ influx upon binding to ENaCs. Within type II taste cells, both umami and kokumi tastes share convergent intracellular signaling cascades, including the elevation of intracellular Ca^{2+} and the activation of TRPM5 ion channels, resulting in depolarization and downstream neurotransmission. Additionally, both sodium and type II taste cells employ a shared neurotransmitter release mechanism that operates through the CALHM1/3 channel [47, 64]. It is widely acknowledged that peptides often evoke more than one taste quality. Salty peptides frequently induce an umami sensation alongside saltiness [17, 22]. Similarly, peptides that exhibit a strong umami flavor also possess salty taste attributes as measured through sensory evaluations [28, 29]. Furthermore, kokumi peptides enhance the overall flavor profile of foods, encompassing both umami and salty perceptions. These peptides appear to modulate multiple receptor types, enabling the co-activation of distinct sensory pathways. However, further studies are required to precisely identify the receptor targets of these peptides and to clarify the signaling mechanisms involved in their modulation of taste perception.

10.5 Applications in the Food Industry

10.5.1 Compound Condiments

Flavor peptides serve as key functional constituents in the formulation of compound condiments. Acting as novel flavor modulators, including salty, umami, and kokumi peptides, these compounds play a pivotal role in enhancing both salty and umami taste perceptions, thereby offering potential alternatives to conventional seasonings such as NaCl and MSG (Figure 10.5a). This replacement strategy contributes to lowering the total sodium (Na^+) content in condiments. Currently, flavor peptides are preferred as essential components in condiment formulations. Leveraging the notable synergistic effects of flavor peptides, various interactions can enhance flavor profiles in compound condiments, including combinations of peptides with amino acids, nucleotides, and cations. For instance, Yang et al. [83] revealed that the combination of kokumi peptides with MSG enhanced umami taste perception by a factor ranging from 0.218 to 1.216 compared to MSG used in isolation. Additionally, Maehashi et al. [84] demonstrated that the inclusion of a blend comprising umami peptides and 0.2% IMP markedly intensified the umami flavor.

In addition to enhancing flavor, flavor peptides have been shown to elevate the overall palatability of food products. Kim et al. [85] identified umami peptides such as Glu-Glu, Glu-Asp, Glu-Ser, Asp-Glu-Ser, and Glu-Gly-Ser as effective modulators of bitterness through their interaction with bitter taste receptors. This indicates their potential in mitigating undesirable taste components. Li and Liu [86] reported that the incorporation of γ-glutamylated hydrolysates into sausage formulations intensified kokumi characteristics and amplified overall flavor while also extending product shelf life. These findings underscore the promising role of

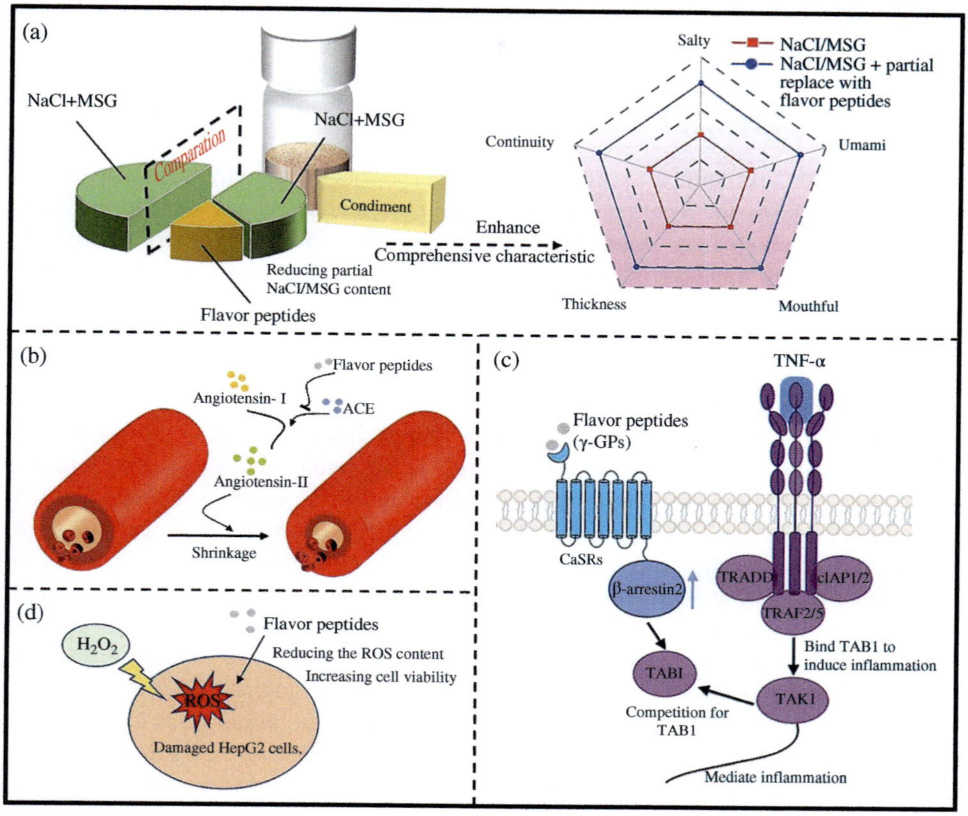

Figure 10.5 Health-promoting attributes associated with flavor peptides. (a) Intensification of overall flavor profiles and facilitation of sodium intake reduction; (b) Modulation of vascular tone via attenuation of vasoconstriction mechanisms; (c) Anti-inflammatory responses elicited by specific peptide sequences. *Source:* Adapted from [47]; (d) Cytoprotective antioxidant action contributing to oxidative stress mitigation in cellular models *Source:* Adapted from [82].

these peptides in the food processing sector. Zhao and Gänzle [44] identified specific γ-GPs, including γ-Glu-Glu, γ-Glu-Leu, γ-Glu-Ile, γ-Glu-Phe, and γ-Glu-Val, as potent flavor enhancers derived from sourdough fermentation, which improved sensory quality without compromising other quality indicators.

From a structural standpoint, these low-molecular-weight peptides possess advantages over traditional sodium-based flavorants. Their relatively lower molecular weight facilitates rapid absorption in the gastrointestinal tract, particularly via the epithelial lining of the small intestine [87, 88], allowing them to enter systemic circulation and exert physiological functions.

Irrespective of their application form, the incorporation of flavor peptides into condiments necessitates adherence to regulatory frameworks and industry norms concerning food additives. As the use of flavor peptides becomes more prevalent, defining and promoting robust quality and safety standards are crucial to meet consumer demands for flavorful and healthier foods with reduced salt content. Ensuring controlled use and avoiding processing errors are vital to preventing potential negative outcomes.

10.5.2 Functional Foods

Functional foods are defined as products that deliver bioactive ingredients capable of promoting health beyond meeting basic nutrient requirements. Among these ingredients, peptides have emerged as promising small molecules. They have been extensively researched for their applications in food production, healthcare products, pharmaceuticals, and skincare. Beyond providing pleasant flavors, flavor peptides have biological roles that should not be overlooked. The broad application potential and favorable socioeconomic prospects have thus positioned flavor peptides as an active research frontier spanning food technology and biomedicine.

Hypertension, a major risk factor for cardiovascular diseases, is frequently associated with high sodium consumption [89, 90]. Despite rising public awareness regarding the health risks of sodium-rich diets, reducing dependence on traditional sodium-based seasonings like NaCl and MSG continues to pose challenges. Certain flavor peptides, such as those delivering salty, umami, or kokumi profiles, can serve as sodium substitutes in condiments, facilitating sodium reduction while supporting physiological balance and minimizing cardiovascular strain. Moreover, these peptides often function as natural inhibitors of angiotensin I-converting enzyme (ACE), an enzyme central to blood pressure regulation (Figure 10.5b). ACE catalyzes the transformation of angiotensin I into the vasoconstrictor angiotensin II [82, 91]. Naturally derived ACE inhibitors from food sources have shown promise in lowering blood pressure with fewer adverse effects than synthetic alternatives. For instance, dipeptides Ala-Phe and Ile-Phe, isolated from sodium-free soy sauce by Zhu et al. [21], demonstrated potent ACE-inhibitory activity, underscoring their potential in dietary strategies targeting hypertension. Notably, these peptides effectively traverse Caco-2 intestinal cell monolayers, suggesting intact absorption and systemic action. Additionally, umami peptides, such as Cys-Cys (CC), Cys-Cys-Asn-Lys (CCNK), and His-Cys-His-Thr (HCHT), have shown significant ACE-inhibitory activity post-digestion, enhancing nitric oxide production and lowering endothelin-1 levels in human endothelial cells, thereby supporting vascular health.

Chronic disease progression is often amplified by persistent inflammation. γ-GPs exhibit flavor-enhancing and anti-inflammatory potential. Their characteristic γ-linkage confers resistance to gastrointestinal degradation, affording high stability and prolonged in vivo half-life (Figure 10.5c) [47]. These peptides are widespread in edible legumes and fermented foods such as cheese and soy sauce [92–94]. Notably, γ-Glu-Val has shown potent anti-inflammatory effects via CaSR-mediated pathways in intestinal cells and has attenuated inflammatory responses in human vascular cells, mouse adipocytes, and rat liver tissue [95–98].

The free-radical theory of disease has further stimulated interest in peptide antioxidants. Beyond ACE inhibition, CC, CCNK, and HCHT protect HepG2 cells from oxidative injury and reduce intracellular reactive oxygen species (Figure 10.5d) [82]. Maillard reaction products, initially valued mainly for flavor generation, also display appreciable antioxidant capacity. Liu et al. [99] produced Maillard-modified peptides that enhanced 2,2-diphenyl-1-picrylhydrazyl (DPPH) radical-scavenging activity while improving flavor, extending food shelf life as a result.

Currently, there is a growing interest in the commercial and scientific exploitation of flavor peptides. Their biological functionality presents compelling opportunities for application in health-promoting foods. However, despite their multifaceted bioactivities, flavor peptides incorporated as functional additives should not be positioned as therapeutic agents. Regulatory caution is warranted when presenting health-related claims on product labels, marketing materials, or advertisements to avoid misleading consumers regarding their intended use.

Abbreviations

ACE	angiotensin I-converting enzyme
$CaCl_2$	calcium chloride
CALHM1/3	calcium homeostasis modulator
CaSRs	calcium-sensitive receptors
DOT	dose-over-threshold
ENaCs	epithelial sodium ion channels
γ-GPs	γ glutamyl peptides
GMP	guanosine monophosphate
GPCRs	G protein-coupled receptors
IMP	inosine monophosphate
IP3R	inositol trisphosphate receptor
K^+	potassium ion
KCl	potassium chloride
$MgCl_2$	magnesium chloride
mGluR1/4	metabotropic glutamate receptor 1/4
MRPs	Maillard reaction products
MSG	monosodium glutamate
Na^+	sodium ion
NaCl	sodium chloride
PLCβ2	phospholipase Cβ2
T1R1	taste receptor type 1 member 1
T1R3	taste receptor type 1 member 3
TMC4	transmembrane channel-like 4
TRPM5	transient receptor potential melastatin 5
UHP	ultra-high pressure
VFT	venus flytrap

References

1 Zhu, C.-Z., Jin, H.-Q., Yin, F. et al. (2021). Emulsion-forming properties of heat-induced pork myofibrillar protein affected by NaCl. *International Journal of Food Science and Technology* 56 (6): 3016–3025.

2 Li, T., Zhong, Q., and Wu, T. (2022). Effects of NaCl on the freezing–thawing induced gelation of egg yolk at pH 2.0–8.0. *Food Biophysics* 17 (1): 106–113.

3 Chen, J., Yang, R., Wang, Y. et al. (2022). Inhibitory effect of D-tryptophan on the spoilage potential of *Shewanella baltica* and *Pseudomonas fluorescens* and its potential application in salmon fillet preservation. *Food Microbiology* 108: 104104.

4 Lindemann, B., Ogiwara, Y., and Ninomiya, Y. (2002). The discovery of umami. *Chemical Senses* 27 (9): 843–844.

5 Thuy, L.N., Salanta, L.C., Tofana, M. et al. (2020). A mini review about monosodium glutamate. *Bulletin of University of Agricultural Sciences and Veterinary Medicine Cluj-Napoca-Food Science and Technology* 77 (1): 1–12.

6 Panzarini, A.C., Menis-Henrique, M.E.C., and Conti-Silva, A.C. (2020). Effect of salt and monosodium glutamate on the sensory characteristics of low-sodium cheese-flavored corn grits expanded snacks. *Journal of Food Processing and Preservation* 44 (12): e14936.

7 DeLalio, L.J., Hahn, S., Katayama, P.L. et al. (2020). Excessive dietary salt promotes aortic stiffness in murine renovascular hypertension. *American Journal of Physiology-Heart and Circulatory Physiology* 318 (5): H1346–H1355.

8 Le, B., Yu, B., Amin, M.S. et al. (2022). Salt taste receptors and associated salty/salt taste-enhancing peptides: a comprehensive review of structure and function. *Trends in Food Science & Technology.* 129: 657–666.

9 Yu, H., Liu, X., Li, Y., and Tian, H. (2023). Research progress of condiment reducing salt and increasing umami. *Food Science* 44 (5): 1–12.

10 Pietrasik, Z., Gaudette, N.J., and Johnston, S.P. (2016). The use of high pressure processing to enhance the quality and shelf life of reduced sodium naturally cured restructured cooked hams. *Meat Science* 116: 102–109.

11 Chen, X.-W., Yang, D.-X., Guo, J. et al. (2018). *Quillaja* saponin-based hollow salt particles as solid carriers for enhancing sensory aroma with reduced sodium intake. *Food & Function* 9 (1): 191–199.

12 Lu, W., Hu, Z., Zhou, X. et al. (2022). Natural biopolymer masks the bitterness of potassium chloride to achieve a highly efficient salt reduction for future foods. *Biomaterials* 283: 121456.

13 Vidal, V.A.S., Biachi, J.P., Paglarini, C.S. et al. (2019). Reducing 50% sodium chloride in healthier jerked beef: An efficient design to ensure suitable stability, technological and sensory properties. *Meat Science* 152: 49–57.

14 Tan, H.-L., Tan, T.-C., and Easa, A.M. (2022). The use of salt substitutes to replace sodium chloride in food products: a review. *International Journal of Food Science and Technology* 57 (11): 6997–7007.

15 Hwang, Y.-H., Ismail, I., and Joo, S.-T. (2020). Identification of umami taste in sous-vide beef by chemical analyses, equivalent umami concentration, and electronic tongue system. *Foods* 9 (3): 251.

16 Rhyu, M.-R., Song, A.-Y., Kim, E.-Y. et al. (2020). Kokumi taste active peptides modulate salt and umami taste. *Nutrients* 12 (4): 1198.

17 Tada, M., Shinoda, I., and Okai, H. (1984). L-Ornithyltaurine, a new salty peptide. *Journal of Agricultural and Food Chemistry* 32 (5): 992–996.

18 Kato, H., Rhue, M.R., and Nishimura, T. (1989). *Role of free amino acids and peptides in food taste.* In: *Flavor Chemistry*, 13rde, vol. 388 (ed. R. Teranishi and R.G. Buttery), 158–174. American: American Chemical Society.

19 Ueda, Y., Sakaguchi, M., Hirayama, K. et al. (1990). Characteristic flavor constituents in water extract of garlic. *Agricultural and Biological Chemistry* 54 (1): 163–169.

20 Wang, H., Suo, R., Liu, X. et al. (2022). Kokumi γ-glutamyl peptides: some insight into their evaluation and detection, biosynthetic pathways, contribution and changes in food processing. *Food Chemistry Advances* 1: 100061.

21 Zhu, X., Watanabe, K., Shiraishi, K. et al. (2008). Identification of ACE-inhibitory peptides in salt-free soy sauce that are transportable across caco-2 cell monolayers. *Peptides* 29 (3): 338–344.

22 Zheng, Y., Tang, L., Yu, M. et al. (2021). Fractionation and identification of salty peptides from yeast extract. *Journal of Food Science and Technology* 58 (3): 1199–1208.

23 Chen, Y.P., Wang, M., Blank, I. et al. (2021). Saltiness-enhancing peptides isolated from the Chinese commercial fermented soybean curds with potential applications in salt reduction. *Journal of Agricultural and Food Chemistry* 69 (35): 10272–10280.

24 Harth, L., Krah, U., Linke, D. et al. (2018). Salt taste enhancing l-arginyl dipeptides from casein and lysozyme released by peptidases of basidiomycota. *Journal of Agricultural and Food Chemistry* 66 (10): 2344–2353.

25 Jünger, M., Mittermeier-Kleßinger, V.K., Farrenkopf, A. et al. (2022). Sensoproteomic discovery of taste-modulating peptides and taste re-engineering of soy sauce. *Journal of Agricultural and Food Chemistry* 70 (21): 6503–6518.

26 Schindler, A., Dunkel, A., Stähler, F. et al. (2011). Discovery of salt taste enhancing arginyl dipeptides in protein digests and fermented fish sauces by means of a sensomics approach. *Journal of Agricultural and Food Chemistry* 59 (23): 12578–12588.

27 Zhang, J.-N., Zhao, M., Su, G., and Lin, L. (2019). Identification and taste characteristics of novel umami and umami-enhancing peptides separated from peanut protein isolate hydrolysate by consecutive chromatography and UPLC–ESI–QTOF–MS/MS. *Food Chemistry* 278: 674–682.

28 Chen, W., Li, W., Wu, D. et al. (2022). Characterization of novel umami-active peptides from *Stropharia rugoso-annulata* mushroom and in silico study on action mechanism. *Journal of Food Composition and Analysis* 110: 104530.

29 Song, S., Zhuang, J., Ma, C. et al. (2023). Identification of novel umami peptides from boletus edulis and its mechanism via sensory analysis and molecular simulation approaches. *Food Chemistry* 398: 133835.

30 Chen, Z., Gao, H., Wu, W. et al. (2021). Effects of fermentation with different microbial species on the umami taste of shiitake mushroom (*Lentinus edodes*). *LWT* 141: 110889.

31 Wang, W., Huang, Y., Zhao, W. et al. (2022). Identification and comparison of umami-peptides in commercially available dry-cured Spanish mackerels (*Scomberomorus niphonius*). *Food Chemistry* 380: 132175.

32 Dunkel, A., Köster, J., and Hofmann, T. (2007). Molecular and sensory characterization of γ-glutamyl peptides as key contributors to the kokumi taste of edible beans (*Phaseolus vulgaris L.*). *Journal of Agricultural and Food Chemistry* 55 (16): 6712–6719.

33 Mukherjee, S., Mandayam Bharathi, H., Weng, W.-H. et al. (2023). Predicting conduction properties of TMC proteins involved in sensory perception in silico. *Biophysical Journal* 122 (3, Supplement 1): 92a.

34 Chaudhari, N., Landin, A.M., and Roper, S.D. (2000). A metabotropic glutamate receptor variant functions as a taste receptor. *Nature Neuroscience* 3 (2): 113–119.

35 Li, X., Staszewski, L., Xu, H. et al. (2002). Human receptors for sweet and umami taste. *Proceedings of the National Academy of Sciences* 99 (7): 4692–4696.

36 Zhu, W., Zhu, L., Yang, W. et al. (2020). Optimization of the enzymatic hydrolysis assisted by ultra-high pressure processing of Alaska Pollock frame for improving flavour. *Journal of Aquatic Food Product Technology* 29 (6): 567–576.

37 Gao, P., Zhang, W., Wei, M. et al. (2022). Analysis of the non-volatile components and volatile compounds of hydrolysates derived from unmatured cheese curd hydrolysis by different enzymes. *LWT* 168: 113896.

38 Zheng, Z., Zhang, M., Fan, H., and Liu, Y. (2021). Effect of microwave combined with ultrasonic pretreatment on flavor and antioxidant activity of hydrolysates based on enzymatic hydrolysis of bovine bone. *Food Bioscience* 44: :101399.

39 Fan, W., Tan, X., Tu, M. et al. (2018). Preparation of the rainbow trout bone peptides directed by nutritional properties and flavor analyses. *Food Science & Nutrition* 6 (4): 925–933.

40 Cai, Z., Ruan, Y., He, J. et al. (2020). Effects of microbial fermentation on the flavor of cured duck legs. *Poultry Science* 99 (9): 4642–4652.

41 Chen, M., Gao, X., Pan, D. et al. (2021). Taste characteristics and umami mechanism of novel umami peptides and umami-enhancing peptides isolated from the hydrolysates of Sanhuang chicken. *European Food Research and Technology* 247 (7): 1633–1644.

42 Bai, W., Liang, J., Zhao, W. et al. (2022). Umami and umami-enhancing peptides from myofibrillar protein hydrolysates in low-sodium dry-cured Spanish mackerel (*Scomberomorus niphonius*) under the action of *lactobacillus plantarum*. *International Journal of Food Science & Technology* 57 (8): 5494–5503.

43 Chen, Q., Liu, Q., Sun, Q. et al. (2015). Flavour formation from hydrolysis of pork sarcoplasmic protein extract by a unique LAB culture isolated from Harbin dry sausage. *Meat Science* 100: 110–117.

44 Zhao, C.J. and Gänzle, M.G. (2016). Synthesis of taste-active γ-glutamyl dipeptides during sourdough fermentation by *lactobacillus reuteri*. *Journal of Agricultural and Food Chemistry* 64 (40): 7561–7568.

45 Luo, X., Xiao, S., Ruan, Q. et al. (2022). Differences in flavor characteristics of frozen surimi products reheated by microwave, water boiling, steaming, and frying. *Food Chemistry* 372: 131260.

46 Zhao, L., Zhang, Y., Venkitasamy, C. et al. (2018). Preparation of umami octopeptide with recombined *Escherichia coli*: feasibility and challenges. *Bioengineered* 9 (1): 166–169.

47 Guha, S. and Majumder, K. (2022). Comprehensive review of γ-glutamyl peptides (γ-GPs) and their effect on inflammation concerning cardiovascular health. *Journal of Agricultural and Food Chemistry* 70 (26): 7851–7870.

48 Huang, W. and BI, J. and Yan, H. (2020). Advances in salty peptides. *The Food Industry* 41 (10): 261–264.

49 Miao, X., Liu, Y., Qiu, C. et al. (2014). Research progress in the structure–activity relationship of flavor peptides. *Science and Technology of Food Industry* 35 (06): 357–362.

50 Cheung, L.K.Y., Aluko, R.E., Cliff, M.A., and Li-Chan, E.C.Y. (2015). Effects of exopeptidase treatment on antihypertensive activity and taste attributes of enzymatic whey protein hydrolysates. *Journal of Functional Foods* 13: 262–275.

51 Zhang, N., Cui, Z., Li, M. et al. (2022). Typical umami ligand-induced binding interaction and conformational change of T1R1-VFT. *Journal of Agricultural and Food Chemistry* 70 (37): 11652–11666.

52 Zhang, Y., Venkitasamy, C., Pan, Z. et al. (2017). Novel umami ingredients: umami peptides and their taste. *Journal of Food Science* 82 (1): 16–23.

53 Zhang, J.-H., Wang, F., and Zhang, Y. (2020). The effect of partial substitution of soybean flavor peptide for flavor nucleotide disodium in chicken essence seasoning. *Food industry* 41 (11): 153–156.

54 Zhang, J.-H., Su, G., Zhao, T. et al. (2022). Preparation, sensory characterization, and umami-enhancing mechanism of novel peptide glycoconjugates. *Journal of Agricultural and Food Chemistry* 70 (26): 8043–8051.

55 Breslin, P.A.S. and Beauchamp, G.K. (1995). Suppression of bitterness by sodium: variation among bitter taste stimuli. *Chemical Senses* 20 (6): 609–623.

56 Reis Rocha, R.A., Reis Rocha, L.C., Ribeiro, M.N. et al. (2021). Effect of the food matrix on the capacity of flavor enhancers in intensifying salty taste. *Journal of Food Science* 86 (3): 1022–1032.

57 Zhang, Y.-H., Zheng, H., Liu, N. et al. (2022). Design of colloidal structures for salt-reduced foods. *Food Science* 43 (01): 213–222.

58 Hino, K., Hirashima, S., Tsuneyoshi, R. et al. (2022). Three-dimensional ultrastructure and histomorphology of mouse circumvallate papillary taste buds before and after birth using focused ion beam-scanning electron microscope tomography. *Tissue and Cell* 75: 101714.

59 Roper, S.D. and Chaudhari, N. (2017). Taste buds: cells, signals and synapses. *Nature Reviews Neuroscience* 18: 485–497.

60 Yang, R., Tabata, S., Crowley, H.H. et al. (2000). Ultrastructural localization of gustducin immunoreactivity in microvilli of type II taste cells in the rat. *Journal of Comparative Neurology* 425 (1): 139–151.

61 Liman, E.R. (2020). Salty taste: from transduction to transmitter release, hold the calcium. *Neuron* 106 (5): 709–711.

62 Noreng, S., Bharadwaj, A., Posert, R. et al. (2018). Structure of the human epithelial sodium channel by cryo-electron microscopy. *ELife* 7: e39340.

63 Diao, J., DeBono, A., Josephs, T.M. et al. (2021). Therapeutic opportunities of targeting allosteric binding sites on the calcium-sensing receptor. *ACS Pharmacology & Translational Science* 4 (2): 666–679.

64 Taruno, A., Nomura, K., Kusakizako, T. et al. (2021). Taste transduction and channel synapses in taste buds. *Pflügers Archiv: European Journal of Physiology* 473: 3–13.

65 Lyall, V., Alam, R.I., Phan, D.Q. et al. (2001). Decrease in rat taste receptor cell intracellular pH is the proximate stimulus in sour taste transduction. *American Journal of Physiology-Cell Physiology* 281 (3): C1005–C1013.

66 Xu, J., Elkaddi, N., Garcia-Blanco, A. et al. (2017). Arginyl dipeptides increase the frequency of NaCl-elicited responses via epithelial sodium channel alpha and delta subunits in cultured human fungiform taste papillae cells. *Scientific Reports* 7 (1): 7483.

67 Chandrashekar, J., Kuhn, C., Oka, Y. et al. (2010). The cells and peripheral representation of sodium taste in mice. *Nature* 464 (7,286): 297–301.

68 Oka, Y., Matthew, B., von, B.,.L. et al. (2013). High salt recruits aversive taste pathways. *Nature* 494: 472–475.

69 Kretz, O., Barbry, P., Bock, R., and Lindemann, B. (1999). Differential expression of RNA and protein of the three pore-forming subunits of the amiloride-sensitive epithelial sodium channel in taste buds of the rat. *The Journal of Histochemistry and Cytochemistry* 47 (1): 51–64.

70 Kasahara, Y., Narukawa, M., Ishimaru, Y. et al. (2021). TMC4 is a novel chloride channel involved in high-concentration salt taste sensation. *The Journal of Physiological Sciences* 71 (1): 23.

71 Shen, D.-Y., Pan, F., Yang, Z.-C. et al. (2022). Identification of novel saltiness-enhancing peptides from yeast extract and their mechanism of action for transmembrane channel-like 4 (TMC4) protein through experimental and integrated computational modeling. *Food Chemistry* 388: 132993.

72 Heck, G.L., Mierson, S., and DeSimone, J.A. (1984). Salt taste transduction occurs through an amiloride-sensitive sodium transport pathway. *Science (New York, N.Y.)* 223 (4634): 403–405.

73 Halpern, B.P. (1998). Amiloride and vertebrate gustatory responses to NaCl. *Neuroscience & Biobehavioral Reviews* 23 (1): 5–47.

74 Bigiani, A. and Rhyu, M. (2023). Effect of kokumi taste-active γ-glutamyl peptides on amiloride-sensitive epithelial Na^+ channels in rat fungiform taste cells. *Biochemistry and Biophysics Reports* 33: 101400.

75 Spaggiari, G., Di Pizio, A., and Cozzini, P. (2020). Sweet, umami and bitter taste receptors: state of the art of in silico molecular modeling approaches. *Trends in Food Science & Technology* 96: 21–29.

76 Zhao, G.Q., Zhang, Y., Hoon, M.A. et al. (2003). The receptors for mammalian sweet and umami taste. *Cell* 115 (3): 255–266.

77 Dang, Y., Hao, L., Cao, J. et al. (2019). Molecular docking and simulation of the synergistic effect between umami peptides, monosodium glutamate and taste receptor T1R1/T1R3. *Food Chemistry* 271: 697–706.

78 Li, C., Hua, Y., Pan, D. et al. (2023). A rapid selection strategy for umami peptide screening based on machine learning and molecular docking. *Food Chemistry* 404: 134562.

79 Rui, G., Chun, C., and Sun-Waterhouse, D. (2022). Research progress on the taste-enhancing mechanism and potential physiological functions of partial kokumi peptides. *China Condiment* 47 (03): 200–205+210. https://doi.org/10.3969/j.issn.1000-9973.2022.03.040.

80 Broadhead, G.K., Mun, H., Avlani, V.A. et al. (2011). Allosteric modulation of the calcium-sensing receptor by γ-glutamyl peptides: inhibition of PTH secretion, suppression of intracellular camp levels, and a common mechanism of action with l-amino acids*. *Journal of Biological Chemistry* 286 (11): 8786–8797.

81 Dellafiora, L., Magnaghi, F., Galaverna, G., and Dall'Asta, C. (2022). A mechanistic investigation on kokumi-active γ-Glutamyl tripeptides—A computational study to understand molecular basis of their activity and to identify novel potential kokumi-tasting sequences. *Food Research International* 162: 111932.

82 Feng, X., Liao, D., Sun, L. et al. (2021). Affinity purification of angiotensin converting enzyme inhibitory peptides from Wakame (*Undaria Pinnatifida*) using immobilized ACE on magnetic metal organic frameworks. *Marine Drugs* 19 (3): 177.

83 Yang, J., Liao, J., Dong, H. et al. (2022). Synergistic effect of kokumi-active γ-glutamyl peptides and l-glutamate on enhancing umami sensation and stimulating cholecystokinin

secretion via T1R1/T1R3 activation in STC-1 cells. *Journal of Agricultural and Food Chemistry* 70 (45): 14395–14402.

84 Maehashi, K., Matsuzaki, M., Yamamoto, Y., and Udaka, S. (1999). Isolation of peptides from an enzymatic hydrolysate of food proteins and characterization of their taste properties. *Bioscience, Biotechnology, and Biochemistry* 63 (3): 555–559.

85 Kim, M.J., Son, H.J., Kim, Y. et al. (2015). Umami–bitter interactions: the suppression of bitterness by umami peptides via human bitter taste receptor. *Biochemical and Biophysical Research Communications* 456 (2): 586–590.

86 Li, Q., Liu, J., Cao, L. et al. (2022). Effects of γ-glutamylated hydrolysates from porcine hemoglobin and meat on kokumi enhancement and oxidative stability of emulsion-type sausages. *Food and Bioprocess Technology* 15 (8): 1851–1865.

87 Haque, E., Chand, R., and Kapila, S. (2008). Biofunctional properties of bioactive peptides of milk origin. *Food Reviews International* 25 (1): 28–43.

88 Ryznar-Luty, A. (2022). Food industry by-products as a source of biologically active peptides. *Przemysl Chemiczny* 101 (3): 186–190.

89 Ertuglu, L.A. and Kirabo, A. (2022). Dendritic cell epithelial sodium channel in inflammation, salt-sensitive hypertension, and kidney damage. *Kidney360* 3 (9): 1620–1629.

90 Sorato, M.M., Davari, M., Kebriaeezadeh, A. et al. (2021). Risk of fatal and nonfatal coronary heart disease and stroke events among adult patients with hypertension: basic Markov model inputs for evaluating cost-effectiveness of hypertension treatment: systematic review of cohort studies. *Journal of Pharmaceutical Health Services Research* 12 (2): 283–302.

91 Lu, Y., Wu, Y., Hou, X. et al. (2022). Separation and identification of ACE inhibitory peptides from lizard fish proteins hydrolysates by metal affinity-immobilized magnetic liposome. *Protein Expression and Purification* 191: 106027.

92 Hillmann, H., Behr, J., Ehrmann, M.A. et al. (2016). Formation of kokumi-enhancing γ-glutamyl dipeptides in parmesan cheese by means of γ-glutamyltransferase activity and stable isotope double-labeling studies. *Journal of Agricultural and Food Chemistry* 64 (8): 1784–1793.

93 Kuroda, M., Kato, Y., Yamazaki, J. et al. (2013). Determination and quantification of the kokumi peptide, γ-glutamyl-valyl-glycine, in commercial soy sauces. *Food Chemistry* 141 (2): 823–828.

94 Liao, D., Cram, D., Sharpe, A., and Marsolais, F. (2013). Transcriptome profiling identifies candidate genes associated with the accumulation of distinct sulfur γ-glutamyl dipeptides in *Phaseolus vulgaris* and *Vigna mungo* seeds. *Frontiers in Plant Science* 4: 1–12.

95 Guha, S., Paul, C., Alvarez, S. et al. (2020). Dietary γ-glutamyl valine ameliorates TNF-α-induced vascular inflammation via endothelial calcium-sensing receptors. *Journal of Agricultural and Food Chemistry* 68 (34): 9139–9149.

96 Salama, S.A., Arab, H.H., Hassan, M.H., and Al robaian, M. M. and Maghrabi, I. A. (2019). Cadmium-induced hepatocellular injury: modulatory effects of γ-glutamyl cysteine on the biomarkers of inflammation, DNA damage, and apoptotic cell death. *Journal of Trace Elements in Medicine and Biology* 52: 74–82.

97 Xing, L., Zhang, H., Majumder, K. et al. (2019). γ-glutamylvaline prevents low-grade chronic inflammation via activation of a calcium-sensing receptor pathway in 3T3-L1mouse adipocytes. *Journal of Agricultural and Food Chemistry* 67 (30): 8361–8369.

98 Zhang, H., Kovacs-Nolan, J., Kodera, T. et al. (2015). γ-Glutamyl cysteine and γ-glutamyl valine inhibit TNF-α signaling in intestinal epithelial cells and reduce inflammation in a mouse model of colitis via allosteric activation of the calcium-sensing receptor. *Biochimica et Biophysica Acta* (BBA) - Molecular Basis of Disease 1852 (5): 792–804.

99 Liu, P., Huang, M., Song, S. et al. (2012). Sensory characteristics and antioxidant activities of Maillard reaction products from soy protein hydrolysates with different molecular weight distribution. *Food and Bioprocess Technology* 5: 1775–1789.

11

Self-Assembling Peptides

Peptide self-assembly is widely found in organisms and is the basis for many life activities and biological functions. It can be defined as the spontaneous and reversible process by which disordered peptides form organized structures. The resulting supramolecular structures include nanofibers, nanospheres, nanotubes, and other nanostructures. Therefore, peptides that can spontaneously form supramolecular structures under certain conditions are considered self-assembling peptides (SAPs). Its development process can be divided into three stages (Table 11.1). SAPs, as one of the most promising biomimetic materials in the 21st century, are reshaping humans' understanding of molecular engineering. This chapter systematically reviews the source, molecular mechanism, safety assessment, and application of SAPs.

11.1 Sources

In the study of SAPs, according to the source of peptides, they can be divided into food-derived SAPs and artificially designed SAPs.

11.1.1 Food-Derived SAPs

In general, food-derived SAPs are mainly protein hydrolysates obtained from food-derived proteins through enzymatic hydrolysis or fermentation. The product is usually a complex mixture, and the amino acid sequence of the peptide fragment is affected by the specific cleavage site of the enzyme and the conditions of action (such as temperature, pH, etc.). Therefore, the study of the sequence regularity of food-derived SAPs is difficult and has not been fully explored. Subsequent separation and purification of the peptide mixture can yield pure peptides with confirmed sequences.

At present, more and more SAPs from food sources have been studied and reported (Table 11.2), such as milk enzyme peptides, milk fermented peptides, egg white hydrolyzed peptides, soy enzyme peptides, etc. When these peptides are found to have self-assembly ability, they are usually found as hydrogel products during protein degradation or food

Food-Derived Bioactive Peptides: Green Preparation, Activity Mechanisms, and Health Applications, First Edition. Shao-Yun Wang, Xu Chen, and Xixi Cai.

Table 11.1 Development history of SAPs.

Time	Research
1980–2000	Natural peptide structure resolution (e.g., β-sheet structure of silk protein)
2001–2015	Manual design (RADA16, MAX series)
2016–present	AI-driven design era (AlphaFold-Pep algorithm)

Table 11.2 SAPs from food sources.

Source	Food-derived SAPs
Canola	DQPLVII, FVRYHE, GFAYVVQ, GLEETI, GLYLP, GLYLPSFF, GLYLPTFF, MVLPK, MVLPQ, PRGPPQSPQDN, PRPFYLAGN, SALRALP, SHWIYN, SHWIYNTGDQPLVE, SLGVPPQLGNA, SPKISYVVQGMGI, SVLRGLPLEVI, SYTLPILQYIRL, TGDQPLLVII
Soy	GVAWWMYNNEDTPVVAVSIIDTNSLE, SILSGFTLEFLEHAFSV, LDIFLSIVDMNEGALLLPHF, ADFLLFVLSGRAILTLVNND, LSSTQAQQSYLQGFSHNILETSFHSEFE
Potato	GGGIKGIIPAIILEFLEGQLQEVD, PALLSLSVATRLAQED, PAFASIRSLNYKQLLLLSLGTGTNSEF, APCANGIFRYNS

processing. For example, Yu et al. found that buckwheat protein peptides can self-assemble to form hydrogels. Under the treatment of hot and acid, the peptide products formed by egg white lyase can form hydrogels under the action of Mg^{2+}. In addition, the 49–101 and 53–101 peptide fragments which have selfassembly ability, are generated by Cys64-Cys80 and Cys76-Cys94 [1].

A mixture of peptides extracted from fermented curd by Manna et al. can aggregate to form hydrogels, but this work did not show which peptides play a key role [2]. Therefore, food-derived peptide mixtures have the ability to self-assemble, but accurate peptide sequence signatures are difficult to verify. This problem can be verified by separating and purifying peptide mixtures to obtain peptides with accurate amino acid sequences. For example, Pouliot et al. isolated peptide β-lg AB[f1–8] (sequence: LIVTQTMK) from whey protein hydrolysate and found that the peptide could form a β-sheet structure [3]. The amphipathic nature of the C-terminal and N-terminal ends of the β-lg AB[f1–8] sequence may explain its gelation ability. In addition, Josefsson et al. [4] found that glycine hydrolysates can aggregate to form fibrous structures, and the important feature of these sequences is that they are highly hydrophobic, which is responsible for the formation of insoluble aggregates.

11.1.2 Artificially Designed SAPs

Most sources of SAPs are artificially designed peptide sequences according to specific rules and are achieved by solid-phase synthesis. Due to the diversity of amino acid chemical functions, the design of SAP sequences can be performed by adjusting the length and type of building blocks. At present, peptide amphiphiles, ion-complementary peptides, β-hairpin peptides, aromatic peptides, and other artificially designed SAPs (Table 11.3) have been widely studied and reported.

Table 11.3 Artificially designed SAPs.

Types	Artificially designed peptides
Peptide amphiphiles (hydrophilic peptide sequence and lipophilic groups or hydrophobic amino acids)	$capA_6R$ $(CONH\text{-}AAAAAAR\text{-}NH_2)$ $K_3(FA)_4K_3$ (KKKFAFAFAFAKKK) $K_6(FA)_4$ (KKKKKKFAFAFAFA) P_5VP_5 $(Ac\text{-}RLIRKVKRILR\text{-}NH_2)$ PA_4 $(C_{18}K_5)$ PA_7 $(C_{18}K\text{-}K_5)$ K-11 $(Ac\text{-}{}^LK{}^LH{}^LH{}^LQ{}^LK{}^LL{}^LV{}^LF{}^LF{}^LA{}^LK\text{-}NH_2)$ KKd-11 $(Ac\text{-}{}^DK{}^DH{}^DH{}^DQ{}^DK{}^DL{}^DV{}^DF{}^DF{}^DA{}^DK\text{-}NH_2)$
Ionic complementary peptides (amino acid pairing with regular charge distribution)	RADA4 (RADARADARADARADA) FFEFK EMK8-II (MEMEMKMK)
Aromatic peptides (a short peptide sequence [commonly dipeptides or a single amino acid] at C-terminus and a synthetic aromatic moiety at N-terminus)	BOC-FF Fmoc-FF
β-Hairpin peptides (β refers to the dihedral angle of the secondary structure of the peptide)	MAX8 $(VKVKVKVKVDPPTKVEVKVKV\text{-}NH_2)$ MAX1 $(VKVKVKVKVDPPTKVKVKVKV\text{-}NH_2)$
Short peptide	FF

Table 11.4 Peptide sequence design by computer-aided design tools.

Tool name	Function	Typical case
RosettaDesign	Energy minimization optimizes the sequence	Coil29 nanotube design
AlphaFold-Pep	Predict assembly paths with structures	AF-Pep12 error < 0.1 nm
MDynaMix	Molecular dynamics simulates assembly dynamics	Analysis of RADA16 hydrogen bond network

Various types of artificially designed SAPs have their own sequence characteristics. Peptide amphiphiles (surfactant-like peptides and lipid-peptide amphiphiles) are characterized by the presence of hydrophobic segments and hydrophilic groups. For example, the pioneering work done by Zhang and his team includes DA_6, DV_6, D_2L_6, KA_6, H_2V_6, K_2V_6, K_2L_6, and so on [5, 6]. Another peptide amphiphile found by Stupp's team [7], the N-terminal of which consists of a hydrophobic tail composed of one or more long alkyl chains, and the C-terminal of which is a multi-domain peptide sequence.

Ion the complementary peptide, the sequence is characterized by amino acid pairs, charge distribution with rules, including type I (+ or + − −) and type II (− − + + or + + − −), and type III (− − − − + + + + + + + + or − − − −). For example, Hong et al. [8] EAK16-II and EAK16-IV have the same amino acid composition and different charge distribution.

Aromatic peptide, whose sequence is characterized by a short peptide sequence (usually a dipeptide or a single amino acid) at the C-terminus and a synthetic aromatic part at the N-terminus. For example, the classical Fmoc-FF is an aromatic peptide.

β-hairpin peptide, which contains a β-hairpin region in its sequence. β is the dihedral angle of the secondary structure of the peptide. This type of SAP was first designed by the team of Schneider and Pochan to self-assemble to form self-supporting hydrogels. For example, MAX1, the β-hairpin is rich in Val on one side and Lys on the other side, and forms a short, fiber-rich β-sheet structure through face and side binding, and finally forms a mechanically rigid hydrogel [9].

SAPs also contain short peptides and cyclic peptides. Therefore, different SAPs can be designed according to the reported types of SAPs. Sequence design can also be performed by using computer-aided design tools (Table 11.4).

11.2 Supramolecular Structure of SAPs

The diversity of SAP sequences promotes the formation of various supramolecular structures, including nanotubes, nanofibers, nanoribbons, nanospheres, nanocups, hydrogels, etc. Therefore, peptide self-assembling supramolecular structures can provide a unique platform for the construction of nanomaterials with controllable structural properties. In general, the supramolecular structure of SAPs can be determined by microscopic techniques such as transmission electron microscope (TEM), scanning electron microscope (SEM), atomic force microscope (AFM), etc. This section discusses several typical examples of peptide nanostructures. Table 11.5 shows the types and characteristics of peptide nanostructures and typical sequences.

Table 11.5 Peptide nanostructure types and characteristics.

Structure	Features	Typical sequence
Nanofibers	β-folded sheet stacks, 5–20 nm in diameter, micron in length	RADA16, KFE8
Nanotubes/vesicles	Hollow cylindrical or spherical structures	FF, A_6K
β-hairpin hydrogel	Dynamic cross-linked network, high water content (>99%)	MAX1, MAX8
Nanosheets/ribbons	2-D ordered arrangement	AC-I_3NGK-NH_2, C16-OVEVE

11.2.1 Nanotube

A nanotube is a thin, long, tube-like structure with certain internal pores. Covalent bond formation of the nanotube structure in the past few decades has received widespread attention; recently, non-covalent bond nanotubes, because of its easy to synthesis, assembly control and high efficiency become the hot research topic. Self-assembling nanotubes have been reported using different structures, including hollow bundle-rod elements, helical structures and stacked rings. Diphenylalanine peptide (FF) is one of the simplest and most common building motifs for molecular self-assembly. It forms a tubular structure of about 100 μm in length. Cyclic peptides and peptide amphiphiles can also self-assemble to form nanotube structures, such as A_6K and I_3K [10]. It can be seen from literature reports that nanotubes are mainly formed by the assembly of designed peptides.

11.2.2 Nanofiber

Unlike nanotubes, nanofibers are fiber structures with a diameter of less than 100 nm and are solid structures. Fiber structures formed by peptides and proteins are common in biological systems, such as amyloid fibrils. Such structures can be formed from many SAPs, including designed and food-derived peptides. There are two main processes in the formation of fiber structures of peptides. First is the formation of nanospheres, leading to fibril aggregation, followed by mature fibril growth, which is relatively slow [11, 12]. In the formation of SAP fiber structure, π-π stacking and hydrogen bonding determine the interaction direction, and electrostatic interaction is conducive to the formation of β-sheet structure [13].

11.2.3 Nanosheets/Ribbons

Nanosheets/ribbons are thin, long, strip-like structures with rectangular interfaces. Cui et al. [14] designed a peptide amphiphile (C16-OVEVE) that self-assembled to form nanoribbons, whose structure would be affected by pH. When the pH increases, the flat nanoribbons turn into groove-like nanoribbons. Wang et al. designed AC-I_3NGK-NH_2 and AC-I_3NGK-NH_2, two SAPs that can form single-layer nanoribbons with a thickness of 3.3 and 3.2 nm, respectively [15]. In addition, the assembly morphology and size can be adjusted by changing the amino acids, such as AC-I_3K-NH_2 [16]. Due to the hydrophobic interaction of Ile side chains, the electrostatic repulsion between Lys side chains is eliminated, and finally, fine nanofibers are formed.

11.2.4 Other Supramolecular Structures

Peptides can also be assembled into other supramolecular structures, such as nanospheres, nanocups, vesicles, etc. For example, Mandal et al. designed cyclic peptide [WR]n (n = 3–5), and the experimental results showed that all [WR]n can self-assemble into vesicular structures [17], and the interaction between [WR]5 molecules is stronger than that of [WR]3 and [WR]4. Peptide self-assembling supramolecular structure can be controlled by adjusting the interaction conditions (temperature, pH, ions, etc.) and peptide sequence to transform the morphology. Table 11.6 summarizes the nanostructures formed by SAPs.

Table 11.6 Nanostructures formed by SAPs.

Name	Sequence(n → c)	Nanostructure
A_6R	AAAAAAR	Sheets, helical ribbon, and nanotube
capA_6R	CONH-AAAAAAR-NH_2	β-sheet fibrils
A_6K	AAAAAAAAAK	Nanofibers
A_9K	$(AAA)_3K$	Nanorods
$K_3(FA)_4K_3$	KKKFAFAFAFAKKK	Nanofibrils
$K_6(FA)_4$	KKKKKKFAFAFAFA	Nanofibrils
P_5VP_5	Ac-RLIRKVKRILR-NH_2	Nanoparticles
PA_4	$C_{18}K_5$	Micelles
PA_7	$C_{18}K$-K_5	Micelles
KK-11	Ac-$^LK^LH^LH^LQ^LK^LL^LV^LF^LF^LA^LK$-$NH_2$	Nanofibril hydrogel
KKd-11	Ac-$^DK^DH^DH^DQ^DK^DL^DV^DF^DF^DA^DK$-$NH_2$	Nanofibril hydrogel
I_4K_2	IIIIKK	Nanofibril hydrogel
Soy peptide	/	Nanoparticles
Egg yolk peptide	/	Micellar nanoparticles
Phg_4	PhgEPhgK	Nanofibers
Soy peptide	/	Nanoparticles
Zein peptide	/	Nanoparticles
RADA16-I	ac-(RADA)4-CONH2	Fiber hydrogel
MAX8	VKVKVKVKV^{D}PPTKVEVKVKV-NH_2	Fiber hydrogel
Meal-derived peptides	/	Nanoparticles
	EDLAALEK	Clusters aggregation
E_WK-5	EIWLK	Nanoparticles/fiber hydrogel
	Pyrene-FEFEF	Fiber hydrogel
T13	LNALEPDNTVQSEAGTIETWNPK	Fiber hydrogel
	FmocFF	Fiber hydrogel

11.3 Mechanism Consideration

Non-covalent interaction forces are involved in peptide self-assembly, including hydrogen bonding, π-π stacking, electrostatic, hydrophobic, and van der Waals forces. The synergistic interaction of these forces determines the formation of thermodynamically stable supramolecular structures of SAPs. It is a spontaneous process driven by thermodynamics and kinetics. Due to their relatively weak nature, these can be precisely and reasonably tuned by various kinetic parameters. By changing the mode of noncovalent interactions with selected kinetic parameters, including pH, temperature, concentration, and solvent, assembled structures can be controlled in metastable states that will eventually develop into thermodynamically favorable states without the intervention of kinetic parameters. Thus, the competitive relationship between the dynamic and thermodynamic states of self-assembly offers the possibility of moving from thermodynamic to dynamic control. In this section, the mechanism of peptide self-assembly is elucidated from the thermodynamic and kinetic perspectives.

11.3.1 Thermodynamics of Self-Assembly

The synergy of various noncovalent interactions during self-assembly is necessary to determine a thermodynamically stable structure. The fact that individual amino acids, such as Phe or Ser, can form self-assemblies (oligomers, clusters, or even fibrous aggregates) in solution through noncovalent interactions hints at the mechanism of peptide assembly and growth trends [18]. At present, the thermodynamic research on peptide self-assembly mainly focuses on two aspects: (i) control/driving force and (ii) the free energy of assembly.

The classical self-assembling dipeptide FF forms a variety of self-assembling structures, such as hexagonal microtubules [19, 20], etc. These structures can realize the transition between structures by adjusting the kinetic parameters. Some studies have found that the microtubule structure may be the most thermodynamically stable in water, and its crystalline part is stabilized by the hydrophobic interaction between the side chains, and the electrostatic interaction between the two ends makes the main chain into a more ordered state [21]. Amphiphilic peptide molecules can also assemble into various supramolecular structures, among which cylindrically shaped nanofibers may be a thermodynamically favorable assembly structure. Hydrogen bonding, π-π stacking, hydrophobic interactions, and electrostatic repulsion are the main sources of capability contributions during the assembly of amphiphilic peptide molecules. Among them, hydrogen bonding, π-π interaction, and hydrophobic interaction promote the aggregation of amphiphilic peptide molecules, while electrostatic repulsion promotes the dissociation of amphiphilic peptide molecules. MD simulation results resolved the self-assembly process of amphiphilic peptides, and the "phase diagram" was obtained by combining the hydrogen bond strength and the hydrophobic attraction strength. The simulation results showed that the "pure" hydrophobic interaction promotes micellar structure formation, while the "pure" hydrogen bond interaction promotes the aggregation of 1D β-sheets [22].

Therefore, the driving force is important in the process of peptide self-assembly. To illustrate more intuitively the free energy involved in peptide self-assembly, the self-assembled peptides P11-2 and butoxycarbonyl-FF (BOC-FF) are discussed as examples [23, 24]. The P11-2 sequence was designed and synthesized according to the amino acid properties.

The forces involved in the sequence are hydrophobic interaction of side chains, π-π stacking interaction of aromatic groups, and complementary charge attraction between charged amino acids. The assembly process of P11-2 was confirmed by layering assembly experiments. The randomly coiled monomer formed by P11-2 first transformed into a β-strand, and the free energy increased to $\varepsilon_{\text{tape}}$, which was mainly due to the loss of configurative entropy. The β-sheet tapes band is then formed, and the binding free energy is generated between each pair of interacting monomers. In addition, the tape has a spiral shape because one side of the tape is conducive to interacting with itself, and the other side tends to interact with the solvent. When the tape pairs due to the hydrotropism and hydrophobicity of the surface, it creates a twisted ribbon, which occurs as the free energy of each peptide changes to $\varepsilon_{ribbon}^{attr}$. As the ribbons pile up, a fibrillary structure is formed, with a free energy change of $\varepsilon_{\text{fibril}}^{attr}$ for each pair of interacting peptides. When the edges of the fibrils are attracted, the fiber is produced and its free energy changes to $\varepsilon_{\text{fiber}}^{attr}$. These twisted aggregates bend for stacking purposes and change their twisted state with the loss of elastic energy $\varepsilon_{\text{elast}}$. The critical assembly concentration of the peptide structure can be determined according to the corresponding calculation formula of the free energy. For the assembly of P11–2, the critical concentration for band formation was 0.1 mM, and the concentration for fibril formation was 0.6 mM. When the concentration of P11–2 increased to about 15 mg/mL, the formation of fiber structure could be detected, and P11–2 could form hydrogel.

For the self-assembly thermodynamics of BOC-FF, the Gibbs free energy difference at each stage of the self-assembly process can be determined by chemical thermodynamics. The Gibbs free energy difference between the monomer and the sphere is $-16.3\,\text{kJ mol}^{-1}$, and the difference between the sphere and the tube is $-19.6\,\text{kJ mol}^{-1}$. In addition, the contribution of the free energy to the entropy (ΔS) and enthalpy (ΔH) can be determined by calorimetric measurement experiments. During the transition from amorphous sphere to transistor in BOC-FF self-assembly, the entropy change is unfavorable.

11.3.2 Dynamics of Self-Assembly

The dynamic control of peptide self-assembly is of great significance for resolving the mechanism of peptide self-assembly and constructing peptide-based dynamic materials. The supramolecular structure of SAPs can be regulated by pH, temperature, electrolyte concentration, and other factors. The study of dynamics includes two important aspects: (i) the regulation of structure by kinetic parameters and (ii) dynamic nucleation and growth processes.

Although self-assembly is an energy-driven process, kinetic parameters can regulate the assembly structure by affecting the thermodynamic interactions between peptide molecules and between peptide molecules and solvents [25, 26]. In the discussion of thermodynamics, it was mentioned that the self-assembly of FF to form a variety of supramolecular structures can be adjusted by kinetic parameters. For example, FF can form flower-like microcrystalline structures in ethanol and nanofibers in toluene. In addition, it is also regulated by the concentration. When FF is diluted in solvent-friendly HFIP and solvent-hydrophobic water, it can form nanotubes, and in the dilution process, the microtubule structure will transform into a spherical vesicle structure [27]. The results of MD also confirmed that the assembly path of FF is concentration-dependent [20]. In addition to the structural transition of the assembly path of FF dependent on solvent and concentration, its supramolecular structure is also regulated by conditions such as pH, temperature, and enzyme [28].

In the study of amphiphilic peptides, Stupp's team found that by regulating pH and concentration, the balance of electrostatic repulsion and hydrogen bond interaction of amphiphilic peptides was destroyed, and the nanoribbons were transformed into twisted forms. Recently, Stupp et al. adjusted the preparation pathway by changing the order of water and HFIP addition. In addition, MD simulations investigated the effect of solvents on the assembly kinetics of amphiphilic peptides. The results showed that when the solution environment of amphiphilic peptides becomes solventphilic, the increase in hydrophobic strength, that is, the decrease in solvent hydrogen bond formation ability, will cause the supramolecular structure of amphiphilic peptides to change from an open β-sheet mesh and nanofiber structure to an elongated micelle structure [29].

The amphiphilic peptide PEP-1 (Asp-Lys-Ala-Ala-Asn-Leu-Leu-Phe) [30] parameters (Figure 11.1). The first is the regulation of the structure of PEP-1 by pH. At pH 5.5, PEP-1 forms short, disordered oval aggregates. At pH 7.4, PEP-1 transforms into a twisted nanofiber structure. At pH 13.0, PEP-1 transforms into a fractal structure with a strong β-sheet conformation. In addition, concentration and temperature can also have an effect on the assembled structure. At pH 7.4, when diluted by PEP-1, the nanofibers with a twisted sheet secondary structure are transformed into undefined spherical nanoparticles, where the peptides adopted a random coil conformation and, when thermally annealed, are further transformed into short nanorods with α-helix structure. At pH 7.4 and high

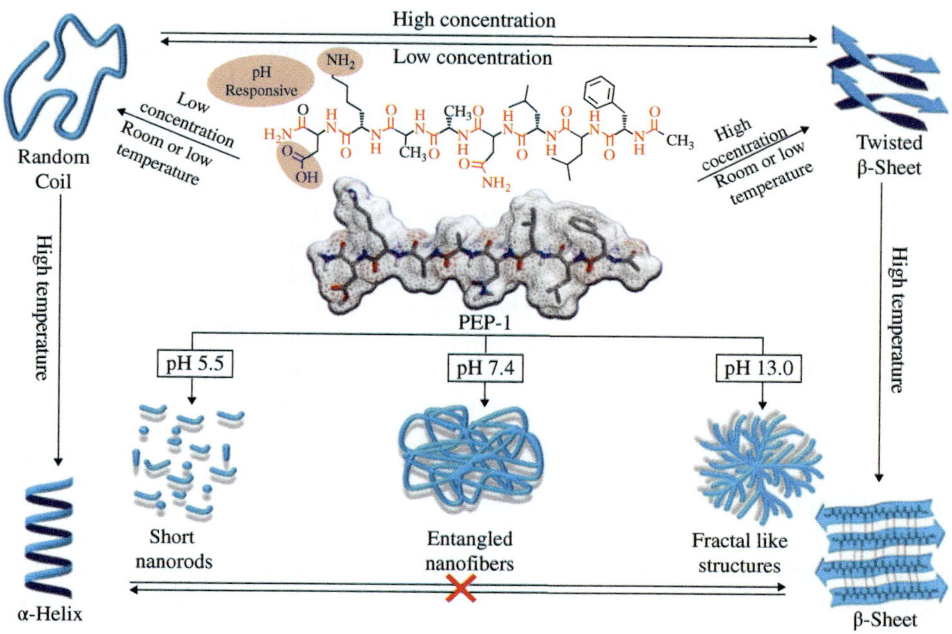

Figure 11.1 Morphological and secondary structural transformations of PEP-1. *Source:* [30]/John Wiley & Sons/CC BY-NC 4.0.

concentration of PEP-1, the twisted nanofibers undergo thermal annealing to form 2D sheets with well-defined β-sheet conformation. Therefore, the adjustment of kinetic parameters alters the interactions between peptide units, resulting in the construction of different supramolecular structures.

For the kinetic nucleation and growth process, it can be learned from the formation of SAP hydrogels, as the process strongly depends on kinetics. In the case of amyloid fibril formation, for example, there are several representative nucleation and growth pathways. First, unstructured peptide monomers with random configurations collide in solution to form nuclei. The nuclei undergo a lag period before reaching a critical concentration in solution [31]. When the nucleus reaches a critical size, fibrils are formed, and peptide monomers are added to the fibril ends to promote the formation of full-length fibers through a dock and lock mechanism [32, 33]. When the peptide concentration is too low, the newly formed fibrils will break or dissociate, and the free peptide polymer will undergo secondary nucleation.

11.3.3 Effects of Kinetic/Thermodynamic Interactions

In many cases, there is no clear line between thermodynamic and dynamic control. For example, the design of a peptide molecule, or the dynamic control of solvent properties, is actually a kind of thermodynamic control as well. Because the balance of the original driving force, or the energy barrier, shifts. Temperature can affect the hydrogen bonds and hydrophobic interactions between peptide molecules. Increased temperature can weaken or destroy intermolecular hydrogen bonds [34], but promote hydrophobic effects [35]. For example, Max1 presents an unfolded state at temperatures below 20 °C, and folding of the peptide occurs when the temperature increases because the hydrophobic surface of the peptide is buried away from water [9].The pH of the solution affects the competitive solvation between the hydrogen bond donor and acceptor sites. The nature of the solvent has a great influence on the solvation of peptide molecules.

Taking FF as an example, Kim et al. [36] found that changing the concentration, temperature, ions, and ultrasound of FF all induced a structural transition between nanowires and microtubules. It is mainly the hydrogen bonds between water and FF molecules that play a role in assembly. Heating the FF assembly system, introducing ions, and ultrasonic treatment all weaken the hydrogen bonds between the peptide molecules. In addition, the concentration of FF regulates the hydrogen bond ratio between FF and water molecules. In this case, the regulation of dynamic factors has an impact on the thermodynamic interaction (hydrogen bonding), driving the structural transition of the nanotubes and nanowires.

11.4 Function Properties of SAPs

Supramolecular structures generated by the self-assembly of peptides possess some functional properties due to their unique structural characteristics. Understanding these functional properties can help researchers obtain desired functions by constructing peptide self-assembly systems. The functional properties of SAPs are closely related to proteins, such as foaming ability, water retention, and gelling ability. However, they also have some unique functions, such as assembly reversibility. In this section, we introduce several typical functional properties of peptide self-assemblies in order to provide some potential applications in the food industry.

11.4.1 Amphiphilicity

For most SAPs, peptide sequences typically contain both hydrophobic and hydrophilic amino acids. If the hydrophobicity of peptide-based fibers is too high, self-supporting hydrogels cannot be formed, and precipitation occurs instead. Thus, hydrophilic residues in the sequence can improve hydrophilicity to solvate peptide-based fibers. The higher the hydrophilicity, the greater the self-assembly free energy. Therefore, if the peptide sequence is too hydrophilic, the interaction between peptide molecules will be significantly reduced, and the continuous existence of self-assembled fibrils cannot be maintained [37]. As a result, SAPs are usually amphiphilic.

Among artificially designed SAPs, ion-complementary peptides and peptide amphiphiles are typical sequences with amphiphilicity. Among them, the hydrophobic amino acids in the ion-complementary peptide form hydrophobic regions on the inside of the fiber, while the charged complementary hydrophilic amino acids are on the outside. For example, when the ion-complementary peptide RADA16–1 forms a stable β-sheet in water, the inner part of the β-sheet structure is composed of a hydrophobic Ala array, and the outer part is formed by the electrostatic interaction of negatively charged Asp and positively charged Arg [38]. The sequence of the surfactant-like peptide in the peptide amphiphile is composed of consecutive hydrophobic amino acid residues and charged amino acids so that the hydrophilic amino acids form the hydrophilic side and the hydrophobic amino acids form the hydrophobic core. For example, ANKs ($n = 3$, 6, and 9) assemble into membranes, nanotubes, and nanosheets, respectively, with the increasing number of Ala. However, their assembled structures all share a common hydrophobic core and a hydrophilic side [39].

In food-derived SAPs, the sequences are also amphiphilic. For example, buckwheat peptide can self-assemble into a 3D gel network structure with good rheological properties, and its sequence is an alternating pattern of hydrophilic and hydrophobic amino acids [40, 41]. Du et al. used amphiphilic yolk peptides to construct nanoparticles with high surface activity, interfacial deformability, and moderate wettability, which can be used for the stabilization of Pickering emulsions [42]. In addition, zein deamination peptides have high carboxyl and hydrophobic residues and can self-assemble into nanospheres [43]. In addition, the amphiphilicity of β-lg AB [f1–8], a peptide mentioned above, confirmed its ability to self-assemble into a β-sheet structure and eventually form a hydrogel [3].

Therefore, in general, peptides with amphiphilic properties tend to cluster the hydrophobic regions of the peptide together to reduce their exposure to the surrounding aqueous environment, while the hydrophilic regions face outward to interact with solvents.

11.4.2 Gelation Ability

Another important functional property of SAPs is gelation, that is, the formation of semi-solid transparent gels with the peptide as a gelling factor by binding to water molecules. Both designed peptides and food-derived peptides can form hydrogel structures. Among them, food-derived peptides include milk hydrolysate, milk fermentation, egg white, [1, 2, 44, 45] etc. The formation process of SAPs hydrogels can be divided into three steps: (i) SAPs molecules are arranged in a certain order to form fibrils under the joint action of various non-covalent interactions (electrostatic interaction, hydrogen bond, π-π bond, hydrophobic interaction); (ii) the fibrils continue to aggregate to form nanofibers; (iii) the nanofibers intertwine with each other and form a network structure with water molecules to form a self-supporting macroscopically visible hydrogel [46]. There have been many reports on the research progress of peptide hydrogels, including different preparation methods of peptide self-assembled hydrogels. [47–49].

Most current studies on SAPs hydrogels have focused on artificially designed peptides, but the gelation ability of food-derived natural peptides has also been reported. Recently, Yu et al. [40] prepared peptide hydrogels, that is, the buckwheat protein was hydrolyzed by alkaline protease to obtain peptides with a molecular weight less than 5 kDa. When the concentration was 4%, the buckwheat protein peptide could self-assemble into a hydrogel with good rheological properties. Similarly, peptide f1–8 at concentrations greater than 2.5 mg/mL and pH 9.0–11.0 form a hydrogel [45]. In addition, the preparation of SAP hydrogels can also be achieved by the addition of metal ions. α-lactalbumin (α-la) peptide nanotubes can form a hydrogel after adding Mn2+, and the higher the Mn2+ content, the lower the hardness of the hydrogel [50]. Egg white lysozyme can also form a hydrogel under the action of Mg2+ after being heated and acid-hydrolyzed [1].

11.4.3 Assembly Reversibility

Since the self-assembly of peptides is based on noncovalent interactions, the supramolecular network structure can be reversibly transformed in response to external stimuli, typically in response to changes in mechanical stress, pH, temperature, ionic strength, and the presence of other types of molecules.

SAP hydrogels benefit from this property and are modifiable, adjustable, and dynamic under external forces. For example, E_WK-5 hydrogel would be shear thinning and flowing when subjected to mechanical shear and could recover its original stiffness after removing stress [51]. In addition, pH regulation can promote the protonation or deprotonation of peptides, which leads to reversible rearrangement of molecular stacking and structural changes during peptide self-assembly by affecting the strength of hydrogen bonds between hydrogels. Moyer et al. [52] designed an amphiphilic peptide that can assemble to form a fiber structure at pH 7.5, while the fiber decomposes at pH 6.5, and the process is reversible. Similarly, Geng et al. [53, 54] demonstrated that the nanostructure of α-la peptide can be transformed by conditions. At pH 7.5, α-la peptide self-assembled into long, straight nanotubes. When the pH was decreased to 4.0, the α-la peptide formed short, curved nanotubes. Temperature affects the interaction between hydrophobic and hydrogen bonds in the self-assembly. Loveday et al. [55] found that β-lg peptide self-assembled into a fibrous structure after heating at 75–120 °C, but prolonged heating at high temperatures destroyed the formed fibers. In addition, Huang et al. [56] found that FF preferentially self-assembled into a stable nanofibrous gel during thermal cooling. With the increase of temperature, a further transformation into a thermodynamically stable crystal structure is observed.

Therefore, peptide self-assembly can be controlled and reversible by changing the external conditions (mechanical stress, pH, and temperature) to adjust the supramolecular structure.

11.4.4 Photoelectric Properties

It has been found that strong Coulomb interactions between highly confined electrons and holes exist in nanoscale quantum confinement structures formed by self-assembled short peptides, conferring optical properties, and therefore, self-assembled short peptides may have good semiconductor properties [57].

FF can be used as quantum dots that self-assemble into dimers, forming a supramolecular semiconductor [58]. This is because the directional aromatic interaction between molecules and the hydrogen bond network leads to the formation of a nanometer structure internal quantum limit domain. When solvents range from water to methanol, the FF peptide monomer formation of the nanotube structure will be broken down into individual dimers and quantum dots. When reduced to the solution, the quantum dots would further self-assemble and once again form the nanotubes; the process is reversible [58]. BOC-FF, an elemental analogue of FF, self-assembled to form a rigid nanosphere structure, and the peak absorption spectroscopy results showed that the quantum dot was composed of two Boc-FF molecular groups [59].

Some rings with 2, 5-2 ketone piperazine dipeptide nanostructures can be self-assembled into quantum constraints, used as building blocks to form dimer quantum dots. Tao et al. [60] designed aromatic cyclic dipeptides that can self-assemble into quantum-constrained microstructures with photoluminescence properties. The experimental results showed that the diameter of the quantum dot is 2.24 nm, which is about twice the size of the monomer. The main driving force for the dimerization of cyclic dipeptides is the π-π interactions between aromatic side chains, especially the indole ring, which is the steric conjugation of the adjacent electron clouds of the indole ring. After the formation of dimeric QDS, the cyclic dipeptide monomer serves as the basic building block for further assembly into supramolecular structures.

11.5 Security Assessment

SAPs' security needs to be assessed comprehensively from four dimensions: molecular toxicity, biocompatibility, immunogenicity, and the environmental of residual risk. Table 11.7 shows the safety assessment of different methods. For SAPs, PuraMatrix® (RADA16-I), an approved clinical product, is currently in a Phase I clinical trial (NCT02503892). The results of the trial showed mild local inflammation (redness and swelling score grade 1) in only 1 of 20 patients with chronic wounds 11. Long-term follow-up for 5 years, no tumorigenic antispasmodic drugs or organ fibrosis were found. However, there were also cases of failure, such as antimicrobial peptide LL-37 nanofibers, which terminated when a high dose (2 mg/kg) of peptide increased C3a level to 200 ng/mL (normal value <50 ng/mL), triggering systemic complement activation 13. Therefore, the safety of SAPs needs to be further evaluated, and several key safety data and examples will be discussed in this section.

Table 11.7 Safety assessment parameters and methods.

Evaluation dimensions	Key parameters	Methods of standardization
Acute toxicity	LD_{50} (median lethal dose), EC_{50} (median effective concentration)	OECD 423 (Rodent test)
Long-term toxicity	Organ histopathology and blood biochemical indicators	Dose toxicity tests were repeated on 28/90 days
Immunogenicity	Cytokine release (IL-6, TNF-α), complement activation	ELISA and flow cytometry
Toxicity of degradation products	Amino acid metabolites, oligopeptide fragment toxicity	HPLC—MS metabonomics analysis

11.5.1 Biocompatibility

The biocompatibility of SAPs can be evaluated from two aspects: cytotoxicity and hemolysis risk. For cytotoxicity, the RADA16-I study found that the viability of L929 fibroblasts was greater than 95% at a concentration of 1 mg/mL (ISO 10993-5), and high concentration (>5 mg/mL) resulted in a decrease in mitochondrial membrane potential (40% reduction in JC-1 fluorescence ratio). For hemolysis risk, studies have shown that when the concentration of the cationic peptides K8 (KKKKKKKK) is 0.1 mg/mL, the erythrocyte hemolysis rate was less than 5% (in line with the ISO 10993-4 standard), and when the concentration is greater than 0.5 mg/mL, the hemolysis rate soared to 30%, mainly because of static electricity damage to cell membranes.

11.5.2 Immunogenicity

Immunogenicity is divided into innate immune activation and adaptive immune response. When the concentration of EAK16-II is between 10 and 100 ng/mL, the β-sheet fibers formed by EAK16-II can activate the TLR2/4 pathway and induce the secretion of IL-6 by macrophages. The MAX8 neutral peptide studies have shown that the immunogenicity was significantly lower than cationic peptides, and IL—1β release a quantity reduced by 80%. As a result, some SAPs can activate innate immune cells. GAGAGS may trigger T cell epitope responses, and low-risk sequences need to be screened by MHC-II molecular binding prediction tools.

11.5.3 Degradation Product Toxicity

The safety of SAPs can be evaluated from the kinetics of enzymatic hydrolysis. The degradation half-life of RADA16 in plasma is about 48 h, and the main product is RADA tetramer, so it is not cytotoxic. Cyclo-modified peptides (e.g., Cyclin-RGD) had a 10-fold increase in antitrypsin ability and reduced degradation products to 20% of wild type. It can also be evaluated from the accumulation of metabolites. The concentration of phenylalanine released by phenylalanine-rich peptides (such as FF-FF) after degradation should be less than 50 μM, which may induce the risk of phenylketonuria.

11.6 Applications

Supramolecular structures formed by SAPs have a wide range of applications in various fields, including food, biomedicine, and environmental science and so on.

11.6.1 Food Industry

11.6.1.1 Nutrient Delivery Systems

In the food industry, fat-soluble nutrients (such as curcumin, carotene, resveratrol, etc.) have high nutritional value, but they have low bioavailability, such as poor water solubility and chemical instability problems [61, 62]. Nutrient delivery systems, such as emulsions, micelles, liposomes, etc., can be constructed by utilizing the amphiphilic and gelation properties of SAPs, with the advantages of a simple process, high encapsulation efficiency, and biocompatibility.

Soybean peptide, corn protein peptide, and the α-la peptide nanoparticles can coat curcumin [43, 63, 64]. Rapeseed meal-derived peptide nanoparticles [65, 66] can encapsulate β-carotene, and also achieve improved stability and water solubility. In addition, whey protein-derived peptide (sequence: VLPVPQ) nanomicelle can coat vitamin D_3, improving the bioavailability of vitamin D_3 in vitro. Peptide self-assembly supramolecular structures can also be used for encapsulation and protection of probiotics, such as amphiphilic peptide A_6K nanocapsule wrapped in lactic acid bacteria, acid environment survival rate increased from 10 to 85% in vitro simulation experiment.

11.6.1.2 Food Detection

Self-assembling peptide nanostructures (such as inorganic quantum dots or carbon dots) have made some progress in the field of food detection. The combination of signals from nanostructures (including fluorescence, chemiluminescence, and electrochemiluminescence) and sensors (such as enzymes, antibodies, aptamers or molecularly imprinted polymer sensors) has been used to realize the detection of agricultural and veterinary drug residues, harmful substances, and metal ions in food.

A biosensor based on modified FF peptide nanotubes can realize sensitive detection of glucose, and the reaction mechanism is that glucose oxidase on peptide nanotubes reacts with glucose to produce hydrogen peroxide [67] peptide nanotubes were used as the coating template of horseradish peroxidase, and were co-fixed with glucose oxidase on the nano-gold modified electrodes to form a kind of nano environmental sensors [68]. In addition to using the material of the structure of peptide self-assembly supramolecular detection, it can also be used for the determination of the freshness of food detection. Anna et al. developed a Cu^{2+}-reactive peptide hydrogel that is sensitive to amines. When amines in food volatilize, they can produce discoloration, which can be used for food freshness detection [69].

11.6.1.3 Other Applications

SAPs have other applications as well. For example, pea peptide P1–8 (sequence: LIVTQTMK) self-assembled into a fiber network, can simulate animal muscle texture (texture profile analysis [TPA] hardness of 2,000 g). The cyclic peptide Cyclo-(FFRGD) can encase caffeine and reduce bitter perception through hydrogen bonding (electronic tongue bitter value decreased from 9.5 to 3.2).

11.6.2 Biomedical Fields

11.6.2.1 Pharmaceutical Effect

Many bioactive peptides have pharmacological effects and are known as therapeutic peptides. Pharmacokinetics of therapeutic peptides is a hot research topic, and the means by which they usually have to add the encoded amino acid and cholesterol or fatty acids combined with [70]. The supramolecular structure of SAPs has a certain effect on prolonging the retention time of therapeutic peptides. Semaglutide (SMG) is a 37 peptide for the treatment of type 2 diabetes [71]. SMG analogues can be designed to prolong pharmacokinetics by altering the sequence. Venanzi et al. [72] substituted the Ala residue at position 8 with Aib to inhibit enzymatic degradation. Lys is at position 26, and two consecutive PEG groups are attached to the side chain; in addition, a C_{18}-OH chain is attached to the Glu side chain, increasing the affinity of SMG for human serum albumin. In aqueous solution, SMG is mainly composed of monomers and oligomers with a nanospherical structure. With the prolongation of maturation time (weeks), a stable aggregation process occurs, and the dendritic structure of the nanorods is formed. This aggregation potential is critical for the design of peptide drugs with optimal pharmacokinetics.

Wang designed a SAP to induce T cell-mediated lysis of cancer cells [73]. The peptide consists of a targeting group and a self-assembling module (G7-RGD) that self-assembles into nanofibers. The targeting group on G7-RGD can specifically bind to the overexpressed integrin receptors on the membrane of tumor cells, and the resulting G7-RGD cluster drives the oligomerization of integrin receptors to realize an in situ self-assembly strategy for cancer immunotherapy. The team also designed a double specificity coupling (D3—the G7—RGD), promoting the CD3 receptor in vitro oligomerization and T cell activation, and eventually led to dissolve T cell-mediated tumors. Chen et al. engineered a pH-sensitive peptide H5K8 (sequence: HVHVHVHVKKKKKKKK) that disassembled to release doxorubicin (DOX) in the tumor microenvironment (pH 6.5), reducing the IC_{50} against MCF-7 breast cancer cells from 1.2 to 0.3 μM. Therefore, applying the self-assembly strategy to the design of therapeutic peptides can achieve a more effective therapeutic effect.

11.6.2.2 Antibacterial Nanomaterials

In some cases, antibacterial effects are enhanced when antimicrobial peptides self-assemble into nanostructures. This is because the structural characteristics of self-assembled antimicrobial peptides increase the positive charge density and high surface energy/surface area-to-volume ratio and have a larger contact and adhesion area with the bacterial surface, which can produce more electrostatic interactions [74]. More importantly, self-assembled antimicrobial peptides have better stability, and their sensitivity to enzymatic degradation, renal filtration, and reticuloendothelial system uptake is reduced [75].

Many SAP supramolecular structures have antimicrobial activity themselves and have been developed as novel antimicrobial agents. Some SAPs with cationic groups have been shown to possess antimicrobial activity. For example, A_6R, A_6K, and A_9K have antibacterial activity against Escherichia coli and Staphylococcus aureus [76]. Among them, there is a positive correlation between the membrane penetration ability of A_9K and the bactericidal ability [77]. Sha et al. designed $K_3(FA)_4K_3$ and $K_6(FA)_4$ to self-assemble into nanofibrils that inhibit the activity of *Escherichia coli*, *Salmonella* [78, 79] and *Xanthomonas axonicum* [80].

Huang et al. used the self-assembly property of RADA16 [81] and the antibacterial activity of antimicrobial peptide Tet213 to construct a new antibacterial nanomaterial (RADA16-AMP), which can inhibit the activity of *Staphylococcus aureus*. Zotting et al. [82] designed a chimeric peptide that can self-assemble into nanorods. Peptide nanorods in combination with the macromolecular adjuvant montanide gel provided complete protection against lethal experimental infection with AH1N1 influenza virus in vivo. Therefore, adding a self-assembly motif to the antimicrobial peptide sequence can also realize the construction and activity enhancement of self-assembled antimicrobial peptide systems.

11.6.2.3 Tissue Engineering

Self-assembling peptide hydrogels, as a scaffold material, can be adapted to advantages such as local therapy by injection into specific tissues, making them ideal for tissue regenerative medicine applications.

Due to the loss of skin integrity and function caused by trauma, the host loses its protective barrier against invading pathogens. A heparin-like peptide-based self-assembling hydrogel scaffold (HM-PA) has facilitated wound healing by improving the key drawback of heparin (its rapid degradation in the biological environment). HM-PA hydrogel increased angiogenic activity during early regeneration and promoted re-epithelialization and granulation tissue formation at later stages of burn injury. Recently, Gao and colleagues [83] designed BQA-GGFF, a peptide-based quinazolinone derivative that mimics the innate immune process of neutrophil extracellular traps. At sites of bacterial infection in vivo, reactive oxygen species produced by inflammation successfully induce peptides to self-assemble into nanofibers in situ, thereby trapping bacteria, preventing infection, reducing inflammation, and promoting faster healing of infected wounds.

Wang et al. [84] designed several SAPs and performed 2D tissue culture studies on osteoblasts implanted on the surface of peptide hydrogels. Their strategy included adding a short bioactive motif (i) ALK (ALKRQGRTLYGF) from osteogenic growth peptides, (ii) DGR (DGRGDSVAYG) from osteopontin cell adhesion motif, and (iii) PRG (PRGDSGYRGDS) from fibritin and other ECM proteins that promote cell adhesion with SAP Ac-$(RADA)_4$-$CONH_2$. In this study, the authors observed that the mixture of an unmodified self-assembled peptide Ac-$(RADA)_4$-$CONH_2$ with a functionalized self-assembled peptide Ac-$(RADA)_4$ -$CONH_2$ with ALK, DGR, or PRG functional motifs resulted in a significant increase in mouse osteoblast MC3T3-E1 cell proliferation and osteocalcin secretion in a hydrogel scaffold, which is a hallmark of osteoblast differentiation.

11.6.3 Other Applications

In recent years, the application of SAPs in biotechnology has been diversified and expanded, such as vascular suture [85]. Recently, Smith et al. reported that a peptide-based gel was able to undergo a multiphase transition, enabling it to greatly facilitate a high degree of anastomosis to ultra-small blood vessels and achieve the suturing of blood vessel [85].

In their design, the SAP forms a supramolecular hydrogel in a syringe, and can be injected into the broken site of the blood vessel and enter the space between the two blood vessels, greatly enhancing the suture process of the blood vessel. After completing the suture, any external gel is washed and the intravascular gel is dissolved by initiating its gel–sol phase transition in which light irradiation disrupts the intravascular gel-hydrogel network, thereby restoring normal blood flow. Yang et al. [86] developed a vaccine nodule by encapsulating dendritic cells and tumor antigens into a RADA16 hydrogel. After subcutaneous injection, the vaccine tumor remains active and biologically functional, enhancing the proliferation of antigen-specific spleen cells and causing cellular immune responses. Therefore, RADA16 hydrogel can be used for immunotherapy of infectious diseases and cancer. Raffaele found that lupin-derived peptides can self-assemble into hydrogels with antioxidant activity without the addition of other antioxidants [87].

11.7 Limitations and Challenges

The current limitations and development directions of SAPs are as follows:

1) The bottleneck of large-scale production: Peptides are mainly produced by chemical synthesis or enzymatic hydrolysis in the laboratory. In solid-phase synthesis, the cost of producing 1 kg of RADA16 is 500,000 US dollars, while that of PLGA is only 10,000 US dollars. Therefore, it has the problems of high cost, complex process, and low yield.
2) Stability: SAPs will undergo enzymatic hydrolysis in the physiological environment, resulting in structural disintegration, especially for SAPs with long sequences and natural amino acids. For example, RADA16 lost 50% of its structural integrity after 24 h in the medium containing 10% fetal bovine serum.
3) Limitation of dynamic monitoring technology: At present, the assembly process of peptide self-assembly cannot be dynamically monitored. For example, cryoelectron microscopy can only capture the static structure and cannot resolve the intermediate state of assembly.
4) Safety: It is necessary to establish a complete system for the safety evaluation of SAPs, including long-term toxicity mechanism, multi-omics integration analysis, and biomimetic immune tolerance design.

References

1 Yang, L., Li, H., Yao, L. et al. (2019). Amyloid-based injectable hydrogel derived from hydrolyzed hen egg white lysozyme. *ACS Omega* 4: 8071–8080.

2 Manna, S., Ghosh, A.K., and S.M.J.o. (2019). Mandal, curd-peptide based novel hydrogel inhibits biofilm formation, quorum sensing, swimming motility of multi-antibiotic resistant clinical isolates and accelerates wound healing activity. *Frontiers in Microbiology* 10: 951.

3 Pouliot, Y., Guy, M.-M., Tremblay, M. et al. (2009). Isolation and characterization of an aggregating peptide from a tryptic hydrolysate of whey proteins. *Journal of Agricultural and Food Chemistry* 57: 3760–3764.

4 Josefsson, L., Cronhamn, M., Ekman, M. et al. (2019). Structural basis for the formation of soy protein nanofibrils. *RSC Advances* 9: 6310–6319.

5 Vauthey, S., Santoso, S., Gong, H. et al. (2002). Molecular self-assembly of surfactant-like peptides to form nanotubes and nanovesicles. *Proceedings of the National Academy of Sciences of the United States of America* 99: 5355–5360.

6 Von Maltzahn, G., Vauthey, S., Santoso, S., and Zhang, S. (2003). Positively charged surfactant-like peptides self-assemble into nanostructures. *Langmuir* 19: 4332–4337.

7 Stupp, S.I., Hartgerink, J.D., and Beniash, E. (2001). Self-Assembly and Mineralization of Peptide-Amphiphile Nanofibers. *Science* 294: 1684–1688.

8 Hong, Y., Legge, R.L., Zhang, S., and Chen, P. (2003). Effect of amino acid sequence and pH on nanofiber formation of self-assembling peptides EAK16-II and EAK16-IV. *Biomacromolecules* 4: 1433–1442.

9 Rughani, R.V. and Schneider, J.P. (2008). Molecular design of beta-hairpin peptides for material construction. *MRS Bulletin* 33: 530–535.

10 Bucak, S., Cenker, C., Nasir, I. et al. (2009). Peptide nanotube Nematic phase. *Langmuir* 25: 4262–4265.

11 Luo, J., Wärmländer, S.K.T.S., Gräslund, A., and Abrahams, J.P. (2014). Alzheimer peptides aggregate into transient nanoglobules that nucleate fibrils. *Biochemistry* 53: 6302–6308.

12 Michaels, T.C.T., Šarić, A., Curk, S. et al. (2020). Dynamics of oligomer populations formed during the aggregation of Alzheimer's Aβ42 peptide. *Nature Chemistry* 12: 445–451.

13 Jalali, S., Yang, Y., Mahmoudinobar, F. et al. (2022). Using all-atom simulations in explicit solvent to study aggregation of amphipathic peptides into amyloid-like fibrils. *Journal of Molecular Liquids* 347: 118283.

14 Cui, H., Muraoka, T., Cheetham, A.G., and Stupp, S.I. (2009). Self-assembly of giant peptide nanobelts. *Nano Letters* 9: 945–951.

15 Wang, M., Wang, J., Zhou, P. et al. (2018). Nanoribbons self-assembled from short peptides demonstrate the formation of polar zippers between β-sheets. *Nature Communications* 9: 5118.

16 Wang, A., Cui, L., Debnath, S. et al. (2017). Tuning supramolecular structure and functions of peptide bola-amphiphile by solvent evaporation–dissolution. *ACS Applied Materials & Interfaces* 9: 21390–21396.

17 Mandal, D., Tiwari, R.K., Nasrolahi Shirazi, A. et al. (2013). Self-assembled surfactant cyclic peptide nanostructures as stabilizing agents. *Soft Matter* 9: 9465–9475.

18 Do, T.D., Kincannon, W.M., and Bowers, M.T. (2015). Phenylalanine oligomers and fibrils: the mechanism of assembly and the importance of tetramers and counterions. *Journal of the American Chemical Society* 137: 10080–10083.

19 Wang, Y., Huang, R., Qi, W. et al. (2015). Capillary force-driven, hierarchical co-assembly of dandelion-like peptide microstructures. *Small (Weinheim an der Bergstrasse, Germany)* 11: 2893–2902.

20 Kim, J., Han, T.H., Kim, Y.-I. et al. (2010). Role of water in directing diphenylalanine assembly into nanotubes and nanowires. *Advanced Materials* 22: 583–587.

21 Reches, M. and Gazit, E. (2005). Self-assembly of peptide nanotubes and amyloid-like structures by charged-termini-capped diphenylalanine peptide analogues. *Israel Journal of Chemistry* 45: 363–371.

22 Cui, H., Webber, M.J., and Stupp, S.I. (2010). Self-assembly of peptide amphiphiles: from molecules to nanostructures to biomaterials. *Peptide Science* 94: 1–18.

23 Aggeli, A., Nyrkova, I.A., Bell, M. et al. (2001). Hierarchical self-assembly of chiral rod-like molecules as a model for peptide beta -sheet tapes, ribbons, fibrils, and fibers. *Proceedings of the National Academy of Sciences of the United States of America* 98: 11857–11862.

24 Levin, A., Mason, T.O., Adler-Abramovich, L. et al. (2014). Ostwald's rule of stages governs structural transitions and morphology of dipeptide supramolecular polymers. *Nature Communications* 5: 5219.

25 Zhou, P., Deng, L., Wang, Y. et al. (2016). Different nanostructures caused by competition of intra- and inter-β-sheet interactions in hierarchical self-assembly of short peptides. *Journal of Colloid and Interface Science* 464: 219–228.

26 Cheng, H., Cheng, Y.-J., Bhasin, S. et al. (2015). Complementary hydrogen bonding interaction triggered co-assembly of an amphiphilic peptide and an anti-tumor drug. *Chemical Communications* 51: 6936–6939.

27 Yan, X., Cui, Y., He, Q. et al. (2008). Reversible transitions between peptide nanotubes and vesicle-like structures including theoretical modeling studies. *Chemistry--A European Journal* 14: 5974–5980.

28 Yan, X., Zhu, P., and Li, J. (2010). Self-assembly and application of diphenylalanine-based nanostructures. *Chemical Society Reviews* 39: 1877–1890.

29 Fu, I.W., Markegard, C.B., and Nguyen, H.D. (2015). Solvent effects on kinetic mechanisms of self-assembly by peptide amphiphiles via molecular dynamics simulations. *Langmuir* 31: 315–324.

30 Ghosh, G., Barman, R., Mukherjee, A. et al. (2022). Control over multiple nano- and secondary structures in peptide self-assembly. *Angewandte Chemie—International Edition* 61: e202113403.

31 Arosio, P., Knowles, T.P., and Linse, S. (2015). On the lag phase in amyloid fibril formation, physical chemistry chemical physics. *Physical Chemistry Chemical Physics* 17: 7606–7618.

32 Lomakin, A., Chung, D.S., Benedek, G.B. et al. (1996). On the nucleation and growth of amyloid beta-protein fibrils: detection of nuclei and quantitation of rate constants. *Proceedings of the National Academy of Sciences of the United States of America* 93: 1125–1129.

33 Lomakin, A., Teplow, D.B., Kirschner, D.A., and Benedek, G.B. (1997). Kinetic theory of fibrillogenesis of amyloid beta-protein. *Proceedings of the National Academy of Sciences of the United States of America* 94: 7942–7947.

34 Legon, A.C. and Millen, D.J. (1987). Angular geometries and other properties of hydrogen-bonded dimers: a simple electrostatic interpretation of the success of the electron-pair model. *Chemical Society Reviews* 16: 467–498.

35 Jaremko, M., Jaremko, Ł., Kim, H.Y. et al. (2013). Cold denaturation of a protein dimer monitored at atomic resolution. *Nature Chemical Biology* 9: 264–270.

36 Kim, J., Han, T.H., Kim, Y.-I. et al. (2010). Self-assembled nanostructures: role of water in directing diphenylalanine assembly into nanotubes and nanowires. *Advanced Materials* 22: 583–587.

37 Li, Y. and Cao, Y. (2018). The physical chemistry for the self-assembly of peptide hydrogels. *Chinese Journal of Polymer Science* 36: 366–378.

38 Yanlian, Y., Ulung, K., Xiumei, W. et al. (2009). Designer self-assembling peptide nanomaterials. *Nano Today* 4: 193–210.

39 Colherinhas, G. and Fileti, E. (2014). Molecular dynamics study of surfactant-like peptide based nanostructures. *The Journal of Physical Chemistry. B* 118: 12215–12222.

40 Yu, M., Lin, S., Ge, R. et al. (2022). Buckwheat self-assembling peptide-based hydrogel: preparation, characteristics and forming mechanism. *Food Hydrocolloids* 125: 107378.

41 Yin, X., Yu, R., Zhang, J. et al. (2023). Characterization, formation mechanism, and curcumin delivery of buckwheat peptide-based emulsion gels. *Journal of Agricultural and Food Chemistry* 71: 8589–8601.

42 Du, Z., Li, Q., Li, J. et al. (2019). Self-assembled egg yolk peptide micellar nanoparticles as a versatile emulsifier for food-grade oil-in-water Pickering nanoemulsions. *Journal of Agricultural and Food Chemistry* 67: 11728–11740.

43 Li, L. and Yao, P. (2020). High dispersity, stability and bioaccessibility of curcumin by assembling with deamidated zein peptide. *Food Chemistry* 319: 126577.

44 Guy, M.-M. and Voyer, N. (2012). Structure and hydrogel formation studies on homologs of a lactoglobulin-derived peptide. *Biophysical Chemistry* 163-164: 1–10.

45 Guy, M.-M., Tremblay, M., Voyer, N. et al. (2011). Formation and stability of nanofibers from a milk-derived peptide. *Journal of Agricultural and Food Chemistry* 59: 720–726.

46 Babu, S.S., Praveen, V.K., and Ajayaghosh, A. (2014). Functional π-gelators and their applications. *Chemical Reviews* 114: 1973–2129.

47 Du, Z., Fan, B., Dai, Q. et al. (2022). Supramolecular peptide nanostructures: self-assembly and biomedical applications. *Giant* 9: 100082.

48 Li, T., Lu, X.M., Zhang, M.R. et al. (2022). Peptide-based nanomaterials: self-assembly, properties and applications. *Bioactive Materials* 11: 268–282.

49 Fu, K., Wu, H., and Su, Z. (2021). Self-assembling peptide-based hydrogels: fabrication, properties, and applications. *Biotechnology Advances* 49: 107752.

50 Fuciños, C., Estévez, N., Pastrana, L. et al. (2021). Biofunctionality assessment of α-lactalbumin nanotubes. *Food Hydrocolloids* 117: 106665.

51 Chen, H., Zhang, T., Tian, Y. et al. (2022). Novel self-assembling peptide hydrogel with pH-tunable assembly microstructure, gel mechanics and the entrapment of curcumin. *Food Hydrocolloids* 124: 107338.

52 Moyer, T.J., Finbloom, J.A., Chen, F. et al. (2014). pH and amphiphilic structure direct supramolecular behavior in biofunctional assemblies. *Journal of the American Chemical Society* 136: 14746–14752.

53 Geng, X., Kirkensgaard, J.J.K., Arleth, L. et al. (2019). The influence of pH, protein concentration and calcium ratio on the formation and structure of nanotubes from partially hydrolyzed bovine α-lactalbumin. *Soft Matter* 15: 4787–4796.

54 Geng, X.L., Bjerrum, M.J., Arleth, L. et al. (2016). Formation of nanotubes and gels at a broad pH range upon partial hydrolysis of bovine α-lactalbumin. *International Dairy Journal* 52: 72–81.

55 Loveday, S.M., Wang, X.L., Rao, M.A. et al. (2012). β-Lactoglobulin nanofibrils: effect of temperature on fibril formation kinetics, fibril morphology and the rheological properties of fibril dispersions. *Food Hydrocolloids* 27: 242–249.

56 Huang, R., Wang, Y., Qi, W. et al. (2014). Temperature-induced reversible self-assembly of diphenylalanine peptide and the structural transition from organogel to crystalline nanowires. *Nanoscale Research Letters* 9: 653.

57 Bingbing, X., Qianli, G., and Ehud, J. (2019). Photoactive properties of supramolecular assembled short peptides. *Chemical Society Reviews* 48: 4387–4400.

58 Kim, S., Kim, J.H., Lee, J.S., and Park, C.B. (2015). Beta-sheet-forming, self-assembled peptide nanomaterials towards optical, energy, and healthcare applications. *Small (Weinheim an der Bergstrasse, Germany)* 11: 3623–3640.

59 Amdursky, N., Molotskii, M., Gazit, E., and Rosenman, G. (2009). Self-assembled bioinspired quantum dots: optical properties. *Applied Physics Letters* 94 (261907): 261907.

60 Tao, K., Fan, Z., Sun, L. et al. (2018). Quantum confined peptide assemblies with tunable visible to near-infrared spectral range. *Nature Communications* 9: 3217.

61 Chen, H., Chen, X., Chen, X. et al. (2022). New perspectives on fabrication of peptide-based nanomaterials in food industry: a review. *Trends in Food Science & Technology* 129: 49–60.

62 Chen, H., Cai, X., Cheng, J., and Wang, S. (2022). Self-assembling peptides: molecule-nanostructure-function and application on food industry. *Trends in Food Science & Technology* 120: 212–222.

63 Zhang, Y., Zhao, M., Ning, Z. et al. (2018). Development of a sono-assembled, bifunctional soy peptide nanoparticle for cellular delivery of hydrophobic active cargoes. *Journal of Agricultural and Food Chemistry* 66: 4208–4218.

64 Liu, B., Liu, B., Wang, R., and Li, Y. (2021). α-lactalbumin self-assembled nanoparticles with various morphologies, stiffnesses, and sizes as Pickering stabilizers for oil-in-water emulsions and delivery of curcumin. *Journal of Agricultural and Food Chemistry* 69: 2485–2492.

65 Lan, M., Fu, Y., Dai, H. et al. (2021). Encapsulation of β-carotene by self-assembly of rapeseed meal-derived peptides: factor optimization and structural characterization. *LWT* 138: 110456.

66 Du, Y., Bao, C., Huang, J. et al. (2019). Improved stability, epithelial permeability and cellular antioxidant activity of β-carotene via encapsulation by self-assembled α-lactalbumin micelles. *Food Chemistry* 271: 707–714.

67 Yemini, M., Reches, M., Rishpon, J., and Gazit, E. (2005). Novel electrochemical biosensing platform using self-assembled peptide nanotubes. *Nano Letters* 5: 183–186.

68 Park, B.W., Zheng, R., Ko, K.A. et al. (2012). A novel glucose biosensor using bi-enzyme incorporated with peptide nanotubes. *Biosensors & Bioelectronics* 38: 295–301.

69 Fortunato, A. and Mba, M. (2021). Metal cation triggered peptide hydrogels and their application in food freshness monitoring and dye adsorption. *Gels* 7: 85.

70 Bech, E.M., Pedersen, S.L., and Jensen, K.J. (2018). Chemical strategies for half-life extension of biopharmaceuticals: lipidation and its alternatives. *ACS Medicinal Chemistry Letters* 9: 577–580.

71 Chao, A.M., Tronieri, J.S., Amaro, A., and Wadden, T.A. (2021). Semaglutide for the treatment of obesity. *Trends in Cardiovascular Medicine* 33: 159–166.

72 Venanzi, M., Savioli, M., Cimino, R. et al. (2020). A spectroscopic and molecular dynamics study on the aggregation process of a long-acting lipidated therapeutic peptide: the case of semaglutide. *Soft Matter* 16: 10122–10131.

73 Wang, M., Lv, G., An, H. et al. (2022). In situ self-assembly of bispecific peptide for cancer immunotherapy. *Angewandte Chemie International Edition* 61: e202113649.

74 Gogos, A., Knauer, K., and Bucheli, T.D. (2012). Nanomaterials in plant protection and fertilization: current state, foreseen applications, and research priorities. *Journal of Agricultural and Food Chemistry* 60: 9781–9792.

75 Hong, W., Zhao, Y., Guo, Y. et al. (2018). PEGylated self-assembled nano-bacitracin a: probing the antibacterial mechanism and real-time tracing of target delivery in vivo. *ACS Applied Materials & Interfaces* 10: 10688–10705.

76 Dehsorkhi, A., Castelletto, V., Hamley, I.W. et al. (2013). Interaction between a cationic surfactant-like peptide and lipid vesicles and its relationship to antimicrobial activity. *Langmuir* 29: 14246–14253.

77 Chen, C., Pan, F., Zhang, S. et al. (2010). Antibacterial activities of short designer peptides: a link between propensity for Nanostructuring and capacity for membrane destabilization. *Biomacromolecules* 11: 402–411.

78 Sha, X., Li, P., Feng, Y. et al. (2020). Self-assembled peptide Nanofibrils designed to release membrane-lysing antimicrobial peptides. *ACS Applied Bio Materials* 3: 3648–3655.

79 Shuai, J., Guan, F., He, B. et al. (2019). Self-assembled nanoparticles of symmetrical cationic peptide against citrus pathogenic bacteria. *Journal of Agricultural and Food Chemistry* 67: 5720–5727.

80 Guo, Z., Wang, Y., Tan, T. et al. (2021). Antimicrobial d-peptide hydrogels. *ACS Biomaterials Science & Engineering* 7: 1703–1712.

81 Yang, G., Tingben, H., Wang, Y. et al. (2018). Sustained release of antimicrobial peptide from self-assembling hydrogel enhanced osteogenesis. *Journal of Biomaterials Science, Polymer Edition* 29: 1–31.

82 Zottig, X., Al-Halifa, S., Côté-Cyr, M. et al. (2021). Self-assembled peptide nanorod vaccine confers protection against influenza a virus. *Biomaterials* 269: 120672.

83 Huang, Z., Liu, Y., Wang, L. et al. (2020). Supramolecular assemblies mimicking neutrophil extracellular traps for MRSE infection control. *Biomaterials* 253: 120124.

84 Gelain, F., Bottai, D., Vescovi, A., and Zhang, S. (2006). Designer self-assembling peptide nanofiber scaffolds for adult mouse neural stem cell 3-dimensional cultures. *PLoS One* 1: e119.

85 Smith, D.J., Brat, G.A., Medina, S.H. et al. (2016). A multi-phase transitioning peptide hydrogel for suturing ultra-small vessels. *Nature Nanotechnology* 11: 95–102.

86 Yang, P., Song, H., Qin, Y. et al. (2018). Engineering dendritic-cell-based vaccines and PD-1 blockade in self-assembled peptide nanofibrous hydrogel to amplify antitumor T-cell immunity. *Nano Letters* 18: 4377–4385.

87 Pugliese, R., Arnoldi, A., and Lammi, C. (2021). Nanostructure, self-assembly, mechanical properties, and antioxidant activity of a lupin-derived peptide hydrogel. *Biomedicine* 9: 294.

Index

Food-Derived Bioactive Peptides: Green Preparation, Activity Mechanisms, and Health Applications, First Edition. Shao-Yun Wang, Xu Chen, and Xixi Cai.

b

c

d

i

S

Printed and bound by CPI Group (UK) Ltd, Croydon, CR0 4YY
24/08/2026
14947823-0003